Dynamics of Development:
Experiments and Inferences

Dynamics of Development: Experiments and Inferences

PAUL A. WEISS

The Rockefeller University
New York, New York

Selected Papers on Developmental Biology

1968

ACADEMIC PRESS New York and London

ACADEMIC PRESS INC.
111 Fifth Avenue, New York, New York 10003

United Kingdom Edition published by
ACADEMIC PRESS INC. (LONDON) LTD.
Berkeley Square House, London W.1

LIBRARY OF CONGRESS CATALOG CARD NUMBER: 68-23476

PRINTED IN THE UNITED STATES OF AMERICA

PREFACE I

As ideas are preserved and communicated by means of words it necessarily follows that we cannot improve the language of any science without at the same time improving the science itself; neither can we, on the other hand, improve a science without improving the language or nomenclature which belongs to it. However certain the facts of any science may be, and however just the ideas we have formed of these facts, we can only communicate false impressions to others while we want words by which these may be properly expressed.

Lavoisier, "Elements of Chemistry," Preface

I have never heard Paul Weiss quote these lines by Lavoisier. No doubt he knows them. But whether or not he has given voice to them, he has lived by them, as the articles that follow attest. His has been a life devoted to improving his science and its language.

In this one book, in the writings of one man, is traced the recent history of ideas in developmental biology—in fact, the very emergence of developmental biology as a unified field of science. There will be readers now, and especially among the students of the future, for whom such a volume should have special meaning, for they are encountering the man and his works for the first time. They must be prepared for an array of ideas unrestricted by the conventional boundaries of scientific disciplines or techniques, ideas generated by a mind trained first in a happy combination of biology and engineering and sharpened by forays into mathematics, physics, and medicine, while pursuing the search for a better understanding of the basic principles of development.

Alfred Hershey has written that one of the great discoveries of modern science is that its goal—achieving an intelligible view of the universe —cannot be achieved piecemeal, by the accumulation of facts. To understand a phenomenon is to understand a category of phenomena or it is nothing. Understanding is reached through creative acts. His thesis is amply confirmed in these pages. Whether Professor Weiss is treating the ordering role of ground substances in morphogenesis, the molecular basis of cell specificity, the mechanisms of growth control, or the far-reaching implications of neuroplasmic flow, he moves freely and naturally from observations and experiments to new concepts, and, in full cycle, again to experimentation. He has always kept the larger goal, the understanding of development as a whole, before him and his readers, being guided by the precept that—in his own words—"Achievement is marked not so much by what one has learned, but by how one is using that which one has learned."

JAMES D. EBERT

Carnegie Institution of Washington
April, 1968

PREFACE II

The Effects of Discovery upon the Discoverer

Most discoverers become to some extent the creatures of what they have found, and the effect on them can be widely varied. Many of the fortunate have at once been overwhelmed with delight. Archimedes' shout "Eureka!" as he rushed forth naked into the street reverberates to this day. Yet the happiness of other discoverers has not infrequently depended on beliefs already existing at the time, or on their personal character. Galileo, though admirable, and loving not only science but music and poetry, barely escaped destruction by bigots, and then on humiliating terms. Isaac Newton was of such a "suspicious and quarrelsome temper" (according to "Chambers' Biographical Dictionary") that despite the loud acclaim of most of the scientists of his day the objections of a few made him "worse than ever." Yet, all in all, it can be said that most discoverers have exulted in their findings. Cold acceptance of these discoveries by them has been rare. Pleasant indeed it is to think of Pierre and Marie Curie going in the night to the cold shed in which they had labored throughout the day just to gaze and gaze upon a huge tub of fluid glowing in the dark that contained the radium that they soon would purify.

Until now, in terms of human time, most of the facts brought to light by purposeful finders had already existed unbeknownst for "ever so long." Searching and finding these facts has been archeology at its best, the new facts coming to life, as one might say, and often adding largely to life's betterment. Their discoverers, if gifted, have eagerly pressed on: in most instances no great harm has come from their further revelations.

Now the state of affairs is astonishingly different. Scientific research has almost abruptly become such a brilliantly rewarding profession that a myriad of able minds previously concerned with other fields have turned to science; and a truly creative era in science has begun, a building of facts upon other new facts with far-reaching results. Discovery has indeed become the most transcending of all man's efforts. It has even enabled him to venture forth into the airless Universe. What elation! But it has also now equipped him with weapons enabling him to destroy whole nations abruptly should he so desire. What horror! Needless to say some of the discoverers who have brought this about feel a sense of guilt; and perceptive statesmen are deeply concerned with what man should or shouldn't do.

Here let us hastily turn for good cheer to what Paul Weiss has found out about the inherent capability of living organisms, including ourselves. It is all for the good. Walter Cannon's book "The Wisdom of the Body," first published in 1932, should be recalled in this connection. Cannon, one of the most revealing physiologists of his time, made plain in this book "the role of the autonomic system in maintaining steady states" of the body. Paul Weiss has dealt with a kindred theme. He has been concerned with what can be termed the rights of the animal organism, its inherent endowment, morphogenesis, developmental processes, and the dynamics of the living systems through which it asserts itself successfully. His discoveries force him almost literally (as one can well understand) to reprint his most telling papers, together with a foreword which stresses what they mean when all are taken together. Hail to him and them!

PEYTON ROUS

The Rockefeller University
April, 1968

FOREWORD

The papers assembled in this book represent about ten percent of my writings published during half a century at the scientific workbench. Then, why republish? As one who has repeatedly expressed his apprehensions about the contemporary "publication explosion," I ask this question in all sincerity. All I can do for an answer is set forth the reasons or rationale that motivated me. Was it paternal pride? Not really. Of course, no one would honestly believe an author to be so inhumanly obtuse or self-effacing as not to relish the narcissistic exhilaration engendered by seeing his productive labors reflected in print. But this is essentially a matter of personal gratification; in fact, just the climax of the elation that comes from accomplished work. Publication, however, as the term implies, is a public function, transcending private satisfaction.

Part of that function is purely practical. The individual items of this compilation have been so widely scattered over such a variety of media, some of them not easily available or accessible, that assembling them in one package renders them more serviceable. Yet, my chief motive was more fundamental; it was the hope that the juxtaposition and overlap would let emerge the basic theme which, like Ariadne's thread, runs as a guiding tenet through the whole series of papers: the theme that a whole contains and conveys more information than does the sum of its unassembled parts.

Attention to this theme is crucial in these days in which research in the life sciences expands at such an explosive rate that the necessary preoccupation with detail in evermore separate channels of specialization exacts a mounting toll of shrinking perspective and disintegrating unity of basic principles. Simultaneous illumination from a variety of disciplinary angles promises to rescue that fading common base from total eclipse. Brought back into full light, it might attract renewed attention and the investigative fervor of future workers.

My potential for such an integrative service stems presumably from the highly diversified diet that has nurtured my scientific growth. My dual training in physics and engineering, on the one hand, and biology and preclinical disciplines, on the other, both technically and conceptually, have set my course: to help in the approach to the vague problems of *morphogenesis* by rigorous *experimental analysis*. Experi-

ix

mentation then was essentially problem-oriented, not technique-guided. Moreover, my closeness to the "Vienna circle" of philosophers convinced me that pure pragmatism—"doing without thinking"—and sheer philosophizing—"thinking without doing"—are equally stultifying. As a result, the urge to relate data always to a conceptual framework, keeping the framework, in turn, adaptable to new experiences, has never since left me; which also accounts for my periodic recourse to mathematical formalism.

Along with my doctoral work on animal behavior—an early attempt to replace micromechanistic thinking by a concept of "systems dynamics" —I concentrated on the analysis of formative growth processes, mainly in regeneration. An accidental discovery on coordination of limb function then prompted me to delve into the properties of the nervous system, which soon impressed me with the profound similarity between the patterning of nervous functions and of developmental processes, reflected in some chapters of this book. Indeed, the more I varied subjects and techniques, the more I came to recognize *an underlying fabric of rules and relationships in common to the manifold manifestations of the dynamics of the living system.* As one to whom an integrated overview from many vistas has revealed inner resemblances and congruities among such diverse phenomena as embryonic development, regeneration, growth control, cell locomotion, and neural functions, I naturally saw some merit in expounding them. And consequently, I view the cardinal function of these papers in their conjunction to be their contribution to the emergence of a unified understanding of the *dynamic order of living systems.*

Such understanding must be approached methodically by the unprejudiced and unrestricted exploration of *all* the aspects which a living system offers, evaluated logically and soberly: through "experiment and inference," as intimated in the title of this book. No single branch of the life sciences holds the master key to the understanding of the common core of principles of life from which it has branched off; claims to the contrary stem from limited perspective, no matter whether they come from the molecular or the evolutionary or the psychological or the ecological or any other branch. I myself have tried repeatedly one-sided "explanations" from limited viewpoints—and have failed. The more comprehensive my experience became, the more remote and unrealistic appeared the prospect of ever comprehending the basic properties of the living system by giving primacy to any one of the various sectors of its study. The universality of features and problems of development, of growth, of form and order, of fit and fitting, in common to the building and functioning of both body and brain, demands that the set of explanations be of equal *universality.*

This calls for a broad sweep of vision and concerted investigative effort, which is no easy task. In fact, it seems that the immenseness of the task and the feeling of inadequacy regarding the tools and methodology for its pursuit tend to induce us to cut the task down to the magnitude of means at hand instead of straining the imagination to fit the means to the magnitude of the task. If continued, this trend would inevitably lead to trading the spirit of inquisitive penetration into the vast unknown for a sense of smug acquiescence in the knowledge of the little that we know. I trust that the content of this book will manifest my effort to steer clear of that trend by glorying not so much in the advances that have been made as by stressing the enormous distance from the goal that remains to be traveled; and above all, that the approach must and can be made from *many* directions. For I believe that it would divert us from a goal-directed course if we were to accept blindly the claims of Pied Pipers contending to know the Royal Road to full enlightenment. One-sided spotlighting does not justify one-sidedness of approach.

It seems to me that current contentions to the effect that the principles of development might be "resolvable" entirely into operations on the *molecular* level are in that class. Without denying the valiance of such claims, the evidence gathered in this book disputes their validity. There is no phenomenon in a *living system* that is *not* molecular, but there is none that is *only* molecular, either. While I admit to my fatherhood of the term "molecular biology" (coincidentally with Astbury) and unreservedly share the general admiration for the spectacular achievements and future prospects of work on that level, I merely meant the term to signify the lowest level in the total hierarchical structure of biological principles, without intending or expecting it to aspire to monopolistic status. Indeed, much of the substance of this book serves to document the need and the legitimacy of giving the phenomena at *all* levels, from molecule, through cell, organ, and individual, to group organization —from "molecular ecology" and crystallinity to the ecology of communities and patterned behavior—equal status. The major problems of morphogenesis lie in domains of "supramolecular biology." They are amenable to disciplined experimental attack, taking advantage of modern progress in the physical sciences (e.g., cytophysics, solid-state physics, etc.), and in their underdeveloped state call for the efforts of the truly curious, imaginative, and resourceful workers who favor the challenge of the wilderness over the safety of the beaten track.

Besides its substantive content of data, conclusions, and propositions, the book contains a medley of observations and practical suggestions which could have been followed up with profit, but have not. There are phenomena and problems outlined in it that have been either overlooked or lost sight of and whose disinterment might spur research effort in new directions. In retrospect, of the ideas scattered through the papers, some have proven fruitful, others untestable, and some, untenable. There are still others, however, which, the reader will find, have been adopted and assimilated so generally that the very identity of my authorship has disappeared. Contrary to value systems in commerce, arts, and letters, the attainment of such anonymity, based on the full incorporation and integration of one's product into the growing body of knowledge, is, and has been to me, a source of deepest gratification, for it is the truest testimony to the validity and utility of the product.

What seems to me unfair, however, is that this anonymity should be shared by the host of collaborators without whose dedicated and resourceful help my ideas and probings could never have become translated into products as effectively. Therefore, I gladly seize upon this opportunity to cite specifically from among the many helpers in the conduct of my work, at least those who by their hands and thoughts have had a major share in it: Cecil Taylor, Murray Rosenberg, Mac Edds, Lee Kavanau, Helen Hiscoe, Aron Moscona, Gedeon Matoltsy, Margaret Cavanaugh, Yvonne Holland, Sam Ferris, Jane Overton, Gert Andres, Agnes Burt, Jean Cummings, Albert Bock, and Ian Linden.

I also express my gratitude to the public and private organizations which have provided substantial financial support for my research, especially the Rockefeller Institute, the American Cancer Society, the Rockefeller Foundation, the National Institutes of Health (U. S. Public Health Service), and the Faith Foundation of Houston, and, last but not least, to Academic Press for having taken on and carried out the task of publishing this book.

For background information on the phenomena and principles of development, to which the subject matter of this book is intimately related, the reader is referred to my text on "Principles of Development," 601 pages, 1939 (Henry Holt & Co., New York), which, after having been out of print for many years, is now in the process of republication by Hafner Publishing Co., New York.

PAUL A. WEISS

March, 1968

CONTENTS

I. ON BASIC PRINCIPLES

II. ON DIFFERENTIATION

III. ON GROWTH AND AGING

IV. ON FORM AND FORMATIVE PROCESSES

V. ON THE DYNAMICS OF THE NERVOUS SYSTEM

I. ON BASIC PRINCIPLES

Reprinted from: ADAPTATION. Edited by John Romano. Cornell University Press, 1-22, 1949.

CHAPTER 1

The Biological Basis

of Adaptation

✦

PAUL WEISS

Professor of Zoology,
The University of Chicago

THE inclusion of a biologist in this volume is a symbolic act: it reaffirms the unity of all life. I, therefore, take it that I am supposed to survey the principle of adaptation in living beings in general. There was a time, not so long ago, when we readily acquiesced in a textbook generality such as this: "Adaptation is a general property of all protoplasm." Modern biology wants to know more concretely and more precisely just what adaptation really consists of. My task, therefore, is not merely to present the concept of adaptation, but to dissect it critically and to classify its members.

Right at the start we stumble over the word. Adaptation, in daily language, means "the state of being adapted" as well as "the process of becoming adapted." To avoid confusion, let us confine the term "adaptation" to the adaptive process only and refer to the adapted state as "adaptedness." Also, while we are on preliminaries, it may be pointed out that the term "adapted" always refers to the relation of one entity to another. No system is adapted as such. It can only be adapted or conform to something else. If this conformance is achieved by direct interaction of the two, we shall speak of "adaptation," otherwise merely of "adaptedness." Adaptation, then, is the fitting, and adaptedness the fitness, through which a system is harmonized with the conditions of its existence.

[3]

Such harmony or fitness is the premise for the endurance of any circumscribed system in nature. The more complex a system, the more precarious is its maintenance, and the more exacting are its adaptive needs. The living organism, being the most complex of systems, must, therefore, also be the most subtly adapted. All organic systems, whether cell or organ or individual or species or population, are composites. They contain innumerable parts of different properties and activities, which interact in a most intricate manner, yet so that the group as a whole retains identity and stability long enough to deserve a name. The system endures beyond the flux of its parts and beyond the variations of its environment. The very fact that these fluctuations do not disrupt the organized existence of the system connotes, for the system, that it is in tune with the world around it and, for its parts, that they are in mutual harmony—in brief, adapted.

The origin of such adaptedness has been a fertile field of inquiry and speculation. The answers have ranged all the way from the pre-established harmony of Leibniz to the direct adaptation by functional use of Lamarck. Neither of these extreme doctrines proved tenable. It is generally conceded now that adaptedness is shaped by evolution. That is, as organisms, species, climates, and environments change, ill-matched and maladapted combinations simply do not survive the changes. On the whole, this evolutionary interpretation of adaptedness has proved satisfactory,

[4]

with perhaps one major reservation. It does not really account for the general and basic properties of living systems as such, but only for their various modifications and recombinations through which particular organic beings have been fitted into particular frames of conditions. Throughout evolution the fundamental features of life seem to have been and to have remained the same. This we deduce from the fact that they are in common to all existing creatures, from the lowliest of bacteria to the highest of animals.[1] All kinds of organisms contain the same classes of biochemical compounds and use the same methods of biosynthesis and of energy transfer, the same structural features of fibers and membranes, the same limited number of amino acids and growth requirements, the same mechanisms of metabolism, respiration, secretion, digestion, contractility, excitation, genic transmission, growth, and so forth.[2, 3] In their intricate interdependence and harmonious interaction, these elementary processes are prototypes of mutual adaptedness; yet for this primordial adaptedness we have no adequate scientific explanation. But if these basic elements of the fabric of life are taken for granted, all further adaptedness can be accounted for in principle by the evolutionary doctrine of variability and selection: fitness prevails because the misfit, unable to survive and procreate, cannot perpetuate its kind. No level of organic existence can escape the rigor of this test for fitness. Naturally, therefore, fitness pervades organic systems of

[5]

all magnitudes, from the molecular ecology of the cell to the group ecology of populations. Structural order and functional co-ordination are the outward signs of the resulting adaptedness.

Let me briefly list samples of adaptedness on the various levels of organization. On the cellular level we note harmonious co-ordination of molecular processes, such as the selective relation of enzymes to substrates or the orderly sequences of steps in a metabolic cycle. On the tissue level we think of the collaboration and interdependence among cells (as well as of cells and medium), such as the nutrient relations of egg cells and nurse cells, or between tissue cells and blood; we think of the mechanical support of limp epithelium by firm connective tissue or even of the immunity of a digestive gland to its own digestive juice. Within the individual the reciprocal fitness among the various constituent organs is abundantly illustrated by their hormonal and neural correlations, by the matching of muscles and skeleton. On the group level we are struck by the remarkable structural and functional complementariness of the sexes, the altruism of brooding care, the selectivity of food habits, and the like; on the interspecies level, by the perfect matching of parasite and host, of bee and flower, of predator and prey; and, above all, by what Henderson has described as the "fitness of the environment." [4]

What we plainly recognize from even such a cur-

[6]

sory listing as this is that there is no common denom-inator to adaptive features other than the sheer fact of their adaptedness. Presenting an almost infinite variety of forms, they must have arisen in an equally large variety of ways. In this light it seems absurd to assume that there may be only a single adaptive mech-anism. By the same token Lamarck's concept of adapt-edness through direct functional adaptation is absurd, not just wrong.

It is instructive in this connection to contemplate developmental stages rather than the completed body. If we disregard for the moment physiological mainte-nance, which is of as much immediate concern to the embryo as it is to the adult, most of what is being developed in the embryo has adaptive significance only in reference to its future function.[5] The eye develops without seeing, digestive enzymes without food, feath-ers without flight, instincts without practice. They all are predesigned, prearranged, prefitted in forward reference to the needs of the mature body. A function that has not yet made its appearance can obviously have no part in the molding of its tools, nor be in-strumental in the fitting of these tools to each other and to their later use. Evolution, therefore, brings about adaptedness not by the purposeful construction of matching designs, but by the routing out from an ever-changing assortment of designs of all but those that happen to match.

As a result, any form now extant owes its adapted-

[7]

ness essentially to the proven success of its developmental formula transmitted faithfully from generation to generation by the mechanism of heredity, preserved because in innumerable tests of survival it has not been found wanting. But here a closer look at the facts gives us pause. It occurs to us that these innumerable tests, which innumerable generations managed to pass successfully, can in reality never have been identical in any two cases. No two individuals, no two cells—in fact, no two natural systems of any description—are ever faced with precisely the same constellation of conditions in space and time. Thus each system in its lifetime passes through a train of exposures to conditions and events whose constellation is unique and, as such, unpredictable in its details. This leads us to a highly important realization. If organisms were rigidly preadapted to fit precisely one particular detailed course of life, their chances of ever encountering just that one expected course, hence, of surviving, would be infinitesimally small. Here, then, is the limit beyond which evolutionary prefitting may not go without dooming its creatures. The precision of fit must not exceed the precision with which future conditions can be predicted, and since predictability is but another expression for the regularity of past recurrences, we realize that what an organism is prefitted for by its evolutionary endowment is merely a statistical norm of conditions, the standard range of which is relatively constant, the individual manifestations of

[8]

which, however, vary at random from case to case. The gross lines are predetermined, but the details are left indeterminate for the individual organism to fill in according to the contingencies it will meet. This is where adaptedness, the property, leaves off, and adaptation, the process, takes over. This is where evolutionary providence, which has provided for the gross outlines, turns the organism over to its own resources to cope with the details. At last we have joined our central issue—the faculty of adjustment left to the organism within the fixed limits set by heredity. The prefabricated responses, which are evolutionary products of the past, appear now in contradistinction to the adjustive responses, which are products of the moment: the stereotype reveals its flexibility. An adaptive act can, therefore, appropriately be described as a preformed standard performance, adjusted to the realities of the moment.

It should be made clear at this point, however, that the ability to respond appropriately to a variety of actual situations is not yet sufficient proof of itself that direct adaptation is at play. Each organic system owes some of its response flexibility simply to the possession of subsidiary preformed response mechanisms. Evolution evidently equips organisms not only for a single set of conditions but endows them with spare mechanisms for recurrent emergencies, that is, for conditions somewhat out of the ordinary, but occurring with sufficient frequency to give the species that can

[9]

meet them a distinct advantage of survival. Wound healing and regeneration after mutilation, encystment and spore formation under adverse conditions, inflammatory reactions, blood clotting, skin color change to blend with the surroundings—all these strategic measures of preservation are spare equipment of old evolutionary design rather than on-the-spot inventions. Of course, here, too, only the general character of the response is prepared, while the details of execution are left to the elements on the spot to solve.

To exemplify the relative roles of prefittedness and direct adjustment, let us choose the case of the skeletal system. Bones are the pillars, braces, and levers of the body, and as such are subject to the stresses of load, pull, pressure, and shear. They are found arranged in just the right manner to sustain those stresses. To cite a spectacular example, the vertebral column of mammals is constructed like a suspension bridge, with the vertebral processes graded in length, shape, and slant exactly the way the engineer would arrange girders to take up the principal stresses.[6] Since the vertebral column assumes this pattern early in embryonic and fetal life, long before load carriage becomes a factor, and since even parts of the skeleton, isolated experimentally from the rest, develop essentially as if they were still integral parts of the whole system, it is evident that each part has a predesigned course and that the eventual bridge structure results from assembled piecework.

[10]

Not so, however, the internal architecture of the individual skeletal element. The fine lamellae and trabeculae in cancellous bone have long been recognized to conform to the stress trajectories. Here the acting stresses do the actual molding. The bony elements remain highly responsive to the stress pattern, and if the latter changes (as in deformities, ossification of joints, etc.), the interior of the bone is remodeled accordingly.[7] The mechanism by which tensions fashion tissue structure has been elucidated in the case of connective tissues. There tension acts primarily by orienting the protein chains of the ground substance, which then in turn act as guide rails for the cells and interstitial fibers.[8] This preferential orientation of the fibrous network in the direction of maximum stress is further reinforced by the proteolytic dissolution of all disoriented cross links.[9] Many instances of direct functional adaptation to mechanical stress, including that of cartilage, have yielded to this explanation. The application to bone is still only indirect,[10] although there is little doubt that bone architecture, too, is the resultant of two opposite effects—deposition along lines of maximum stress and resorption in stress-free directions.

Just as its interior structure, so the detailed surface configuration of a bone is subject to adaptive molding.[11] Muscle attachments, for instance, are in their gross outlines prepared on the skeleton, but adjust their finer details directly in accordance with the

[11]

actual pull transmitted. Perhaps even more instructive is the formation of joints. When the primordia of the two skeletal members of a joint are reared separately, each develops grossly a joint surface of appropriate curvature.[12] However, brought together, the two independently developed surfaces fail to match. Direct contact interaction is needed to produce a precise fit. Similarly, the interlocking of upper and lower jaws and teeth is crudely provided for in their respective courses of differentiation, but the proper meshing is achieved by direct mutual interactions during growth and functioning.

There are countless examples of this type that could be cited from all branches of biology. They include such common phenomena as the correspondence between phasic changes of organic systems and the rhythmic fluctuations of their environment, such correspondence again being based in part on inherited and inherent adaptedness and in part on direct adaptation. For example, many seasonal and diurnal cycles of growth and function (photoperiodism in plants, hibernation, breeding cycles, diurnal activity fluctuations, and so forth) are grossly preformed in the physiological machinery of the living system, but attain their synchronization with the external world through a direct adjustive response to stimuli of light or temperature.

All these cases confirm the general principle of the duality of fitness—approximate fitness by evolution-

[12]

ary predesign, elaborated to greater perfection and precision of fit by direct on-the-spot adjustments. Direct adaptation thus is conceded only a very restricted sphere of action, although in this statistical world this narrow latitude means the margin between death and survival.

Considering the narrow scope of adaptation, it is hardly surprising to find that it operates mostly by a mere quantitative regulation of activities that owe their specific character to preadaptation. It manifests itself as "more or less," "faster or slower," "earlier or later," "in one direction rather than another"—always of a given established activity. Muscle is caused to build more muscle in response to increased load and, conversely, to lose substance in disuse; vessels to widen with increased blood flow and to shrink with reduced pressure; breathing to speed up or slow down depending on oxygen demand.

It has often been maintained that the capacity for such quantitative regulations is a natural attribute of all self-preserving systems.[13, 14] For any system to persevere in stationary equilibrium implies, almost by definition, that any forcible change in one part be offset by a corresponding change of opposite sign in some other part, leaving the state of the whole unaltered. The self-preservative tendency to counter a change from without by appropriate response from within is, therefore, truly a basic property of all such systems, from the automatic rebound of an elastic

[13]

body after deformation to the automatic spurt in growth of an organism after a period of starvation.

By way of generalization one might say that a system under stress automatically develops means to resist the stress (that is, up to the limit of tolerance). However, as for organic systems, adaptation by such systemic resistance of the elastic type is only half the story. The other half is what we might call adaptation by accommodation of the plastic type. Instead of bracing itself against enforced change, the system yields, adopts the change, and submits to its consequences. Of course, thereafter it is no longer quite the same system, but it has retained its identity and saved itself from disruption. An accommodative adaptation leaves behind a direct impress of the external factor, comparable to the molding of a plastic substance. Whether a system will adapt by resisting or by yielding is an empirical question. Local pressure on tissue may lead to resistive toughening—for instance, by increased collagen deposition—or to accommodative softening —for instance, by enzymatic resorption. The rigors of cold may lead to the reactive growth of denser fur or to the accommodative departure for warmer regions; a food shortage to the reactive stepping up of foraying or to the accommodative cutting down of expenditure of energy. A stretched muscle may increasingly resist extension or adopt the new length, as in plastic tone. No array of facts could be more convincing than these in proving the futility of postulating or searching for

[14]

any single mechanism common to the various phenomena of organic adaptation.

To add further to this already impressive variety, let us finally bring up what seem to be the most specific of all direct adaptations, namely, those affecting the very structure of organic molecules in living systems. The prototypes of this category are the specific antibodies and the adaptive enzymes so-called, molecules that assume specific characters to match foreign antigens or foreign substrates, respectively. The relation between such complementary pairs of molecules is usually and very aptly compared to the fitting between key and lock. In fact, a good case has been made for the theory that the molecules concerned do actually interlock by virtue of their steric conformance, in true key-lock fashion. We may profitably use this simile to place those phenomena in their proper context in our present discussion.

There are two ways of unlocking a lock. One can either sample a large assortment of available keys until he finds one that fits, or one may start with a dummy key and fashion it appropriately until it fits, either by trial and error or by making a cast of the lock. It is currently assumed that, in antibody formation, certain protein molecules of standard shape, the gamma globulins, serve as the dummy keys upon which the "lock" molecule of the antigen then impresses its peculiar serration.[15, 16] If correct, this would be strictly a case of direct adaptation on the molecular level.

[15]

In other instances the complementary matching of molecular pairs more likely takes place by selection from among existing molecular varieties rather than by the creation of new ones; it is a molecular case of multiple preadaptedness rather than direct adaptation. It has been suggested that some such molecular mechanisms operate in the cellular interactions in growth, differentiation, and tissue formation,[17] but the evidence is still very tenuous.

We are on safer ground in the case of the so-called adaptive enzymes. Enzymes in general are prefitted keys unlocking matched substrates. It has been found, however, that certain bacteria, yeasts, and fungi can "learn" to adapt themselves to unfamiliar substrates that they normally cannot attack, evidently because of lack of the proper key enzyme. In order to be able to adapt, they must be left exposed to the new substrate for some time. Then, suddenly, enzymes of the proper structure appear on the scene. In some cases this is undoubtedly the result of spontaneous mutations which happen to turn up, perchance, an occasional fitting enzyme and whose carriers are subsequently favored by selective survival and propagation. The adaptation is then a feat of the strain rather than of a given individual. There are other cases, however, in which the adaptation is a directly substrate-related response of the individual cell, a true adaptation. It is possible, of course, that it consists merely of the unmasking or rendering active of molecules prefitted to

[16]

the new substrate that were formerly inactive, but there are strong indications that a direct molding effect of the substrate on the configuration of a master enzyme or precursor is involved.[18] By and large, the case is similar to antibody production.

Although we are dealing here with direct adaptive interactions on the molecular level, it is significant to note that this need not imply an adaptive benefit to the organism concerned. Rated biologically, the effect may be good or bad. There may result increased refractoriness to the foreign agent, as in immunity, or, conversely, increased sensitization, as in anaphylactic shock—acclimatization as well as decreased tolerance.

This is just another illustration of the ambiguity inherent in the term adaptation. An adaptive mechanism in the analytical sense need have no adaptive value in functional regards.

Let us now briefly take stock of the situation in which this analysis of adaptation leaves us. I have presented to you some of the pieces that we can isolate and identify in the conglomerate that is covered, and often enough covered up, by the term adaptation. I hope to have demonstrated that while we may legitimately recognize a universal principle of adaptedness in living systems, based on the supreme rule of harmony as a prerequisite for continued survival, the mechanisms for achieving such harmony are numerous, diverse, and variable, and that no single formula can ever embrace them. This leads us to the eminently

[17]

practical, if sobering, realization that it is illusory to hope for such an all-inclusive formula. The only scientific way to deal with adaptation is to get the facts for each case. Only after the facts are known is it possible to tell just how much of the adaptedness of a given phenomenon is due to inherited evolutionary prearrangements, and how much to direct adjustive interactions. This ratio varies greatly and unpredictably from species to species, from function to function, from unit to unit.

There is one generalization, however, that can safely be drawn despite all this indefiniteness, and that is that the latitude left to direct adaptation is extremely narrow as compared to the wealth of inherited adaptedness of evolutionary origin. The only system in which this margin has reached a maximum of truly imposing dimensions is the brain. Here our remarks assume a direct bearing on the topic of behavior to be dealt with by subsequent papers in this volume.

Examining the nervous system, we note that in growth and performance it displays the same variety of adaptive aspects that we have encountered in other biological systems. We note the same preadaptedness, blueprinted in the hereditary endowment of the germ, then elaborated by laws of growth and maturation to a state of functional adequacy that is reached prior to, hence without the guidance of, function as such. The whole vital response repertory, from the simplest reflex to the most complex instinct, is thus prefitted for

[18]

the needs of individual and species. We note the same provision of subsidiary mechanisms for emergency situations, the same strict confinement of the organism to the limited scope of operations circumscribed by the limited repertory of preformed instrumentalities. There are then the various direct adaptations to actual needs, beginning with the simple regulation of the size and number of neurons in accordance with their terminal fields and functional load, and culminating in the ingenious devices of the mind in solving problems. As in other biological phenomena, the adaptation may take the course of merely quantitative changes— e.g., facilitation, inhibition, acceleration, compensation, and conditioning—or it may involve changes of structure and pattern—e.g., invention, memory, and re-education. Again, the direction that an adaptive change will take defies a priori prediction. Stress may be responded to by increased resistance or by accommodation, by greater effort or by escape; repetitive stimulation, by sharpening or by dulling of the response. Lastly, even the highly specific adaptations of molecular configurations of the antigen-antibody type may be conjectured to be operative in the nervous system (in establishing the specific correspondences between peripheral and central elements, as well as among the latter).

In conclusion, the nervous system shows no more uniformity in its adaptive features than do other biological systems. If the study of the latter has pointed

[19]

up the futility of grandiose generalizations about adaptation, the same holds for the nervous system and behavior. Concepts of a minutely set predeterminism at one extreme, or unlimited plasticity and adaptability at the other, are equally unrealistic. It is no longer a matter of predetermination versus plasticity, but of just how much of one and how much of the other there is in a given phenomenon, psychological, neural, or plainly biological. Biological experience has settled the case by reconciling the opponents. As a biologist, I have presented the case to you. I am now resting my case—not without re-emphasizing, however, the immensely greater powers for direct adjustive action given to our brain, as compared to other organic systems; powers that are our challenge and opportunity, if we only develop and use them wisely.

REFERENCES

1. Williams, R. R. Social implications of vitamins, *Science*, 94: 471–475, 502–506, 1941.
2. Baldwin, Ernest. *An Introduction to Comparative Biochemistry*. Cambridge, Eng., 1940.
3. Florkin, Marcel. *Biochemical Evolution*. New York, 1949.
4. Henderson, L. J. *The Fitness of the Environment*. New York, 1913.
5. Weiss, Paul. *Principles of Development*. New York, 1939.

[20]

6. Thompson, Sir D'A. W. *On Growth and Form.* Cambridge, Eng., 1942.

7. Wolff, Julius. *Das Gesetz der Transformation der Knochen.* Berlin, 1892.

8. Weiss, Paul. Functional adaptation and the rôle of ground substances in development, *American Naturalist,* 67: 322–340, 1933.

9. Weiss, Paul, and Taylor, A. C. Histomechanical analysis of nerve reunion in rat after tubular splicing, *Archives of Surgery,* 47: 419–447, 1943.

10. Murray, P. D. F. *Bones; A Study of the Development and Structure of the Vertebrate Skeleton.* Cambridge, Eng., 1936.

11. Washburn, S. L. Relation of temporal muscle to form of skull, *Anatomical Record,* 99: 239–248, 1947.

12. Braus, H. Gliedmassenpfropfung und Grundfragen der Skeletbildung, I. Die Skeletanlage vor Auftreten des Vorknorpels und ihre Beziehung zu den späteren Differenzierungen, *Gegenbauers Morphologisches Jahrbuch,* 39: 155–301, 1909.

13. Weiss, Paul. Tierisches Verhalten als "Systemreaktion": Die Orientierung der Ruhestellungen von Schmetterlingen (Vanessa) gegen Licht und Schwerkraft, *Biologia Generalis,* 1: 167–248, 1925.

14. Bertalanffy, Ludwig von. *Theoretische Biologie.* Berlin, 1932.

15. Landsteiner, Karl. *The Specificity of Serological Reactions.* Cambridge, Mass., 1946.

16. Pauling, Linus. Theory of structure and process of formation of antibodies, *Journal of the American Chemical Society,* 62: 2643–2657, 1940.

[21]

17. Weiss, Paul. Problem of specificity in growth and development, *Yale Journal of Biology and Medicine*, 19: 235–278, 1947.

18. Monod, Jacques. The phenomenon of enzymatic adaptation and its bearings on problems of genetics and cellular differentiation, *Growth*, 11: 223–289, 1947.

I. On Basic Principles

Reprinted from THE MOLECULAR CONTROL OF CELLULAR ACTIVITY.
Editor: J. M. Allen, McGraw-Hill Co., New York, 1-72, 1961.

CHAPTER 2

From Cell to Molecule*

PAUL WEISS

The Rockefeller Institute, New York

INTRODUCTION

The privilege of introducing this series of lectures on *The Molecular Control of Cellular Activity* is all the more precious to me because it provides me with an opportunity to recenter the object whose "activities" are to be "controlled" — the cell — from the increasingly off-center, out-of-focus position which it has assumed in current thought. Of the twelve lectures of the series which are to follow, all twelve deal with important fragments of the molecular inventory of cells, and seven alone with nucleic acids. This is a true reflection of current hopes — or illusions — that it might be possible to pinpoint in the cell a master compound "responsible" for "life" — an obvious reversion in modern guise to animistic biology, which let animated particles under whatever name impart the property of organization to inanimate matter. Therefore, lest our necessary and highly successful preoccupation with cell fragments and fractions obscure the fact that the cell is not just an inert playground for a few almighty masterminding molecules, but is a *system*, a hierarchically *ordered* system, of mutually interdependent species of molecules, molecular groupings, and supramolecular entities; and that life, through cell life, depends on the *order* of their interactions; it may be well to restate at the outset the case for the cell as a *unit*. A unit retains its unity by virtue of the power of subordination which it exerts upon its constituent elements in such a manner that their individual activities, instead of being free and unrelated, will be restrained and directed toward a combined unitary resultant. In short, the story of "molecular control of cellular activities" is bound to remain fragmentary and incomplete unless it is matched by knowledge of what makes a cell the unit that it is, namely, the "cellular control of molecular activities."

*Introductory lecture delivered March 1, 1960, in the lecture series on "The Molecular Control of Cellular Activity" of The Institute of Science and Technology, University of Michigan.

Some of the research results used as examples in this address have been obtained in investigations that were aided in part by grants from the American Cancer Society and the National Cancer Institute (National Institutes of Health of the Public Health Service).

It is on this principle then that I shall concentrate here, for it is as poorly understood as it is neglected. Or rather, it suffers not so much from outright neglect as from being frozen into literary symbols (e.g., "control," "organization," "information," "coordination," "regulation," etc.), the resolution of which into objective terms has been bypassed far too long. The problem of the cell as a unitary system may easily escape notice by those whose practical experience is confined to a limited sector of cellular activities, or it may be recognized but relegated to the mental attic as uncomfortable or untractable. Yet unless it is restored explicitly and courageously to a central place commanding universal recognition, it will not receive the concentrated attention and investigative effort which it deserves and without which the concepts of the cell — more broadly, of organisms — will remain an incongruous mixture of solid factual descriptions and vacuous anthropomorphisms.

FROM CELL TO MOLECULE

Analysis — the way from cell to molecule — appears a relatively easy road when one compares it with the uphill task of getting back from molecules to cells. Can what we destroy on the way down be replaced in reverse order and the system resynthesized from its shambles? Retracing such synthetic courses in our minds, we rely on verbal crutches, — "reconstitution," "reintegration," and the like. Do these abstractions have concrete counterparts in our actual experience with living things? Or, phrased differently, is it conceivable that we could resynthesize a cell from its fragments, if we only knew how to put the parts together stepwise one by one; and how to keep the intermediary partial assemblies from collapsing before the culminating self-sustaining state has been attained?

Recent developments in cell biology might seem to encourage positive hopes for an affirmative answer. Less than half a century ago, when I studied biology, something called "protoplasm" was globally endowed with all the properties required for a cell to live and function — reproductive capacity, growth, respiration, excretion, contractility, and excitability. Since then all these performances have been successfully allocated to identifiable components — genes, ribosomes, mitochondria, Golgi nets, acto-myosin threads, and polarized membranes, respectively — and some of them have actually been obtained from the respective components in isolation. What is often overlooked, however, is that in order to obtain performance by an isolated part, the experimenter must provide it with accessories which themselves are products of cellular activity, such as enzymes. Isolation thus connotes by no means the cessation of dependence upon activities of other cell systems. In fact, it is precisely this indissoluble interdependence among its various component operations which marks the cell as an entity in its own right and, as I shall indicate presently, discourages prospects for a "synthetic cell."

In passing, we might examine the reason why this qualifying fact is often overlooked. It lies, I believe, in our stage of scientific development to which future historians of science will perhaps refer as the "age of identification." We are looking for the doer and forget about the deeds. There seems to be some predilection for indentifying agents and charging them with "responsibility" for actions, the mechanisms of which we do not seem to be equally intent on exploring. For instance, we have identified, purified, crystallized, and even synthesized some of the hormones, but how they exert their selective effects on cells of various types has as yet been revealed in no single case. We state that the hormone "controls" a given cellular activity, and we mostly let it go at that. In fact, belief in identification as the real object of research accounts for the presumptuous habit of attaching labels of identity to unknowns which have not even been identified at all. For instance, how often does one hear the gratuitous assertion that in a given unknown mechanism "an enzyme may be involved," without any real evidence that the process does "involve" enzymes, or how it does, let alone which enzymes are involved? Of course, identification is defended as merely the first and indispensable step to understanding, but a first step leads nowhere unless the next step follows, and furthermore, understanding of an activity does not necessarily presuppose knowledge of the identity of the agency, as is well illustrated by the fact that the basic laws of optics were established long before the electromagnetic nature of light waves became known.

Accordingly, unless we follow through with the second step, the historian of the future will rightly blame us for partial blindness. This second step, which we have been slack in taking, is the study of the orderly interdependence of the partial mechanisms of the cell of which the identified molecules or particles are the tools.

Life is a dynamic *process*. Logically, the elements of a process can be only elementary *processes*, and not elementary *particles* or any other static units. Cell life, accordingly, can never be defined in terms of a static inventory of compounds, however detailed, but only in terms of their interactions — with stress on *inter-*, for as I indicated before, to credit compounds with "actions," "responsibility," "control," and other personifying traits of spontaneity is nothing but old-fashioned animism in disguise.

There is an obvious disproportion between the relative wealth of information about molecular and supramolecular entities as such and the dearth of knowledge about their complex interplay. Undoubtedly, the study of their interactions should command a larger share in the blueprint of future research than is evident in the contemporary scene. But, as the following commentary will indicate, even on the basis of current knowledge, more can be said about those interactions than is commonly made explicit.

In analyzing cell content by progressive fractionation, proceeding downward on the scale of magnitudes, we get from the microscopically visible units (e.g., nucleus, chromosomes, fibers) through submicroscopic particles to macromolecules, and further to simple molecules and radicals, ending up with a selection of the atomic elements, ubiquitous in nature.

Figure 1 exemplifies the scale of sizes of organic units between the cellular and macromolecular levels, which extends across the border zone between living and nonliving systems; even the entities within the lower range have never as yet been proved to be reproducible except through the mediation of a cell. One finds, in fact, the same gamut of units of different size orders represented within each living cell. By physical fragmentation (or its mental counterpart), we can effect, or at least visualize, the decomposition of a cell, or more generally, of units of the upper range, into piles of units having the characteristics of the lower range, that is, being definitely incapable of continued life independently of cells. The question then is this: just what is it that is lost in this degrading process from cells to nonliving constituents, and that would therefore have to be restored if ever one were to get back in ascending order from the elements to the whole?

There has been no loss of mass, the total of the fragments equaling the mass of the unfragmented cell. The number and proportions of atomic elements contained in both have likewise remained the same. What has been lost are some of the specific kinds of combinations and constellations in which the fragments had existed in the living cell and which are vital for those specific interactions on which the integral existence of the cell depends. This network of interrelations between components is what is meant by "complexity." Since the interacting components are in themselves of different compositions, complexity at the same time implies "inhomogeneity." And these two attributes are often all one has reference to when one describes a cell simply as "an infinitely complex system of heterogeneous molecular species."

However, this description misses the most important point of cellular organization, which is that the heterogeneous mixture of components combined in the complex system operates within a framework of *order*, the stability of which contrasts sharply with

Diameter or
width X length in mμ

Red blood cells	7500
B. prodigiosus (Serratia marcescens)	750
Rickettsia	475
Psittacosis	270
Myxoma	230 x 290
Vaccinia	210 x 260
Pleuro-pneumonia organism	150
Herpes simplex	130
Cytoplasmic virus (Tipula paludosa)	130
Rabies fixe	125
Newcastle disease	115
Avian leucosis	120
Vesicular stomatitis	65 x 165
Polyhedral virus (Bombyx mori)	40 x 280
Influenza	85
Adeno	75
Fowl plague	70
T2 E coli bacteriophage	65 x 95
Chicken tumor I (Rous sarcoma)	65
Equine encephalomyelitis	50
T3 E coli bacteriophage	45
Rabbit papilloma (Shope)	45
Tobacco mosaic and strains	15 x 300
Cymbidium (orchid) mosaic	12 x 480
Genetic unit (Muller's est of max size)	20 x 125
Southern bean mosaic	30
Tomato bushy stunt	30
Coxsackie	27
Poliomyelitis	27
Turnip yellow mosaic	26
Tobacco ringspot	26
Yellow fever	22
Squash mosaic	22
Hemocyanin molecule (Busycon)	22
Foot-and-mouth disease	21
Japanese B. encephalitis	18
Tobacco necrosis	16
Hemoglobin molecule (Horse)	3 x 15
Egg albumin molecule	2.5 x 10

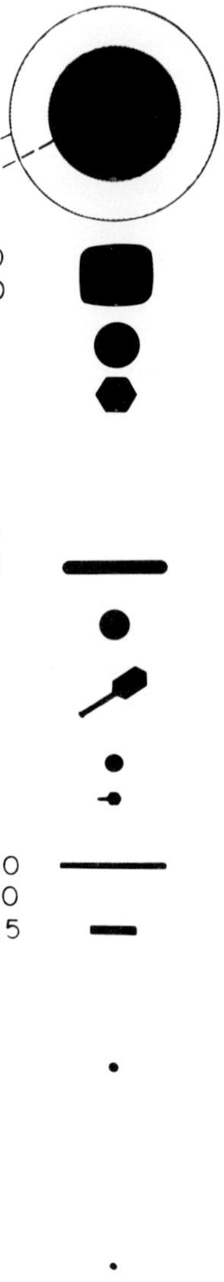

FIG. 1. Size spectrum of organic bodies. (*Revised in* 1958 *by R. C. Williams from W. M. Stanley* "Chemical Studies in Viruses," *Chem. Eng. News,* 25:3786-3791, 1947.)

the potential randomness of the individual component events if these were not subject to some over-all control. This *stability* of order manifests itself in two basic facts — first, that the individual cell as such remains recognizably similar to itself, i.e., essentially invariant, despite the incessant turnover and reshuffling of its content; and second, that the countless specimens of cells of a given kind remain recurrently similar to one another although in detail the content of each has a unique and nonrecurrent fate. Considering the cell as a population of parts of various magnitudes, the rule of order is objectively described by the fact that *the resultant behavior of the population as a whole is infinitely less variant from moment to moment than are the momentary activities of its parts.* Despite the continual flux of components, both as to composition and location, the system *as a whole* preserves its character. Small molecules go in and out, macromolecules break down and are replaced, particles lose and gain macromolecular constituents, divide and merge, and all parts move at one time or another, unpredictably, so that it is safe to state that at no time in the history of a given cell, much less in comparable stages of different cells, will precisely the same constellations of parts ever recur.

By contrast, however, one does not find this uniqueness, hence unpredictability, of the precise state and distribution of components reflected in the resultant total system, whose over-all pattern and behavior (or what one usually refers to as "organization") remain relatively unaltered, hence predictable. This forces us to conclude that although the individual members of the molecular and particulate population have a large number of degrees of freedom of behavior in random directions, the population as a whole is a system which restrains those degrees of freedom in such a manner that their joint behavior converges upon a nonrandom resultant, keeping the state of the population as a whole relatively invariant. It is this property of directive restraints, then, that is the most essential loss a cell suffers in the process of analytical disintegration, and since it is a property of a *collective,* we cannot observe its manifestations when we study the members of the population *singly* in isolation. Examples to illustrate this proposition will be presented later on.

In its more strictly morphological past, biology tended to ascribe the stability of the organization of the total behavior pattern of a cell to a rigid frame of fixed structures within the cell — a cytoskeleton supposedly exempt from the metabolic and motile changes to which the rest of the cell content is constantly subjected. This notion has arisen from the preoccupation with microscopic pictures of fixed dead cells, and even in our day it is still sometimes carried over to the submicroscopic realm of electron microscopy. But as will be explained presently, stable structures that are demonstrable in the living cell, other than chromosomes, have mostly turned out to be secondary derivatives, rather than primary carriers, of cellular organization. Just as the turnover of radioactively labeled compounds has revealed the flux of cell composition on the molecular level, so the microscopic observation of the living cell in action, particularly with the optical speed-up of time-lapse cinemicrography, has revealed such incessant reshuffling of the cell content that even the thought that at least the supramolecular units (particulates) might be linked into a stable framework can be safely dismissed. In motion picture films of single cells in tissue culture (a sample film taken under the phase-contrast microscope was shown at the lecture), one can directly observe how cell contour, intracellular fiber systems, and granules of various descriptions change their configurations and positions continuously, thus ruling out the presence, or at any rate, the relevance, of a consistent three-dimensional cytoskeleton. Two-dimensional continua are present in the various surfaces and membrane systems, but their repeated disruptions likewise fail to impair the essential integrity of the cell.

Yet despite the absence of an orderly static frame, the various activities of all parts remain coordinated in the maintenance of a standard pattern of order in any given cell. It is an order of relations rather than of fixed positions. Lacking a static foundation and barring sham explanations by extraneous vital agencies, we evidently must seek the source of this order in the *population dynamics* of the cellular constituents of various magnitudes. That is to say, the cell (as well as any of its subsystems) is not only made up of heterogeneous parts, but the various segments of this molecular and particulate population are so constituted that they assume the proper mutual space and functional relations simply by virtue of their own activities, rather than by passive allocation within a fixed framework. Some of the interacting subsystems are in such cooperative interdependence — symbiotically as it were — that neither can proceed without essential contributions from the other; in other cases, the dependence will be unilateral, comparable to parasitism; and in still other instances, there will be mutual interference or incompatibility between adjoining processes. In this dynamic concept, organization rests on properties which, to be sure, are inherent in the individual members of the heterogeneous population, but which can find expression only in collective interaction. By analogy to human populations, one could compare this emergent ordering process with the "self-structuring" of a community — an ecological simile to which we shall return later.

In the light of these somewhat ponderous but relevant prefatory remarks, our earlier question about the possibility of recompounding a fragmented cell must be rephrased more articulately. As the cell is not simply a random array of molecules, but a hierarchy in which molecular groupings are combined into macromolecules, macromolecules into particulates, and particulates into organelles, the question must evidently be posed separately for each level of this hierarchical organization, leading to such specific queries as the following: If we start with a random mixture of selected molecular species, we must ask just how high a system in the ascending scale of order would they be able to build up by sheer free interaction, without the intervention of a living cell or of cell products? This implies a sharp logical distinction between de novo synthesis, i.e., actual compounding, from the atomic level up on the one hand, and recombinations of (or with) compounds which themselves had a cellular origin, on the other. If we start from the opposite end, the intact cell, we must ask just how far can a cell be broken up and yet again be restored to integrity from reassembled pieces? To find the answers to such questions is a purely empirical problem and no longer a matter of abstract speculation. Yet, by the same token, "synthetic" success at any one level does not automatically spell success for any other level, and any generalization remains likewise a matter of empirical tests, rather than just confident assertion. Therefore, to predict whether or not cells will ever be synthesized from scrambled molecules is in that generality not only an idle but a logically unsound undertaking with more emotional and cultural overtones than scientific foundations.

In conclusion, once we have acknowledged that the cell is nothing but the systematically organized community of molecular populations in dynamic interaction, the *dynamic organization* of the system becomes our central problem. And solely by learning more and more about the isolated pieces, we can never hope to gain understanding of the higher degrees of order to which the pieces are subordinated in those collective groupings which we know as cells, and whose continuity as organized systems has been passed down uninterruptedly through the whole course of evolution.

To sum up, even though we have now some fairly good road maps for the analytical trip "from cell to molecule," most of those roads are still one-way, and the reverse trek "from molecule to cell" takes us into uncharted land.

FROM MOLECULE TO CELL

By way of dramatizing our problem, I am showing in Fig. 2, side by side, a 6-day-old chick embryo immersed in liquid, before (Fig. 2a) and after (Fig. 2b) having been homogenized by crushing. As no

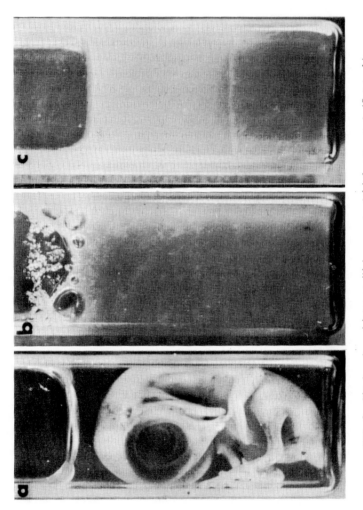

FIG. 2. Chick embryo (*a*) intact (*b*) homogenized (*c*) fractionated. (*Original.*)

DYNAMICS OF DEVELOPMENT

substance has been lost or added during the procedure, the content of the vial before and after is the same in weight and composition. An inventory of molecules, if it could be taken, would likewise reveal no change (disregarding, for the moment, mechanical breakage of some macromolecules, whose fragments are, of course, still there). What has been lost is structural organization from the highest level of the organism down to the order of whatever subsystems were small or consistent enough to have escaped the disruptive force of our crushing technique and whose ordered heterogeneity has therefore failed to become "homogenized." As pictured here, organs and tissues have been broken up, so also have the individual cells, their membranes, nuclei, and cytoplasmic systems. To get from a to b was easy. How to return from b to a—that is our uphill problem. Of course, if one considers the change from a to b as nothing but a general transition from heterogeneous to more homogeneous distribution, a change in the reverse direction can readily be brought about; for instance, by subjecting the homogenate of Fig. 2b to centrifugation or other separatory measures by which the molecular or particulate scramble can be partly unscrambled into distinct fractions, as shown in Fig. 2c. But the result is not a step back toward the lost old order, but rather a step toward a new and artificial order, which bears no more resemblance to the original than a neatly stacked assortment of spare parts bears to an intact machine. The different components are all there, but the specific structural order on which the functional capacity of the whole assembly depends is lacking.

Now, as stated before, this order comes in hierarchic steps, hence it needs to be considered separately for each level. The very fact that cellular subsystems, functionally specialized, are interposed between the levels of organization of the cell and of the molecule has long been recognized. But it has remained for the electron microscope to resolve the details of those subunits. The examples presented in the following pages will outline certain common properties of such cell "organelles," as well as their bearing on the problems of cellular organization.

For practical purposes, one might draw a distinction between continuous subsystems, such as membranes or fibrous networks, and discontinuous systems, consisting of such discrete units as chromosomes or mitochondria. In some sense, however, this may be misleading, for the real cell is a physical continuum, no part of which can be considered as truly separate from the rest. Moreover, both in life and during fixation, continuous structures may break up into fragments, while conversely, discrete bodies may coalesce. These qualifications should be borne in mind in the following account.

Our first example shows some of the common implements of

cells (Fig. 3). Surrounding the nucleus (n) concentrically, one observes a system of more or less parallel double-contoured lamellae, called "endoplasmic reticulum" or "ergastoplasm" (e), which per-

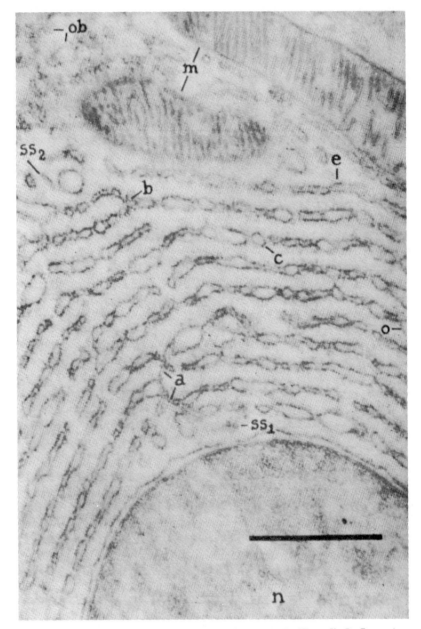

FIG. 3. Electron micrograph of section of gland cell. (*From K. R. Porter.*)

vade the cytoplasm as a branched and anastomosing network. Conspicuousness and dimensions of this system vary greatly, not only according to cell type, but what is more important in the present context, with the age and functional activity of any one cell. This latter fact proves that the lamellar system is not a permanent and rigid fixture formed once and for all, but is a "dynamic" system which undergoes reductions and restorations while nevertheless preserving its basic pattern.

It is the nature of this pattern which concerns us here. One notes at once two salient features, namely (1) the rather strict orientation of the lamellae along well-defined planes, and (2) the tendency for these planes to be equidistant in any one cell. These features can serve as clues to the dynamics of the formative processes to which the lamellar system owes its specific pattern. That is to say, the observed geometric regularity is but an index of some singular constellation of underlying physical and chemical conditions. The presence of an overt lamella in a given plane reveals that prior to the appearance of the lamella that particular plane must have been distinguished as a uniquely favorable site at which lamella-forming elements could persist long enough to become detectable. Just what the locally favored process is will vary from case to case; it may be the synthesis of lamellar material, or the assembling and orderly deposition of material from a dispersed state, or the active inflow of formed material [1] along invisible interfaces — at any rate, some process which is responsive to the physical constellation of its environment. The formed lamellae thus become the visible traces of the configuration of an invisible field of interactions, figuratively comparable to the iron filings which trace the lines of force of a magnetic field.

Accordingly, the fact that the planes occupied by the lamellar system are separated by a standard distance (e.g., 1,300 A in Fig. 3) can be taken to signify that some such processes as the following are involved: either (1) the material itself is produced in rhythmic waves spreading at a constant rate from a few basic planes (e.g., the nuclear surface); or (2) the material occurs ubiquitously, but once a layer has formed it inhibits the deposition of a like layer within its range of action, comparable to the phenomenon of rhythmic Liesegang rings; or (3) the ground substance of the cytoplasm, appearing structureless under electron microscopic inspection, may in reality be composed of shells or laminae of a given thickness of which the lamellar system would outline the borders. Alternative (1) seems to be ruled out by the frequent branchings and intercalations (e.g., at points *a* and *b* and at the asterisks, respectively, in Fig. 3). No decision can be made at present between points (2) and (3). Both are equally conceivable, although they differ crucially in that the former assumes direct interaction among the folds of the lamellar system itself, while the latter refers the lamellar pattern back to a lamination of the ground substance. This issue is a fundamental one and will recur in our later discussion.

A similar rhythmic pattern on a small scale is encountered in the internal structure of the mitochondria (*m*, Fig. 3), whose diagrammatic interpretation by Sjöstrand is given in Fig. 4. As in the preced-

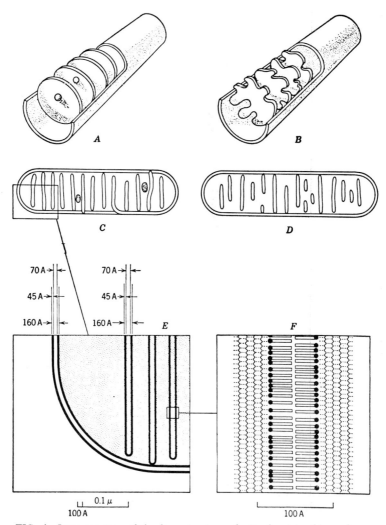

FIG. 4. Interpretation of the fine structure of mitochondria. (*According to F. Sjöstrand.*)

ing case, the matrix, which is enclosed in a double-contoured sheath, contains, in its interior, lamellar plates and folds, called "cristae." The distances between these lamellar plates are again relatively uniform, indicating that there is either mutual interaction or an underlying periodic structure in the matrix. The double contour of the lamellar walls (Fig. 4E, F) has been explained quite plausibly as resulting from the self-arraying of mixtures of protein and lipid molecules by virtue of the latter's hydrophilic and hydrophobic ends — a principle well demonstrated by the formation in vitro of polar monomolecular layers of stearates and proteins (according to Langmuir and others) and first applied successfully to biological systems in the interpretation of the lamellar fine structure of the myelin sheath of nerves. In our present context, however, it must be stressed that even though the molecular basis of the fine structure of the individual lamellae be known, this does not of itself lead to an understanding of the higher organization of a mitochondrium as a whole. Observations of living cells have shown mitochondria to be highly mobile and morphologically unstable units, given to subdividing, coalescing, branching, and anastomosing [2]. In view of this lability, it would seem impossible to account for the structural regularities noted in the electron micrograms of fixed cells otherwise than on the premise that they are but momentary records of formative dynamics which continuously remakes and adjusts the structural details, while maintaining the integrity of the mitochondrial pattern as a whole.

Once this fact has been realized, a series of new questions comes to mind, for which no answer is as yet available, partly because the questions have never been raised explicitly. For instance, what determines and maintains the average distance between the cristae? Mitochondria, though varying in length, seem to be of fairly uniform diameters for given cells and conditions; to what equilibrium condition do they owe this relative stability of girth? Why does their growth stop when standard unit size has been reached? Within what limits does the unit size adapt itself to metabolic and functional conditions, and by what means? Some of these questions touch on a most basic problem of morphogenesis, which is: Why are discrete subunits of cells produced in quantal steps, rather than in a continuous spectrum of sizes? Does not this fact itself suggest a dynamics of the integral higher system of such a kind that at a particular size range a singularly stable state for the whole assembly of subsystems would be reached?

Another example of cell organelles with lamellar fine structure are the chloroplasts of plant cells (Fig. 5a), which show further specialization in the fact that at certain points the lamellae are compacted into more tightly packed "grana" (two of which are seen in

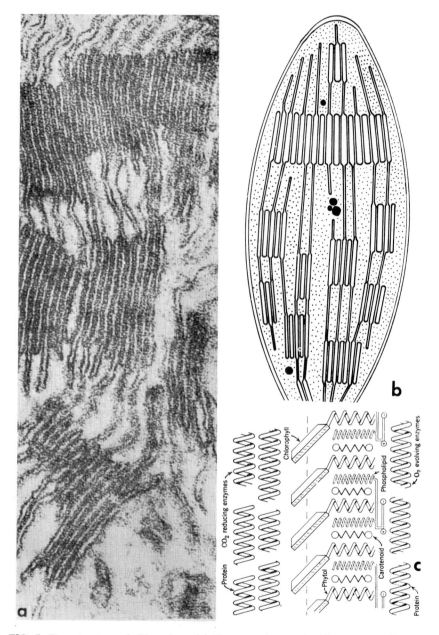

FIG. 5. Fine structure of chloroplast. (*a*) Electron micrograph. (*Courtesy von Wett-stein.*) (*b*) Diagrammatic reconstruction of the lamellar arrangement. (*After von Wett-stein.*) (*c*) Hypothetical molecular organization of the lamellar system. (*After Calvin.*)

the illustration). From combined electron-microscopic, biochemical, and X-ray studies, a hypothetical concept of the molecular composition and architecture of the individual lamellae has been constructed (Fig. 5c) which, regardless of validity in detail, indicates the degree of orderly complexity which one must postulate for even the elemental constituents of this structure. But as with the mitochondria, the supraelemental order remains to be explored and explained. Why, for instance, do all the lamellae of a given group change suddenly along a sharp line from the loose to the compact packing, or why does a chloroplast, on reaching a liminal size level, subdivide instead of keeping on to enlarge?

Sometimes the cell surface itself contains specialized subunits of rather constant size. For instance, the underside of larval amphibian epidermis cells, illustrated in Fig. 6, shows the plasma membrane of the cell dotted with regular bobbin-shaped bodies, each consisting of two lipid-rich discs connected by a hydrophilic neck, which, according to experimental evidence, serve as adhesive devices [3]. As one can readily see, both their dimensions and spacing are so regular as to raise again the question of how the shape, size, and arrangement of such complex supramolecular entities are regulated to a standard norm. The question is all the more cogent, since this is one of the instances in which the lability of the pattern could be directly demonstrated: in a cell which has been mobilized by wounding the skin, the adhesive discs undergo resorption, but once the cell has settled down again, the discs are formed anew in the same typical size, shape, and distribution as before. The conditions for the formation of this surface pattern are, therefore, ever present in those cells.

As another surface differentiation of unit character and fleeting existence, I may cite certain filamentous projections which are observed in electron micrographs of single cells cultured in vitro (Fig. 7); these projections from the cytoplasmic margin are supported by one or several rather rigid core filaments ("microspikes") of variable lengths, but of a standard diameter of slightly less than 1,000 A. Since in these highly motile cells the margin changes its composition and configuration incessantly, it must be taken for granted that the microspikes are continually renewed, yet always of the same unit width.

Many of the more permanent structures subserving specialized cell functions likewise display unit character. The most common examples are cilia and flagella. All of them, whether of protozoans, plants, animal tissues, or spermatozoa, are built alike. Even the visual elements of the retina contain abortive cilia, presumably homologous to the light-sensitive spot of flagellates, which is likewise a modified cilium. Cilia, shown in cross section in Fig. 8, con-

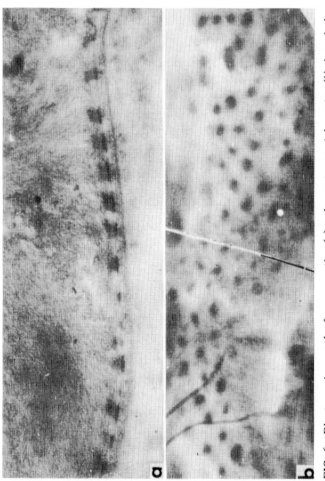

FIG. 6. Electron micrograph of cross section (*a*) and near-tangential section (*b*) through the basal end of an epidermis cell (in larval amphibians), showing the adhesive "bobbins." (*From Weiss and Ferris.*)

DYNAMICS OF DEVELOPMENT

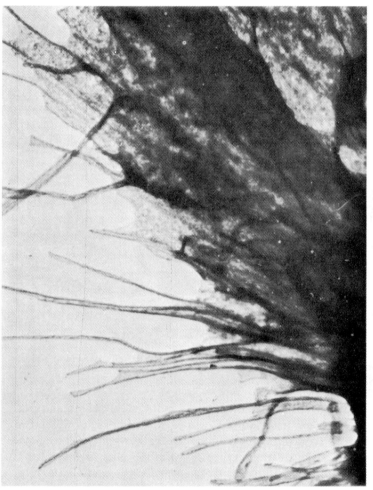

FIG. 7. Electron micrograph of free margin of single cell grown in tissue culture. (*Original, Weiss and Robbins.*)

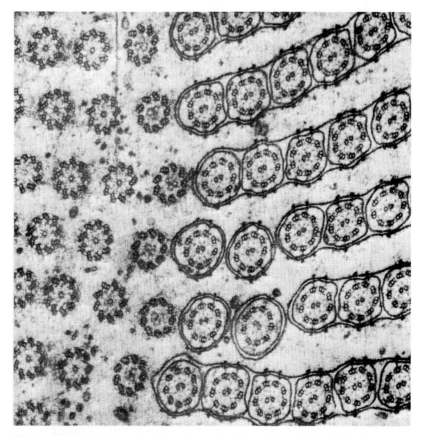

FIG. 8. Electron micrograph of cross section through the ciliary field of a proto-zoan. (*Courtesy I. Gibbons.*)

sist of a matrix, which is enclosed in a cylindrical sheath and con-tains in the interior invariably eleven parallel fibers, of which two lie in the center, while the other nine are rather evenly distributed about the periphery; in some types, each of the nine is a doublet, sometimes with an additional odd appendage. The basic pattern is remarkably constant. Over the whole range of species, cell types, and stages, measurements of the diameter of individual cilia, of the order of about 0.2 micra, seem to vary by no more than a factor of two. The occurrence of the number 9 is puzzling and unaccountable by ordinary rules of symmetry; in favorable cases (Fig. 8), they can be seen to be arranged pinwheel-fashion, rather than in a plain circle, while the two central fibers define a plane of bilateral symmetry for each cilium, the planes of symmetry for all cilia of a given tract ap-pearing to be essentially parallel. Each cilium grows forth from a basal body localized beneath the cell surface, but just how this takes place is still quite obscure. Depending on whether these production sites are spaced at random or are themselves arranged in definite patterns, the resulting ciliary fields are either irregular or ordered in grids, as exemplified in Fig. 8. A cell can repeatedly re-form a ciliary field of the typical pattern after resorption or loss of a primary field (e.g., after the fission of ciliate protozoa); consequently, the conspicuous morphological features of ciliation must again be viewed not as singular products of some single occurrence in the developmental history of the cell, but as the tangible manifestations of an ever-present and active dynamics of the cell.

It should have become abundantly clear by now that this dynamics operates hierarchically: macromolecules are linearly linked to filaments, filaments are laterally compounded to fibers of definite diameter, fibers in fixed numbers (9 + 2) are combined with matrix in a fixed proportion and distribution to make up a sheathed cylindrical cilium of fixed diameter, and rows of cilia compose a ciliary field. Each level of this hierarchy is thus characterized by its peculiar constants of composition, dimensions, proportions, and configuration [3a].

As a second example of an intracellular system of great structural regularity, we may cite the contractile apparatus of the cross-striated muscle fiber. Its major elements on the molecular level are chains of the proteins myosin and actin which are combined with other components of the muscle cell (sarcoplasm) into myofibrils, which in turn are assembled into muscle fibers, which then are grouped and wrapped by fibrous sheaths into still larger units, the muscle fascicles, a given number of which constitutes a muscle. For every one of these unit classes, the sizes, numbers, and arrangement of its subunits seem to be held within such close statistical limits that one must conclude that each one is subject to a separate regulatory mechanism. Little is known beyond the sheer fact that such an order exists, and only on the two lowest levels have studies of the fine structure presented us with some details on the degree of order. According to those studies, leaving aside all controversial interpretations, each myofibril is a tandem array of identical segments ("sarcomeres") cemented together by transversal discs. Each segment consists of large numbers of parallel protein filaments (Fig. 9a) extending from one cementing disc to the next. In the middle portion of each segment, another protein (presumably myosin) appears in close association with the continuous filaments and so exactly in register among all the molecular chains of the sarcomere as to give rise to the sharp "anisotropic" band of microscopic "cross striation." To this strict sequential order along the axis of each com-

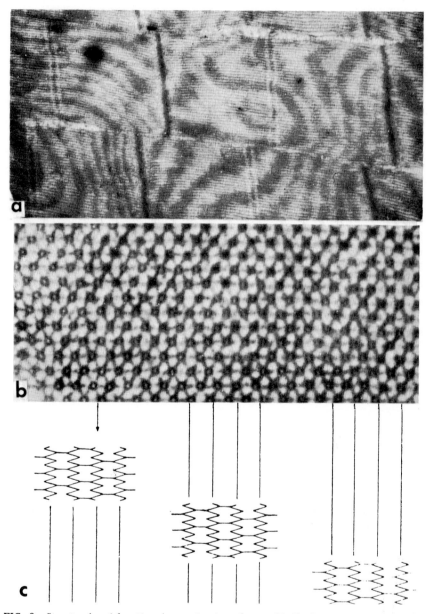

FIG. 9. Structural and functional organization of a myofibril of cross-striated muscle. (a) Electron micrograph of longitudinal section, X23,500. (*From A. Hodge.*) (b) Electron micrograph of cross section. (*Courtesy A. Hodge.*) (c) Phase coordination of contractile wave. Polar displacement of a liquid "plasma" strand. (*After Frey-Wyssling.*)

pound molecular chain is then added a higher intermolecular order in the lateral direction, of which the register within the anisotropic band is probably a collateral expression. In cross section (Fig. 9b), each individual compound molecular chain appears as a circle, of which the light center has been assumed to be the actin core, with the dark ring around it representing the myosin jacket. As one can see from the picture, these elements, without being stacked in direct lateral contact, are spaced in a regular hexagonal grid pattern.

Although the dynamic basis of this pattern is still obscure, it points to a strong lateral group interaction between the component filaments, hence it is of decidedly supramolecular order. The same inference is reached on physiological grounds, for the effective contraction of a muscle fiber presupposes that the contractile wave be synchronized for all the constituent chains. If contractility is based on the propagated alternation of the folding and unfolding of protein chains, as has been contended, all the elements of a unit at any given transverse plane would always have to be in the same phase, as indicated in the diagram of Fig. 9c. If, as others have proposed, contraction results from a lengthwise slippage of chains relative to one another, this likewise would have to be cross-synchronized. Either theory requires strict lateral coordination among units which are separated by distances of several hundred angstroms. This functional cross linkage may be mediated by structural cross connections [4], but such periodic cross structures themselves must have been preceded by a process which staked out the respective planes in the first place.

The muscle cell is particularly impressive in demonstrating the inexhaustibility of the faculty of cells to produce substructures of invariably the same pattern. The "hypertrophy" observed in muscles as a result of exercise or endocrine stimulation involves in the main the enlargement of individual muscle fibers by the production from within their cytoplasm of additional myofibrils, built exactly like the old ones, with which they are made to line up. The basic property of a muscle cell, therefore, is not its possession of a given complement of myofibrils, but its capacity to produce and reproduce myofibrils almost indefinitely.

Many more similar examples could be listed. Undoubtedly the rapid expansion of fine structural research will enlarge the list at a phenomenal rate. As a result the facts and problems of supramolecular order will become more familiar and attractive. But even the scanty evidence thus far available, as sampled in the foregoing, permits us to draw a few general conclusions as guides to our thinking and further research as follows:

THE HIERARCHY OF ORDER

Any description of cell constitution and cell behavior couched solely and directly in molecular terms misses the basic fact of the existence of intermediate supramolecular entities of unit character. To be fully consistent, one might as well assert that a cell consists of atoms in various combinations, which of course is true but meaningless. As different atoms are compounded into molecules with specific chemical properties, so cells contain, besides free molecules, larger units in which diverse species of molecules are combined in definite numerical ratios and specific mutual space relations to yield composite structures of rather uniform dimensions, proportions, and architecture. Although the de novo origin of these supramolecular subcellular entities has not yet been observed directly, the empirical fact that cells can keep on producing and reproducing them proves that they can be formed whenever and wherever conditions are right. The more complex the units are, the more specific prerequisites must evidently be fulfilled in the same place at the same time in order that the crucial event of complexing be able to take place.

The fact that each kind of these supramolecular units appears mostly in characteristic uniform dimensions at all cell ages indicates strongly that the particular molecular assemblies composing a given unit can exist solely in that single unique combination and configuration, and that partial or intermediary combinations are too unstable ever to be discovered. This is the obvious explanation of the fact that no one seems ever to have seen a fractionally completed mitochondrion, cilium, or sarcomere. Plainly, unless the conditions in a given spot at a given time are absolutely correct for all the components of such a unit to fall into their proper places in the right numbers and proportions all at once, no structure is formed to leave a record. Consequently, such structures are known only in quantal units, as it were, and in their multiples, but are never found in fractions.

On the other hand, in stressing uniformity of size and proportions of such composite units, one must remain aware of the degree of latitude left to the individual specimen. Although there are no comprehensive measurements and statistical data available on the actual variances among the dimensions of different cilia, mitochondria, or comparable subunits, the fact that they are subject to some degree of variation is clear from ordinary observations. That is to say, the microprecision of molecular composition ascribed to biological macromolecules (e.g., amino acid composition of proteins; nucleotide composition of nucleic acids) does not prevail at higher organizational levels. The numbers and ratios of macromolecules going into the formation of an organelle are unquestionably not exactly identical in each instance, but are merely kept within a certain standard range. This only strengthens our conclusion that the unit size of the complex is not the passive outcome of the stacking up of a preassigned number of molecules but rather that some *equilibrium level for the unit as a whole* determines the approximate number of subunits that it can accommodate.

Let us consider, for example, a cilium. Whether each of the nine peripheral filaments is a bundle of exactly the same number of molecular chains is not known, but it is highly doubtful. The dimensions of the two central filaments definitely indicate a composition different from the peripheral ones. As for the amorphous matrix, the number of constituent molecules is decidedly an open one, as one can readily see from the fact that as the cilium grows forth from its basal body, the matrix increases steadily. By contrast, however, the number of filaments is rigidly fixed. As for the distances between them, only the average value seems constant, leaving the individual measurements to fluctuate about that value. The diameter of the whole cylindrical structure again seems to be far more constant than one could expect in view of the indefiniteness of the mass of the amorphous matrix, which raises the possibility that the circumference of the sheath may be the critical equilibrium value which secondarily determines the amount of matrix as that quantity which it can hold within its confines. And going on to the spacing of the individual cilia in a ciliary field, it is quite evident that however regular the rows and distances of their generative basal bodies may appear on the supramolecular scale, it would be absurd to assume that this could be due to their being separated always by precisely the same number of molecules.

In summary, patterns of supramolecular order result not directly from freely interacting *molecular* units, but from the interactions of *supramolecular* entities of various degrees of complexity and various orders of magnitude, which behave as relatively constant units. Such unit groups possess properties and faculties that are characteristic of the particular collective, but are not manifested by the individual components in the unassembled state. Although the rules of the dynamics of such group behavior are still largely unexplored, much could be learned about them from a proper evaluation of their resultant morphological expressions.

This presupposes that we consistently cultivate the mental habit of viewing morphological patterns, such as microscopic or ultramicroscopic structures, as merely indices and residues of the patterns of processes by which they have been formed — rhythmic structures, of processes with periodicity; polar structures, of polarizing processes; etc. Dynamic process is the foundation of static form, rather than the reverse. Processes are not visible, only their effects are. Which brings to mind the words of Robert Louis Stevenson:

> Who has seen the wind?
> Neither you nor I.
> But when the trees bow down their heads,
> The wind is passing by.

FROM THREAD TO FABRIC

In the realization that the way "from molecule to cell" leads stepwise through a hierarchy of levels of increasing ordered complexity, we can now turn to examining on which of those levels the compounding of higher units from a free mixture of lower-order constituents has been observed, or better still, has actually been accomplished outside of cells. Each step, as stated earlier, presents a problem of its own. Since the study of the origin of cellular subunits is in its infancy, our case will have to rest on scanty evidence. We shall not consider here subunits of the so-called "self-reproducing" kind, where a new unit springs directly from a preexisting unit of the same level of organization, although they grade over into the category of "replications after a model," on which we shall touch later. Rather we shall concentrate on the crucial case in which, by the sheer free interaction of elements, a more complex unit of a higher order is synthesized even in the absence of a corresponding model unit, that process referred to earlier as true "compounding," as molecules are compounded from ions.

As the simplest instance, one could cite the formation of linear polymers, the self-stacking of mixtures of polar molecules along interfaces, and the three-dimensional self-sorting and ordering observed when polar molecules combine into coacervates. In all these cases, what started out as a random mixture of molecules ends up in a less random condition, manifested as segregation, orientation, alignment, and the like, attained either solely by the mutual interaction of the molecules concerned or with the aid of physical guidance from the environment, as will be discussed further below. All of these are still relatively primitive systems, not confined to the organic, but illustrating at least in rudimentary form the principle of emergent order in groups.

Going on to the specifically biological objects, however, one finds oneself at a loss for truly crucial examples to cite, except for one outstanding phenomenon — fibrogenesis, especially of collagen. The formation of the collagen fiber has been one of the very few processes whose analytical study has been driven far enough to document both the fact and the manner of the compounding of genuinely biological fine structures [5]. The main conclusions seem to be sufficiently firmly established, even though some of the detailed assumptions on which they rest are still hypothetical and in part controversial.

The elementary molecule of collagen, "tropocollagen," is a protein of a molecular weight of about 300.000; it consists of a triple-stranded chain of amino acids with a particularly high ratio of glycine and hydroxyproline. The length of this macromolecule has been determined as about 2,600 A and its width as 14 A. In the body, the molecules are known only as polymerized chains, "protofibrils," which in turn are bundled up into larger fibrils. Electron microscopy and X-ray analysis of native collagen fibers have revealed a characteristic axial periodicity in the fibrils (Fig. 10), with a major cross band repeating itself every 640 A and minor cross bands spaced in an aperiodic sequence in between (see Fig. 11), resulting in a definite morphological polarity of all segments. The fact that a fibril, which is an aggregate of hundreds or thousands of molecular strands, shows these over-all regularities, proves that the molecules have not aggregated at random, but in a definite linear and lateral order. Since the serial order of the minor bands is the same in all segments, polymerization must have taken place with the head end of one monomer always linking up with the tail end of the next throughout the whole chain. The cross striation of the assembled fibril in turn reveals that the constituent molecular chains have become joined laterally in such a way that the homologous fractions of all chains have come to lie in strict register; any slippage out of register obviously would blur the band pattern. The current view is that there are only four possible equilibrium positions in which the protofibrils can join up laterally and that the typical band pattern (Fig. 11) is the result of the statistical summation of the four staggered positions — each period of 640 A representing the ends of one fourth of the population of unit segments 2,600 A long. As we shall elaborate presently, a further element of order lies in the fact that many fibers fall into definite size classes, reminiscent of the more elaborate subcellular units discussed in the preceding chapter, which means that in the cases concerned there is a finite upper limit beyond which lateral aggregation of protofibrils cannot continue. So much for the collagen fibers encountered in the body.

Yet under appropriate conditions, precisely the same type of organized fiber can be compounded from the component molecules outside the body in vitro. Nageotte had discovered that collagen from connective tissues (e.g., tendon) can be brought into molecular solution by treatment with weak acids. Upon adding salts to this solution, or better, by dialyzing against salt solutions, a colloid precipitate would form that had all the then known properties of native collagen.

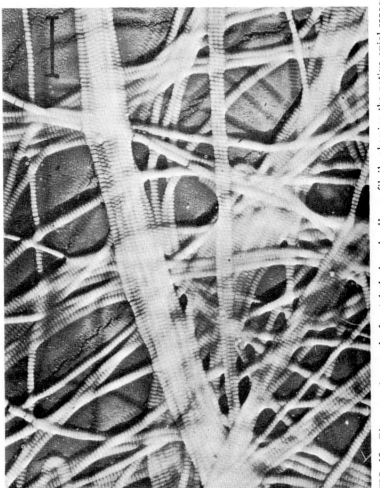

FIG. 10. Electron micrograph of metal-shadowed collagen fibrils showing the native axial repeat period at 640 A. (*From F. O. Schmitt.*)

DYNAMICS OF DEVELOPMENT

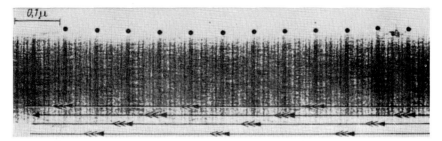
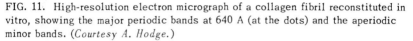

FIG. 11. High-resolution electron micrograph of a collagen fibril reconstituted in vitro, showing the major periodic bands at 640 A (at the dots) and the aperiodic minor bands. (*Courtesy A. Hodge.*)

It was this observation which the subsequent electron-optical, X-ray, and biochemical studies were signally successful in elaborating. Fibers reconstituted from such solutions in vitro showed electron-microscopically (Fig. 11) the typical 640-angstrom periodicity, and in several other critical tests gave every evidence of being the equivalent of the native fibers from which the molecular solution had been derived. As in the native structure, the molecules had become polymerized in correct polar order into protofibrils and the protofibrils had aggregated in register to form cross-banded fibers.

These results are a patent documentation of the fact that when identical molecules of given constitution come together in an appropriate environment they do not remain a random pile, but snap into mutually compatible group arrays, which are characterized by common orientations, alignments, and size rules of higher degrees of order than were present in the original molecular scramble. Here then is a crucial case in point to substantiate the thesis of "order emerging in compounds," which we have set out to test.

The issue gained a novel and even more exciting aspect when it was shown that the native collagen array (with the 640-angstrom periodicity) is only one among several possible group constellations which collagen molecules can assume, including some quite artificial patterns not known — or at least not yet observed — in nature. By allowing the dissolved collagen molecules to precipitate in different ionic and otherwise altered environments (e.g., in the presence of adenosine triphosphate), the following modifications in the form of aggregation in vitro could be obtained: (a) fibers without periodicity (abolition of lateral register); (b) fibers with 200-angstrom repeat pattern (abnormal staggering); (c) fibers with a 2,600-angstrom periodicity, but with polarized segments (Fig. 12; only a single instead of four possible stacking arrays); (d) fibers with a 2,600-angstrom periodicity but with symmetrical banding patterns within each segment.

I. On Basic Principles   51

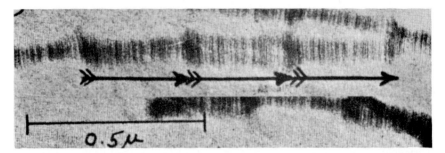

FIG. 12. Electron micrographs of aberrant assemblies of collagen reconstituted in vitro, forming periodic segments of a length of 2,600 Å. (*From F. O. Schmitt.*)

These are the facts. It must be admitted that the detailed nature of the energy relations and force constellations that limit the combination of macromolecules to just a very few specific group configurations is scarcely known and awaits further analysis. But in the spirit for which I have been pleading, the acceptance and exploitation of these facts for the explanation of biological structure formation is not necessarily contingent on a final resolution of that problem, and we may take it for granted from here on that the kind of self-ordering of macromolecules in groups, which we have illustrated here, is a reality, whatever its physicochemical basis may turn out to be.

Proceeding from this foundation, we can now take a further step in our ascending course, which leads to an even higher degree of ordered complexity. It concerns the process of ossification, or more generally, of mineralization of tissues [6]. Bone is a tissue which derives its mechanical strength from the incrustation of its organic matrix with the inorganic crystals of hydroxy-apatite. The matrix, as in all varieties of connective tissue, consists of a base of mucopolysaccharide interlaced with collagen fibers of the native type just described. If the crystals were simply scattered freely through the ground substance, they would be no more than grit; to act as reinforcements for the fibrous framework they must be firmly linked with it, like microbraces. This is achieved by having the crystals incorporated in the collagen fibers. Since these fibers form prior to the mineralization process, one may expect the pattern of crystallization to be guided by the collagen pattern. In confirmation of this expectation, it was actually observed [7] that at the onset of osteogenesis in the embryo, the first deposits of crystals visible in electron micrographs appeared along the collagen fibrils at constant intervals corresponding to the 640-angstrom repeat period of the fibrillar cross bands (Fig. 13*a*). Evidently some specific condition prevailing at those particular sites of the collagen fiber initiated or enhanced

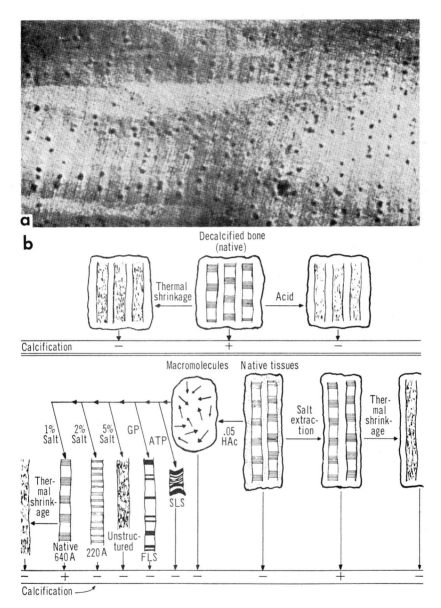

FIG. 13. Localized combination between collagen fibrils and hydroxyapatite crystals in ossification. (a) Electron micrograph, showing crystal deposition (dark dots) at periodic repeat bands of collagen fibrils in the initial stage of embryonic bone formation. (After S. Fitton Jackson.) (b) Diagram summarizing the success (+) or failure (−) of calcification of different varieties of collagen in metastable solutions of calcium phosphate. (From Glimcher.)

either the formation or the early growth of the crystals. In their later stages, the elongate crystals are found to lie, on the whole, parallel to the axis of the fibrils with which they are associated, thus revealing a continued directive dependency.

It was at this point that the success in reconstituting collagens of various forms in vitro offered a promising, and as it turned out, most profitable approach to a more penetrating study of that dependency. The task was to find out whether collagen in vitro could actually induce the formation of apatite crystals in a metastable solution of calcium phosphate. And this did take place if reconstituted collagen fibrils of the native 640-angstrom periodicity were used as inducers, even when the particular collagen had been extracted from normally not calcifying tissue, such as tendon, skin, or swim bladder. The crystal germs reappeared at sites corresponding to the 640-angstrom segments of the fibers so the conclusion was compelling that some peculiar configuration of the fibrillar structure at those levels, in contact with the ambient solution, initiated the formation of crystal nuclei. But not until collagen patterns of other than the 640-angstrom variety were tested could the degree of specificity of this effect be assessed. As was mentioned before and is diagrammatically represented in Fig. 13b, the same molecular solution of tropocollagen can be made at will to yield in vitro fibrils of either the native type with the 640-angstrom repeat period, or several other types of fine structure. Now, in exposing each of these varieties, as well as variously treated native tissues, to metastable calcium phosphate solutions under otherwise identical conditions, it was established that crystal nucleation occurred only in the presence of fibrils showing the 640-angstrom repeat period, and in no other combinations (Fig. 13b). Since neither the simple tropocollagen molecules, nor their linear polymers, nor any larger aggregates of other than the native stacking pattern has ever led to nucleation, it is evident that the effect depends strictly on the specific group arrangement which characterizes the macromolecular collective in the native-type fibril.

There exists thus a matching relationship between certain steric properties of the native collagen aggregate on the one hand and the crystalline state of the hydroxyapatite on the other, which predisposes these two systems to combine in a definite space pattern and thereby to "compound" a new system of higher-order complexity. The fact that native collagen from ordinary connective tissue will become calcified in vitro, but not in the body (except under certain pathological conditions), points to crucial differences in the local microenvironments of the fibrils at different tissue sites, only some of which would be favorable for crystallization; but this is another matter. The basic lesson to be drawn from this remarkable story is that the

principle according to which two otherwise unrelated and heterogeneous components can be conjugated into a higher-order union if there
is some steric or other specific correspondence between them is not
confined to intermolecular reactions, for which it has been generally
acknowledged, but is equally valid in the realm of large supramolecular systems. The discovery that virus particles, which consist of a
nucleic acid core and a protein jacket, can be dissociated into these
two components, and that if the two separate fractions are mixed again,
they will combine again in the proper spatial arrangement [8] points
to the same conclusion of "complementarily prematched systems."

Whether or not this is to be interpreted as a veritable "missing
link" in the chain "from molecule to cell" is something we shall
consider later. For the moment, it serves to illustrate that in order
to become integrated into a higher-unit system, the elements that
come together must have properties that prefit them for each other;
then, if they are of proper fit, they apparently fall into place instantaneously. The mutual fitting may be based, as in this last example,
on some configurational correspondence; but it may also reside in the
fact that the time courses of the two component events are harmonious ("resonance"). Since the configurational model is the simpler
one, we shall take the next step in our ascent from there.

COMPLEX ORDERED FABRICS

If the foregoing has acquainted us with properties of a multimolec-
ular complex, the collagen fibril, we now pass on to consider a system of higher order, in which the fibrils are assembled to fibers of
rather uniform size, and the fibers are arranged in a definite geometric pattern in a continuous matrix, or "ground substance," of
mucopolysaccharide. This dual composition at once raises the question of whether, as in the simpler case of calcification, the structural
order might not be a conjoint product of both component systems — in
this case, collagen and ground substance.

The issue is presented most clearly by the basement lamella of
the larval amphibian skin [9]. This is a membrane which lines the
underside of the epidermis and separates the latter from the subjacent loose connective tissue. As can be seen in Fig. 14a, it is a
laminated structure. Notwithstanding certain variations according
to species, age, and body region, its basic construction is essentially
constant, consisting of an electron-optically amorphous sheet of matrix, about 4 micra thick, in which the collagen fibers lie embedded,
like steel cables in reinforced concrete. A higher-power electron
micrograph (Fig. 14b) illustrates the regular geometric pattern which
the courses of these fibers describe, as well as other constant fea-

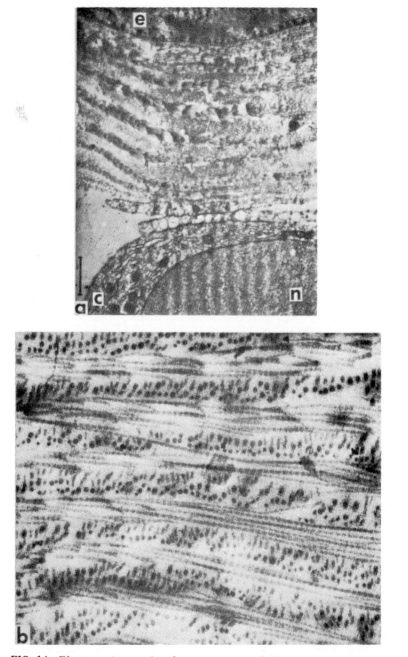

FIG. 14. Electron micrographs of cross sections of the laminated basement lamella of larval amphibian skin. (*a*) Low-power view (ca. 10,500X); *e*, epidermal cell; *n*, nucleus, and *c*, cytoplasm of fibroblast. (*b*) Part of the lamella at higher magnification (ca. 27,000X.) (*Originals.*)

tures, as follows: (1) The fibers run in layers of about 2,500-angstrom width each. (2) All fibers within a given layer are parallel. (3) The direction of the fiber axes alternates between adjacent layers by an angle of 90°; that is to say, fiber orientation is the same for all even layers, as well as for all odd layers, and a surface projection makes the fibrous fabric appear as an orthogonal grid, while a favorable transverse section, such as that of Fig. 14b, shows the fiber systems alternately in side view (as beaded strings) and in cross sections (as circular dots). (4) Along the trunk, the two systems of fibers run diagonally with regard to the main body axis. (5) The individual fibers are cylindrical, with relatively uniform diameters of 500 to 600 A. (6) The fibers are not densely packed and may be separated by distances of several hundred angstroms (control experiments have excluded the possibility that this separation might be a hydration artifact arising during fixation of the tissue). (7) Despite the lack of lateral contact between fibers, the periodic bands of all of them within a given layer appear to be in register, defining thus a periodic system of parallel planes perpendicular to the fiber axes.

We are faced here with a fabric of great over-all regularity, in which the individual fibers, themselves supramolecular units, function as subordinate elements. Evidently, such an object affords a unique opportunity for studying the manner in which higher-order patterns develop. We used for this study the process of wound healing (Fig. 15a). Small holes made in the skin, which penetrate the basement lamella, are quickly covered by epidermal cells moving over the lesion. The gap in the underlying basement lamella is repaired more slowly, so that its stepwise reconstruction could be followed. At first the gap is filled with a structureless exudate of undetermined nature. Then fibroblasts appear underneath and deposit collagenous material in this matrix. The precise mode of fibrogenesis is still debated; but the prevailing view is that the fibroblasts extrude fibrous precursors, which on interaction with the outer medium become polymerized and aggregated into fibrils. At any rate, the young fibers in the wound in the vicinity of the fibroblasts already show the typical periodic banding of native collagen. Significantly, however, they show no preferential orientation whatever and lie helterskelter in the matrix, like cordwood dumped from a delivery truck (Fig. 15b). Moreover, these fibers are much thinner than the mature collagen fibers and, of course, much sparser. The hole in the old basement lamella is thus filled at first by a primitive patch of ground substance containing an irregular feltwork of matted immature collagen fibers. This, incidentally, corresponds to the terminal stage at which the development of the lamella is normally arrested in certain areas of the body surface.

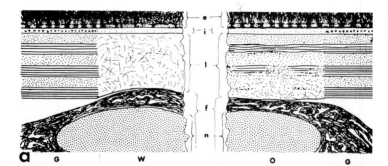

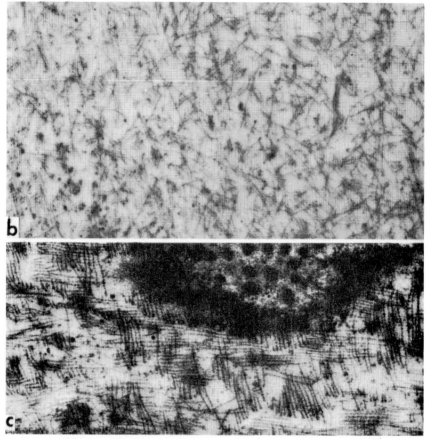

FIG. 15. Healing of wounds in the basement lamella. (*a*) Diagram (cross section) of restoration of lamella (*l*, only six layers are represented). G, old intact region; W, O, wound region during first and second week of healing, respectively; *e*, epidermal cell; *f*, fibroblast. (*b*) Electron micrograph of cross section of wound region with unordered early fibers. (*From Weiss and Ferris.*) (*c*) Electron micrograph of tangential section through wound region adjacent to epidermal underside (island dotted with "bobbins") in second week of healing, with beginning ordering of fibers. (*From Weiss and Ferris.*)

Secondarily, however, this erstwhile irregular texture is superseded by the typical architectural order described above, which is
characteristic of most skin regions. The ordering process starts
from the epithelial surface and sweeps downward. It manifests itself in a marked reorientation of the fibers nearest the epithelial
underside; instead of the earlier random directions these assume
orientations (a) in a common plane parallel to the epithelial sheet,
and (b) parallel to one another within that plane, their axial direction
being determined by the stumps of the old fibers at the former wound
edge. The same process then repeats itself in the next subjacent
layer, but with axis orientation perpendicular to that of the preceding
layer. Then a third layer appears, and so on down, always with alternation axis orientations (Fig. 15c), until the whole lamella has again
become laminated. At about the same time, the fibers acquire mature dimensions.

The unexpected lesson from these observations has been that the
imposition of geometric order upon the fabric is a secondary process,
or rather a series of processes: preexisting fibrous units are being
rearranged and restacked, comparable to the orderly crosswise layer-
ing of cordwood from a dumped heap. This is expressing the matter
in oversimplified terms, especially since there may be a recasting of
fibers involved, comparable to recrystallization, that might have escaped detection; yet basically, the simile is correct. The question
then arises whether such a physical reordering process can be accounted for in terms of interactions among the structural components.
A closer study of the disposition of the fibers during the early phase
of rearrangement furnished some clues, as can be seen from the ex-
ceptionally favorable tangential section pictured in Fig. 15c. The
section lies in a plane roughly grazing the underside of the epidermis,
a sliced-off cap of which (with bobbins) is seen on top. The basement
lamella having been hit at a very low angle, the consecutive layers of
fibers in it appear almost side by side, with only partial overlap. Thus
seen in surface view, the following details could be discerned: (1) Each
layer contains as yet one row of fibers only. (2) Within each layer,
the fibers have already assumed a common parallel orientation and,
most significantly, with their periodic bands aligned in lateral register, just as in the original undamaged membrane. (3) The individual
fibers are only about one-third as thick as fibers of the mature tissue, but they are regularly spaced at lateral distances from center to
center of 500 to 600 A. (4) The fibers of each layer intersect the projections of those of adjacent layers approximately at right angles
(making allowance for distortion during fixation). Since the lateral
distance between the fiber axes is of the same order as the axial repeat (~ 600 A), the projections of overlapping adjacent orthogonal

sets of fibers describe squares, many of which are plainly evident
in Fig. 15c.

The fact that the segments of the parallel fibers within any given
layer lie in register despite the lack of lateral contact calls for some
principle of alignment operating across the intervening spaces of
several hundred angstroms. One can conceive of two different hypotheses both of which would satisfy this condition. The first is outlined
in the diagram of Fig. 16. It goes as follows: The eventual consolidated fiber grid bears the same relation to the earlier random mat-
ting, which it supersedes, as does the crystalline to the amorphous
state of a substance. The "crystallinity" of the ordered system would
consist in the fact that certain critical constellations of the collagen
fiber, morphologically expressed by repeat bands, could become stabilized only at equidistant points, roughly 600 A apart. The first
row of fibers, laid down in a plane defined by the epidermal underside
and polarized by the linear extensions of the "crystal" axes of the
severed old fibers around the wound, would then constitute a grid with
a square pattern that would serve as foundation. A second story of
fibers would then be set on top, and if we assume again an equilibration distance of some 600 A, we arrive at a cubic system of nodal
stabilization points. There are two line systems to define the short-
est connection between these points on the second story of the lattice
(Fig. 16), one parallel (*A*) and the other (*B*) normal, to the basal
grid. If one assumed that symmetry relations would make *B* superior to *A* energywise, the orthogonal alternation from row to row
would be explained.

In all of this, the geometric order of the final fabric would be a
result of the self-positioning of the fibrous units relative to one another into a distribution and orientation which would be the only permissible one for a collective body of the given constitution in the given
physicochemical environment. In this version, the hypothesis would be
but a logical extension to the next higher level of the principle underlying, for instance, the formation of the orderly aggregate of a fiber

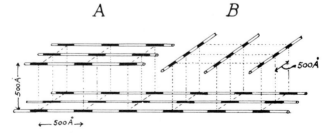

FIG. 16. Hypothetical cubic lattice for the stabilization of collagen fibers in the basement lamella. (*From Weiss.*)

from a pool of free molecules. The fibers, by virtue of their interactions with each other and with their environment, would determine a field of forces with energetically distinguished equilibrium points spaced in the indicated cubic lattice pattern. The pattern of the emergent system of higher order thus would result from the fact that the interacting units themselves have a distinctly nonrandom, patterned constitution. At any rate, in this view, the active role in the performance is ascribed to the collagen units.[10]

However, an alternative hypothesis, which also deserves serious attention, would assign a much more crucial role to the ground substance. It would attribute the requisite property of crystallinity with a cubic lattice and a lattice constant of around 600 A to the ground substance, and it would postulate that the nodal points are sites where physicochemical conditions are singularly suited for the linkage between ground substance and collagen. In other words, the periodic collagen chains would settle on grids of lines staked out by acceptor sites of the matrix which coincidentally are spaced in the same steric pattern. The resulting higher order would then be based on a property of neither the ground substance alone, nor of collagen alone, but on the fact that both systems share a fundamental steric property. This, as one realizes now, would truly tie this case conceptually to the lower-order one of mineralization discussed before, and more generally, to the broader biological thesis of specificity as based on the interaction between systems mutually attuned by pairwise matching properties.

This latter hypothesis assumes that the supposedly "hyaline" ground substance in reality possesses definite structural order analogous to "crystallinity." This assumption might seem gratuitous in view of the fact that the electron microscope, at least after the conventional procedures of tissue fixation, has as yet disclosed no direct evidence of fine structure in the matrix. But there are some observations that ought to caution us against accepting the negative evidence of the electron microscope as conclusive. Two of these observations pertain directly to our present object — the basement lamella — hence, will be related here. A third one, more involved, will be dealt with later.

It will have been noted in Fig. 14a and b that the boundaries between the domains of consecutive layers of the lamella are absolutely sharp and straight, even though the planes one can lay through the outermost row of fibers in each layer are quite erratic; to visualize this, one need only connect the dots marking fiber cross sections in Fig. 14b. This in itself suggests that the lamination is a feature of the ground substance, notwithstanding the absence of clear optical delineations between laminae.

I. ON BASIC PRINCIPLES   61

But there is more direct proof that the ground substance is actually built of discrete layers, plywood fashion. At metamorphosis, the cell-free larval membrane is suddenly invaded by mesenchyme cells from the underlying connective tissue. As these cells move in, they do not drill through the lamella in arbitrary directions, but can be seen to extend preferentially along invisible cleavage planes between the individual layers, wedging lamina from lamina. A cross section through the exfoliating membrane at this stage (Fig. 17a) shows quite clearly that each lamina is a cleanly separable slab. Because cells have been known to prepare invasion routes by trail-blazing enzymes lytic for ground substance, and because the matrix in our object could by analogy be judged to be a mucopolysaccharide of the hyaluronic acid class, we tried to duplicate the metamorphic exfoliation by placing larval skin directly into hyaluronic acid. As Fig. 17b shows (by comparison with Fig. 14b), this treatment actually dissolved the ground substance between the collagen fibers (causing the latter to become more densely packed in between), but as in metamorphosis, the enzymes spread faster and more effectively in the planar direction and along the invisible borders between adjacent layers (in the horizontal sense in the illustration).

Thus there is concrete evidence of a planar pattern of organization in the ground substance; that is, there are structural discontinuities which set off layer from layer, although these have become discernible only in biological tests, not in electron optical inspection. On the other hand, our hypothetical proposition of an additional structural order in the third dimension, as presented above, can derive from these tests, at best, some logical, but surely not yet actual, support; it is still highly speculative, as is much of this excursion into obscure and uncharted land.

Yet there is one recent instance which positively implicates the ground substance in the determination of architectural features of tissues. This case deals with the development of cartilage by chondrogenic cells in vitro. At a certain stage of their development, mesenchymal blastemas destined to form cartilage, which have been removed from the embryo and explanted into nutrient media, will continue to develop in vitro and form skeletal pieces characteristic of the embryonic sites from which they were taken [11]. Thus mesenchyme from avian limb buds, grown in vitro, will form limb cartilages; that from the ventral midline, sternum; that from the eyeball, sclera; each with the characteristic site-specific architecture. Later it was found [12] that when cells of embryonic limb mesenchyme were separated from one another by trypsin treatment and the suspension was placed in a culture medium, the reaggregated cells would develop into cartilaginous nodules. The identification of the develop-

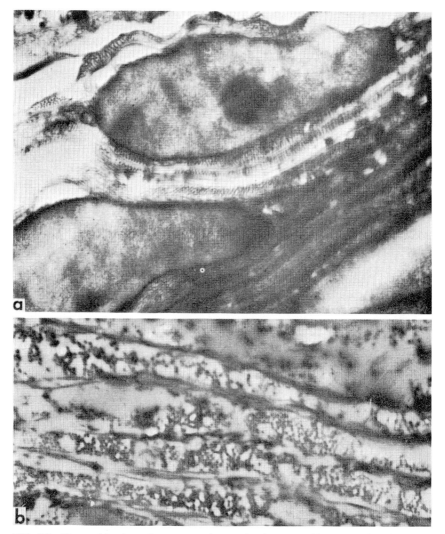

FIG. 17. Signs of laminated structure of ground substance of basement lamella. (a) Invading mesenchyme cells. Two nuclei, one with dark nucleolus, are seen in section; cytoplasm, extremely thin and indistinct, split the lamella along preformed cleavage planes. (*From Weiss and Ferris.*) (b) Electron micrograph of cross section of larval lamella, fixed after immersion of live skin in hyaluronic acid, showing faster dissolution of ground substance in the horizontal than in the vertical direction. (*Original.*)

mental result rested largely on the appearance of intercellular ground substance typical of normal cartilage, in which the cells became entrapped. Evidently the faculty of generating specific cartilaginous ground substance was inherent in each individual cell of the poten-

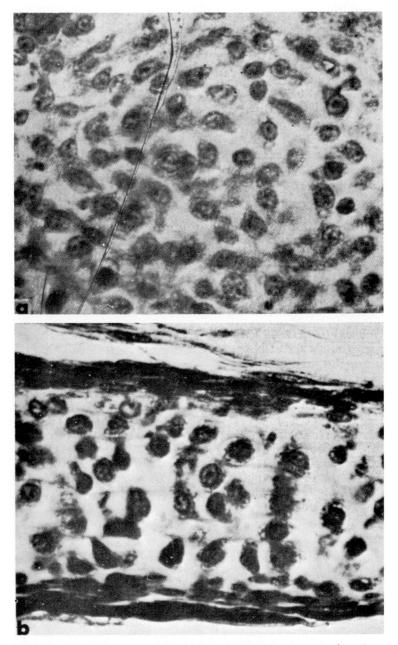

FIG. 18. Transverse sections through cartilages formed in vitro from dissociated and reaggregated chondrogenic blastemas of limb bud (*a*) and sclera
(*b*). (*From Weiss and Moscona.*)

tially chondrogenic tissue, rather than in some property which only the integral blastema would possess.

But a further extension of the cartilage experiments indicated a far more differentiated architectural role of the ground substance. As just indicated, cartilages allowed to develop in vitro from undisrupted chondrogenic blastemas bear certain distinctive features characteristic of their original destinations. Thus limb cartilage, *in situ* as well as in vitro, tends to grow rather massively in concentric whorls, whereas scleral cartilage in both conditions is laid down in the form of a slab of rather uniform thickness. It was surprising to find [13] that these distinctions were still clearly displayed by cartilages developed from blastemas that had been broken up completely into their component cells. Mesenchyme cells from the limb bud and from around the eye, respectively, dissociated, scrambled, reassembled, and then cultured under identical conditions, produced two distinctly different kinds of cartilage, which, as one can judge from Fig. 18, are readily recognizable as of the massive whorllike limb pattern (Fig. 18*a*) and of the flat plate-shaped scleral pattern (Fig. 18*b*), respectively.

This is no longer just a sign of the ability of chondrogenic cells to secrete a generalized cartilaginous material — chondroitin sulfate and collagen — but is an index of far subtler properties. For obviously, although it takes a multitude of cells to build collectively either a nodule or a slab, each individual cell of either tissue must have "known" how to contribute its distinctive share to the erection of its special type of cartilage structure when it became rejoined to other cells of like character. Contrary to man's technology, where culverts or cathedrals can be built from the same kind of bricks, the cellular bricks for different skeletal structures are differently constituted and carry, as it were, the blueprint for the respective construction procedures within themselves.

This is, of course, a matter difficult to visualize and even harder to translate into analytical terms. However, in order to make at least a beginning, I have suggested that perhaps each cell type produces a ground substance of different "crystalline" characteristics, which would determine the specific mode of its accretion in the different dimensions of space, with the radial direction dominating in the case of the limb, but a two-dimensional extension in the case of the sclera. The cells, enclosed in their own product, would then assume conforming positions within this framework, and the architectural integration of the "compound" would thus be completed. There seems to be sufficient variety in the family of mucopolysaccharide ground substances [14] to make such a hypothesis at least reasonably possible. In combination with the preceding account of the architecture of the basement lamella it assumes even greater plausibility. If confirmed, it would bring the evidence for the "compounding" of higher-order systems by the free interaction of lower-order units clear up to the level of tissue structure. Let it be remembered, however, that in all these instances the integrated pattern of the group interaction is prepared — built in, as it were — in properties of the interacting units, which, as observers who know the outcome, we recognize as anticipatory in character.

SOME ORDERING PRINCIPLES

From this survey of ordered or regulated interactions of units at all levels from macromolecules to tissues, one can extract a number of general conclusions which are instructive in two respects: in one sense, they give greater precision to our initial thesis about "the cellular control of molecular activities," and secondly, they lead to the posing of more concrete and succinct questions to which research on "cellular organization" could be directed than would be feasible from the worm's-eye view of molecules alone. It is for the purpose of setting the sights, rather than with any presumption of finality or comprehensiveness, that I venture to set forth the following somewhat arbitrary selection of thoughts on the subject.

One result which emerges clearly from a survey of our illustrative examples is the almost commonplace realization that the operation of a system cannot be inferred from its composition. A description of composition is a static record of the content of a given unit: a catalogue of all the items in it, whether they operate or not. No matter in what terms we list the component items, whether as molecules or higher complexes, no item as such ever "acts." It can only "interact" with another item [15]. Yet such interaction is, of course, contingent on the opportunity of both to come together. If we are dealing with a body in which all components can move freely and at random, the probability for any two of them to meet is merely a matter of time; homogenization is a device to approximate this condition. However, in any system whose components are restrained in some degrees of freedom of motion, there will be correspondingly fewer probabilities for interaction from unguided pairwise collisions. In other words, of the total number of possible constellations and combinations that could conceivably be formed from the elements of a given inventory, were they free, only a very limited fraction has actually a chance to materialize within the framework of an organized unit system. One could vaguely compare the static chemical equipment of a cell to a chemist's store with chemicals on shelves confined in bottles: the mere presence of the bottles on the shelves as such has no effect whatever. However, when by design and choice some are uncorked and their contents mixed, predictable reaction products are obtained — as the outcome of "organized behavior." Homogenization, by contrast, is comparable to the smashing of all bottles, spilling and mixing freely the contents of them all.

Therefore, unless the listing of the content of a system ("composition") is supplemented by a description of the set of conditions which limit the opportunities for the occurrence of random interactions, the operation of the system remains incomprehensible. Although a plain truism, this rule is so commonly ignored that it deserves to be reiterated.

Two recurrent mistakes may illustrate the point. One is the tacit conversion of enzymatic "activity" observed in a living cell into terms of enzyme "concentration." Evidently, any enzymes bound to specific sites (hence not distributed ubiquitously throughout the cell) can be "active" only in proportion to their accessibility to substrate and to the fraction of active groups actually exposed to the substrate, and in reference to the physicochemical environment at that particular microsite. Collision frequencies between enzyme and substrate molecules, which are what we register as "activity," are further reduced if substrate cannot diffuse freely but is channeled into limited compartments. For any given localized group of fixed enzyme molecules, on the other hand, "activity" will be enhanced if all of them are oriented alike so as to expose the maximum number of reactive groups to the passing stream of substrate, particularly as parallel orientation would also permit the closest stacking, hence the accommodation of the maximum possible number of enzyme molecules per unit of available surface. Again, we encounter structure, in the broadest sense, as one of the major factors "controlling molecular activities" in the cell, so that homogenates, unleashed from its restraining influence, can never portray truly the operation of the structured system.

A second related set of errors arises from inadequate attention to barriers of penetration by agents to which a cell or cell part is exposed (e.g., hormones, antibodies, drugs, etc.). Such situations are often dealt with as if the ability of the agent in question to permeate the cell and all its compartments could be taken for granted. Admittedly, the barrier of the cell membrane is rather generally respected, but it is not equally well recognized that every interface, microscopic or submicroscopic, must be considered to be a potential screen for the passage of substances or the transmission of influences. Moreover, there is nothing to justify the expectation that any given agent that can pass through a given cell compartment will emerge from it in the same form as it has entered. For instance, in order for a chemical externally administered to an embryo to be able to alter the genic response of certain cells, and hence to cause developmental alterations, it must reach the genes. But in order to reach them, it must pass through cells, intercellular spaces, body fluids, and finally in the target cell, through cytoplasm, nuclear envelope, nuclear "sap," and chromosomal matrix — a complicated journey through spaces metabolically far from inert, during which it is more likely than not to suffer some alterations. The same is true of the traffic in the reverse direction, from genes to body products. This is not to say that substances may not get to distant destinations without essential changes, as often they will, but merely to caution against assuming such inertia as a matter of course.

These two examples serve well to illustrate the sort of structural limitations of the degrees of freedom of interaction which distinguish the organized system from its randomized homogenate. The constructive side of this principle lies in the fact that as it restrains ubiquity of interaction, it also creates uniquely favorable conditions for special reactions to occur at now restricted sites. I have already referred to the potentiation of the effectiveness of enzymes resulting from their adsorption from solution, where they had n degrees of freedom of orientation, to an interface, where they are stacked in parallel. In general, structural organization will be effective only to the extent to which, in restraining wasteful ubiquity and randomness, it establishes guides for optimal utilization of the available energy; just as a combustion engine turns explosive energy into useful work. In the cell as in the machine, structural provisions route energy from random, undirected, dissipation into useful channels.

A common morphological expression of such directive guidance, as well as one of its major tools, is a uniform (or at least nonrandom) orientation of polar elements. As we shall show below, tool and product are here in a reciprocal relation, which is perhaps most simply illustrated by the grooving action of water flowing downhill — the river fashioning its bed and the bed confining the river. Processes oriented in a given direction (e.g., shear, hydraulic, or electric flow) can result in structural orientation, which then, in turn, will channel further flow. Orientation, therefore, is a device for, as well as an index of, increased efficiency.

In this general perspective, it seems highly doubtful whether a living cell could operate efficiently if substance traffic in it depended chiefly on the relatively inefficient process of free diffusion, particularly in view of the narrowness of the capillary spaces involved. Just as higher organisms have expedited intertissue traffic by the institution of circulatory systems, so intracellular traffic may be enhanced by convection currents much more widely than is commonly envisaged.

Aside from the rather conspicuous streaming observed in cell bodies of radiolarians, amoebae, plant cells ("cyclosis"), and fish eggs, the possibility of "directed traffic" inside the cell has received no more than casual attention, and certainly no systematic study. In a speculative mood, one might add that such "traffic direction" need not be confined to the familiar device of channeling and pumping liquid flow, but that the interfaces of linear bodies in liquid pools may also be able to propagate substances down their length faster than would be feasible by free diffusion in the pool, hence, would act like everted channels. This would be particularly pertinent if oriented molecular chains should turn out to have the faculty of acting as "bucket brigades" in conducting electrons or protons down the line at rates far in excess of those attainable in solution [16]. If this should come to pass, one might be led to conclude that all the truly relevant directive processes in a living system are carried out by molecular arrays that form a continuous, though labile, network with the properties of a solid, rather than by the liquid diffusion pools, which are subject to random agitation.

However, to return from these conjectures to familiar ground, there are enough established facts to argue for the dominant role of molecular orientation as one of the basic principles through which biological organization becomes effective. The following primitive models, which I presented a decade ago, may serve to symbolize some elementary ordering processes in this category. They are to convey the realization that even processes that give an aspect of elementariness, like "molecular orientation," may still be truly composite, occurring in several steps "controlled" by different and often quite unrelated factors. The fact of such compositeness makes explicitly clear why, as was postulated in the preceding sections, there must be "integration" of all those unrelated factors on a higher level if they are to be rendered interrelated for the conjoint act of constituting and preserving a higher unit system.

Figure 19 shows the progressive ordering of a linear array. We start at *a*, with a random pile of filamentous macromolecules (e. g., of a fibrous protein) with different polar end groups. On introducing (in *b*) an appropriate linkage, we obtain a linear polymere chain (in *c*) with polarized segments, but neither straight nor oriented (ignoring, for simplicity, attractive and repulsive forces between side chains). To straighten the chain, we introduce an external force, for instance, mechanical tension (in *d*), which if applied to a whole group of chains, turns all of them into a common direction (in *e*), thus imparting common linear order to the collective. In this situation, further order emerges from the interactions among the formerly separated elements, which as a result of their parallel orientation have been brought into closer proximity, hence, mutual lateral interaction, conducive to aggregation into larger bundles (ignoring, again for simplicity, more complicated group configurations such as the formation of tactoids). An ulterior ordering step ensues if some equilibrium condition for the group prescribes that the ends of the monomeres of neighboring chains be as closely approximated to one another as possible, which would cause them to slip from random axial positions (*e*) into strict register (*f*). It will be noted that this series of steps is essentially a portrayal of such phenomena as the collagen fibrogenesis described above.

Let us now proceed to a model of planar order. For this, the model experiments of Langmuir and of Harkins on monomolecular layers can serve as a point of departure. They show the orderly adsorption from solution of macromolecules with polar end groups along an interface dividing media of different affinity for the respective end groups. For instance, molecules with a hydrophilic group at one end and a hydrophobic group at the other, when entering the boundary between oil and water, will form a picket fence, the ends of whose stakes are each immersed in the medium appropriate to it (Fig. 20*b*). On such a polar monolayer, a second, inverted, layer can then be deposited; on this a third layer, and so forth. In other words, the exposed ends of an adsorbed monolayer act like a new interface. Consequently, a hydrophilic protein can be deposited on top of the hydrophilic face of a lipid layer, and in this manner, a mixture of both species (Fig. 20*a*) can be segregated as a double foil (Fig. 20*c*). For practical examples, we need only refer to our earlier description of some lamellar systems within cells (Figs. 4 and 5*b*). As one can see, this scheme could be extended to allow for much more specific and exclusive affinities as determinants of the molecular combinations to be accomplished than is possible by the mere binary choice among the terminal charges on a dipole. Once a ground layer has been established of molecules with end groups of specific steric configuration, all of which stick out into the medium, these would function as acceptors or traps for any passing molecules that have a complementary configuration, thus leading to the building on of a second layer of different composition. And if these latter molecules were then to act, on their part, in a similar manner, still a third molecular species with affinities to the exposed ends of the second would be fished out from the common medium (Fig. 20*d*).

As one can sense, in deriving the rationale for this model of "stacking according to steric correspondence," I have leaned on the phenomena of immunology. The model is as hypothetical as is the steric basis for antibody-antigen binding, but not much more so. If it should prove untenable, some other model for the building up of

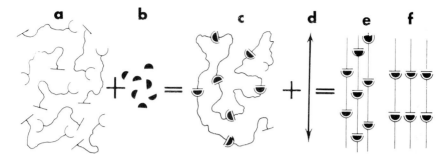

FIG. 19. Diagrammatic model of stepwise ordering of linear array of macromolecules. (*From Weiss.*)

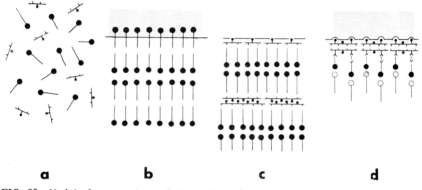

FIG. 20. Model of progressive ordering and stacking of macromolecules along interfaces. (*From Weiss.*)

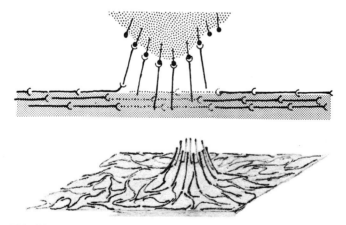

FIG. 21. Hypothetical generation of surface breaches and leaks as a result of molecular reorientation toward a specific stimulus. (*From Weiss.*)

I. ON BASIC PRINCIPLES

the third dimension in an organized system would have to be invented. For one important consideration that must be taken into account in explaining three-dimensional organization of supramolecular systems is that the interior of such systems initially is often liquid, hence randomized. The only spatial patterns of sufficient stability to provide firm bearings for any subsequent organized consolidation of the interiors, therefore, reside in the boundary layers, which by virtue of their interfacial positions act as two-dimensional solids. As such, in contrast to the liquid contents, they can hold in position space mosaics of diverse molecular populations in a typical nonrandom mapping.

Given such primary surface mosaic, interactions of its settled molecules with the disordered content would then secondarily extend order into the interior, vaguely comparable to the erection of a building on its orderly foundation. For instance, I proposed long ago that the basis for the localization of the first differential steps in embryonic development is to be sought in a surface mosaic of areas with specifically different molecular populations, blocked out in the cortex of the egg, and all the experiences gathered since have strengthened this concept. It would seem plausible to broaden it by assuming that the organization of other cell and subcell systems likewise resides primarily in planar patterns. Growing support for this assumption comes from the prominence of smectic systems in the cell (myelin sheath, cell membrane, endoplasmic reticulum, etc.) revealed by the electron microscope. The secondary elaboration of a planar pattern in depth adds not only body to the system, but most significantly, a further dimension of order in the progressive chemical diversification along the new third axis, much as primary differences in soil composition find secondary expression in the pattern of vegetation.

For further details of these concepts, I must refer to earlier accounts [17]. Indeed, without such supplementary commentary the static frame of the system described in the preceding pages would seem to be incompatible with the lability and flux of structural organization emphasized in the earlier sections of this paper. First, I labored the fact that outer and inner membrane systems of the cell are in a state of continual flux and remodeling, and now I have introduced a topographic mosaic of conventional stability as the fixed basic reference system from which further organization is to proceed. Let me, therefore, briefly summarize the formula by which I have in the past attempted to reconcile the two propositions. It is implicit in what I have termed the concept of "molecular ecology" [18].

It considers molecular populations in the living cell essentially analogous to populations of organisms which are adjusted to their ecological environment. Each kind of population is capable of existing and surviving only in an appropriately fitting set of external conditions, comprising chiefly the physicochemical milieu and the activities and products of other populations in the joint territory. We know that macromolecules of the kinds to be found in living systems require for their formation and maintenance highly specific constellations of metabolic conditions, such as the adequacy of concentration, proportions, and local availability of the requisite building blocks, of the energy supply, of essential catalysts and other cofactors, of the right temperature range, pH, and so forth, as well as the sort of kinetic enhancement by proper alignment and orientation which was mentioned above. The more complex the molecular species in question, the more particular are its existential prerequisites, and the more factors must combine cooperatively at a single site in order to satisfy them. Once formed, a given macromolecular species still can only lodge at a site compatible with its continued survival, where it is protected against metabolic extinction. Conversely, it will be positively concentrated at a site for which it has chemical affinity.

Because different molecular species with different requirements must coexist and function in a common space without being insulated from one another by preformed mechanical compartments, the activities of each are of critical concern to the survival of its neighbors. Any localized chemical activity in a given spot inside a cell consumes, transforms, and gives off substances and energy in a characteristic time course and thus constitutes part of the fluctuating environment of other centers in the vicinity, just as the activities and products of the latter reciprocally impinge upon the former. In this dynamic interdependency, no localized system can become established and persist unless what the neighboring systems are doing is in every respect compatible with it. The inner organization of a cell thereby becomes "self-organization," which becomes stationary whenever the various chemical domains have assumed a distribution relative to one another such that maximum mutual harmony and complementariness of operation are guaranteed. The crucial point is that this state is brought about and actively maintained by the interacting subsystems themselves, rather than by their fixation in a rigid mechanical framework — their grouping being the result of the dynamics of the populations involved, rather than of passive allocation to given positions by outside forces.

In all of this, different macromolecular species may appear either in cooperative, mutually supporting ("symbiotic") roles or in competitive relations. Consequently, when a mixed molecular population is faced with a variety of possible sites for adsorption, or chemical combination, or otherwise preferential location, it will sort itself out into its separate components, as each species occupies the site most appropriate to its kind. In preempting that site, it automatically excludes other species from it. If either a site or a species residing there changes in character, some other species, previously barred, suddenly acquires thereby favorite status and consequently crowds out the original occupant. The patterns of relative distribution, segregation, and localization among the different segments of a mixed molecular population thus sustain themselves dynamically as a result of the "ecological" conditions which the various species create for one another in a bounded physical space. In this sense, structure and ordered activity appear again as merely different phases of the same phenomenon, as the "primary" structural patterns of localization in surfaces, to which we ascribed above a guiding role in cell organization, presumably originate, themselves, from the dynamic self-sorting and territorial segregation of molecular species in competitive interaction.

Although stated in this generality, these concepts of molecular ecology must sound rather vague, they could be liberally documented but for the lack of space. Nor is it as important for the moment to be convincing as it is to introduce the ecological point of view into the molecular realm, where it can overcome serious conceptual barriers to an objective treatment of biological organization. Needless to say that in line with the hierarchical scheme of organization, there also will have to be created a macromolecular ecology, an ecology of supramolecular units, of organelles, of cells, and of cell populations.

As will be shown later, the ecological scheme is, in fact, a more concrete version of the more formal "field" concept in so far as it endows geometric field parameters such as "central," "peripheral," "marginal," "axial," etc., (as well as references to "position" or "distance") with physical meaning in terms of differential effects of forces and limiting conditions. I mention this in the present context because our earlier discussion of molecular orientation as a dominant ecological factor in the regulation of molecular behavior is a case in point. I have indicated its bearing on structure formation and on enzyme efficiency. To this I shall add a bried remark on its instrumental role in cellular selectivity and specificity.

Cells possess the discriminatory capacity of selectively admitting or excluding specific physical stimuli, chemical agents, and traffic with fellow cells (e.g., in permeability, food ingestion, drug reactions, hormone responses, parasitic infections, immune reactions, fertilization, virus penetration, sensory perception, or nerve excitation). While it has been generally assumed since Ehrlich's days that in at least some of these phenomena the specificity resides in the interlocking of molecules of matching configuration between the two interacting partners, the relation between the specificity and the dynamics of such interactions (transfer of substance and energy, current flow, cementing, and so forth) has remained undefined.

As a tentative approach to the problem, I have recently [19] proposed a "dualistic" hypothesis based on the following assumptions. (1) A major fraction of the cell surface is occupied by a network of filiform macromolecules in essentially planar (surface-parallel) array, barring substance passage ("barrier position"); (2) certain of these molecules have end groups of specific configuration as selective acceptors for complementary groups; (3) carriers of complementary end groups approaching the cell from the environment, in attracting and combining with matching surface groups, thereby turn the respective molecules from their erstwhile tangential into radial ("open-gate") positions (Fig. 21); (4) for molecules with an axial (length:diameter) ratio of 100, this reorientation implies the uncovering of 99 per cent of a formerly covered surface site, in other words, the opening of local "breaches" or "pores" as channels for secondary outflow or inflow across the surface; (5) local electrostatic disturbances produce surface "leaks" in similar, but unspecific, fashion.

According to this concept, specific molecular interactions at a surface serve merely to unlock passage for less specific and dynamically more powerful transport and transmission mechanisms. What has been said for the cell surface applies, of course, equally well to other membrane barriers within the cell. Although the validity of the idea remains subject to further tests, the main point in presenting it on this occasion is to focus attention on the crucial role of macromolecular orientation in the ecology of the cell and its subunits. It is noteworthy that this type of effect would remain concealed to ordinary chemical determinations, as it involves no change in the composition of the molecular population concerned.

DIFFERENTIATION

Our considerations thus far have been confined to a cell which can be regarded as stationary, immutably maintaining its essential character as a whole despite the incessant variations of content, distribution, shape, and behavior of its constituent elements (other than perhaps the genes). This is permissible as long as we contemplate only a relatively short sample period of its life span. A new class of problems along the way from molecule to cell emerges, however, as soon as we widen our scope to encompass the full life cycle of the individual cell, including its growth, differentiation ("cytodifferentiation"), and possibly its transmutation, and aging; or broader still, the development of the whole protoplasmic continuum of many cell generations, of which the individual cell is merely a small probing sample. Again, at the hurried pace of this cursory survey, I can do no more than sketch the gist of the problem [20] as follows:

As indicated above, in molecular terms one can ascribe the progressive specialization of the interior of a cell to the competitive segregation of selected molecular species into various and separate "ecological niches" favorable to their kinds, followed by a series of interactions between these settled species and those of the still roaming population; the nature of the interactions varies with the nature of the species which have monopolized the critical positions. It does not matter in this context whether or not the particular key species exercise their master functions as clusters of enzymes or as nucleation sites (or anchoring points) on which to build structural elements. The salient feature is that one can understand in principle how an orderly mosaic of chemical properties, once it has been initiated in the cell, can be progressively elaborated to yield the specialized equipment displayed by the "differentiated" cell, examples of which have been given in an earlier section. I have also indicated before that the master positions of given segments of the population need not be permanent; but that if conditions change, some ruling groups could be supplanted by totally different ones, entailing rather thorough modifications of the behavior, products, and appearance of the whole cell [21]

This picture of cytodifferentiation, which is restricted to the individual cell during its individual life span, presents nothing new that could not be fundamentally covered by our ecological cell model. Yet if we turn from the cell individual to the family of cell lines, the problem of differentiation becomes truly vexing. For then the problem arises of how two cell strains, derived from one common ancestral cell — or to put it more pertinently, derived from two cells of exactly identical composition and constitution and fully interchangeable — can acquire the radically different properties which characterize them in later stages of development, often irreversibly. Considering the contingencies of cell life, random differences between two originally equivalent cells could arise and then be transmitted, and even be amplified in their transmission, to the descendant cells. It is not difficult to visualize how in this manner a cell population could undergo random diversification. In fact, this is apparently what takes place among cell strains cultivated in vitro for prolonged periods. However, in the organism the diversification is not fortuitous, but specific as to time and place, and this requires an extension of our cell model. We must assume that if two cells are destined to give rise to different kinds of progeny, this is achieved by inducing a different set of compounds in each to occupy the critical master positions, thus initiating radically different sequences of subsequent interactions in the two lines.

Figure 22 illustrates diagrammatically an extreme case, in which such a differential dichotomy between two equivalent cells is induced by their exposure to different environments whose molecules are prematched to different complementary key fractions in the cell. In this case, the discriminate response is of the "resonance" type. Supposedly, however, there are other instances in which the same dichotomy of cell fate can be evoked by environmental differentials of less specific character (e.g., different metabolites favoring different members of competitive pairs of metabolic processes). What conditions determine the switch in one direction or the other must be decided empirically for each separate instance of cell-type differentiation. It also remains an empirical task to determine whether the course of events following the initial critical dichotomy will or will not deprive a given cell of its faculty to return, in response to changed conditions, to the state at which it was at the time of the original dichotomy and thereafter to transform into something else [22].

In ascribing the primary dichotomies of cell lines to the differential effects of outside conditions, we do not intent to obscure the fact that the differential courses of cell development thus initiated from the cell periphery will, at the central end, result in differential reactions and activations of genes, with telling effect on the subsequent transformation of the cell and of its progeny into a given specialized type. In line with our earlier comments, however, the genes are rated as reactive rather than active participants in differentiation. The studies of experimental embryology and developmental genetics have furnished abundant and compelling evidence to support this view.

The total repertory of reactions and productions a given cell could possibly display at a given stage is limited and defined not only by its initial physicochemical endowment, but by all the modifications through losses, gains, transformations, and dislocations, which this endowment has undergone in the prior ontogenetic history of that cell. Yet of the remaining faculties only a very limited fraction is ever given a chance to materialize, depending on which portion of the molecular population attains controlling dominance, as symbolized by our model, in response to conditions outside the cell. It cannot be emphasized too strongly that the critical "conditions" in question — which are popularly referred to as "inductive" — are different for each specific type of differentiation, and that tendencies to discover and identify a single universal "inductive agent," or at least a common class of such agents, are not only illusory, but in outright contradiction to many established facts of development. On this point,

I. ON BASIC PRINCIPLES   77

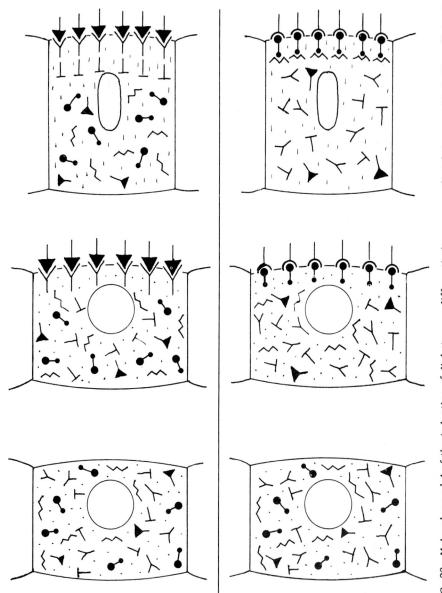

FIG. 22. Molecular model of the induction of dichotomous differentiation among cells of the same kind. (*From Weiss.*)

DYNAMICS OF DEVELOPMENT

the actual concrete phenomena of development, as they have become known through the disciplined and precise studies of the past, are in marked contrast to some current notions of development as an abstract generality. This cautionary remark seems necessary in view of the express danger that the growing and enthusiastic collaboration modern cell biology is receiving from the physical sciences might be misdirected toward some false and fictitious notion of a cell and of what really is involved when cell fate is turned in one direction rather than another.

Perhaps because of this ambitious search for an illusory over-all master solution, there is as yet hardly a single case in which we have detailed factual information on, let alone control of, how a given cell with alternative reactive potencies is actuated to take one specific course to the exclusion of another. There are certain hopeful beginnings, but that is all. One case in this category deserves mention here because of its exemplary simplicity and clarity. Epidermis of the chick embryo, explanted into a medium of blood plasma in vitro, produces the usual stratified epithelium of skin, the outer layers of which transform into typical keratin. But if the medium is enriched with vitamin A, the cells take an alternative course and form a columnar epithelium, which secretes mucus; and even a brief bath in vitamin A, followed by rearing in unenriched medium has the same modifying effect on the cells that have been exposed [23]. Although the intimate mechanism of this spectacular deflection of a cell line from one course into another is still unknown, here at least is one instance in which the initiation of the conversion can be linked directly with a known difference in the molecular environment [24].

This whole line of argument, however aptly it may apply to the progressive orderly diversification among cells, does not, of course, answer the familiar question of how the first critical differentials may have come about when the egg was still a single cell, uniformly exposed, as it is in many cases, to an environment devoid of the kind of systematic differentials to which one could ascribe a differentiating effect. For an answer to this problem, one must turn to our earlier statement that there are major regional differences in the molecular composition of the egg surface, which, in being parceled out directly among the cleavage cells, leaves them, right from the start, with specifically different endowments in accordance with the particular parcel of egg surface each has acquired [25].

Despite their sketchiness, these remarks on differentiation should have sufficed to indicate that in development, as in the stationary mature cell, more catalogues of what is present in the cell, and what *might* come of it, can yield no understanding of the orderly sequence of actual events, that is, of just what part of the inventory

becomes operative, when and where; and what really *does* take place
in the system. That understanding presupposes knowledge of the
principles which govern the nonrandom, typically patterned, distribution and discriminative activation of specific segments of the inventory. Unfortunately, such knowledge is still very scanty. It might
grow faster if more attention were paid to problems of this sort.
And perhaps more attention would be focused on the problems if they
were stated in explicit, rather than symbolic, terms. The models
presented in the foregoing should be regarded in this spirit — more
as aids in the phrasing of questions, than as answers.

GROUP DYNAMICS

The key theme of our entire discussion has been the extent to
which the order observed in a unit of higher order can conceivably
be derived from the ordered group dynamics of its interacting elements of lower order. We have encountered this issue at levels of
all magnitudes, from the complexing of macromolecules to the cooperative building of typical tissue architecture by randomly aggregated cells. If it were not for the fact that the level of the cell was
set as arbitrary cutoff mark for the topic of my address, the theme
could readily have been further expanded upwards to the order-
determining group dynamics of organisms, populations, and species,
which is the subject of the discipline of ecology. The fact that its
basic tenets are formally so similar throughout the whole range of
biological magnitudes, has led me to the concepts of "molecular
ecology" and "cellular ecology," as well as earlier to a "field"
theory of development [26]. Of course, labels like these do no more
than identify the problem and point to its nature; its resolution still
depends on our success in determining precisely the rules of interaction that mark the patterned effect of group activity as of a higher
order than the sum total of the effects of the individual constituents
operating separately. So what is needed is not a new set of noncommittal terms (and I include among them the much misused symbol of
"information"), but factual descriptions of how specific kinds of interactions among elements in given environments can yield order in
the group. Carried out methodically, such studies should lead to a
consistent science of "dynamic morphology," which could subsume
under a common principle, such phenomena as, for instance, the
dendritic patterns of electric discharges, of snow crystals, of nerve
cells, of lichens, and of trees. As yet we do not even have a nucleus
for such a systematic science. But in order not to dwell on abstractions, I would like to refer briefly to two concrete examples.

The first one is a phenomenon which I have called the "two-center effect." It appears in situations in which two separate centers of activity reside in a common medium in relative proximity. In such cases, the radial symmetry of the field of possible effects emanating from a single center is distorted in the direction of the connecting line between the two centers so that there emerges, instead of two partly overlapping radial patterns, a single novel pattern with axial symmetry, as illustrated in Fig. 23. In this manner, the independent scalar and undirected actions (e.g., chemical emanations) of each of the two centers yield, by virtue of the fact that they operate in a common matrix, a vectorial effect with unique geometric features of great consequence; in the place of the isotropic spaces around the independent centers, we now find a structured space connecting the two centers directly, establishing a preferential channel of traffic between them, hence terminating their former independence.

This description has deliberately been kept in the most general terms so as to be applicable to a wide variety of manifestations of the "two-center" principle. The practical case from which it was first derived concerned the interaction between two separate tissue fragments cultured in a common blood plasma medium [27]. The chain of events went like this (Fig. 23): Growing tissue dehydratized its colloidal environment (syneresis); water was lost from the

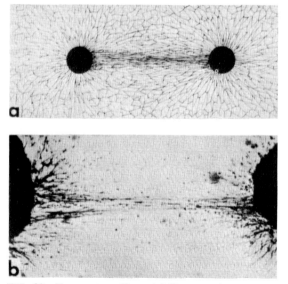

FIG. 23. Two-center effect. (a) Diagram of orientation of molecular chains in the fibrous matrix around two active centers. (b) Cell bridge formed between two explants (spinal ganglia) in thin plasma membrane. (From Weiss.)

plasma-fibrin sponge; the fibrin net therefore contracted, engendering radial tensions; two centers of radial tension in a common net
produced maximum strain along the connecting line; fibrin chains
consequently became oriented in this direction (Fig. 23a); cells migrating from the centers thus found a preferential oriented track; as
a result, not only did the two centers become connected and united
by a cell bridge (Fig. 23b) but the continued syneretic activity of
these growing cells amplified the effect progressively in positive
feedback fashion. The point is that once the asymmetry has been
initiated, it becomes further accentuated automatically by the chain
of sequelae. One can readily extrapolate the picture to the simultaneous activity of more than two centers. Evidently the effects will not
just summate arithmetically, as would be the case if only stoichiometric chemistry were involved, but they will create characteristic
patterns (see Fig. 24 for three centers). Thus, in conclusion, we have
a practical case in which order of a sort emerges demonstrably from
the interaction of separate independent bodies.

This is only one example from among many obeying the same underlying principle. Let me just mention one other variety. Supposing a
cell emanates some agents which can affect the surface constitution
of other cells. Naturally, unless other cells are present in the vicinity to serve as indicators, the effect cannot manifest itself. But if
two cells are present in close vicinity, a polarizing group effect may
emerge, as the accumulation of the agent on the inner side of the pair
sets up a steady differential of concentrations to which the cells are
exposed on their inward and outward facing sides. In liquid media,
there is a critical distance above which no such surface asymmetry

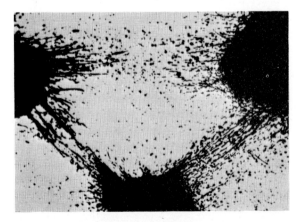

FIG. 24. Two-center effect exerted pairwise between
three separate cultures. (*From Weiss.*)

could develop because of the leveling action of thermal agitation. But below that threshold distance, cells can mutually polarize each other by their products. The functional result of this polarization will vary depending on whether the agent in question softens or stiffens the surface, resulting in the protrusion of cell content toward the opposite cell or away from it, respectively. In the former case, one may observe an extension of cell processes or actual movement of the cells toward each other, and in the latter case a movement of the cells away from each other [28]. It is important to bear in mind that even very slight and inconspicuous but systematic asymmetries of this kind can establish differentials which in the course of subsequent chains of interactions are not only perpetuated, but amplified. The two-center effect in its various expressions illustrates one general principle of morphogenesis which prevails from the tissue level down, presumably to the subcellular units (e.g., mitotic spindle), and perhaps lower. Its salient feature is the emergence of novel patterns from interactions within the confines of the group of elements without pattern-determining directives from outside the group, in other words, "self-organization" [29].

The second example to be cited is of even broader application, although less well documented. It has to do with the emergence of crucial differentials within a mass of initially fully equivalent units, producing patterns in which the kind of reaction manifested by a given unit is demonstrably related to the position of that unit within the whole group. Such "position" effects have been observed at various levels, from the position of a gene in the chromosome to the position of an embryonic cell within the germ. As indicated earlier, "position" evidently signifies the prevalence at that particular site of characteristic physical and chemical conditions, which depend upon and vary with, the constellation of the whole system.

In order to instill some concrete sense into this abstract formulation, I shall give an elementary example of how equal units can suddenly become unequal, depending on their position within the group. Let us take a system the existence of which depends on an equilibrated exchange with its environment along a defined boundary (Fig. 25a). Let us then have a small number of such units, regardless of whether the multiplication has come about by division or aggregation (Fig. 25b); every one of these components still shares in the surface along which the exchange takes place, though in a smaller measure than the original body, hence perhaps quantitatively restrained by having become part of a collective. But carrying the numerical increase (by division or aggregation) still further, a condition abruptly arises (Fig. 25c) in which some elements are no longer in contact and communication with the outside medium, but

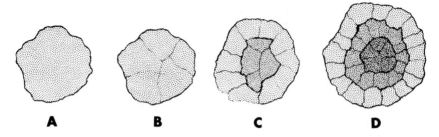

FIG. 25. Diagrammatic illustration of the emergence of differentials in groups of equivalent units. (*a*) Undivided unit (*b*) subdivided unit (*c*) further subdivision, yielding inner and outer units (*d*) increasing complexity by interaction of outer and inner units. (*Original.*)

are completely surrounded by their former fellow units, which henceforth act as mediators and screens from and to the outside medium. Among the units which initially were all "outer," a new positional category has sprung up, which now is "inner." Cut off from their former medium, they are suddenly deprived of all interactions with it, and become instead subject to a new set of interactions, namely, those with their new envelopment — the girdle of outer units. A simple example of such a positional differentiation is the separation in the solid blood island of the chick embryo of the outer cells, which form the endothelial tube, from the inner cells, which become blood cells. Now, let us increase the number of units still further so that a belt of units intermediate between the former outer and inner units emerges (Fig. 25*d*). Then these will obviously come under a dual interactive influence from both adjacent layers, as a result of which they will become a third variety [30]. In this manner, it is possible to conceive of progressive diversification of units in accordance with a relatively invariant over-all pattern of distribution and proportions irrespective of absolute size and dimensions [31].

Considerations like these may at least point the way in which an abstract "field" concept could be translated into operational terms. The actual translation, however, remains to be carried out in yet a single practical case. This is a program for the future. Basic to it is the realization that in a body composed of identical and interchangeable units, the reactions of individual units and hence their future fates, can be crucially different depending on where they are located in the body, whether in the surface or in the interior, or for that matter nearer or farther relative to the source of any agents with which they can interact. It has been the historic merit of the gradient theory of Child to have claimed a dynamic basis for geometrical "position" effects. It fell short of the goal by unnecessarily confining the dynamics chiefly to a single monotonic scale of values of metabolic intensity; for in doing so, it ignored the specificity and selectivity of intermolecular reactions which we now recognize to be among the foundations of organized structures and processes. And for this same reason, the crude scheme of "field" differentials presented above would lose its meaning unless it is coupled with a concept of "ecology of reactive elements," which makes the specific qualitative response of any element a determinate function of the constellation of physical-chemical conditions in its microenvironment at that particular locus; that is, of its coordinates within the integral field. Just what those conditions are which can evoke or suppress a particular response from among alternative responses, will vary for macromolecules, for macromolecular complexes, for cell organelles, and for cells, and within each level, will vary from type to type.

If this seems like a forbidding task for study, it becomes even more formidable when one remembers the introductory proposition about organized systems, which specified that all elementary events, however much they may fluctuate individually, must be so coordinated and controlled that their total combined effect does not impair the integrity, unity, and relative invariance of the system of which they are the constituents. Therefore, molecules can contribute to the "control of cellular activity" only insofar as "cellular control" prevails over their individual activities.

In order to illustrate concretely, once more and lastly, the fact that this "control" is not exercised by a single monopolistic master agent, but is reinsured by a multiplicity of cooperative and synergistic devices — many more than Spemann or Lehmann had envisaged when they spoke of "double insurance" or "combinative unitary performance"[32] — let me briefly refer to one means by which cells seem to exert control over one another's growth.

GROWTH CONTROL

It had become apparent that the growth of different body parts is regulated not only by the gross conditions of space, accessibility to nutrients, activity, and the like, but by much subtler and more specific means of harmonization. The hormone system constitutes one such device. But it, in itself, might be only an evolutionary specialization of a more general chemical control system inherent in the growth process. Proceeding from this general premise and a number of specific facts, including the experience that the artificial reduction of the cell population of a given organ system (e.g., liver, kidney, blood) is followed promptly by an automatic compensatory growth reaction of the residual part of the very same population, no matter how dispersed, I proposed a molecular "control" mechanism operating as follows [33] (Fig. 26):

1. Each specific cell type reproduces its protoplasm, i.e., it grows, by a mechanism in which key compounds that are characteristic of the individual cell type (symbolized by large circles and triangles in Fig. 26) act as catalysts. The postulated cell-specific diversity of compounds is the chemical correlate of the differentiation of cell strains. Growth rate is proportional to the concentration of these intracellular specific catalysts (or "templates") in the free or active state. Under normal conditions these compounds remain confined within the cell, where some become switched into nonreproductive differentiation products (stippled in Fig. 26).

2. Each cell also produces compounds ("antitemplates," small full circles and triangles in Fig. 26) which can inhibit the former

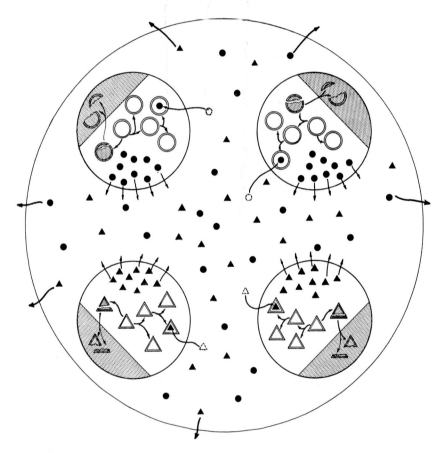

FIG. 26. Model of organ-specific molecular control of organ growth. Explanation in text. (*From Weiss and Kavanau.*)

species by combining with them into inactive complexes. These may be turned out as direct by-products in the process of protoplasmic reproduction or may be secondary differentiation products. They may be steric complements to the former or matched to them in some other fashion. The only prerequisites are that (a) contrary to the specific templates, they can leave and reenter the cell and get into the extracellular space and into circulation; (b) that they carry the specific tag of their producer-cell type, which endows them with selective affinity for any cell of the same type; and (c) that they are in constant production so as to make up for their extracellular katabolic decomposition and final excretion.

3. As the concentration of antitemplates in the extracellular medium increases, their intracellular density, hence inactivating effect on corresponding templates, will likewise increase; in short, growth rate will decline in all cells belonging to that particular strain bathed by the common humoral pool. When stationary equilibrium between intracellular and extracellular concentrations is reached, incremental growth will cease. This mechanism results in a sigmoid growth curve for the total mass of each organ system, and the familiar sigmoid curve for the whole organism is then essentially an aggregate of similar curves for the individual constituent organ systems.

This general concept of a "negative-feedback" regulation of growth offers a rational explanation both for the self-limiting character of normal growth in a confined medium (organism or culture), and for the homologous organ-specific growth reactions after experimental interference or pathological alterations. As can readily be seen, each interference or alteration will have to be examined in a dual light as to its effects on the concentration of both templates and antitemplates, since it is the ratio of both that determines growth rate. The following conclusions can immediately be deduced from this scheme:

A. Removal of part of an organ system removes part of the sources of corresponding types of templates and antitemplates. Since the former, according to our premise (1), have been in intracellular confinement, neither their former presence nor their recent loss can be noticed as such by other cells of the system. This is not so for the antitemplates, which are in circulation and a reduction of whose production source would soon register by their lowered concentration in the extracellular pool. According to points (2) and (3), this would shift the intracellular ratio of templates to antitemplates temporarily in favor of the former, causing automatic resumption of growth till a steady state is restored, resulting in a "compensatory" growth reaction.

B. Addition of a part should have opposite effects, depending on whether or not its cells survive, or rather on the ratio of surviving to disintegrating cells. If all cells survive, the net effect would be an increased concentration in the circulation of the particular anti-templates, hence a reduction in growth rate of the corresponding host system, provided it is still in a phase of growth. If all cells disintegrate, they release into the extracellular space a contingent of specific templates that would otherwise never have escaped. Assuming that these, according to point (2), combine with, or otherwise trap, homologous antitemplates, their presence in the pool will entail a temporary lowering of antitemplate concentration, hence again a spurt of growth in the homologous cell strains of the host. The simultaneous release of antitemplates from the disintegrating cells would have to be assumed to be insufficient to cancel this effect because of their faster metabolic degradation, point (2c). An alternative possibility is that the templates freed from cracked cells are directly adopted by homologous cells, where they would temporarily increase the intracellular concentration of growth catalysts, hence growth rate [34]. In either scheme, the release of cell content would accelerate homologous growth by increasing the intracellular ratio of templates to antitemplates — in the former case, by reducing the denominator, in the latter case, by increasing the numerator. It can be seen that in terms of this interpretation, partial necrosis of an organ will have the same effect as partial removal, and that implantation of a fragment, followed by some degeneration, as well as the injection of cell debris, are merely further variants of the same procedure.

When the mathematical formulation of this theory was subjected to quantitative tests, it reproduced with accuracy the observed time course of normal growth (e.g., Fig. 27); but above all, when electronic computers were programmed to derive from our equations the time course of changes to be expected in a given system after artificial reduction or augmentation of its mass, the records showed not only the automatic return to the original equilibrium mass, but its attainment by a series of damped oscillations, as is characteristic of systems with negative feedback regulation [35].

If further substantiated, this theory would lead to the conclusion that organ-specific features of molecules (e.g., organ-specific antigenicity) are not only products and indicators of differentiation, but are instrumental devices for the maintenance of intercellular balance and harmony in a multicellular system [36]. It might well be that the concept could also be scaled down to apply to intracellular regulation carried out by complementary pairs of molecules, one compound fixed to structures (analogous to the cell-bound template system), the other freely mobile in the liquid spaces (comparable to the freely

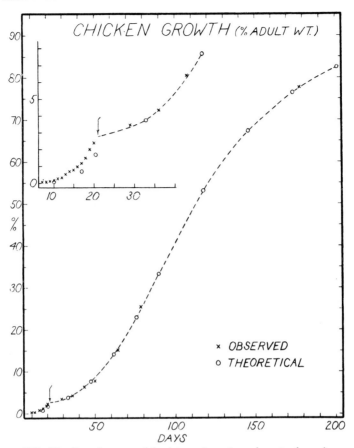

FIG. 27. Growth curve of fowl in embryonic and postembryonic phase. Crosses represent observed values according to Schmalhausen and to Landauer; circles are values calculated from mathematical growth model of Weiss and Kavanau.

diffusible inhibitors). This would bring us back to the vital interplay between the two interpenetrating systems of the cell emphasized in an earlier section — the ordered solid-state arrays and the liquid diffusion pools bathing them. But this is treading on wholly unexplored ground.

EPILOGUE

With these remarks on growth control, we have arrived at last at a point which comes rather close to what might be defined as "molecular control of cellular activity," only to discover that the "controlling" molecules have themselves acquired their specific configurations, which are the key to their power of control, by virtue of their membership in the population of an organized cell, hence under "cellular control." And this indeed has been the whole purpose of my long discourse; to document by practical examples that the distinction between molecular control of cellular activity and cellular control of molecular activity is based on the semantic ambiguity of the term "control," hence fades in the light of true understanding of the phenomena involved. A cell is nothing but the population of component entities that constitute it. But these entities are not just of molecular rank, nor can their ordered behavior in the group be fully appreciated and understood solely by studying them in isolation, out of context. As I have tried to carry the principle of self-organization of higher organizational units by the free interaction of elements of lower order as far as present factual evidence — not hopes, nor beliefs — would honestly entitle us to do, I had to add at every turn that elements endowed for such ordered group performance have always been prefitted for it by properties previously imparted to them as members of just such an organized group unit, whether cell parts, cell, or germ. This circular argument contains one of the most fundamental truths about the nature of organisms, and as one can readily understand, does not augur well for an eventual affirmative answer to our introductory question, in its naïve form, of whether a cell will ever be synthesized de novo without the active intervention of another cell.

On the other hand, I have also tried to document the fact that there are now available practical and constructive approaches to the gradual replacement of symbolic references to "organization" by true insight into the dynamics involved. Our knowledge of the dynamics is rudimentary and spotty. But it is consistent enough for us to realize that almost every mechanism in a living system employs a combination of dynamic principles, rather than just a single kind. There was a time, not long ago, when biologists would proclaim monopolies for certain forms of energy as key to life: electrical, mechanical, and chemical theories vied for primacy. The modern unified concept of energy makes this sound obsolete, and so it is. Of itself, scalar energy cannot define, in the sense of a precise description, a single process in a living system (except perhaps its disintegration). While energy input is needed to create and maintain nonrandom diversity, the difference between just *any* nonrandom state on the one hand, and the repetitive and conservative order of diversity in organisms on the other, is not spelled by the scalar values of energy, but by the vectors of its channeling; just as it is the ordered channeling that makes a given amount of energy fed into a machine yield useful work, instead of dissipating itself in an explosion. If energy is needed for a cell to move, what makes the difference between this coordinated effect and merely a diffuse warming up? And this is where the old questions of forms of energy involved are still as fully pertinent as ever. The answers no longer select favorites, but by experience have come to admit all forms as being used in various combinations, depending on the specific mechanisms concerned. No generality can exempt us from the effort of determining the workings of each mechanism — including "control" mechanisms — separately in its own right.

How often do we hear in the discussions of biological problems pronouncements to the effect that this or that event is "biochemical." Such statements are platitudinous unless they are accompanied by indications of how the particular reaction is conditioned by the physical setting in which it occurs and how its effects, in turn, modify the physical settings for subsequent reactions. In this broader perspective, ordering process and ordered structure become a single continuum, determining and limiting each other in endless sequences of activities, so that as a given chemical event may control (that is, condition) the appearance of a particular physical array or "structure," the latter then will go on to control (condition) the next chemical transaction, which in further consequence may again alter the prior structure, and so forth almost *ad infinitum*.

Since examples of such chains are scattered throughout this text, there is no need of belaboring the point further. But one important lesson should be reemphasized. Although the word "structure" evokes primarily a picture of such gross mechanical functions as are subserved by tension cables, weight supports, confining envelopes, cross ties, etc., we must not let this limited aspect dominate our thinking. In extrapolating into the future the picture I have presented, one can foresee the ever-growing importance that structure will assume in furnishing the chemical systems of the cell with expediting channels or inhibiting barriers, creating or abolishing, respectively, opportunities for interactions.

More work along this line is badly needed. Not only more work, but also more penetrating thinking about the problems of cellular organization in unfragmented, undiluted, uncorrupted form — the real problems, not some sham versions — so that our minds may gradually acquire a mature and soundly structured concept of cellular organization, as a guide to well-directed further exploration and interpretation of the phenomena of cellular control that make the cell the integrated operating unit it is, and that at the present time are still in deep obscurity.

I. On Basic Principles   91

NOTES AND REFERENCES

1. H. S. Bennett, Membrane flow. *J. Biophys. Biochem. Cytol.*, Suppl. 2, 99-103 (1956).
2. This is, for instance, strikingly illustrated in the motion picture made by F. Frédéric and M. Chèvremont, *Arch. Biol.*, 63:109 (1952).
3. For the evidence, see P. Weiss, *Harvey Lectures, 1959-1960*, Academic Press, Inc., New York, 1961.
3a. Recently, a most stimulating discussion of structural hierarchy, extrapolated from studies of the ciliary system, has been published by C. F. Ehret, *Science*, 132:115 (July, 1960), which in many respects is closely related to the topic of this lecture and, but for lateness, would have deserved a more explicit review in this place; readers interested in our points of contact would do well to consult the original article.
4. K. R. Porter and G. F. Palade, *J. Biophys. Biochem. Cytol.*, 3:269 (1957) have proposed transverse extensions of the endoplasmic reticulum as coordinating communication system.
5. The following brief account is based for the most part on the work and interpretations of F. O. Schmitt and his group (see *Rev. Mod. Physics*, 31:349, 1959) at the Massachusetts Institute of Technology, although, of course, many workers and laboratories have had a share in this success story.
6. The following discussion rests essentially on the excellent review of the subject by Glimcher, *Rev. Mod. Physics*, 31:359 (1959), who has also made some of the most spectacular advances in knowledge of this virgin field.
7. By S. Fitton-Jackson, *Proc. Roy. Soc.* (London) B 146:270 (1957).
8. See, for instance, Fraenkel-Conrat, *The Viruses*, F. M. Burnet and W. M. Stanley, eds. Academic Press Inc., New York, 1959, p. 429.
9. Only the features most relevant in the present context are reviewed here. For details and literature, see P. Weiss and W. Ferris, *Proc. Natl. Acad. Sci. U.S.*, 40:528 (1954).
10. Some observations on primitive higher-order patterning among fibers of collagen reconstituted in vitro (J. Gross, *J. Biophys. Biochem. Cytol.*, 2: Suppl. 261, 1956) would seem to encourage such a concept.
11. H. B. Fell, *Arch. Exp. Zellforsch*, 7:390 (1928). P. Weiss and R. Amprino, *Growth*, 4:245 (1940).
12. A. and H. Moscona, *J. Anat.*, 86:287 (1952).
13. P. Weiss and A. Moscona, *J. Embryol. Exp. Morphol.*, 6:238 (1958).
14. Karl Meyer, E. Davidson, A. Linker, P. Hoffman, *Biochem. Biophys. Acta*, 21 21:506. (1956).
15. Oftentimes our impression of "action," even spontaneity, signifies merely that we have inserted ourselves as observers into the process too late to recognize its prior inception from "interaction."
16. Some of these possibilities were contemplated in an unpublished joint conference of biologists and solid-state physicists, held at the Rockefeller Institute in 1959; the growth of a systematic trend in this direction is documented by the recent conference series bearing on the subject held at the Department of Biology, Massachusetts Institute of Technology (Fast Fundamental Transfer Processes in Aqueous Biomolecular Systems, June, 1960).
17. For instance, P. Weiss, *J. Embryol. Exp. Morphol.*, 1:181 (1953).
18. P. Weiss, "Differential Growth," in: *Chemistry and Physiology of Growth*, A. K. Parpart, ed., Princeton University Press, Princeton, N.J., 1949, p. 135.
19. P. Weiss, *Proc. Natl. Acad. Sci. U.S.*, 46:993 (1960).

20. For a fuller discussion see P. Weiss, *Yale J. Biol. Med.*, 19:235 (1947); *Quart. Rev. Biol.*, 25:177 (1950); *J. Embryol. Exp. Morphol.*, 1:181 (1953).

21. I have designated as "modulations," in *Principles of Development*, Holt, Rinehart and Winston, Inc., New York, 1939, the various expressions which a cell can assume with the same basic molecular equipment by virtue of the alterna-· tive functional dominance of different portions of that equipment, in contradistinction to true differentiation, which connotes unidirectional changes—gains or losses—of equipment. Both the distinction and the term have proved useful in articulating the problem of differentiation more precisely, and consequently, have been rather widely adopted.

22. It is remarkable to note the manner in which this fundamental problem of the degree of reversibility and irreversibility of "differentiation" is often slighted in current literature. If differentiation were an objective scientific term with identical connotations to all its users, the solution of the problem would, of course, be simply a matter of finding the facts. However, since the term means different things to different authors, which they often do not bother to specify, we are treated to a display of quite discordant and contradictory professions of faith, rather than statements of fact, pertaining to a common word—"differentiation"— which, since it covers such a wide array of disparate phenomena, lends itself readily to sham arguments based on semantic confusion. For example, one common source of misunderstanding is the outdated, though not yet outmoded, habit of equating the state of differentiation with features discernible under the microscope, thereby confining the living object to the limited detecting power of a particular instrument. As a constructive step toward clarification, I have made a modest effort to identify and list the various kinds of processes and phenomena that one hears usually referred to as "differentiation" (*J. Embryol. Exp. Morphol.*, 1:181, 1953), but despite some success in engendering a more critical and realistic attitude toward the issue, the matter is still often dealt with as an article of faith rather than as a subject for sober evaluation of facts. In factual terms, the question of whether "differentiation" is or is not reversible is plainly nonsensical. All depends on what sort of "differentiation" the questioner has in mind. There is reversibility and there is irreversibility in cellular processes, and nothing short of determining just when, where, and how much of either is connected with a given transformation, will ever make scientific sense.

23. The original experiment by Fell and Mellanby (*J. Physiol.*, 119:470, 1953), in which skin fragments were cultured in a vitamin A-enriched medium, was later modified by Weiss and James (*Exp. Cell. Res.*, Suppl. 3, 381, 1955) in that a suspension of separated skin cells was merely given a brief exposure to the vitamin. But the question of whether the metaplastic effect was produced entirely during the actual exposure or as a result of residual traces of the vitamin absorbed by the cells, has not been crucially decided.

24. Other pertinent examples of recent date are the deflection of prospective muscle cells into a cartilaginous course by influences emanating from the spinal cord (C. Grobstein and G. Parker, *Proc. Soc. Exp. Biol.*, 85:477, 1954; H. Holtzer, in *Regeneration of Vertebrates*, C. S. Thornton, ed., University of Chicago Press, Chicago, 1956, p. 15), and of ectoderm cells into true pigment cells by exposure to a pigment precursor (C. E. Wilde in *Cell, Organism and Milieu*, D. Rudnick, ed., The Ronald Press Company, New York, 1958, p. 3).

25. For a basic discussion of these embryological problems see P. Weiss, *Principles of Development*, Holt, Rinehart and Winston, Inc., New York, 1939; *Analysis of Development*, B. H. Willier, Paul Weiss, and Viktor Hamburger, eds.,

W. B. Saunders Company, Philadelphia, 1955; C. H. Waddington, *Principles of Embryology*, George Allen & Unwin, Ltd., London, 1956; C. P. Raven, *An Outline of Developmental Physiology*, McGraw-Hill Book Company, Inc., New York, 1954, among others.

26. My first explicit reference to a "field" principle in development was made in regard to phenomena of regeneration (*Naturwiss, Jg.*, 11:669, 1923), but was soon extended to encompass embryology as well (*Morphodynamik. Abhandl. z. theoret. Biol. H.*, 23, 1926, *Morphodynamische Feldtheorie und Genetic. V. Intern. Gen. Congr.*, Berlin; *Z. Indukt. Abstammungsu. Vererbungslehre* II, 1567, 1928); for a summary see P. Weiss, *Principles of Development*, Holt, Rinehart, and Winston, Inc., New York, 1939.

27. First described in 1929 (P. Weiss, Roux' *Arch. Entwicklungsmech. Org.*, 116; 438, 1929), the phenomenon was further analyzed in later work (see *Chemistry and Physiology of Growth*, A. K. Parpart, ed., Princeton University Press, Princeton, N.J., 1949, p. 135), and definitely discounted as possibly due to "chemotaxis" (P. Weiss, *Science*, 115:293, 1952).

28. Such polarization effects have actually been observed. Protoplasmic protrusions oriented toward each other by cells have been reported in rootlet formation of seaweed eggs (D. M. Whitaker, *J. Gen. Physiol.*, 20:491, 1937; D. M. Whitaker and E. W. Lowrance, *J. Gen. Physiol.*, 21:57, 1937); and in the outgrowth of nerve fibers from a circle of neuroblasts (A. Stefanelli, *Acta Embr. Morph. Exp.*, 1:56, 1957). Movement of cells away from each other (giving the illusion of "repulsion") has been recorded for pigment cells in vitro (V. C. Twitty and M. C. Niu, *J. Exptl. Zool.*, 125:541, 1954). Similarly symmetrical spindle cells, when making mutual contact with their ends, blunt and immobilize each other on the near sides so that they assume turnip shapes, pointing in opposite directions, and move apart (P. Weiss, *Intern. Rev. Cytol.*, 7:391, 1958; M. Abercrombie, M. L. Johnson, and G. A. Thomas, *Proc. Roy. Soc.* (London), B136:448, 1949).

29. If it were within the scope of this article to enlarge upon supracellular organization, one of the most dramatic manifestations of the "field" principle in the "self-organization" of organs would be dealt with in this place. It concerns the ability of a scrambled suspension of single cells from a fairly advanced embryonic stage to reconstitute themselves without specific "inductive" guidance from the environment, into amazingly complete and harmoniously organized organs; e.g., liver, kidney, feathers, of the typical morphology and functional activity (P. Weiss and A. C. Taylor, *Proc. Natl. Acad. Sci. U.S.*, 46:1177, 1960) and considerably beyond the histiotypic reorganization previously reported (e.g., A. Moscona, *Proc. Soc. Exp. Biol. Med.*, 92:410, 1956; C. Grobstein, *J. Exp. Zool.*, 124:383, 1953).

30. This model becomes even more pertinent if one takes into account the existence of activity gradients and cellular response thresholds (see Figs. 8 and 9 in P. Weiss, *J. Embryol. Exp. Morphol.*, 1:181, 1953).

31. Since every compound structure requires a certain minimum number of component units for its execution, it is obvious that a completely proportionate and harmonious organization of form can be achieved only above a critical size minimum (e.g., N. J. Berrill in *Analysis of Development*, Willier, Weiss and Hamburger, eds., W. B. Saunders Company, Philadelphia, 1955, p. 620; G. Andres, *J. Exp. Zool.*, 122:507, 1953; C. Grobstein, *J. Exp. Zool.*, 120:437, 1952).

32. See H. Spemann, *Embryonic Development and Induction*, Yale University Press, New Haven, Conn., 1939; F. E. Lehmann, *Einführung in die physiologische Embryologie*, Birkhauser Verlag, Basel, 1945.

33. The general proposition, based on experiments on the effects of antibodies (P. Weiss, *Anat. Rec.*, 75: suppl., 67, 1939) or organ transplants (P. Weiss and H. Wang, *Ant. Rec.*, 79: suppl., 52, 1941) on the growth of the corresponding em bryonic organs, was first set forth in a symposium in 1946 (P. Weiss, *Yale J. Biol. Med.*, 19:235, 1947), later expanded by further experimental evidence (P. Weiss in *Biological Specificity and Growth*, E. G. Butler, ed., Princeton University Press, Princeton, N.J., 1955, p. 195), and finally formalized in a workable mathematical model (P. Weiss, and J. L. Kavanau, *J. Gen. Physiol.*, 41:1, 1957; J. L. Kavanau, *Proc. Natl. Acad. Sci. U.S.*, 46, 1960).

34. Intriguing evidence for this view has been brought forward by Ebert in *Aspects of Synthesis and Order in Growth*, D. Rudnick, ed., Princeton University Press, Princeton, N.J., 1954, p. 69.

35. See P. Weiss and J. L. Kavanau, *J. Gen. Physiol.*, 41:1 (1957); J. L. Kavanau, *Proc. Natl. Acad. Sci. U.S.*, 46:1658 (1960). Experimental support for the theory is accumulating (see A. D. Glinos in *A Symposium on the Chemical Basis of Development*, W. D. McElroy and B. Glass, eds., Johns Hopkins Press, Baltimore, 1958, p. 813).

Reprinted from PROCEEDINGS OF THE ROBERT A. WELCH FOUNDATION
CONFERENCES ON CHEMICAL RESEARCH, $\underline{V}$, 5-31, 1961.

CHAPTER 3

STRUCTURE AS THE COORDINATING PRINCIPLE
IN THE LIFE OF THE CELL*

PAUL A. WEISS, *The Rockefeller Institute, New York, New York***

The concept underlying this series of annual conferences on Chemical Research has been a steady progression from the elementary to the complex, heading toward an anticipated climax—the chemical resolution of life, or rather of its universal agency—the cell. The program of the present conference centers on resolving *fragments* of the chemical endowment of the cell.

The ultimate objective is, of course, to relate these components to the unitary system of which they form part—the *integrated* cell. This objective still lies far ahead in a dim future. Words like coordination, organization, information, and so forth, just label the problem but bring no insight; nor can exclusive pre-occupation with the component activities in isolation— the worm's eye view of "*molecular* biology"—shed light on the pattern of their cooperative behavior in the integral system, unless it is supplemented by the bird's eye view of the cell in its integrity—by "*cellular* biology." To be compatible, both views must be constantly adjusted to each other.

I take it that this has been the reason for featuring on the program a prologue on the *cell* as the true target to which all biochemical detail must ultimately be related: the *real* cell, not some vague verbal symbols or fictitious models. The bridge between the cell and its constituents can only be built by convergence from both ends. My view today will be from the end of the cell. Considering the width of the gap and the urgency of closing it, my presentation will be more in the nature of a preview of future needs and potentialities than of a review of past accomplishments. In this portrayal of the cell, which of necessity will be crude and sketchy, I shall confine myself to an over-all perspective, so as to bring into view a number of fundamental questions, as yet barely raised, let alone answered, but answerable once they have been clearly recognized.

First, who is the cell? To define it, as our textbooks do, as a body of protoplasm, is meaningless. For what is protoplasm? In a recent monograph

---

*An address presented before "The Robert A. Welch Foundation Conferences on Chemical Research. V. Molecular Structure and Biochemical Reactions," which was held in Houston, Texas, December 4-6, 1961.

**Original work reported in the article has been supported in part by the American Cancer Society and the National Cancer Institute (National Institutes of Health of the Public Health Service).

DYNAMICS OF DEVELOPMENT

on biocolloids[1], we still read: "Protoplasm is a very complicated *mixture**** of organic and inorganic *substances****." Since mixture implies randomness, this version is either incomplete or wrong, for the salient feature of cellular processes is their orderly *non-randomness*. The authors continue: "The living cell is the *seat**** of anabolic and katabolic processes. . . ." These quotes are symptomatic in that they reflect a rather common view of the cell as a domicile or container, distinct from its inhabitants or content, which is that so-called "mixture" of molecules and molecular groups. Perhaps this notion stems from one-sided attention to truly encased specimens of cells, the plant cell in its cellulose wall, the red cell in its envelope, or the bacterial cell in its capsule. Whatever its origin, it is utterly misleading. For obviously, the cell is nothing but the organized community of its molecular constituents.

Then why should "cellular biology" not be wholly resolvable to terms of "molecular biology?" Because individual fragments of the molecular population behave differently when studied separately in isolation than they do as members of the organized system of the cell, where they mutually restrain and direct one another so as to yield joint unitary resultants. The nature of cellular organization lies in the complex web of interactions and interdependencies among its constituents[2]. The prodigious progress of biochemical analysis is giving us a rapidly growing catalogue of what can and may happen in a cell. But let us keep in mind that in the living cell, not all of this does happen ubiquitously, the same at all points at all times.

The distribution of chemical events in the cell is decidedly non-random. The fundamental difference between the same molecular population of the cell, when studied in its organized non-random configuration on the one hand, and as a random mixture on the other, is clearly reflected in the opposite aims of cytochemistry and homogenization, the one trying to localize actual operative sites, and the other, to bring out the total inventory.

Platitudinous as these preliminary comments may sound, they seem in need of repetition as antidote to the recurrent tendency in some contemporary literature to speak of cells as if they were essentially plastic bags filled with solutions of enzymes, substrates, and subsidiary compounds, masterminded somehow from a central intelligence agency—the nucleus. Accepting then the fact that diverse chemical processes are spatially segregated in the cell, what, one must ask, keeps them from mixing? And what does the system gain by keeping them apart? The answers to both questions are related.

Solutions can be kept apart by physical partitions. A physical compartmentalization of the cell space has been both postulated and documented by

---

****Footnote:* Italicizing mine.

microscopists. But even in the absence of parcellation, chemical domains can be staked out by the selective binding and adsorption of molecular species from an ambient medium to solid centers dispersed in it; the granular and fibrillar elements of cells can serve in this capacity. Membranes in fact can act in dual functions, as both partitions and adsorptive (or binding) sites. The common feature of all of these devices is their solid state, if we define for the purpose of this discussion as solid any molecular array, whether linear, planar or massive, which resists disruption by thermal agitation within the physiological range. Since such arrays are commonly called "structures," one has come to regard structural order as the critical tool of organized, as against random, behavior.

It is only natural that the static mentality of classical morphology has endowed this concept of structure with a degree of mechanical stability, rigidity, and fixity, which is no more tenable than is the opposite extreme of a structureless cell. In the meantime, the development of electron microscopy has carried the resolution of fine-structural details in fixed cells two orders of magnitude beyond the microscope. It has not only verified and extended the main concepts of micromorphology, but in its combination with biochemical techniques, has started to clarify the role of structure as a major tool of chemical dynamics.

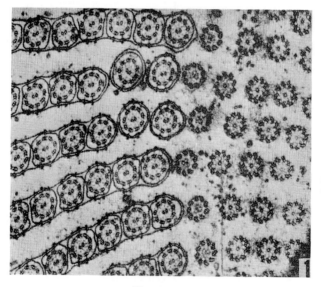

FIGURE 1

Electron-micrograph of thin cross-section through the ciliary field of a protozoan. Magnification: ca. 40,000 times. (By courtesy of I. Gibbons.)

A few familiar examples may serve as illustrations. They represent two different sorts of cellular equipment. The first series will show specialized functional gear peculiar to a given cell type.

Figure 1 shows cilia in cross section. Each cilium, covered by a sheath, contains a pair of fibers in its center, surrounded by a ring of nine doublets of fibers arranged pinwheel fashion, the inner fibers purportedly subserving impulse conduction, the outer fibers, coordinated contraction. This is the standard pattern of cilia[3], whether of plants, protozoans or metazoan cells, and it even appears as an evolutionary relic in the visual cells of the retina; the centrosomes, which are the foci for the formation of the mitotic spindle for nuclear division, are also akin to cilia, consisting of nine tubes around a cylinder[4].

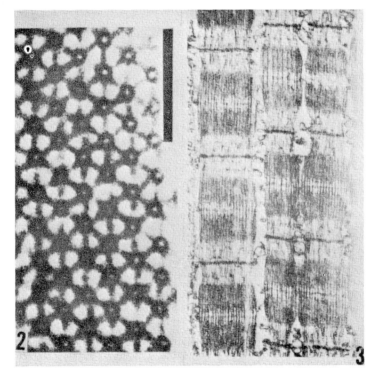

FIGURE 2

Electron-micrograph of ultra-thin cross-section through a muscle fibril. X 300,000. (From A. Hodge.)

FIGURE 3

Electron-micrograph of longitudinal section of three myofibrils, glycerol extracted. X 27,000. (From Huxley and Hanson.)

The myofibrils of muscle fibers, shown in Figure 2 cross-sectioned in their regular hexagonal array, are longitudinally (Fig. 3) seen as composed of a thinner core filament, accompanied at segmental intervals by a heavier jacket. These two structural components have been identified with the two proteins involved in contractility—actin and myosin—, and the relation between fine structure and function in this object seems well along the road to final clarification[5].

So much for special differentiations. They are all built according to patterns of exquisite structural regularity, which reflects corresponding patterns of chemically diverse substructures essential for the execution of their particular functions. They all show notable constancy of composition, of macromolecular architecture and of dimensions. During development, as well as regeneration after loss, they emerge as integral compounded units, and no one has ever seen them in unassembled pieces.

Turning from thse products to the cellular production machinery, we

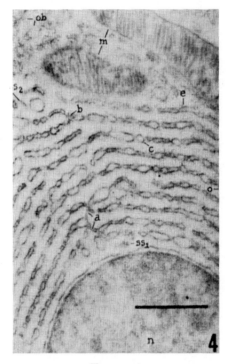

FIGURE 4

Electron-micrograph of section of a gland cell. X 20,000. n, nucleus. e, endoplasmic reticulum with anastomoses at a and b. m, mitochondria. (From K. R. Porter.)

meet the chromosomes, the nucleolus, and various cytoplasmic structures. To single out a few: the ergastoplasm or endoplasmic reticulum (Fig. 4), a system of membranous sacs or vesicles, continuous or disconnected, depend-' ing on circumstances; a population of granules (Fig. 5), named "micro-

FIGURE 5

Electron-micrograph of a portion of the cytoplasm of a gland cell showing the population of microsomes associated with the walls of the endoplasmic reticulum, as well as the structural organization of a mitochondrion below. (From G. Pallade.)

somes" by size, but "ribosomes" if loaded with RNA, either attached to the ergastoplastic wall or loosely scattered; the mitochondria (Fig. 4) with the characteristic inward foldings of their surface membrane; the Golgi complex; lysosomes; and other, unidentified, inclusions[6].

Each of these various structures is involved in some special chemical activity. For instance, there is a direct relation between certain puff-shaped discharges from specific loci of chromosomes in given cells and the chemical function of that particular cell type or stage (Fig. 6)[7]; between the ergastoplasmic-microsomal system and the manufacture of specific macromolecules[8]; between the mitochondria and oxidative phosphorylation[9]; and so forth.

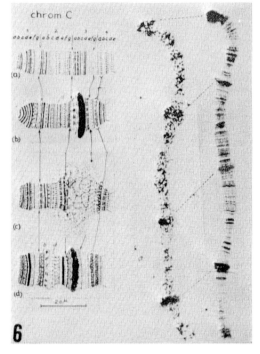

FIGURE 6

Banding pattern of the same giant chromosome in insects at different stages of development, showing the different activity patterns (puffs) of a given locus. The right half of the picture shows paired photographs of the whole chromosome as seen under the microscope (right) and in a radioautogram (left). (From Beermann.)

Does this now mean that chemical order in the cell rests on the rigid parcellation of autonomous local production sites, in the sense of the old microscopists? Evidently, such a concept would be far too static to fit the living cell; for the integrated household of the living cell is based precisely on the mutual dependency of those components in a network of cooperative interactions which require constant and regulated communication and exchange[2].

Which at once brings up the question of the medium of this communication network. Is it essentially fluid? Do these structural components float like islands in an unstructured pool to which they abandon their products for random dissemination by ordinary diffusion? Since this would be infinitely less efficient than if transport went along tracks or channels connecting sources and destinations, one wonders whether cell physiology might not have overstressed the relevance of traffic by diffusion, instead of viewing the whole cell as a structured continuum, though not of static, but of stationary, character. In terms of chemistry: Is not the cell perhaps pervaded with solid structures designed for stepping up the probability, hence frequency, of strategic molecular encounters way above the chance frequency of collision by sheer thermal agitation? Evidently, the effective concentration of any given number of molecules is far greater if they are gathered along a linear track, or even spread in a plane, than if they are dissipated in bulk. In *theory*, at least, structural order seems to be a postulate for maximum efficiency of the chemical machinery of the cell.

For the moment, however, let us return to the purely *factual* question of just how little or how much of a consistent structural framework the living cell contains. For an answer, we may examine living cells in action, recorded cinematographically.

(The motion pictures of cells in culture shown at the meeting were all produced in my laboratory by Dr. Taylor, Dr. Rosenberg, and Mr. Bock; other films shown came from Warren Lewis' laboratory. Most of them are time-lapse films with accelerations on the screen of several hundred times.)

> *1st Scene:* The film sequence shows phase-contrast micrographs of two cells of a strain of human liver (Chang) kept in continuous culture for many months. Note (1) the great variability of the motile border of the cells; (2) the continual changes in the configuration of the cytoplasm and in the disposition of cytoplasmic granules; (3) the relative constancy of spatial relations between the nucleoli, indicative of solidity of the nucleus; and (4) rotation of the nucleus within the cytoplasm.

> *2nd Scene:* The sequence shows parts of two cells from a human conjunctiva strain, the diagonal light band being their boundary. The rotation of the left nucleus can be told from the shift of the axis of the dumbbell-shaped nucleolus. Note also the filamentous mitochondria in the cytoplasm.

The extensive shuttle traffic of particulates in the cytoplasm reveals a high degree of free mobility. Just how the particles move, is not yet understood. They are of course subject to buffeting by Brownian motion and passive convection by streams of liquid. But many of them also move actively and directionally, as shown clearly by Fréderic[10] for mitochondria. Sharply contrasting with the cytoplasmic mobility is the rigidity of the nucleus, whose formed elements are ostensibly linked into a firm gel. Therefore, in view of the low mobility and high adsorbability of macromolecules in such gels, the vital exchange between nucleus and cytoplasm can hardly be just

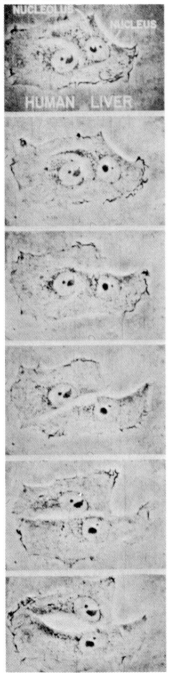

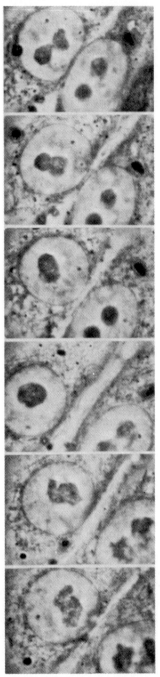

SCENE 1   SCENE 2

a matter of diffusion. Moreover, the cytoplasm, too, often congeals, as has been shown, for instance, in a film record of a fish egg during the first cleavage stages by W. H. Lewis and Roosen-Runge.

> *3rd Scene:* (Shown at the meeting, but not reproduced here. It demonstrated the rhythmic changes of the cytoplasm from solation, with its granular inclusions moving rather freely and irregularly in an amoeboid sort of motion, to gelation, with all particles totally stopped, and back to solation, in the rhythm of cleavage; the actual cell division occurs during the gelled phase.)

Let us thus keep in mind then as a fact that in many critical stages, not only the nucleus, but the whole cell appears solidified, with the free mobility of its molecular populations severely restrained.

Some recent observations in our laboratory[11] have added to the pertinence of this fact. Lowering of the pH in the culture medium below 6.0, either by buffering or by increase of CO_2, freezes cells in their tracks and abolishes their internal mobility, without interfering with their vitality; the phenomenon is fully reversible if the cells are returned to normal pH 7.4 within two hours. Raising the pH of the medium, on the other hand, increases the intracellular mobility, as well as the contractile forces in the cell cortex, entailing detachment from the substratum and rounding-up of the formerly spread cell.

> *4th Scene:* Effect of raising the pH by perfusion with pure oxygen of the erstwhile CO_2-containing normal medium. Note the vigorous retraction of pseudopodia and rounding up of the formerly spread cell bodies, with residual fibrous attachments (frames 3 to 5). Upon restoration of CO_2 supply and return to normal pH, cells spread out again and resume normal shape and activity (frames 6, 7).

> *5th Scene:* Effect of lowering of pH in the medium by increasing CO_2 concentration. Cells spread in normal pH (frames 1, 2) congeal at the low pH values (frames 3, 4), and after return to normal pH (frames 5, 6) resume normal motility. Note the rigid retention of the filamentous cell processes, as well as of the constellation of cytoplasmic granules in frames 3 and 4 as evidence of loss of free mobility.

Such effects are not all-or-none for the whole cell, for if one takes an elongated cell and acidifies (by means of micropipettes; experiments with Dr. Bruce I. B. Scott) one of its ends only, this end alone becomes paralyzed, while the other end keeps right on moving.

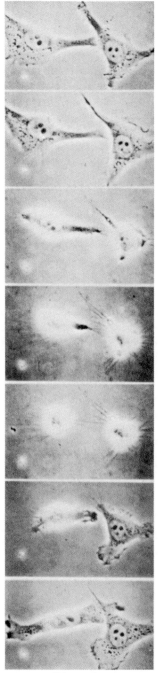

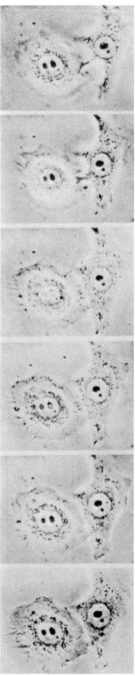

SCENE 4   SCENE 5

These results have important implications; for if in the course of metabolic activity, one spot in a cell develops low pH and thus congeals, this obviously will reduce the rate of chemical exchange with its surroundings, hence produce a transitory chemical insulation of that local domain. It is this type of topographic inequalities and inhomogeneities in the cell, with their functional chain reaction sequelae, with structures forming and dissolving, which is so crucial for our topic, yet has received so little attention and study compared to the dealing with bulk and averages.

If chemical traffic is to proceed efficiently even in the gelled cell, a diffusion mechanism alone will not do. The narrow fluid spaces of such a system, though passable for small molecules, will be the less trafficable for macromolecules, the larger and more adsorbable the latter are, and the more their shape, orientation and clustering tendency reduce their mobility. I am not aware that this fundamental issue has as yet been properly faced. Most biochemical model experiments are carried out in solutions and their lessons are then automatically transferred to some mental image of a relatively fluid cell, in which there is apparently no traffic problem. And consequently, the ways in which the real cell copes with the problem have rarely been looked for. There are, however, a few positive clues.

Some years ago, I found that a nerve fiber is not a stationary structure, but consists of a continuously moving column of material synthesized in the nucleated cell body and advancing downward toward the periphery at a rate of the order of a millimeter per day. A special conveyor belt feeds products from production sites to destinations[12]. One realizes the necessity of such a flow if one considers how inordinately long it would take, for instance, for neural transmitter substances of cerebral origin, such as acetylcholine or noradrenalin, to get from the brain cell of a giraffe to their sites of business in the body through a viscous filament which is about one hundred thousand times as long as it is wide, if they were moving merely by diffusion along a concentration gradient. Fortunately, we do have direct proof of an active driving force being involved. The case is not unique.

The classical example is the continuous protoplasmic streaming in plant cells, called cyclosis[13]. Cyclosis is ostensibly a stirring device to promote encounters among chemical elements way above what a stagnant system could accomplish. How the protoplasm is kept in coordinated motion, is still as enigmatic as ever. Metabolic energy, of course, is needed. But why would a given amount of energy not just raise the temperature of the cell, but be so channelled as to do useful work? We find ourselves right back at our initial proposition, that cellular order is based on the orderly channelling, that is, the systematic restriction of degrees of freedom, of energy distribution for the attainment of maximum efficiency.

I. On Basic Principles

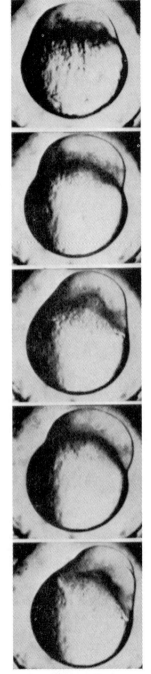

SCENE 6

This order is a space-time order. Structures can only account for the spatial part. Unless they are coupled in time, too, cell behavior would still be uncoordinated; just as discharges of nerve cells at random yield convulsions, instead of coordinated movements. Space-time coordination presupposes that the population of structural elements be sequentially synchronized; that it pulsate. Such timing mechanism no longer rests on structural devices, but is a matter of dynamics.

In the best investigated case, the nerve fiber, one knows in principle that the molecular activities of the excitatory wave are synchronized by an electro-chemical process. It is becoming increasingly evident, however, that the rhythmicity of the neuron is but a special example of a general property of cells; all cells, not just the familiar rhythmic beating types, like cilia, heart cells or cyclic gland cells. It seems that waves synchronizing phases of chemical activities sweep over cells in regular succession. If the cell happens to contain contractile units, such waves register visually as coordinated contraction-relaxation oscillations. Let me present a few examples.

> *6th Scene:* Rhythmic contractions in a whitefish egg (from film by Lewis and Roosen-Runge). The clear cap on top of the more granular yolk mass is the embryonic blastoderm. In normal development, rhythmic contractions of the yolk hemisphere pump substance into the cap in synchrony with the cleavage steps. In the egg here reproduced, actual cell division was suppressed by partial asphyxiation, but the pumping motion kept on sweeping over the egg nevertheless. Its phases can be recognized from the changes in outline and proportions of the yolk mass and the blastoderm. (Pictures of pulsating normal fish eggs were also shown at the meeting.)

To realize that this sweep has a purely physiological basis, that is, does not reside in a preformed structural differentiation as in muscle, one need only look at our free cells in tissue culture, which are so variable in their surface and internal configurations that any notion of a fixed coordinating structure would be absurd. Yet, they show the same sort of contractile rhythms, which can be visually recorded whenever the wave happens to circle in the optical plane.

> *7th Scene:* This film sequence, taken from a culture of monkey kidney cells, shows an isolated cell in the labile state between attachment and detachment. Evidently, as the contractile wave sweeps over the surface of the cell, the particular involved segment of its motile edge becomes detached and retracted, but re-extends and becomes reattached as soon as the wave has passed it. As a

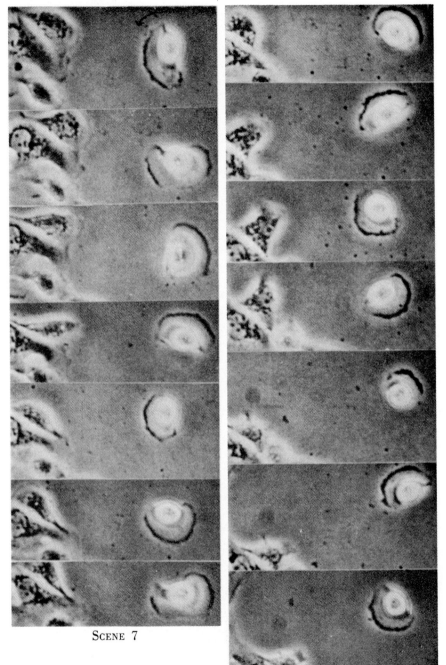

Scene 7

Scene 7A

result, a roughly semi-circular ruffle is seen to travel counter-clockwise around the stationary central portion of the cell. (The pictured sequence shows three full revolutions.)

These examples carry two lessons. In the first place, they demonstrate the coordination of chemical energy release in the cell over large areas. The underlying dynamics are plainly unknown, but whatever they are, the mechanism as such becomes a serious candidate for the role of pump for intracellular transport, forcing fluid both through organized channels (e.g., the endoplasmic reticulum) and through capillary interstices in the manner of a microscopic pulse wave. Devices of this sort lift the chemical machinery of the integrated cell above the limitations of dependence on diffusion. The search for such devices is overdue.

One other such device is presumably the rotation of the nucleus illustrated in the first and second film scenes above and described previously by Hintzsche[14] and by Pomerat[15]. We have found it in several cultured cell types, though never constantly; it may be more common inside the body and simply be disfavored in vitro by the extreme flattening of the cells. Its function may be conceived of as follows. The nucleus is gelled. To the best of our meager evidence, its products are extruded at local sites as vesicles or solid deposits, rather than oozing out all over its surface. Clearly, by cycling a nucleus can distribute its products more widely and uniformly than if it were stagnant. So, here is another stirring device. It even suggests an explanation for the revolving mechanism. The nuclear membrane of many cells is dotted with a sort of micro-manholes shown in the electron-micrographs of Figure 7[16]. If these are nozzles for nuclear discharges, each would produce a jet effect, and the asymmetry of these multiple releases would yield a resultant momentum that would set the nucleus spinning.

Yet the release mechanism itself presents us with another basic problem of cell dynamics. Cross sections reveal the ringlets in the nuclear surface (Fig. 7) to be closed by membranes. This being the case, how could they serve as pores for massive extrusion or intake? A static bulkhead is either tight, in which case it bars transit, or it has leaks, in which case it is useless as a partition. If the nuclear membrane is tightly shut, how can large masses get across? Conversely, if it is full of holes, how can it keep the contents of nucleus and cytoplasm from mixing freely? To remove this apparent contradiction, we must give up our trust in the structural fixity of membranes and adopt a much more elastic and statistical concept. Any fixed microscopic preparation can, after all, only reveal the molecular constellation at the moment of fixation. But in the living cell, the same set of filamentous molecules which at one time bars a passageway by lying crosswise, could at another moment, simply by turning 90 degrees, open a gate for traffic to go through. So-called pores would then be facultative openings, rather than permanent orifices.

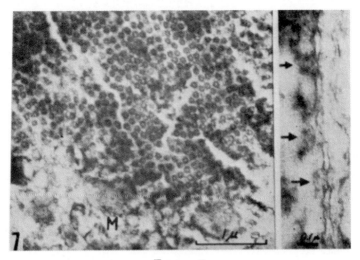

FIGURE 7

Electron-micrographs of tangential (left) and cross-wise (right) sections through the nuclear membrane of an oöcyte; arrows indicating the structures which in surface view appear as ringlets. (From Afzelius.)

I have recently[17] suggested such a mechanism as a possible resolution of the paradox that in many surface interactions in and among cells which involve specificity, relatively small numbers of molecules can initiate incommensurately potent effects, as in fertilization, drug action, phagocytosis, virus infections, synaptic transmission, tissue formation, etc. The idea is this (Fig. 8). The cell surface contains both lipid and protein macromolecules. The latter presumably lie flat in the surface forming a barrier screen to penetration by larger units. However, when a cluster of molecules with specific end groups approaches this fabric, it would attract sterically complementary ends of the macromolecular threads. This would turn them into radial positions, opening a breach for secondary massive non-specific influx or outpouring. This hypothetical construct is unquestionably oversimplified, and it may indeed be wholly wrong. Yet, it serves as a model of the basic, but neglected, roles which molecular shape and orientation must play in establishing, in the living system, physical—or let us call them "structural" —conditions which will permit, facilitate, modify, impair or block the course of chemical events. Too often have such structural features been regarded as morphologically fixed and rigid,—as occasionally, indeed, they are,—

while it is clear now that in the living system they mostly are but statistical groupings of molecules of changeable configuration. The physical manoeuvers of the cell membrane in the selective uptake of large molecules and particles, as in Holter's[18] work on pinocytosis—drinking by cells—illustrates the altered outlook.

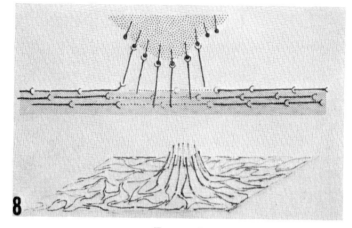

FIGURE 8

Diagrammatic representation of the re-orientation of filamentous surface molecules by strong polar attraction from an approaching outside source, resulting in transitory local perforation of the surface covering. (From Weiss, 1960.)

The more anisodiametric and asymmetrical molecules are, the better, on the whole, they lend themselves to ordering into higher-order assemblies, either transient or permanent. In linear array, they constitute fibrils; in planar array, membranes. Let us consider then how such arrays can affect the chemical processes in their domain: first, linear structures. By polymerization and lateral aggregation, with or without register among chains, sizable filaments arise along lines often determined by tension or by flow. Their mechanical functions in the architecture of cells and tissues are obvious; their chemical significance is less well recognized. Their enormous aggregate surface must bind or adsorb a major contingent of macromolecules from ambient pools. Such molecular coats will not be deposited as random matted felt, but rather like orderly fur, anisodiametric polar molecules assuming common orientations. For enzymes this would potentiate their effectiveness as catalytic agents over that in solution because aligned a much larger number can settle per unit area, and in those with active groups, all the business ends, instead of pointing every which way, would come to face the circulating medium containing reactive substrate. In addition, cooperative phenomena among the ordered molecular collective might increase the efficiency of interaction.

The significance of structural carriers to expedite enzymatic processes in cells is therefore self-evident. Yet, the experimental evidence is still meager. Mazia and Hayashi[19] noted a marked increase of enzymatic efficiency when monomolecular layers of a pepsin-albumin mixture were compressed into fibers, that is, stepped up in geometric order from planar to linear orientation; but since the substrate itself was trapped inside, the full advantage of enzyme orientation could presumably not show up. Bodine's group found years ago[20] that the formation of melanin in insects by the tyrosinase-tyrosin system required an "activator," which turned out to be simply an adsorptive surface. Similar observations have since become rather commonplace in enzyme work. McLaren[21] is carrying on investigations most relevant to this problem of enzyme action in structurally restricted systems; and there are a few more scattered reports. But on the whole, I wonder whether the meaning of these results for the living cell is fully realized and exploited throughout the vast field of enzymology.

There is perhaps an even more significant aspect to linear molecular arrays, too vague still to be mentioned except in passing. It pertains to the idea of molecular bucket brigades acting as tracks for electron, and perhaps proton, transfer. (A recent symposium held in F. O. Schmitt's laboratory

FIGURE 9

Electron-micrograph of thin cross-section through the basement lamella underlying larval amphibian skin, revealing regular laminated structure in which layers of parallel collagen fibers alternate in their directions by 90°, hence appear alternately in profile and in cross-section. Magnification: X 13,000. (From Weiss and Ferris.)

dealt with the subject.) Since the entire problem of the convection of matter along interfaces on any scale is still quite nebulous, all one can say is that the role of fibrils as chemical traffic routes deserves more active study. A model experiment, which we just carried out, gave suggestive results. The thick membrane at the base of the skin consists of a solid matrix of mucopolysaccharide, in which layers of parallel fibers are firmly embedded at alternating angles of about 90 degrees; the fibers are submicroscopic (Fig. 9) or microscopic (Fig. 10A).

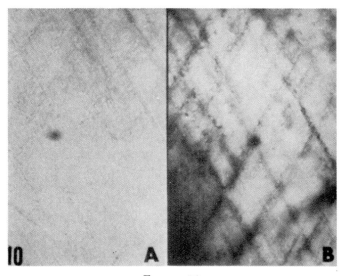

FIGURE 10

Microphotograph of fish skin showing the two intersecting fiber systems. (A), a few millimeters distant from the injection site; (B), in the vicinity of the site of injection of a microdrop of a dye, 24 hours after the injection. The dye can be seen to have expanded preferentially along the interfaces between the fibers and the matrix in which they lie embedded. (Original.)

Vital dies injected as markers into such a membrane were actually found to travel more rapidly along the fibers than in the spaces between (Fig. 10B). Using microdrops of uranil nitrate, the selective conduction of this marker along the linear pathways could be verified under the electron microscope (Fig. 11). Projecting such observations downward into the cell, one feels encouraged in assuming that chemicals may often be guided along tracks, rather than left to scatter in bulk.

The planar structures in the smectic state of matter share of course with the linear systems the faculty of surface catalysis, which they exert mostly

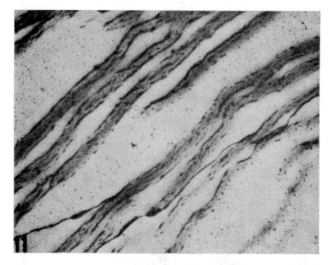

FIGURE 11

Electron - micrograph of a near tangential ultrathin - section through dermal layer of fish skin in the vicinity of site of injection of a microdrop of uranyl nitrate. The metallic compound can be seen to have advanced predominantly in the direction and along the fibers. (Original.)

indirectly by adsorbing in an ordered array a layer of active molecules, thus enabling the latter to operate at the heightened efficiency in the sense just outlined. The distances over which this superiority of geometric order over random dispersion can manifest itself are still wholly problematic. The evidence for truly long-range forces advanced by Rothen has been discounted on theoretical and technical grounds. However, a seemingly fool-proof case of long-range interactions discovered recently in my laboratory by Dr. Murray Rosenberg has brought the whole issue again to the fore.

Here is the gist of it. Round cells spread on different surfaces at different rates characteristic of the particular substance. Spreading on quartz is prompt and extensive (Fig. 12). If one then stacks on a quartz slide monolayers of stearate-stearic acid from Langmuir trays and tests the spreading of the cells on them, one finds not, as expected, a sudden drop of the spreading force, but a gradual decline in proportion to the numbers of layers piled on top of each other. This effect of the distant quartz base on cells through the mediation of ordered layers of stearate molecules has been measured up to about 80 layers,—a range of roughly 2000 Ångstroms. The essential integrity of the stearate film was checked by ellipsometry before and after the tests. The test is so sensitive that cells can distinguish between the two levels of a terrassed stearate deposit by accumulating at the lower level

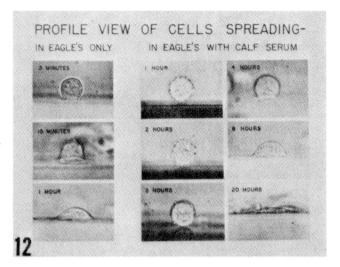

FIGURE 12

Profile views of the spreading of single cells isolated from tissue cultures or embryonic organs on a glass surface exposed to different media. The presence of traces of serum (right) slows the spreading about ten times above that in salt solution alone (left). (From Taylor, 1961.)

(Fig. 13), even if the floor is several hundred Ångstroms from the base and the difference to the top is only another few hundred. To say more, would be departing too far from my subject. Nor do we know whether biological surfaces can act similarly, although we surely must envisage that possibility.

All my examples have centered on one basic principle: that chemical order in the cell is mediated by orderly, non-random, arrays of molecules, which we call "structures," with liquid pools serving primarily as reservoirs and flush lines. This postulates that in the living cell, there is always present a continuous, though changeable, network of interlocking molecular chains and layers which offer solid-state connections from any point to any other point uninterrupted by liquid breaks subject to thermal randomization. Although supermolecular organization thus appears as one of the crucial links between cellular and molecular biology, if this were all, its bearing on the topic of this conference—"Molecular Structure"—would be marginal. There are, however, signs of a much more direct relation. The trend in the science of organic colloids has clearly been away from simple mechanical interpretations to explanations in terms of chemical bonding of various degrees of specificity. What has been learned from the cellular end, goes to accentuate this trend; for supermolecular arrays in cells are being disclosed more and more not as haphazard conglomerations, but as selective combinations of matching molecules, the match being often based on steric fitting.

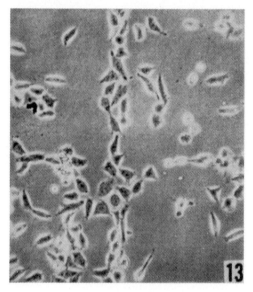

FIGURE 13

Accumulation of cells of a loose single-cell suspension along cross grooves of a terrassed stearate deposit on quartz. (Original, M. D. Rosenberg.)

In a recent review of this problem[2], I have cited specific examples, such as the classic work of F. O. Schmitt and his collaborators on the characteristic patterns in which collagen monomeres aggregate in forming filaments and fibers; the recent demonstration[22] that only in one of these constellations will collagen combine with hydroxyapatite, as in bone formation; or the restoration of a whole tobacco mosaic virus by the selective reunion of its nucleic acid core with a fitting protein jacket after their artificial separation[23]. These are striking examples of the building up of higher order structures by the selective combination of heterogeneous, but mutually fitting, elements.

Let me add a further illustration of this principle of progressive structural complexing from molecule to cell: the grana of plant chloroplasts—sites of photosynthesis. Electronmicroscopically[24] they appear as regular lamellated bodies (Fig. 14). From chemical data, Calvin[25] has proposed a molecular architecture for each lamella as shown in Figure 14, postulating a specific non-random order in the line-up of the molecules. The lamellae themselves, however, seem to originate from the coalescence of uniform bead-like units[26] (Fig. 15).

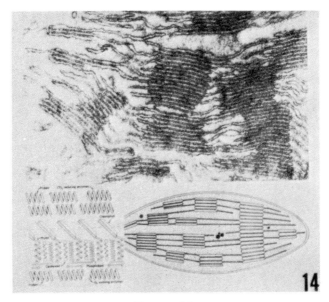

FIGURE 14

Laminated fine structure of the chloroplast of plant cells. (Top and bottom right, from von Wettstein; bottom left, after Calvin.)

The constant size of these units strongly hints at some singularly stable condition of aggregation for molecules in just that particular combination, constellation and quantity. And this is plainly a property of the *group dynamics* of the heterogeneous molecular collective, no longer deducible from the behavior of the *individual* molecular species studied separately.

The inference is obvious. The bridge from molecule to cell needs a mid-stream pillar—the collective behavior of molecular populations as ordering step. Given such a foundation, a new dynamic submicromorphology could rise, which would explain why cell organelles or viruses appear and operate as well defined structural entities, and how new higher patterns of order emerge from the free interaction of those unit bodies. The elementary beads of the grana (Fig. 16, bottom), for instance, have been pictured as forming a lattice pattern with a constant of about 300 Ångstrom[27]. The top portion of Figure 16 pictures a similar geometric scheme proposed[28] for the ringlets in a nuclear membrane. I have been led to the same general conclusion for the regular fabric of collagen fibrils in the lamella under the skin[29]: they seem to be organized about a cubic lattice of stable anchor

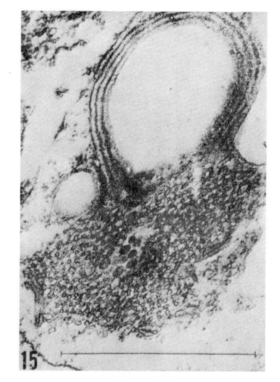

FIGURE 15

Electron-micrograph showing the transition of a tubular lattice structure (lower portion) into the mature laminated structure (upper portion) of a developing plastid. X 53,000. (From Granick.)

points, separated by about 5-600 Ångstroms (Fig. 17). In this light, would it be too farfetched to envisage an ordering principle in cells and cell components akin to crystallinity, with supramolecular units, instead of ions, staking out their respective domains and establishing spatial order by mutual interactions?

This evidently is leaving the firm ground of demonstrated facts and stretching our imagination way into the future. Yet, the temerity for doing it seems justified by a certain timidity which in the past has kept molecular and cellular biology apart in fact, if not in words. Let us face it: The study of cell life presents us with certain postulates, which we cannot satisfy from the limited store of facts and concepts of present-day biochemistry alone. Broadening is needed from both ends. I have tried to illustrate the need, as well as feasibility, from the end of the cell. Incomplete as my presentation has been, I hope it has helped to set the problems of "structure as coordinating principle in cell life" in proper focus. It opens vistas into what is practically virgin territory—let us call it "cytophysics" or "cytodynamics"—calling for more attention and more work commensurate to its importance as link between cell and molecule.

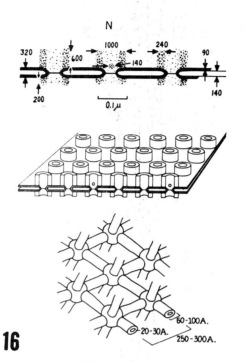

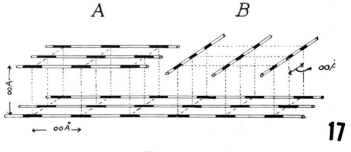

FIGURE 16

FIGURE 17

REFERENCES

[1]Bungenberg DeJong, H. G., and Booij, H. L., in: Protoplasmatologia, Handbuch der Protoplasmaforschung, **Vol. II**, Springer Verl., Wien, (1956).

[2]Weiss, Paul, in: The Molecular Control of Cellular Activity, J. M. Allen, ed., McGraw-Hill, New York, 1962, pp. 1-72.

[3]Fauré-Fremiet, E., Biological Reviews, **36**, 464-536, (1961).

[4]Bessis, M., and Breton-Gorius, J., Compt. Rend. Acad. Sci., (Paris), **246** (8), p. 1289, (1958).

[5]Huxley, H. E., and Hanson, J., in: The Structure and Function of Muscle, Vol. **I**, G. H. Bourne, ed., Academic Press, New York, 1960, p. 183.

[6]Novikoff, A. B. in The Cell, Vol. **II**, J. Brachet and A. E. Mirsky, eds., Academic Press, New York, 1961, pp. 424-481.

[7]Beermann, Wolfgang, in: Developmental Cytology, 16th Symposium of the Society for the Study of Development and Growth, Dorothea Rudnick, ed., Ronald Press, New York, 1957, p. 83.
Breuer, M. E., and Pavan, C., Proc. 9th Intern. Congr. Genet. 1953. Caryologia **6** (Suppl.), 758, (1954).

[8]Siekevitz, Philip, in: The Molecular Control of Cellular Activity, J. M. Allen, ed., McGraw-Hill, New York, 1962, p. 143.

[9]Lehninger, A. L., Rev. Modern Phys. 31, 136, (1959).

[10]Fréderic, J. and Chèvremont, M., Arch. Biol., **63**, 109, 167, 259, (1952).

[11]Taylor, A. Cecil, J. Cell Biol. **15**, 201-209, (1962).

[12]Weiss, Paul, in: Regional Neurochemistry, S. S. Kety and J. Elkes, eds., Pergamon Press, Oxford, 1961, pp. 220-242.

[13]Heilbrunn, L. V., An Outline of General Physiology. (3rd ed.) W. B. Saunders Co., Philadelphia, (1952).

[14]Hintzsche, E., Ztschr. f. Zellforsch., **43**, 526-542, (1956).

[15]Pomerat, C. M., Expt. Cell Res., **5**, 191-196, (1953).

[16]Bernhard, W., Expt. Cell Res., Suppl. **6**, 17-50, (1958).

[17]Weiss, Paul, Proc. Nat. Acad. Sci., **46**, 993-1000, (1960).

[18]Holter, H., and Marshall, J. M., Jr., Compt. rend. trav. Lab. Carlsberg, Sér. chim. **29**, 7, (1954).

[19]Mazia, D., and Hayashi, T., Arch. Biochem. Biophys. **43**, 424, (1952).

[20]Bodine, J. H. and Carlson, L. D., Proc. Nat. Acad. Sci., **40**, (6), 513-515, (1955).

[21]McLaren, A. D., Enzymologia, **21**, 356-364, (1959).

[22]Glimcher, M. J., Rev. Mod. Physics, **31**, 359-393, (1959).

[23]Fraenkel-Conrat, H. L., in: The Viruses, F. M. Burnet and W. M. Stanley, eds., Academic Press, New York, 1959, p. 429.

[24]Hodge, A. J., Martin, E. M., and Morton, R. K., J. Biophys. Biochem. Cytol., **3**, 61, (1957).

[25]Calvin, M., Brookhaven Symposia in Biol., **11**, 160, (1958).

[26]von Wettstein, D., in: Developmental Cytology, 16th Symposium of the Society for the Study of Development and Growth, Dorothea Rudnick, ed., Ronald Press, New York, 1957, p. 123.

[27]Granick, S., in: The Cell, Vol. II, J. Brachet and A. E. Mirsky, eds., Academic Press, New York, 1961, 489-595.

[28]Afzelius, B. A., Expt. Cell Res., **8**, 147-158, (1955).

[29]Weiss, Paul, Proc. Nat. Acad. Sci., **42**, 819-830, (1956).

Reprinted from ICSU REVIEW 5, 185–193 (1963);
Elsevier Publishing Company, Amsterdam

CHAPTER 4

THE CELL AS UNIT

PAUL WEISS

The Rockefeller Institute, New York, N.Y. (U.S.A.)

It is plainly impossible to deal with the problem of 'the cell as a unit' more than perfunctorily in the limited space of a brief essay. The most one can hope to achieve is to set the problem in the proper perspective by an examination, however cursory, of whether 'cell biology' has any factual claim to the status of an autonomous discipline in the organizational hierarchy of nature, or whether it is merely a temporary station on the way to its complete resolution into terms of 'molecular biology'. Since I myself have been responsible for introducing this categorical distinction, I welcome the opportunity to review its merits, which are based on both facts and logic.

Rated by its composition, a cell is obviously no more than the sum total of the molecules composing it, and these in turn can, if one wants to, be described in terms of their component atomic and subatomic elements. But since this process of progressive decomposition yields essentially the same result for a live and for a dead cell, and indeed even for the homogenate of a physically disintegrated cell, one realizes that such a reductionist description loses some highly relevant 'information content' on the way down; it loses the criteria which distinguish the live cell from the dead one, the dead one from its homogenate, the structured macromolecule from a random scramble of its constituent atoms and ions, and so forth. In trying to derive the more complex systems from their elements, therefore, one must make up for this deprivation somehow by restoring the lost properties. The practice of doing this through verbal symbols, such as 'organization' or 'integration', is an old one, but seldom explicit as to whether these symbols are meant to be final logical postulates to compensate for the limitations of pure reductionism, or merely provisional promissory notes that they will ultimately yield to analytical resolution.

Having been batted around for ages in an odd mixture of scientific reasoning and emotional preconceptions, the argument between these two alternatives is at last losing some of its steam under the critical scrutiny of modern 'operationalism'. One honestly cannot deny that hierarchical 'order' and 'organization' as superordinated principles *sui generis* would have gained scientific acceptance more readily if they had not been suspect of theological implications. Divested of all such bywork, what then is the true operational meaning of the above alternatives? Does it not simply lie in the difference between a *mental* reconstruction of a higher system from symbols representing elements on the one hand, and the *physical* reconstruction of such a system from separate components on the other, — in either case, without the cheating intervention of an already ordered 'model' or 'organizer' as integrator?

In other words, the true test of a consistent theory of reductionism is whether or not an ordered unitary system (a cell being such a system) can, after decomposition into a disordered pile of constituent parts, resurrect itself from the shambles by virtue solely of the properties inherent in the isolated pieces. If not, the symbolic terms, which permit us to execute in mental imagery what physical relity is impotent to reproduce, would acquire the logical and scientific respectability of axioms. Conversely, spectacular recent progress in achieving true 'synthesis' of higher-order systems from lower-order elements, in general with the input of energy, has rather fanned the hopes of believers in the eventual triumph of an absolute reductionism. These latter no longer doubt *that* it will be possible to 'synthesize' a living cell from a mixture of molecules; they just ask *when* it will come to pass, and some pretend the feat to be just around the corner.

Unfortunately, such optimism is mostly in direct proportion to the lack of first-hand and penetrating acquaintance with the living cell *as a whole*, which is a unit, rather than a sheer summative assemblage or conglomerate. For however familiar and expert one may be with one particular feature of a cellular system, be it genic replication, contractility, respiration, selective permeability, impulse conduction, enzyme action, membrane formation, or what not, he misses the essence of the problem of cellular unity unless he takes due account of the indispensable *cooperative coexistence of all these features;* that is, that every single one must contribute to the maintenance and operation of all the others in such a way that collectively they achieve a relatively stable and durable group existence. Just bear in mind, for instance, that while models of contractility must fall back on ordered macromolecular structures as synchronizers and coordinators of enzyme activity, enzyme action, in turn, is instrumental in the establishment of structural assemblies. While the mechanisms of respiration and photosynthesis have been convincingly connected with arrays of macromolecular complexes in definite sequential order, this very order depends for its establishment and maintenance on photosynthetically or respiratorially provided energy. As membranes perform selective screening functions between the media to either side, their very formation and growth depend on the uninterrupted presence within the cell of highly discriminative powers as to what it lets in, retains, converts, assimilates, compounds and localizes. Thus, by the time we have laid out the pattern of the reproductive and functional performances of a cell in a total, rather than sectorial, view, we recognize that the basic criterion of cell life lies in the intricate web of interactions and interdependences among *all* of its component activities. True, each one of these particular components can be successfully analyzed in its own right in relative sectional isolation, but only if one takes for granted and then borrows some ready-made cellular derivates, such as enzymes, membranes, chromosomes, from the other sectors.

In contrast to the analyzing scientist, however, the living cell does it all by itself in its own household; its know-how, with only an 'uninformed' environment to draw upon, residing not in the static composition of its chemical endowment, but in the dynamics of its interactions harmoniously coordinated.

Coordinated by what? Entelechy? Feedback? Information? Fields? Should we not desist from coining allusive labels until we have described in sober operational terms the factual content of the phenomena thus to be labelled? Translated to such terms, what is the meaning of 'interdependence' of component events in a system such as a cell? It evidently refers to a relation between a group of events, a, b, c, . . ., in a physical continuum such that the omission of any one of them would preclude the occurrence of all the others. Since in dynamics events are merely spot samples of continuous processes, our formulation must be expanded to imply dependencies among con-current processes, that is, between time courses $a' \to a'' \to a''' \to \ldots, b' \to b'' \to b''' \to \ldots, c' \to c'' \to c''' \to \ldots$ so that at no point on the time line will any one of these series be out of correspondence with the others. All the component processes must mesh like gears in a machine or civic activities in a community.

The crucial alternative raised above can now be phrased more succinctly: Can such interlocking systems be taken apart and put together again *stepwise*, like a machine or jigsaw puzzle, by adding one piece at a time, or is the very existence of the system as a whole predicated on the *simultaneous* presence and operation of all components? In the former instance, an eventual 'synthesis' of artificial cells could be envisaged; in the latter case, it could not. Let us look at the record.

A few years ago the U. S. National Academy of Sciences held a symposium, orga-nized by F. O. SCHMITT, on 'Biomolecular Organization and Life Processes', in which the conditions for the complexing first of molecules into macromolecules, and then of macromolecules into viruses, were reviewed[1]. In the concluding address on *The Compounding of Complex Macromolecular and Cellular Units into Tissue Fabrics*, I presented experimental evidence that would permit us to extend the principle of stepwise assembly of ordered complexes to even higher levels of organization[2]. Just as individual connective tissue fibers are not just random aggregates of molecules of the protein collagen, but strictly ordered strands of such molecules aligned in tandem and register, so the population of fibers need not remain a random feltwork, but in combination with a mucopolysaccharide matrix, can assume the regular geometric arrangement of an ordered fabric; and even a population of cells that had been isola-ted from complex organs, dispersed and scrambled, can rearray itself upon reassem-bly into the typical pattern of the organ of origin. Here then are striking examples that 'organization on a higher level may emerge from ordered interactions of orga-nized elements'[3].

Although our scant factual information on this kind of ordered complexing of heterogeneous elements is largely empirical, certain underlying principles can be dimly perceived. One is the thesis that the orderly grouping of macromolecules into 'macromolecular assemblies'[4] might be based on the steric matching of complemen-tary sites among the components; failing this, some other ordering mechanism would have to be postulated to account for the apparent strict sequential order in which macromolecules (*e.g.*, enzymes) are packed into subcellular structures. Another principle to emerge is what I have called 'macrocrystallinity', that is, the self-ordering of mixed macromolecular populations in definite grids, or space lattices, with periodicities of several hundred Ångströms[5], determined presumably by the optimum equilibrium positions of the domains of the interacting components in those particular combinations. Perhaps one day both principles will yield to a description in terms of 'minimum free energy' of the resultant compound systems with regard to a given common environment, suboptimal for the stability of any one of the participating components alone.

Yet despite the intriguing support which the indicated 'synthesis' of subcellular systems from separate elements seems to lend to an extension of the concept to the cell itself, and hence to the prospect of a reductionist explanation of cell life, there are two major and fundamental objections to its uncritical acceptance. The first one is the qualification that in order for macromolecules to be able to congregate in higher-order patterns, they must themselves possess conforming patterns of organization, that is, properties which pre-match them for mutual conjugation. What this implies, I have stated in the introductory chapter, 'From Cell to Molecule', of a book on '*The Molecular Control of Cellular Activity*'[6]. I quote here from the conclusion:

"We have arrived at last at a point which comes rather close to what might be defined as 'molecular control of cellular activity,' only to discover that the 'controlling' molecules have themselves acquired their specific configurations, which are the key to their power of control, by virtue of their membership in the population of an organized cell, hence under 'cellular control'. And this indeed has been the whole purpose of my long discourse: to document by practical examples that the distinction between molecular control of cellular activity and cellular control of molecular activity is based on the semantic ambiguity of the term 'control', hence fades in the light of a true understanding of the phenomena involved. A cell is nothing but the population of component entities that constitute it. But these entities are not just of molecular rank, nor can their ordered behavior in the group be fully appreciated and understood solely by studying them in isolation, out of context. As I have tried to carry the principle of self-organization of higher organizational units by the free interactions of elements of lower order as far as present factual evidence — not hopes, nor belief — would honestly entitle us to do, I had to add at every turn that elements endowed for such ordered group performance have always been prefitted for it by properties *previously imparted to them as members of just such an organized group unit*, whether cell parts, cell, or germ. This circular argument contains one of the most fundamental truths about the nature of organisms, and as one can readily understand, does not augur well for an eventual affirmative answer to our introductory question, in its naïve form, of whether a cell will ever be synthesized *de novo* without the active intervention of another cell."

A second reservation regarding the concept of 'self-assembly' stems from the essentially *static* character of the examples we have chosen to prove that true 'compounding' without systemic guidance does occur; for we have concentrated our attention on *structural* features, neglecting the inseparable complementarity between *structure* and *process* in the living system, in which processed structure is but an outcome of structured processes. The fact that diverse activities of a definite pattern can coexist and go on concurrently in the space continuum of the cell even in the absence of tight compartmentalization, reveals that although only a fraction of the cellular estate is strictly structured in a mechanical sense, there still is coordination among the diverse biochemical processes, which evidently must remain relatively segregated and localized. So, here we are back again at the question asked before: Coordination, how and by what?

Now, it is intriguing to speculate in the interest of consistency that perhaps the structured portion of the cell might itself also subserve this function of coordinating the unstructured fraction of the cell content by establishing and maintaining differential topographic distributions within the otherwise unsegregated molecular populations of the intracellular pools. I myself happen to be quite partial to this view, and have strongly advocated it in the opening address on '*Structure as the Coordinating Principle in the Life of the Cell*', delivered at the Welch Foundation Symposium in 1961. The trend of thought is this: Ordered patterns of cellular and subcellular structures are definitely capable of inducing a corresponding patterning in the adjacent layers of the ambient liquid space by selective adsorption, chemical bonding and the concomitant local physico-chemical changes; obviously the fact that differential enzyme localization is part of this process, would explain that not only the segregation, but even the regulation of the kinetics of the heterogeneous activities, could likewise be referred back in the last analysis to some ordered structural mosaic.

Most of these considerations would seem to be consonant with current notions of 'coding' of cellular activity. The modern transliteration of the older term 'organization' to 'information' is acceptable, though inconsequential, for this discussion; after all, both are just words. The simile of a 'code' for non-random sequential order, as of letters in a word, likewise is satisfactory to illustrate certain aspects of an organized system. Evidence presented in the preceding articles has shown that genes can be viewed as coded sequences of nucleotides (deoxyribonucleic acids); that different segments of these linear chains can be faithfully copied onto 'messenger' molecules; that these can transfer the 'code' to still other molecules (both ribonucleic acids); and that the sequential code of the latter can act as the model for a corresponding specific sequence of amino acids in the construction of a specific protein. Up to this point, the scheme provides us with a plausible mechanism for the transmission, hence conservation, of specific space order.

Just in passing, it is worth noting a major distinction between this manufacture of macromolecules and the elementary kind of molecular 'synthesis' with which we are familiar from simple inorganic systems. The former requires the presence of organized end products as models, 'templates' or 'primers', whereas, for instance, H and

Cl combine to HCl unaided. This distinction may of course turn out to be fundamental, but this need not concern us here.

By far more critical is the further problem: Suppose we do know how genes beget *proteins* — and surely, this knowledge is a spectacular achievement — how do we get from there to the knowledge, let alone the synthesis, of a *living cell*? In principle, by just more of the same? The standard affirmative answer, that after all proteins in the form of enzymes do hold the key to the synthesis of all other, non-protein, compounds in the cell, begs the question; for it still leads only to a random bag of compounds, instead of the highly coordinated chemical machinery that is the cell. Now, as I indicated before, the cell is neither this sort of random scramble of molecules, nor, at the other extreme, a rigid stereotyped composite of microstructures, but something in between; part fluid, part consolidated. And the fact that it can exist at all, considering the enormous variety of molecular species and groupings it contains, cannot be simply passed off by just referring to the progressive complexity of molecular interactions, but calls for an exact specification of the principle that 'coordinates' these interactions so that their combined performances will insure, as if by concert, a high degree of stability in the total system. Logically, this 'coordinating' principle cannot be of the same categorical order as individual reactions themselves — just one more of them.

The common habit of personifying compounds by calling them 'regulators', 'integrators', 'organizers', or what not, and crediting them verbally with the 'regulatory, integrative and organizing' effects which one observes but cannot explain analytically, either intends to endow chemicals with spiritual powers up and above their ordinary properties, or else is wholly meaningless. To state it bluntly, it would be rather a reversion to the prescientific age if on observing, for instance, the spinning of a whorl of fluid, one were to invoke a special compound as 'spinner'.

By reasons of logic and scientific honesty, we must therefore acknowledge the problem of *coordinated unity* of the cell as a real one. It cannot be hedged by assuming that starting from gene reduplication and the first steps of protein synthesis, all further developments would run off collaterally and uninterrelated for this would imply that once having been mapped out microprecisely down to the most minute details, they would then actually be capable of pursuing with absolute rigidity their individual courses so predesigned as to yield blindly, but unfailingly, a viable product, — a modern version of Leibnitz' 'prestabilized harmony'. The unpredictability of the vicissitudes of the environments in which those courses materialize rules out any such concept of absolute predetermination as utterly unrealistic and absurd: 'environment' is to be regarded as anything in the world surrounding given items — organisms, cells, organelles, molecules, etc. — which can interact with the latter or affect their mutual interactions.

In fact, the whole mental picture of cells as if they were stamped out identically like tin soldiers is false and misleading. Cells of the same type vary far more widely in detailed composition, configuration and activity, among one another and from moment to moment, than is usually realized and taken into account. Therefore, the actual course of a given train of interactions cannot be predicted by either the cell or its observer deterministically, but can be defined only in terms of the probability of its going to lie somewhere within a given range. Thus, obviously, if there is a very large number, n, of independent processes, each subject to a large number, m, of random serial excursions from an average course, the cumulative variance among them would become increasingly greater as time goes on. This would render it highly improbable that any two such bundles of processes would ever lead to recognizably similar results, or even that any one of them would retain essential identity with itself for any length of time of activity. However, notwithstanding this measure of relative indeterminacy of the component processes and the metabolic flux of their substance, living cells of a given kind do resemble each other and do retain essential invariance for long periods of time. We can express this fact symbolically in the formula:

$$V_c < (v_a + v_b + v_c + \ldots v_n),$$

where V_c denotes the total variance within a population of cells of a given type (or between successive stages of the same cell), whereas v_a, v_b, v_c, ... are the variances of component cell activities. The formula represents an 'operational' description of what it is that makes the cell as a unit 'more than the sum of its parts'.

In order that this formula be satisfied, one must evidently postulate that the component processes, when operating in the common integral system, are *interdependent* in such a manner that as any one of them strays off the norm in one direction, this entails an automatic counteraction of the others. In electric control systems, such compensatory stabilization devices are built in as 'negative feedback' loops. But other kinds of systems achieve the same effect without circuitry. The cell is one of these. The principle involved is often referred to as 'homoeostasis.' The question now remains whether it is conceivable that such a system of cooperative dynamics could be assembled from its parts in steps, one at a time, or whether it can exist only in its entirety or not at all.

To be specific, let us consider the requirements a number of biochemical processes a, b, c. . . ., etc., must satisfy in order to be able to go on side by side, as they do in the cell. If they are to coexist, they must either be mutually supporting, each yielding (and receiving) needed products and energy to (and from) the surrounding ones, or at any rate be at least mutually compatible, none yielding products or effects that would interfere with the others. Such families of processes must be harmoniously adjusted to each other not only as to the kinds of reactions, but as to their rates and time courses as well, and furthermore they must keep adjusting continuously to the fluctuations of environmental conditions referred to a while ago.

Now, it is relatively simple to set up a model for a *pair* of reciprocally matched processes a and b chosen to have such properties that whatever a needs in specific compounds and energy will be furnished on schedule by b, and *vice versa*. In essence what a laboratory biochemist does when he reproduces an isolated metabolic reaction (*e.g.*, an enzymatic one) *in vitro*, is nothing but playing the part of *b* for *a*, or of *a* for *b*, to the best of his ability, providing each reaction with the necessary conditions and ingredients (*e.g.*, substrates, accessories, pH buffers, etc.) *ubiquitously* and optimally. In the cell (or any organized fragment recovered from a cell), the same reaction depends of course on whether or not the same conditions and sources are made available *locally* in the immediate micro-environment as the result of commensurate neighboring reactions.

Now, it can readily be seen that the feasibility, in principle, of compounding such coupled interaction systems according to the scheme of $1 + 1 = 2$ ends with a doublet, for if there are more than two components depending integrally upon one another, such a system presents us with logical attributes akin to the 'many body' problem in physics. If *a* is indispensable for both *b* and *c*; *b* for both *a* and *c*; and *c* for both *a* and *b*; no pair of them could exist without the third member of the group, hence any attempt to build up such a system by consecutive additions would break down right at the first step. In other words, a system of this kind can exist only as an entity or not at all. Operationally, the cell falls in this category; to call it a 'unit' is merely a shorthand reference to this operational description. By implication, this also reaffirms the principle of unbroken *continuity of organization* in living systems, which I once expressed as 'omnis organisatio ex organisatione', with the understanding, that higher degrees of organization can emerge from the free interaction of *organized* and *prematched* systems of lower order of complexity (*e.g.*, specific macromolecular assemblies from pools of macromolecules; organs form dispersed cell populations).

In conclusion, on the basis of facts and logic, it seems to me unwarranted, and indeed unsound, to expect that it will ever be possible to describe cell behavior solely in reductionist terms of properties of its component elements, that is, without giving a due account of its 'system character'. Hopefully, a scientific systems theory and methodology, as currently applied to the interpretation of brain function, group behavior, engineering, communication, economics, and so on, will also provide us with conceptual tools for describing and treating the cell as a system more rigorously than heretofore. At any rate, the disciplined study of the *systemic* properties of the cell — of '*cell biology*' — that is, the manner in which its molecular components, which are the prime objects of '*molecular* biology', are subordinated to ordered group coexistence in a system of 'molecular ecology', is one of the major challenges and tasks of modern science. To meet it, we must face it. To face it, we must see it. To see it, we may even at times have to put blinders on so as to reduce the dimming effect of contrast engendered by all the brilliant light that emanates from 'molecular biology', so ably set forth in the preceding essays.

REFERENCES

[1] *Proc. Nat. Acad. Sci. U.S.*, (11) 42 (1956).
[2] PAUL WEISS, *Proc. Nat. Acad. Sci. U.S.*, 42 (11) (1956) 819–830.
[3] PAUL WEISS, *Proc. Nat. Acad. Sci. U.S.*, 42 (11) (1956) 830.
[4] F. O. SCHMITT, *Devel. Biol.*, 7 (1963) 546.
[5] PAUL WEISS, *Proc. Nat. Acad. Sci. U.S.*, 42 (1956) 825.
[6] PAUL WEISS, in *The Molecular Control of Cellular Activity*, edited by J. M. ALLEN, McGraw-Hill Book Co., New York, 1961, pp. 1–72.

II. ON DIFFERENTIATION

Reprinted from "Proceedings of the First National Cancer Conference, 50-60, 1949"

CHAPTER 5

THE PROBLEM OF CELLULAR DIFFERENTIATION*

by PAUL WEISS, PH.D.

University of Chicago, Chicago, Illinois

What is needed in the study of growth and differentiation is not only the accumulation of more data, but a more precise description and more penetrating analysis of the data in a more rigorous conceptual frame of reference. This need calls for a firmer trend away from obscurant verbalisms toward objective scientific formulations of the problems involved. We cannot hope to develop a better understanding of the phenomena of growth, differentiation, organization, induction, control, harmony, and so on, unless we first obtain a realistic picture of just what factual content these various labels cover. By way of example, such an operational analysis is attempted here for the problem of cellular differentiation. It is a condensed and combined version of results reported and thoughts expressed in three previous publications.

WHAT IS DIFFERENTIATION?

The term "growth" may be reserved to designate the increase in protoplasmic mass, that is, the production, or rather reproduction, of more units of a given kind. This may or may not be associated with a further subdivision (by cell divisions) of the growing mass. As long as the resulting fractions retain essentially the same constitution and appearance, we may speak of pure growth. "Differ-

* Original investigations referred to in this paper have been aided by a grant of the American Cancer Society on recommendation of the Committee on Growth of the National Research Council and by the Abbott Memorial Fund of the University of Chicago.

entiation," on the other hand, connotes the appearance of systematic differences among parts that were originally of the same kind. If "growth" means more of a kind, "differentiation" means more kinds. In cellular terms, true "differentiation" then implies the real, not only apparent, diversification (i.e., divergence of character) of cells or cell strains that initially were alike, rather than just looked alike.

DOES TRUE DIFFERENTIATION EXIST?

Whether differentiation in this strict sense really exists, is an empirical question. The answer hinges on the reliability of our testing methods, which furnish the criteria by which to tell the likeness or unlikeness of cells. Historically, undue faith in microscopic criteria has confused the issue. Cells that looked alike were rated as similar, and cells that looked different were considered to be intrinsically different. We must rid ourselves of this ingrained, but utterly fallacious, habit of judging by appearances. A cell is a going concern, in constant interplay with its environment. Its microscopic equipment is merely the cumulative record of its reactions to this environment, reactions of a specifically constituted system to the physicochemical conditions prevailing in the surrounding space. All optically or otherwise discernible characters of a cell are the results and residues of anteceding formative processes. They are indexes, at best, of certain cellular activities only; namely, those that do express themselves morphologically. Since many cellular reactions leave no morphological trace, differentiation of cell character cannot be defined in terms of microscopic criteria. A complete characterization of a given cell would have to include a complete inventory of the molecular species present, their combination and distribution in space, and a list of all possible reactions and manifestations of which the system is capable under any conceivable conditions (the "response repertory"). This seems unattainable in our present state of knowledge. Yet, we can at least exploit to a greater extent than heretofore those tests of cell behavior that are practicable. If we then compare the behavior of two cells or cell strains under identical conditions, e.g., in a common medium, and note constant differences in their behavior in one or more regards, we must conclude that they are of different character, even if they appear alike under the microscope. Conversely, if two strains that differed markedly in appearance while in the organism behave identically when brought into the same medium, this proves conclusively that they have been of one type and had merely displayed different portions of their response repertory in the face of different local environments. The latter process, exemplified by the transformation of fixed histiocytes into macrophages, has been termed "modulation," in contradistinction to "true differentiation" which implies an irreversible change in constitution.

On the basis of such behavioral tests, the occurrence of true differentiation during the development of all higher animals must be affirmed. Tissue culture has shown that cell generations derived from different organs, in spite of assuming similar appearances, retain many of the specific properties of the original strains. This means not only that they had differed in their protoplasmic constitution at the time of explantation, but were able to persist in synthesizing the same specifically different protoplasms without reverting to common type. On the other hand, since the evidence of Experimental Embryology proves that such different cell strains have originated from a common stock, the intrinsic diversification of cell strains during development is an incontrovertible fact. Since this diversification occurs by degrees, it is incorrect to speak of any cell as being "undifferentiated" or "embryonic." Differentiation is not an all-or-none reaction, but a long chain of progressive transformations, so that any cell we are considering has reached a certain point along that line. Also, at each point of the line, the cell is capable of a variety of reactions only part of which are compatible with harmonious development. At any step, an abnormal contingency may provoke a response that may throw the further course of the cell strain off balance and lead to pathological effects.

THE RESPONSE REPERTORY

If, according to the foregoing, visible criteria of shape, arrangement, and so on, are unreliable and incomplete tests of differentiation, what other means of detection do we have available? There are first the chemical products of cells, such as fibers and secretions. Different products are often valid indicators of intrinsic differences in the production plant. Histological stains also sometimes provide sensitive microchemical tests of cellular differences. Histochemical techniques, although still crude, likewise demonstrate specific differences among cells. Further constitutional differences may be revealed by the differential reactions of cells to drugs, hormones, or radiations, provided possible errors due to unequal exposure can be excluded. In this manner, much subtler differences can be detected than are morphologically indicated; for instance, constitutional differences among cells on different branches of the vascular tree, among different types of nerve cells, among different areas of an epithelium, or different regions of the connective tissue.

If this constitutional divergence among cells is based on, or at least associated with, the appearance of distinctive cell proteins, it might be possible one day to trace it by immunological methods. Antiserums against extracts of sperm, lens, kidney, or reticulo-endothelial tissue have been shown to have a selective action on homologous organs. We have seen evidence of similar effects in the embryo. We are now testing the possibility of selective absorption of organ antiserums (tagged by C^{14}) by the embryonic precursors of the homologous organs as a means of tracing back the first stages of biochemical divergence. Still other methods are on the horizon. Combined, such methods will give us more pertinent information on when the crucial steps in the differentiation of a cell strain take place. Only then will our attention become more properly focussed on the process of differentiation rather than on the products of differentiation with which we are mostly preoccupied at present.

MOLECULAR ECOLOGY OF THE CELL

The study of cell behavior in development, in immunological reactions, and in the response to drugs, has drawn increasing attention to the cell surface as the seat of specificity of interaction between cell and environment. Similarly, interfaces in the interior appear as seats of specific interactions in the intracellular, intranuclear, etc., spaces. It seems that mere colloid-physical considerations do not provide reaction mechanisms of sufficiently subtle specificity to account for the highly specific and selective interactions recorded in these fields. Interest, therefore, has turned toward intermolecular forces producing bonds of varying strength depending on the configurational fitting ("steric conformance" of a key-lock type) between the respective molecules (e.g., Pauling). Cell relations would be controlled by the interlocking of complementary compounds. The response repertory of a cell would be limited by the number of such key species present in the cell. Not all of these, however, will become effective. In order to operate or combine, conditions for their operation or combination must be favorable. Interfaces offer such favorable conditions by adsorbing, concentrating, and orienting molecular films. The main point is that in this they act selectively. That is to say, depending on the physical and chemical conditions along the interface, certain segments of the molecular population will be selectively attracted to the exclusion of others. Two cells, otherwise identical, confronted with differently constituted interfaces, therefore develop surfaces of qualitatively different composition, and as a result diverge in their subsequent reactions. It should be clear, even from this very condensed comment, that what really counts in determining cell fate is not just the type of chemical compounds present but which of them are in operative condition. And it is in this respect that the physical constellation in the system becomes of paramount importance in setting the stage for the biochemical events. Disposition in space becomes as significant as chemical composition. The field of study investigating this complex but orderly behavior of molecular populations in cells may be termed "Molecular Ecology."

WHERE DOES DIFFERENTIATION OCCUR?

In the light of the concept just outlined, the first step in the differentiation of a cell may be envisaged as consisting of the selective concentration in its surface of certain specific key compounds. which then by virtue of their unique position and orientation act as anchor points for further molecular apposition and also perhaps as catalysts of specific reactions. A period of lability gives way to gradual consolidation, marking the appearance of irreversible features. The evidence of genetics indicates that the genic equipment of the cells of all tissues is and remains the same throughout development, hence, is not subject to differentiation. Differentiation seems confined to the extragenic protoplasm. At the same time, the number and type of differentiations a cell can undergo is strictly limited by the hereditary endowment of the species. Yet, once a cell has attained a given state of differentiation, it can pass this on without attenuation to generations of descendent cells, as is evidenced by tissue-culture experiments. These seemingly conflicting statements can be readily reconciled on the basis of the preceding remarks, leading to the following concept. 1. The numbers and kinds of key compounds that can be synthesized by a given cell are determined by the genic endowment. This assortment is the material basis of what we used to call cell "potency." 2. In any given case, only a fraction of this is "activated," that is, given opportunity to become effective, by being adsorbed to a surface or otherwise enhanced. It becomes the molecular master population of that cell. 3. This master population would then impose its pattern on the further course of synthesis of protoplasm.

A cytoplasmic master compound could perpetuate its kind in one of two ways, depending upon whether the genic apparatus generates the full assortment of terminal products or merely gives rise to more primordial compounds from which the terminal molecular specialties have to be derived by secondary degradations and conversions. In the former case, the cytoplasmic master compound would build on ready-made units of similar kind, while in the latter case, it would impose its own pattern upon the primordial compounds, template fashion. Thus, although at their source in the nucleus, all basic protoplasmic units may be identical in all cells, they would, on contact with the differentiated populations of the cytoplasm, assume the special characters of the latter. The stuff, in this view, is furnished by the nucleus, but reshaped by the differentiated cytoplasm acting as model. The perpetuation of cytoplasmic specificity is thereby insured as long as the nuclear production site remains trapped inside the cytoplasm. Suggestive support for this view is found in the work of Caspersson on the nuclear production site of proteins, as well as in our own demonstration that the synthesis of protoplasm in nerve fibers occurs exclusively in the nucleated central portion of the neuron. The demonstration by Sonneborn of self-perpetuating, although gene-dependent, cytoplasmic bodies in *Paramaecium* seems capable of a similar interpretation and may, as he suggests, have a bearing on the mechanism of cellular differentiation in higher forms.

In conclusion, differentiation is a process in which different parts of the cell system play different roles, and further research is needed to identify and clarify the component processes involved.

CONTACT HARMONY

Among the least patent criteria of differentiation is the sum total of properties that permit a cell to live in harmony with its neighbors of the same or other types. This is due not merely to the rapid elimination of all disharmonious combinations if and when they occur, but to the subtle preadaptation of cells to one another's prerequisites. One of the most striking examples is the ability of cells to form tissues by (1) aggregating with their own kind, (2) combining with complementary types (e.g., epithelium and mesenchym; nerve fibers and sheath cells), and (3) rejecting association with foreign types. That such associations and separations are not simply the accidental result of proliferation of continuous masses from common centers, but involve active selectivity, is clearly brought out by the selective fusion of identical and complementary tissues in regeneration, wound healing, transplantation, and in the reorganization of cell groups after forcible dissociation. New investigations on this problem are being carried on in my laboratory. Evidently, cells possess a high degree of discriminatory ability in recognizing each other. Their means of recognition must be situated on the surfaces, for this is where they make contact. Contact may occur between naked protoplasts or through an intervening coat of exudate. As suggested earlier, the "means of recognition" may consist of the specific shapes of molecules exposed to the surface. Specific links between conforming molecules could then establish a cohesive union between adjacent cells. Secondarily, such groupings may be consolidated by the formation of fibrous skeletons and membranes. While calcium ions seem to play an important part in promoting adhesiveness between cells, they can evidently not be the determining factor in the selectivity of the associations.

Contact affinities and disaffinities between cells develop in the course of development as corollaries of differentiation (Holtfreter). Thus, two cell strains from different sources can independently develop complementary characters that predispose them for a later union. In other instances, cell types may become mutually adapted after coming in contact. In still other cases, one more advanced cell type may force a less consolidated neighbor into a conforming state, as occurs in the "inductions by contact" to be discussed presently. The final outcome is always harmony of interactions between contiguous cells, as well as between cells and their medium. Establishment of such harmony terminates cell locomotion. Disturbance of the harmonious state sets the cell on the move again. This explains the restlessness at the free edge of an epithelial sheet and similar phenomena. Cells keep moving until all their specific surface contacts are properly matched. As contact harmony depends upon a great many factors, so disharmony may result from a variety of causes. It may originate in a change in the cellular environment or in a change in the cell itself. If the latter change is of such a kind as to reduce the discriminatory acuity that rules in the association among normal cells, inconsistent and abnormal groupings may ensue.

The bearing of these matters on problems of cancer, particularly invasiveness and metastasizing faculty, is evident. To the extent to which these properties mark misdirections of the differentiation process, cancer is a problem of differentiation rather than of growth. The concept outlined in the preceding section explains in principle how a cytoplasmic deviation, once it has occurred, can be perpetuated throughout the subsequent growth of descendent cell generations.

CONTACT INDUCTIONS

The work of Spemann and others has shown that a more mature tissue may influence the course of development of a less differentiated tissue with which it comes in contact. The nature and specificity of this action has long been in the center of interest. That chemical interaction is somehow involved has long since become clear. The underlying mechanism, however, is still as obscure as ever. The immediate effect does not seem to be of the diffusible kind, for it requires that the affected cells be in intimate contact with the inducing cells or, at least, with some cell debris. In the study of one such pair of dependencies, namely the induction of the lens by the eye cup in the chick embryo, we have noted that the first visible trace of the effect consists of a sudden orientation of the prospective lens cells relative to the inducing retina, with the axes of the former becoming perpendicular to the contact surface. Since this reorientation coincides with the area of contact, it must be interpreted as a transcellular contact effect. It is presumably but a sign of the reshuffling and segregation of the molecular population in the epidermal cells in the sense already indicated, that is, attraction of key molecules to the new contact area followed by oriented adsorption, the building on of oriented chains of molecules, and consequent redisposition of the chemical systems of the cell. It is noteworthy that we have seen a similar reorientation of epidermal cells in larval amphibians within a few days after local exposure to an implanted crystal of the carcinogen, methylcholanthrene. Many other instances of contact induction are equally suggestive. Whether or not the particular explanation, attempted here, will prove to be tenable, it shows at least the type of approach from which an eventual resolution of such terms as "induction" into physicochemical realities may be expected. Many pathological processes have long been known to be associated with peculiar cell arrangements, e.g., "pallisading," but little has been done to exploit the descriptive-morphological facts as clues to an understanding of the underlying molecular processes.

A further experimental analysis of the relation between differentiation and contact action seems feasible by exposing cells to surfaces of defined constitution. Nageotte has demonstrated the ability of alcohol-fixed cartilage to induce additional cartilage formation in the surrounding tissue. I have described the specific induction of new cartilage along the surface of implanted cartilage that had been devitalized by quick-freezing and drying, as well as induction of cornea and nerve sheath after similar treatment of the corresponding tissues. The specificity of tissue inductions following injection of tissue extracts, as described by Levander and others, remains to be confirmed. Yet, more intensive work along these lines holds promise of valuable data of general significance for the problem of differentiation.

In all such cases, one will have to distinguish clearly between truly differentiating effects, which push a cell into one of several alternate courses, and mere realizing effects, which permit an already single-tracked cell to express some previously latent character. Early embryos mostly furnish examples of the former category, older animals of the latter. A good case in point are the hormones, which in most instances act by merely promoting or repressing in the target cell a course of differentiation the character of which, including hormone-dependence, has been determined by prehormonal influences.

CONCLUSION

This brief survey may suffice to give an idea of the highly composite nature of the phenomena described as "differentiation" and the multiplicity of factors involved. If a cell deviates in its course of behavior from our expectations, we call it pathological. But we realize that the number of points at which such deviations may occur is as great as the number of steps in the process of differentiation. At each step, the deviation may be due to an aberrant change within the cell system or outside of it. Within, it may originate in the cytoplasm, in the nucleus, or in the genes; outside, it may arise in the immediate contact environment (another cell or surrounding matrix) or may come from a distance (as in the case of diffusible agents). It may imply a trivial deficiency easy to repair, or a profound constitutional alteration doomed to permanency. It may affect chemical composition or simply the realization of physical conditions necessary for given chemical systems to become operative. It may act at the surface or in the interior, affect equilibriums of concentration or conformance of configuration, production of compounds or their distribution, electrical or mechanical properties, mobility or adhesiveness, and so on. Realizing this diversity of factors, nothing could be farther off the mark than trying to embrace differentiation—or for that matter, its pathological variations—in a single formula, as do those who search for, or speculate about, "the" mechanism of differentiation, as well as those who speak of "differentiated" and "undifferentiated" cells, as if these were just two sharply delimited stages. This habit of dealing with "differentiation" as a rather vague generalization has seriously handicapped the breaking down of the general problem into concrete and tractable parts which would lend themselves to experimental attack. The purpose of this paper has been to present illustrations of the feasibility and advantages of a less abstract and more factual approach.

REFERENCES

1. WEISS, PAUL: The problem of specificity in growth and development. *Yale J. Biol. & Med.* 19: 235-278, 1947.

2. WEISS, PAUL: Differential growth. *In* Parpart, A. K., Ed.: The Chemistry and Physiology of Growth. Princeton, N. J. Princeton University Press. 1949.

3. WEISS, PAUL: Growth and differentiation on the cellular and molecular levels. (International Congress for Experimental Cytology, Stockholm. 1947.) *Exper. Cell Research.* In press.

II. ON DIFFERENTIATION   145

Reprinted from JOURNAL OF EMBRYOLOGY AND
EXPERIMENTAL MORPHOLOGY, 1, 181-211, 1953.

CHAPTER 6

Some Introductory Remarks on the Cellular Basis
of Differentiation

by PAUL WEISS[1]

University of Chicago

I. INTRODUCTION

BY request of the committee which organized the conference on 'The Cellular
Basis of Differentiation', I undertook to open the meeting with a sort of topical
outline that might serve as a guide in the subsequent free and informal discussion. In view of the diverse meanings attached to the term 'differentiation' in
various quarters, from development in general in the broadest sense to the production of visible specialized cell structures in the narrowest sense, it seemed
desirable to circumscribe at least the scope within which the conference was to
hold itself. The instructions of the organizing group were to focus the discussion
on the cellular transformations which lead to the final functional state of the
mature cell. This limitation is taken into account in the following outline. At the
same time it is clear that terminal cell specialization is part of a continuous process, hence cannot be neatly separated from antecedent developmental processes
leading up to it. Therefore the difference between this presentation and others
dealing with early embryonic phases lies in the emphasis rather than in the
nature of the objects. In fact, it is becoming ever more evident that the study
of special histogenesis and organogenesis may cast as much light on the interpretation of early embryonic processes as the study of early events, e.g. the formation of the fertilization membrane or of the sperm tail, contributes to the
understanding of specific terminal cytodifferentiation. Accordingly, the main
objective of the following outline is to set the problem of cellular differentiation
into proper focus and perspective within the broad problems of ontogeny.

The term 'differentiation' has long been used vaguely and loosely; longer than
is profitable for either the description and interpretation of facts or the formulation of problems or the design of experiments. The results of a conference such
as this should be gauged by its success in reducing this vagueness and looseness.

[1] *Author's address*: Department of Zoology, University of Chicago, Chicago 37, Ill., U.S.A.
The research of the author has been supported by the Wallace C. and Clara A. Abbott
Memorial Fund of the University of Chicago and by a grant-in-aid from the American Cancer
Society upon recommendation of the Committee on Growth of the National Research Council.

Since science benefits not only from the discovery of new facts but also from the elimination of inconsistencies and misconceptions caused by unfamiliarity with, or disregard of, known facts, a first prerequisite seemed to be for this conference to have before it a complete inventory, as it were, of all the many facts and facets normally subsumed under the single term of 'differentiation'. Such a table of contents of the problem was therefore prepared and distributed to the members of the conference in advance of the meeting. It is reproduced below. A second task that seemed called for was the following.

As on past occasions (1947, 1949, 1950) to which the reader may be referred for further details, I found it necessary in order to deal profitably with the problem of cellular differentiation, to operate with a more realistic and workable conception of 'cell' and 'protoplasm' than is normally implied in these symbolic terms. I have sketched this conception previously under the name of 'molecular ecology'. For brevity and pregnancy, I shall condense some of its major aspects into diagrams presented here for the first time. They are admittedly greatly over-simplified models, intended merely to help us to visualize the phenomena and to formulate problems more cogently. They permit us to phrase specific questions which can be answered decisively. As the answers come in, those models will be validated or invalidated or, most likely, appropriately modified. In the meantime they will have served as a temporary scaffolding for the erection of a more tangible and less non-committal notion about 'differentiation' than we now possess. Therefore no more than this tentative pragmatic value need be attached to them.

Differentiation is not something that should or could be defined, at least not in our present state of ignorance and conflicting opinions. It is a summary term for a great many diverse experiences and evidently its content will grow and change as more experiences are gathered. Hence whatever definition would be attempted would have no finality. To strip the term of its vagueness would not require definition so much as circumscription, and this is what will be attempted in the following. Instead of engaging in the illusory task of defining the properties of some non-existing entity, called 'differentiation', we shall simply list those properties of differentiating living systems that have been observed. This list is a mere sample with no claim to being either categorical or exhaustive. It tries to break down the general problem into a number of specific issues which can serve to bring some preliminary order into the enormous mass of scattered data bearing on the topic.

A Programmatic Inventory of Cytodifferentiation

A. Introduction to the problem of differentiation

Progressive changes in cell *strains* versus the terminal specialization of cell *individuals*. Changes in composition versus changes in distribution and arrangement. Which parts of the cellular system (*a*) change with time in all cells equally? (*b*) change differentially according to cell type? (*c*) remain unchanged throughout differentiation?

II. On Differentiation   147

B. Criteria of cytodifferentiation

(By what signs can we recognize differentiation?)

1. *Morphological criteria* (What signs of differentiation are either immediately visible or can be rendered visible by appropriate tests?)

(*a*) Microscopic characters (intracellular and intercellular structures; their sizes, aggregations, localizations and modifications).

(*b*) Submicroscopic characters (polarization- and electron-optical identification; elemental particulates; orientation; patterned versus irregular deposition; relation of artifacts to reality; microstructural basis of cell shape).

(*c*) Microchemical organization (differential staining reactions, including electron stains; formed secretions; enzyme localization; pH patterns; &c.).

2. *Direct physiological criteria* (What signs of differentiation are revealed by changes in the physiological properties of cells and their discharges?)

(*a*) Functional activities (metabolic activity; secretions; phagocytosis; contractility; selective absorption and storage, as of iodine by the thyroid cell).

(*b*) Changes in composition (chromatography; ionic composition; changes of identifiable macromolecular compounds, including proteins, enzymes, antigens, nucleic acids, &c.).

(*c*) Changes in chemical composition and distribution determined by ultraviolet and infra-red absorption.

3. *Indirect physiological criteria* (What signs of differentiation not directly observable can be deduced from differential reactions of the living cell?)

(*a*) Selective reaction to drugs, hormones, viruses, &c.

(*b*) Organ-specific serological interactions.

(*c*) Self-sorting of cells according to type (e.g. following dissociation; in wound healing; &c.).

C. Mechanisms and causative agents of cytodifferentiation

(The detailed story of just where and how the various criteria of differentiation arise, and of the intracellular and extracellular factors and conditions upon which they depend).

1. *Origin of specialized cell products*

(*a*) Sites of specific synthesis and conversion of protoplasmic systems (role of nucleus, genes, cytoplasmic granules, mitochondria, centrosomes, &c., in the manufacture of specialized cell products; continuity of microscopic and submicroscopic 'self-duplicatory' bodies).

(*b*) Secondary rearrangements (coacervation; polymerization; fibril and membrane formation; formation of cilia, sperm tails, insect hairs and scales, keratinization, &c.).

(*c*) Extrusion of cell products (formation of intercellular cements, ground substances, myelinization, chitinization, &c.).

(*d*) Regeneration of cell parts (restoration of intracellular structures in functional cycles and after injury, including merocrine gland cells, protozoans, neurons, muscle-fibres, &c.; fate of differentiated structures during and after cell division).

2. *Metabolism of cytodifferentiation* (What are the general metabolic requirements for the functional and structural specialization of the cell?)

(a) Thermodynamic considerations and energy requirements for differentiation (relating the metabolic patterns in maintenance, growth, differentiation, and functional operation of the cell).

(b) Comparative aspects (relating the metabolic requirements of different cell types to their specialized features, such as cell size, cell life-span, nucleo-plasma ratio, elongation, cell shape, &c.).

3. *Specific accessories to cytodifferentiation* (What specific contributions and aids from the environment are prerequisite for the expression of given cytodifferentiations?)

(a) Specific nutritive accessories (vitamin A in retinal differentiation; ascorbic acid in connective tissue differentiation; pterins in insect pigments).

(b) Hormones (general hormonal requirements for cytodifferentiation; specific hormone-dependence of hormonal end-organs; differentiative interdependence of endocrine glands; neural influences on differentiation).

(c) Biochemical genetics of defective cytodifferentiation (e.g. albinism, chondrodystrophia, sickle-cell disease, &c.).

(d) Cellular interactions in differentiation (effects of organ-specific discharges other than hormones; contact influences transmitted from cell to cell; homoeogenetic inductions; 'infective' propagation of differentiation, &c).

(e) Physical factors (orienting and aligning forces in the cellular environment in their relations to cell polarization, locomotion, and structure formation; pressure and tension as factors in muscular, skeletal, and connective tissue differentiation; diffusion and hydrodynamic factors in differentiation; the role of interfaces, &c.).

4. *Pluripotency of differentiation* (What is the basis for the fact that most cells prior to their terminal stages are capable of several alternative courses of cytodifferentiation?)

(a) Multivalency of cellular equipment (all-or-none principle of differentiation; physical basis of 'threshold', 'dominance', and 'pacemaker' functions, &c.).

(b) 'Position' effects producing cellular dimorphism (e.g. terminal secretory versus duct cells in glands; endothelial versus blood cells; egg versus nurse cells; ganglion versus satellite cells, &c.).

(c) Proliferation and differentiation (antagonistic relations between reproduction and specialization; occurrence and significance of differential cell divisions; endomitosis in relation to cytodifferentiation, &c.).

D. GENERAL PROBLEMS AND CONCLUSIONS

1. *'Dedifferentiation' and 'redifferentiation'* (To what extent does the loss of certain criteria of differentiation signify (a) loss of type specificity, and (b) true reversion to pluripotent condition, implying the capacity to differentiate in a variety of new directions?)

2. *Genes, nucleus, and cytoplasm in differentiation* (it is essential to proceed beyond the trivial assertion that they all are involved in the process of differentiation, and to

examine just how and when each component participates; including the hypothetic role of 'plasmagenes', somatic mutations, polyploidy, &c.).

3. *Comparative aspects* (What lessons regarding cytodifferentiation can be gained from the study of differentiation in bacteria, slime moulds, protozoans, fungi, and higher plants?)

4. *Cytodifferentiation and morphogenesis* (How cytodifferentiation determines tissue and organ formation).

5. *Cytodifferentiation and carcinogenesis* (Is the primary deviation of the cancer cell a matter of aberrant growth or rather of defective cytodifferentiation? If the latter, which ones among the multiple features of the complex differentiation process enumerated above are predominantly involved?)

With this index as a background, we shall now proceed to single out a number of specific points for discussion.

II. CONNOTATIONS OF 'CYTODIFFERENTIATION'

Ordinarily the term 'cytodifferentiation' is used interchangeably in at least four different meanings. It designates:

1. The variety of terminal characters distinguishing the cells of a developed organism; see, for instance, the frequent statements to the effect that 'a given organism shows such-and-such differentiations'.

2. The process by which a cell of more generalized microscopic appearance assumes the more specialized aspect referred to in point 1 by elaborating distinct 'differentiation products' characteristic of its kind, declaring colour, as it were; e.g. the differentiation of myofibrils in a muscle cell, of melanin granules in a pigment cell, of hemoglobin in an erythrocyte, of colloid in a thyroid cell, of neurofibrils and Nissl bodies in a nerve cell, of keratin in an epidermal cell, &c.

3. The process by which the cells expressing themselves visibly in this fashion develop the basic equipment to do so; e.g. the differentiation (or 'maturation') of a myoblast into a muscle cell; of a melanoblast into a chromatophore; of a hemocytoblast into a blood cell; of a thyroblast into a thyroid cell; of a neuroblast into a neuron. Evidently point 3 grades into point 2, and the empirical distinction between them merely recognizes the fact that even the 'mature' muscle-fibre can still form new myofibrils (as in hypertrophy), the 'mature' pigment cell can still produce pigment, and so forth.

4. The process by which each cell type labelled here as '-blast' acquires the peculiar physico-chemical machinery that enables it to manufacture the specialized protoplasm referred to under point 3, and in many cases not in just one single set, but in an unlimited number of identical sets, as myoblasts proliferate myoblasts, melanoblasts more melanoblasts, and so forth. Whether the proliferating cells are strictly localized ('germinal cords' or 'germinal layers', e.g. in the vertebrate skin, the blood-forming centres, the intestinal mucosa), or whether

they are widely scattered, the basic question remains of how they have come by their differential faculties of reproducing each a characteristic type of protoplasmic descendants identified by the specific performances listed (chronologically reversed) in points 1 to 3.

Point 4 thus raises the question of the origin of divergencies among initially identical cell strains. From the testimony of Experimental Embryology and Pathology we know that the same cell can be the progenitor of a variety of derivative types—the more the earlier we test it—hence diversification among descendants of common precursor cells is a fact. The problem of how this diversity comes about is of a different order than that of the subsequent steps 3, 2, and 1, which merely elaborate an already existing diversity. Even so, since the phenomena 4, 3, 2, and 1 are continuous in time, it seems valid for purposes of our discussion, and probably quite generally, to identify differentiation with the whole series of physico-chemical changes described by these four points. Consequently we can state that differentiation (a) is a complex (as opposed to unitary) phenomenon; (b) is a stepwise process, not an abrupt event; and (c) can be meaningfully referred to only in terms of the stage which that series of processes has reached at a given time. To call a cell either 'undifferentiated' or 'differentiated' without further specification is not only inaccurate but scientifically meaningless.

Ideally one would have to follow a cell and its descendants through development, sample it at various stages along this course, and describe its properties objectively at each sampled stage: the changes of properties registered over the whole course would then add up to a complete record of 'differentiation'. Regardless of whether or not such objective description is ever attainable, we can at least come closer to it by including in our description as many properties as we can possibly identify, instead of limiting ourselves arbitrarily to those few properties which happen to reveal themselves optically or in some other overt manner. Any test, however indirect, and any reaction, however much delayed, that helps us to distinguish between two cells (or the same cell at different times) is a pertinent index of distinctive cellular 'properties'. The above Inventory of Cytodifferentiation illustrates the great variety of covert signs of differential cellular constitution that could properly be used for this purpose. The study of differentiation thus resolves itself into (a) the identification of criteria of the state of a cell; (b) the comparison of these criteria, revealing systematic differences between them; and (c) the exploration of the origin of these differences as they arise between (α) a given cell at an earlier stage and the same cell at a later stage, (β) a given cell and its descendant cells, and (γ) different descendant lines stemming from equivalent (or 'equipotent') precursor cells.

Differentiation thus connotes the appearance of *true* differences of constitution —'progressive transformation'—within a protoplasmic continuum extending along the time line, regardless of whether or not this continuum undergoes further subdivision in space by successive cell divisions or remains a continuous mass. *Identical* courses from identical starts can produce differences only *within*,

II. ON DIFFERENTIATION   151

but not between, protoplasmic strains. The emergence of true differences *between* strains derived from identical sources is due to the fact that their transformations have taken *divergent* courses.

III. TESTS OF DIFFERENTIATION

The term 'true differences' implies a distinction between characters inherent in the cell and features reflecting its environment. Now, evidently a cell without an environment is a fiction; hence no property or manifestation of a cell can be divorced from a consideration of the environment in interaction with which it has occurred or been displayed. However, by comparing the same cells in dif-

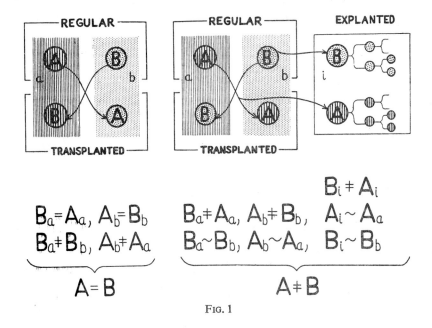

$$B_a = A_a, \quad A_b = B_b$$
$$B_a \neq B_b, \quad A_b \neq A_a$$
$$\underbrace{}$$
$$A = B$$

$$B_a \neq A_a, \quad A_b \neq B_b, \quad B_i \neq A_i$$
$$\qquad\qquad\qquad\qquad A_i \sim A_a$$
$$B_a \sim B_b, \quad A_b \sim A_a, \quad B_i \sim B_b$$
$$\underbrace{}$$
$$A \neq B$$

Fig. 1

ferent environments, and different cells in the same environment, their relative contributions to a given reaction, transformation, and eventually form, can be assessed. These behavioural tests of cell *character*, in contradistinction to environmental expression, are illustrated in the diagram, Fig. 1. It shows in the left third, two cells, *A* and *B*, in their regular environments, *a* and *b*; the conformance between cell and environment being indicated by identical markings, that is, stripes for *a* and dotting for *b*. Cell *A* is then transplanted into environment *b*, and cell *B* reciprocally into environment *a*. The same operation is carried out in the middle diagram. Yet the subsequent behaviour of the transplanted cells is radically different in the two diagrams. In the left one the transposed cells *A* and *B* conform entirely to their new environments, proving that the erstwhile distinctions had not resided in any ingrained differences between the two cells but had been expressions of different behaviour of a single kind of cell in different environments. In the middle diagram, by contrast, the cells after transposition continue to show properties referable to their origin and do not adopt each other's former properties and appearances. Such behaviour is incontestable proof that some intrinsic differences had existed between A and B prior to their transfer. This does not exclude, of course, that these differences might have been conferred upon them by their previous prolonged residence in those environments. Of this we shall speak later. Nor does this test imply that the transposed cell would not show some response to the new environment. In most cases it will, but the salient feature remains that A in b does not become like B in b, and B in a does not become the like of A in a.

It can readily be seen that this scheme is at the base of most classical experiments on embryonic transplantation, in which a change of behaviour from that of the left to that of the middle diagram has been ascribed to a process of 'determination'. In the conviction that determination is based on real physico-chemical changes, and since on the other hand we have extended the term of differentiation to include all indices of transformation, and not just the accidentally visible ones, there seems to be no further justification for retaining these two separate categories on the cellular level. Determination then is but the earlier part of the differentiation process, which is less directly discernible. But power of discernment is a property of the observer and his tools, and not of the observed system.

It is evident that these diagrams of behavioural tests apply equally well to the relation between genotype and phenotype. If A and B are taken to be eggs (or seeds) raised in media (or soils) of different compositions, the left diagram evidently represents phenotypic variation, while the middle one reveals genotypic differences. It is important, however, to bear in mind that inherent differences, which in the zygote state would ordinarily indicate differences of genotype, are manifested among somatic cell strains despite their supposedly identical genic equipment (see below).

This latter fact is best demonstrated by the most cogent test of differential constitution, namely, that diagrammed in the right-hand panel of Fig. 1, representing transfer of cells A and B into a common environment, i. As is well known, somatic cell strains transferred to tissue culture will undergo considerable changes, including the resorption of much overt specialized equipment, yet at the same time show the following signs of the preservation of inherent differences. (1) They retain certain gross morphological and physiological distinctions, such as differences of size, of nutrient requirements, of growth rates, of viability, &c. (2) Returned to conditions appropriate for the restoration of specialized equipment, each strain will develop products characteristic of its own original type or a related type. (3) When being transplanted, after prolonged stay *in vitro*, into a foreign environment in a host organism, they will behave essentially according to the middle rather than the left diagram. (4) While in the *in vitro* environment, each strain may proliferate and give rise to countless descendant cells, all of which will bear the marks of the parent strain, hence likewise

II. On Differentiation   153

behave according to points (1), (2) and (3); that is to say, while B in small i will behave differently from B in b, and A in i differently from A in a, the differentials between B and A are preserved and passed on to their cellular progeny throughout the subsequent processes of growth and cell division without depletion or attenuation.

Consequently the apparent simplification, or as it is commonly called 'de-differentiation', of cells in tissue culture signifies merely a loss of external criteria such as referred to in section II. 2, without loss of type-specificity. It does not imply reversion to a common type. The wider range of responses of cells kept in different media only proves the plurivalency of cells even in advanced stages

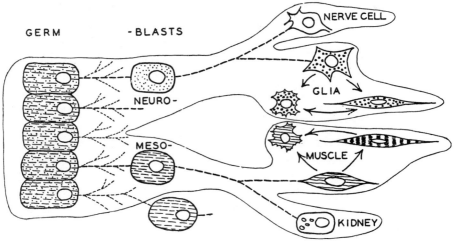

GERM   - BLASTS

NERVE CELL

NEURO-

GLIA

MESO-

MUSCLE

KIDNEY

Fig. 2

of differentiation, rather than a recovery of omnipotency. To determine the breadth of this range, which varies from type to type and from stage to stage, is a purely empirical task. The crucial fact to remember is that this range undergoes progressive narrowing within each sector of the protoplasmic time-continuum. In conclusion, the true, that is inherent, properties which connote cytodifferentiation reside within the cell boundaries and are of such nature that they can be reproduced true to type in unlimited amounts during continued proliferation. No consideration of differentiation that confines itself to the cell individual can thus be complete. A complete account must include the prior history during which equipotential cells have acquired their properties of myoblasts, chondroblasts, nephroblasts, &c., respectively, as well as the subsequent history during which these various cell types can continue to reproduce each its own kind differentially even in an indifferent common medium.

These facts are summarized in the two diagrams, Figs. 2 and 3. In Fig. 2 the segregation of omnipotent cells of the early germ by a series of events into a neuroblastic and a mesoblastic strain is depicted, with the former branching into nerve cells and glia cells, the latter into muscle cells and kidney cells, among other specializations. This diagram takes into account that, even in advanced stages, cells can still assume a variety of expressions; for instance, both muscle and glia cells can appear as either spindle cells or macrophages (see arrows), which are functional conversions, called 'modulations', and indicate latitude of expression within a given type rather than instability of the type as such.

In Fig. 3 one particular cell line is singled out to show the transformations its protoplasm continues to undergo even after type specificity has been established. Stage *a* could represent a medullary plate cell or an epidermal cell maturing (with concomitant growth and division) into stage *b*, then passing on into stage *c*, at which there appears for the first time a separation into reproductive and non-reproductive groups. The reproductive ones may remain segregated in special

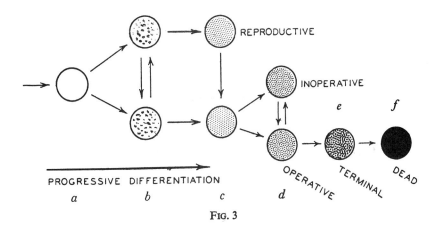

FIG. 3

'germinal' layers or cords, such as the neural epithelium lining the brain ventricles or the Malpighian layer of the skin. Knowing that the former can no longer give rise to anything but neural cells and the latter to nothing but epidermal derivatives, it is hardly proper to call them 'undifferentiated' as is done in common usage. While the upper cell in stage *c* keeps turning out more cells of the particular type, the lower one has become a terminal cell individual. As such it produces additional specialized equipment of the sort usually used for its identification.

It has frequently been asserted that there is a certain general inverse or antagonistic relationship between proliferation and terminal specialization. This rule, however, has by no means universal validity and would be weakened even more if instead of cell division the process of protoplasmic reproduction were used as criterion of growth. Neurons, for instance, keep growing throughout life yet hardly ever divide, and from what is known about cellular hypertrophy, one can make a similar case for other cell types. Perhaps the general absence of mitotic activity in terminal stages of differentiation is due to the diversion of

II. ON DIFFERENTIATION   155

materials or energy resources that would be needed for the mitotic apparatus into the building of specialized equipment.

Even the non-reproductive cells of stage *c* are not necessarily single-tracked. Depending on what type they belong to, they can go on to a variety of functional states. Most common among these are the alternatives of inoperative and operative phases which might be reiterative, either cyclically or aperiodically, or represent a singular event. Familiar examples are the active and inactive states of glands, the castrate or hormonally stimulated states of secondary sex characters, the fixed or mobilized forms of reticulo-endothelial cells, and the like. An operative cell which has become single-tracked may then assume a terminal expression (state *e*), as in the case of neuron or sensory cell, often eventuating in death, as for instance in keratinized cells of the horny layers, red blood cells, secretory cells or holocrine glands, &c. (stage *f*).

The course of events here schematized could be compared to the conveyor belt of an assembly line in an industrial plant in which a raw product is gradually transformed into the finished product. The analogy is correct in that, at every step of the process, additional factors must enter to permit it to proceed to the next step towards completion, and also in that the process may be arrested at any one stage. Thus many cells never reach the maximum possible terminal expression. Contrary to a machine assembly, however, the progress of cellular differentiation is marked less by the stepwise addition of new components than by the reorganization and selective rearrangement of existing ones within the system.

IV. WHICH PARTS OF THE CELL DIFFERENTIATE?

Having thus circumscribed the criteria and the nature of differentiation, let us now turn to the question of what parts of the cell take part in it. We may immediately exclude from our consideration elementary molecular constituents that travel freely between the cell and its environment, such as water, electrolytes, and most small organic molecules, and restrict our question to those organic systems which occur solely within the cell space. For simplicity we may lump them into five categories: (1) the genome; (2) the non-genic parts of the nucleus, or the nucleome; (3) the continuous cytoplasm; (4) the cytoplasmic inclusions, among which we may distinguish two kinds, (4a) those in common to a great variety of cell types, e.g. mitochondria, microsomes, the Golgi system; and (4b) specialized products peculiar of a given cell type, e.g. myofibrils, neurofibrils, secretion granules, pigment, &c.

The last class, (4b), obviously differs widely between different cell types. From what we have said before, particularly in section II. 3 and 4, this presupposes the existence of differential production machineries in the respective cytoplasms. Consequently, at least part of system (3) must be assumed to be subject to differentiation. As for the cell organs of class (4a), the answer is uncertain. Except for trivial differences of size, configuration, and density, they are usually considered as equivalent in all cells. It is quite possible, however, that beyond their universal functional similarities (e.g. concentration of respiratory enzymes on mitochondria, lipid character of the Golgi system, ribonucleic acid accumulation in microsomes), they show finer biochemical distinctions corresponding to their respective cell types. The extragenic nucleome (2) must be considered as differentiated, not only from general cytological appearances, but also because of the reported origin of certain specific cell products (e.g. secretions) within the nucleus. The genome (1), on the other hand, is generally assumed to retain its identity in all the various somatic cell types, at least as far as the quality of its composition is concerned. The evidence rests essentially on genetic data. Occasional attempts to connect differentiation with quantitative changes in the genome, such as polyploidy, can be discounted in view of (a) the normal cytological and histological differentiation of animals with haploid as well as polyploid (from triploid to octoploid) chromosome sets; (b) the regular occurrence in some tissues of cells with multiple chromosome sets (e.g. mammalian liver, insect scales) affecting solely cell size, but not basic cell character; and (c) a simple consideration of the very large number of qualitatively different cell types in the higher animals.

Even after narrowing the issue of differentiation to this point, it is still vague and intractable because we have recognized no further inner distinctions within the various subsystems (2) to (4) to which we concede differentiation, treating them as if they were homogenous substances of identical composition and properties throughout. In order to go beyond verbal generalities and to confront the living system with more realistic and analytical questions, we must try to form a more concrete idea of just what the microcosm of a cell really looks like, consists of, and how and by what forces it undergoes its progressive transformations. This calls for replacing the common notion that protoplasm is a 'substance' by a more realistic representation which takes into account its character as a 'system' composed of populations of molecular species of various properties and groupings, interacting with one another within the ordered facilities, as well as limitations, of the space they occupy. With this in mind, the molecular model of differentiation, presented in the following section, was constructed.

V. CYTODIFFERENTIATION IN MOLECULAR TERMS

Any attempt to formulate a molecular concept of cytodifferentiation is bound to remain, for the time being, highly speculative. However, the free flight of fancy can be considerably restrained by paying rigorous attention to certain principles of cell behaviour in differentiation that have been derived from countless observations and experiments. In an earlier publication I have labelled three of these basic principles as 'discreteness', 'exclusivity', and 'genetic limitation'. Discreteness of cytodifferentiation means that cell types fall into rather sharply delimited classes without intergradations. There are, for instance, no transitions between muscle cells producing actin and myosin, thyroid cells producing thyroxin,

Langerhans cells producing insulin, and nerve-fibres producing myelin. This argument is not weakened by the fact that many cell types may contain or produce common components (for instance, collagen or melanin), much as all living cells must be able to reproduce common equipment for their basic physiological functions such as respiratory enzymes. But with regard to other parts of their endowments, different cell types differ radically. Therefore, the principle of discreteness and the absence of a continuous spectrum of transitions point clearly to the fact that differentiation between strains is based on qualitatively different chains of chemical reactions. The second principle, exclusivity, expresses the fact that a cell cannot follow more than one of the several discrete courses originally open to it, at a time. Once it has become definitely engaged in one course, alternative courses are automatically suppressed. Evidently the cell in differentiation behaves as a unit. This principle resembles the principle of complete dominance established in genetics. The third principle, genetic limitation, expresses the empirical fact that the various ontogenetic courses open to a given cell are strictly circumscribed by the genetic endowment of the species to which it belongs. Combined with the principle of discreteness, this means that the finite, although very large, repertory of reaction types of the various descendants of a zygote is strictly limited from the start by the chemical equipment of the genome.

Heeding these clues and from a general critical interpretation of ontogeny, we arrive at the following concept of differentiation. The genome of the zygote endows all descendant cells with a finite repertory of modes of reaction. What is commonly called 'differentiation potency' may be interpreted as a finite assortment of chemical entities. These entities, of course, must not be viewed as direct precursors of any final characters, but as a reactive system, the constant interaction of which with systems of the extragenic space will only gradually yield the later specific characteristics of the various cell strains. If we envisage the response repertory of a cell as a system of alternative chains of reactions permitted by the original genic endowment, then differentiation involves the selective triggering off of certain of these chains to the exclusion of others. Divergent differentiation between two cell types thus is due not to differences of native composition but to the activation of different parts of the common equipment.

The first divergent activations in a germ arise presumably from preformed regional differences in the chemical composition and configuration of the surface layer of the egg; for blastomeres enclosing one particular sector of this surface mosaic will confront their genome with a different reactive background than will those that have received another sector. The ensuing reactions, further diversified by interactions among neighbouring parts, lead to the next steps of activation from the still multivalent, if already somewhat restricted, reaction repertory, and so on down into the late stages pictured in Fig. 3, in a continuous sequence of interactions. Since the extragenic space, i.e. the genic environment, is thus undergoing progressive transformation, it is evident that every new reaction must be viewed in terms of the cellular system in its actual condition at that particular stage, moulded by the whole antecedent history of transformations and modifications, rather than solely in terms of the unaltered genes at the core. Incidentally, keeping this in mind ought to stop the confusing practice of labelling all intrinsic properties of a cell at an advanced stage as 'genetic', but those brought out by still later interactions with neighbouring cells or diffusible agents as 'environmentally' or 'hormonally' induced, forgetting that no cell develops independently, but that all of them have gone through a long chain of similar 'environmental' interactions with neighbouring cells and the products of distant ones.

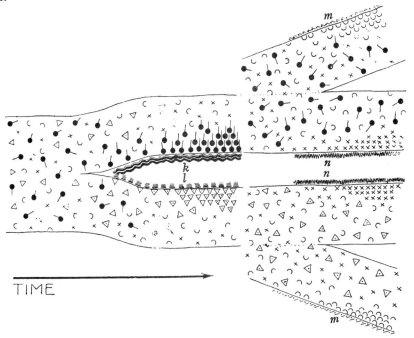

FIG. 4

This view of differentiation as a chain of chemical reactions is, of course, not new, but then it is not sufficiently tangible and specific either. It offers no model for the process of selective activation; no explanation of how, despite the great diversity of possible reactions, their systematic order in time and space, underlying functional organization, can be maintained; in general, it is too non-committal to guide deeper experimental penetration. In an effort to fill it with more specific meaning, I shall sketch in the following a concept the gist of which is incorporated in the diagram, Fig. 4.

The left end of the figure represents a sample of protoplasm (similar models apply to all subsystems listed above in IV. 2–4). We ignore chemical constituents of ubiquitous occurrence and concentrate on those molecular species characteristic of the particular cell. Though their building-blocks are of general occurrence, the peculiar pattern of their assembly into larger systems is unique and apparently reproducible only in the presence and with the aid of pre-existing similar patterns. In order not to encumber our model with unverified assumptions, we shall make no attempt to identify these molecular species chemically as to whether they are proteins, nucleic acids, lipids, polysaccharides, or higher-order combinations of such. Their number may be very large but we shall symbolize them only by four different representatives, depicted as crosses, triangles, crescents, and pins. If the chosen sample were part of the egg cytoplasm, these would be part of the primordial molecular endowment. If the sample represents a cell in a more advanced stage, this would be a derivative population, modified by the past phases of ontogeny.

It is characteristic of living systems that a state of random dispersion like that in the extreme left of the diagram would be unstable and gradually give way to orderly segregations effected and maintained presumably with the aid of metabolic energy and other factors to be mentioned presently. As I have outlined in earlier publications, the mixed molecular populations will sort themselves out according to the specific physical and chemical conditions that prevail in the different regions of the cell space; moreover, as a result of their different localizations, the segregated populations will reciprocally contribute to the establishment of similar ordering conditions for subsequent processes. The resulting organized behaviour of mixed molecular populations in the living cells, their 'molecular ecology', contrasts sharply with the behaviour of the same populations in the random dispersed state existing in homogenates. Only those chemical compounds in a cell are of relevance that are given opportunity either to interact or to alter the conditions for the interaction of others. Now it is generally recognized that the more complex a biochemical system is, the more subtle and specific are the prerequisites for its operation and maintenance. That is to say, in order for any one of the symbolic compounds of our model to have enough stability to be demonstrable, the conditions for its existence in that locality must be uniquely favourable. Such conditions may pertain to its synthesis or local accumulation or simply protection against breakdown and dissipation. They will include physical factors as well as chemical requirements such as the proper concentrations of constituent compounds, accessory factors, energy-yielding reactions, &c. Non-random distribution of key compounds thus signals the existence in different parts of the cell space of different sets of conditions favouring different types of reactions. Consequently, just as we have come to recognize in the visible range that morphological differentiation is merely an index of antecedent differentiating processes, so on the molecular scale we may now consider the demonstrable localization of given molecular species as merely an index of underlying physico-chemical conditions favouring either the reproduction or the accumulation of that particular species in that particular sample of protoplasmic space.

Among the conditions favouring selective molecular grouping, interfaces deserve special attention for the following reasons. Adsorption to interfaces can stabilize a molecular array against disruption by thermal agitation or liquid convection. Anisodiametric molecules at the same time will be adsorbed in a definite spatial orientation and may thus be ordered and aligned with the result that (a) if they are enzymes, their activity will be increased because of their closer packing and the common orientation of their active groups; and (b) if they are structural units, their orderly assembly into larger structures will be facilitated. Factors conferring this organizing faculty upon interfaces are not only differences of electric potential and the purely physical conditions favouring the formation of monomolecular films, but in a subtler sense, the chemical bonding between sterically matching or otherwise corresponding chemical compounds to either side of the interface. Compounds may thus be trapped in the interface by their affinity to sterically interlocking compounds already there. Although models of this type of interaction are currently popular as explanations of enzyme-substrate relations and antibody-antigen binding, and although their extension to phenomena of gene reproduction, protoplasmic replication, and surface inter-actions among cells seems highly suggestive, our model is independent of any such special interpretation. All it assumes is that a condition k, indicated in the upper branch of the diagram by a wavy line, representing a certain physical and chemical constellation along that interface, favours the selective accumulation from the interior of the pin-shaped compounds, whereas the condition l pre-vailing in another interface, symbolized by the broken line in the lower branch, promotes the adsorption and concentration of the triangular species.

Thus the same protoplasm, faced with two different conditions, will acquire surfaces composed of radically different compounds, hence qualitatively different. Let us call such populations which have assumed singular controlling positions 'master populations'. To explain subsequent development they must satisfy two demands—first, they must have a governing influence on the further reaction pattern of the cell, and second, they must stimulate the reproduction of their own kind. Qualitative effects on metabolism by border populations are indicated by the fact that (a) they can control selectively the substance traffic between the system and its medium in the way of a 'living membrane'; (b) if endowed with enzymatic activity, potentiated by their ordered state, they will catalyse characteristic chains of reactions; and (c) as structural elements, they constitute foundations for the anchoring and stacking up of other selected com-pounds (see later). Moreover, if these master compounds in the surface, either directly or through some of their derivative effects, were to monopolize certain metabolic resources to the exclusion of potentially competing compounds, which do not assume equally favourable positions, the latter would gradually be starved out and disappear irretrievably. This is indicated in the model by the dotting and later omission of triangles in the upper branch, and pins in the lower branch, of the diagram.

According to this model, the earliest steps towards differentiation involve

II. ON DIFFERENTIATION   161

primarily changes in the *disposition* of existing compounds, some of them being shifted to, and concentrated in, preferred interfacial positions. Only in further consequence, and with the passing of time, will their controlling functions in these preferred positions entail changes in the substantial *composition* of the systems to which they belong. This distinction between disposition and composition is fundamental, as the former is reversible, whereas the latter is not. So long as their contents in molecular species have not changed, two systems, even though they may have displayed different segments of their molecular populations in master positions, hence manifested different morphological and physiological aspects, can still be returned to a common equivalent state, if the respective key species can be dislodged from their controlling positions. Such reshuffling of the molecular population may come about as the result of solvatization, protoplasmic streaming, or other unstabilization of protoplasm; for instance, following either transfer of a cell into a foreign medium or changes in the inner milieu. as in inflammation or other pathological states.

On the other hand, once a given selected master species of molecules has occupied controlling positions long enough to have caused the competitive depletion of certain other species, then obviously a return to a common condition is no longer possible even though the key compounds may still be displaced from their controlling positions and others be induced to occupy their places. This condition is exemplified by the right-hand part of the diagram. In this the two molecular populations of the upper and lower branches of the left half of the diagram are assumed to have been first thoroughly stirred up by some factor mobilizing the cellular content and then confronted with new conditions, one symbolized as *m* and the other as *n*. Surface condition *m* traps the crescent molecules while the cross molecules aggregate in the *n* surface. Since both of the original cell types still possess both crosses and crescents, their reactions to conditions *m* and *n* are similar, as will be recognized by comparing the two inner branches and the two outer branches in the right half of the figure with each other. It should also be noted, however, that these resemblances do not connote identity and that the erstwhile differences in composition between the strains derived from the upper and the lower branches of our original protoplasmic strain have persisted. The rearrangements indicated in the right half of the diagram are the molecular version of the cellular phenomena previously described as modulations. They represent adaptations of cells to different conditions without change of fundamental equipment, but also without implying that the cells of different strains, which have undergone similar or parallel adaptations in response to identical media, have thereby become constitutionally alike.

According to our model, all irreversible differentiations can be said to arise by way of an initial reversible modulation. The other principles of cellular differentiation listed before can likewise be readily translated into terms of this model. The principle of discreteness is the result of the presence in the molecular repertory of the particular protoplasmic strain of a limited number of discrete key compounds that can assume controlling master positions. The principle of genetic limitation reflects the fact that the number and character of these key species is determined by the genic endowment of the zygote. The principle of exclusivity expresses the complete dominance in the determination of consecutive cell transformations by the molecular master species selected for surface occupancy over other molecular species not so favoured, hence excluded from exercising a determinative role.

Ordering processes of this type, spreading from an interface into the interior, are instruments of progressive organization. They are agents of selective activation and specific communication by which a given surface state can gradually evoke a conforming response from the enclosed parts. It should be borne in mind that in this progressive interaction all other interfaces can act as sites of selective screening and conversion according to their own molecular occupancy, so that one cannot take it for granted that any complex chemical entity can pass through a series of such boundaries (nuclear surface, surface of chromosomes, genes, cytoplasmic particles, &c.) without major modifications. In reverse, the actions thus activated within each enclosed system can alter conditions in the whole hierarchy of systems in ascending order. Accordingly the process we have modelled here crudely for a single protoplasmic fragment must be envisaged as repeating itself in manifold variations as each system interacts with its adjacent space. In this light, one could ask whether what we normally call 'activation' of certain components of the genome in cellular differentiation does not likewise consist of the selective segregation into controlling or active master positions of the appropriate fraction of the molecular repertory of the genes.

In conclusion, this model epitomizes the broadest statement that can be made about the biochemistry of the living cell in its organized state, as contrasted with its homogenized or disorganized condition, namely, that what determines the activities of the system is not the totality of chemical compounds it contains but the specifically selected assortment of compounds that have an opportunity to interact or otherwise operate, this being only a relatively small fraction of the total. Thus knowledge of the content of a protoplasmic system is of interest only in that it limits the possibilities of what can happen. However, in order to know just what will happen in a given case requires knowledge of just what part of the content will be placed in the appropriate conditions where it can operate. This is merely another and more explicit description of the property we usually refer to as 'organization'. To take it into account, biochemistry will have to develop a special field of 'topochemistry'.

We have assumed in this model that, in order for a protoplasmic strain to undergo divergent differentiation, the two branch lines have to be exposed to specifically different external conditions k and l. Similar branching points will arise later leading to further subspecialization. Some such dichotomies may be merely in the nature of modulations, as in the divergent expressions in media m and n. In modulation, the particular type of organization will last only as long as the respective conditioning environments are actually present, whereas in true differentiation a permanent residue of the response to a particular environment has become fixed in the cell so that it can continue itself even in the absence of the organizing environment. The descendants of a modulating cell may, of course, carry permanent and irreversible characters reflecting the particular state of the mother cell during which they were procreated.

Clearly the tacit assumption underlying this model has been that the original protoplasmic system, unless subjected to either condition k or l, would have remained stationary and unchanged. However, the validity of this assumption is open to question. It may be doubted whether any living system, even when left entirely to its own devices in a stable environment, could remain unaltered over

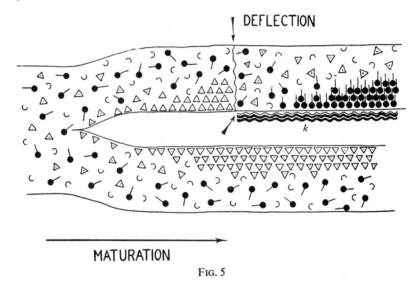

FIG. 5

prolonged periods of time. Slow progressive changes are known to occur in unfertilized eggs as well as in 'ageing' protozoan and metazoan cells. Moreover, any sequence of protoplasmic transformations that greatly outlasts the duration of the condition that has set it off will give the appearance of autonomous intrinsic change.

This being the case, our model may have to be amended in the sense of Fig. 5. The lower branch shows what would happen to the protoplastic system if subject to no additional differentiating influences. It can be seen that a progressive segregation of molecular species occurs, but in this case 'autonomously', that is, by virtue of a course of events initiated much earlier in the cell's history. This course, however, can now be deflected into a different direction by the appearance of a single differentiating condition k, provided this new condition (a) remobilizes the molecular populations, (b) dislodges the master species of triangles, and (c) replaces it by a master population of pins with greater affinity (meaning perhaps better steric conformance) to the inducing category k.

According to this model, divergent differentiation within a given cell strain would require exposure not to two different sets of conditions but to only one, while the other would simply continue an intrinsic pattern of 'maturation'. Let us quote some examples. Divergent differentiation of secondary sex characters has often been described as the switching of a neutral cell form into either the male or the female direction by corresponding sex-differentiating factors, including hormones. This implies double-switch action. On the other hand, it has also become clear that in many forms of animals the differentiation of one type of sex character is actually identical with the assumed neutral condition, with the opposite sex development being actively enforced by appropriate hormonal deflexion

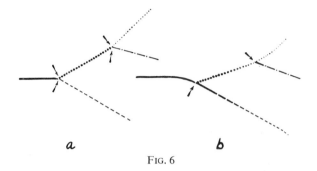

a   b

FIG. 6

from the original course. This then is single-switch action. Similarly, when tapetum of the developed urodele eye proves capable of reproducing retinal cells of the optic layer, but not vice versa, one could interpret this to mean that tapetal differentiation represents the autonomous course from which the cells would have to be positively diverted in order to produce retinal derivatives. There are innumerable examples in experimental embryology and pathology amenable to similar interpretations. The two variants of our model, corresponding to Figs. 4 and 5, are summarized in Fig. 6a and b, mainly to show that in practice the decision between them may often be difficult to make, particularly if the branching points, that is, the events causing divergencies, follow each other in quick succession. At any rate, these are empirical questions, and the models are merely intended to help phrasing them in a realistic light.

VI. HOW IS THE DIFFERENTIATED STATE PERPETUATED
AND TRANSMITTED DURING GROWTH?

Our models depict the progressive transformations of a given molecular population in time, but they do not take into account the continued increase of this population, which we call growth. Since the symbolic molecules used in our models are by definition peculiar to the particular type of protoplasm, hence are found only inside the cell, the mechanism of their reproduction must be looked for entirely within the cellular system. Despite the splendid upsurge of work on protein synthesis and particularly on the role of nucleic acid systems in cell growth, we still lack the major keys to the understanding of just what goes on in protoplasmic reproduction. It may be well, therefore, to outline certain basic considerations which any future theory of somatic growth must take into account.

1. Specialized cell protoplasm of a given cell strain can continue to propagate its own kind, as set forth in section III.

2. There is no evidence that this reproduction of type-specific protoplasm can be referred to the existence of corresponding type-specific differentiations among the genomes of the various somatic cell strains (see above, section IV).

3. To reconcile these two points, one might assume that during differentiation the capacity for self-reproduction or self-replication has been conferred upon some of the molecular key species that distinguish different cell strains according to our models.

4. The autonomy, that is independence from the genome, of such hypothetical self-reproducing cytoplasmic units is contradicted by experimental results in protozoans and yeasts. As for somatic cells, actual observations on suitable objects have made even the very concept of self-reproducibility highly questionable. Our own demonstration of the fact that neurons are in perpetual growth and that this growth proceeds solely from the nuclear territory of the cell space, supported by the recent cytochemical work on the substantial role of the nucleus in cytoplasmic synthesis, indicates strongly that the actual production sites in the process of growth are located in the nuclear territory. Therefore it is not unlikely that the so-called self-reproducing cytoplasmic particles derive the substance for their 'growth' essentially preformed from the nuclear (or more restrictedly, genic) space, hence represent stations for the deposition and possible type-specific conversion, rather than for the synthesis, of the basic protoplasmic compounds. Their growth thus would be by accretion.

5. Considering the fact that the extragenic nuclear space ostensibly undergoes differentiation (see section IV), it would be equally plausible, of course, to assume that the conversion of primordial genic products into type-specific variants occurs already within the nucleus itself.

An effort to bring these various considerations to a common denominator leads to the following integrated concept. The various specialized high-molecular key-compounds in the cytoplasm would not really possess the faculty of catalysing the synthesis of more of their own kind from elementary constituents, as implied in the concept of 'self-reproduction', but would only have the role of models in the reshaping or converting of more complex primordial compounds, furnished from the genic space, into conforming patterns. For this reason these model compounds may be given the purely descriptive name of 'templates'. By imposing their pattern upon other compounds, they would perpetuate their kind without being themselves involved in the process of multiplication. They therefore have the faculty of 'self-perpetuation', not self-multiplication. Evidently, if the type-specific master compounds which we have assumed to characterize differentiated strains act in this template fashion, the continued reproduction of the particular type of protoplasm would be ensured. Myoplasm would engender more myoplasm, nephroplasm more nephroplasm, neuroplasm more neuroplasm, and so forth, despite the identity of the genome.

In passing, it may be pointed out that this concept, suitably expanded, also furnishes the clue for the harmonious growth relations between different parts of the same organism and for their automatic regulation upon disturbance. This growth control, explained more fully in previous publications, operates on the basis of the following premises: (a) Some of the master compounds selectively sorted out in different strains act as models for their own multiplication and thus for the perpetuation of the strain. (b) Production of primordial genic (species-specific) compounds in each growing cell is superabundant. (c) The rate of their conversion into type-specific protoplasm is proportional to the number of extragenic master compounds free to act as templates. (d) Complementary compounds combining with templates render the latter ineffective, hence inactivate or veritably sterilize their template function. (e) Compounds of such complementary combining power are being thrown off as by-products of the type-specific remodelling process. In a very crude picture we might visualize them as the chips coming off as a primordial compound is whittled down to the shape of the template model in the replication process. Again with a purely descriptive term, we might call these small units of a configuration complementary to the templates 'antitemplates'. (f) Because of their small size, the antitemplates can diffuse from the cell and pass freely between the cell and its exterior, whereas the templates, because of larger size or conjugation, remain confined to the inside. (g) By virtue of the diffusion gradient between their intracellular production sites and the large extracellular liquid space in blood, lymph, and tissue juices, the antitemplates will leave the cell at a given rate. (h) As their concentration in the outside medium increases and the gradient flattens, the rate of diffusion from the intracellular to the extracellular space will decrease, until finally equilibrium is reached. Correspondingly, as their cellulifugal diffusion falls off, their relative intracellular concentration increases. (i) Since the proportion of free templates to inactivated templates in the intracellular space will decrease as the concentration of antitemplates inactivating them increases, the reduction in the rate of outward diffusion of antitemplates will automatically produce a reduction in the number of free template molecules, hence, according to (c), an automatic retardation in the reproduction of the type-specific protoplasm which we measure as growth. Whether this, in turn, rebounds on the rate of primordial genic synthesis or merely switches the utilization of the primordial products from the reproduction of basic protoplasm to the elaboration of specialized cell products, thus accounting for the observed interference of cell proliferation with specialized cell function, is problematical. Further complications are introduced into this scheme by the consideration that in the intracellular space both templates and antitemplates will be metabolized, whereas in the extracellular space the antitemplates will be katabolized without resynthesis, which presumably would cause a steady cellulifugal drift.

It can be readily seen that this system permits all protoplasm of a given kind, however widely dispersed throughout the body, to retain intercommunication and regulate its total growth. Let us consider, for instance, that part of a given type of growing protoplasm is artificially removed, thus reducing the production of templates and antitemplates of that particular type by a given amount. To the residual part of the body the loss of templates will not be perceptible since, as strictly intracellular entities, they had not been in circulation. The only information of the changed situation will come from the sudden drop of the corresponding antitemplate species in the extracellular pool. As a result of this drop of extracellular concentration, the rate of their diffusion from all residual cells of the same kind will increase, thus leaving uncovered intracellular templates for renewed growth catalysis. This will register as an automatic spurt of growth in all tissues having the same characteristic as the removed one. We have experimental evidence to show that this is at least a major part of the mechanism of so-called compensatory hypertrophy and compensatory hyperplasia.

A second way by which to reduce the effective concentration of a given antitemplate population in the extracellular pool, hence to cause an automatic growth response in the homologous cell types, is to release into the extracellular space free templates from their intracellular confinement. As these combine with their specific antitemplates in the pool, the intra- to extracellular concentration gradient of antitemplates will steepen and their growth-inhibiting effect will be correspondingly diminished. Thus, injury to a tissue, by bringing cell content into circulation, will have the same effect on homologous tissues as has partial removal. On the supposition that the 'growth-promoting' effect of embryonic extract in tissue culture is due precisely to this mechanism, we have obtained experimental support of highly suggestive, if not yet fully conclusive, nature. Even stronger support has come from experiments carried out in the embryo itself in which organ growth can be influenced in the expected direction by spilling cell content into the vascular or extra-embryonic spaces.

Whether the templates and antitemplates are to be conceived in terms of steric fitting, like antibodies and antigens, is wholly conjectural and by no means crucial for the scheme as here presented. The envisaged mechanism views the organism as a vast system of chemically intercommunicating differentiated protoplasts. Communication by special hormones appears as merely a more highly adapted and specialized version of this more general principle, the difference being that hormones are specialized cell products turned out in terminal cell phases (section II. 1 and 2), whereas we are here concerned with the underlying cell substance itself (section II. 3 and 4).

VII. THE PRODUCTION OF STRUCTURAL ORDER

Differentiation in molecular terms implies unscrambling and selective localization of molecular populations, setting the stage for consecutive reaction chains. The result is over-all structural order. We are satisfied that thermodynamically the production and maintenance of a non-random condition requires the constant input of energy but the actual mechanisms by which chemical processes are translated into orderly physical structure are for the most part still obscure. They

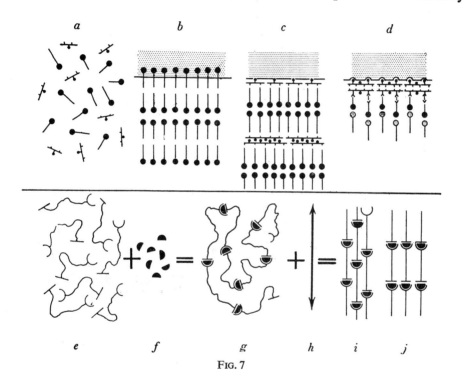

Fig. 7

are ostensibly of many diverse kinds, each to be subject to separate analysis in its own right. A few common examples are presented in the following.

One is based on the ordering effect of interfaces. It starts with the formation of a monomolecular film of oriented molecules, yielding a first-degree planar order, which then constitutes the floor plan, as it were, for a higher degree of order attained by the orderly stacking up of additional layers in the third dimension. The simplest cases are schematized in the upper part of Fig. 7. Let us assume a population of two molecular species in random dispersion (Fig. 7a), the pin-shaped kind with a lyophilic group at one end and a hydrophilic group at the other, and a water-soluble kind, e.g. a certain protein. If only the former were present at an oil-water interface, oriented molecular fixation of the kind shown in Fig. 7b would take place. In a more general sense, we may replace the oil-water system by any diphasic system with regard to which different end groups of the same molecule would show differential selective affinities. In the presence of the second species, the mixed population could sort itself out according to the diagram, Fig. 7c. Current concepts of the cell membrane and of the myelin sheath of nerve-fibres support this model. By introducing additional molecular species we can construct more complex systems such as in Fig. 7d, in which the third dimension no longer shows repetitive structure as in b and c, but displays qualitative variety. It is important to keep in mind that the adsorbed surface population maintains its stable arrangement in the face of convections and thermal agitation of an otherwise liquid system. Through the stacking on of additional layers the stable organized crust gains in width and may become microscopically distinct as gelated exoplasm. If this is true of the cell surface, then similar processes must also be conceded to the interfaces along genes, chromosomes, nucleolar and nuclear surfaces, and particulates in general.

In its application to the primordial differentiation of the egg, this scheme implies that if the egg surface contains a mosaic of molecular species of different properties segregated in different sectors, this topographical pattern would retain its stable localization despite the movements of the egg content during cleavage or the experimental reshuffling by centrifugation or stirring. It therefore provides firm bearings for the subsequent changes in individual blastomeres as outlined above in section V.

Differences are thus initiated in different parts of the original protoplasmic mass which may not become effective or manifest until at a much later stage. Yet as soon as differences appear, they present an emergent condition of further differentiation for the various interacting cell strains, for the condition of any one cell is at the same time an environmental factor for its adjacent cells. In consulting our model, Fig. 4, it is clear that if conditions k and l, after having produced differential effects in the respective cells, subside and the two cells come in apposition, each constitutes a new outside condition for the other. Depending on the state of consolidation or responsiveness reached, either cell may now react to the other by a further step of transformation. The interaction is mutual, yet whether or not a response will materialize is a matter not only of the presence or absence of adequately responsive units but also of the degree of mobility and displaceability.

One readily recognizes in this a general model of processes of induction by cellular contact. As I have indicated on earlier occasions, the neuralization of ectoderm by subjacent cell layers, or layers of organic molecules deposited on its surface, might be a case in point, although the evidence that this type of induction is of transmissive rather than transportative nature is by no means conclusive. On the other hand, in the case of lens induction by contact with the retinal layer, signs of cellular orientation signalling molecular regrouping of the requisite kind have been observed.

A second example of typical unrandomization of protoplasmic components is found in the formation of linear complexes with definite orientation in space, best illustrated by the fibrous proteins but presumably applying to all kinds of elongate and anisodiametric molecules. A simple sequence of steps is represented in the lower half of Fig. 7. In *e* we see constituent molecules with specifically configurated end groups; the addition of molecules of fitting properties (*f*) will link the original elements into chains (*g*). Polymerization and coacervation *in vitro*, blood clotting, protoplasmic coagulation, &c., furnish examples for such linear compounding. The resulting chains (*g*) are linear but not straight. Save for true crystallization and the formation of tactoids, such filaments require extraneous vectors in order to be straightened out. Such vectors (*h*) may consist of physical tensions, convection currents, and perhaps strong electrostatic fields. The resulting arrangement (*i*) leads to the building up of straight fibrils, fibrillar bundles, and fibres by progressive condensation. A still higher degree of order is produced in those cases in which, in addition to the lengthwise alignment, the constituent elements fall in lateral register (*j*), producing cross-banding, as in chromosomes, collagen fibrils, myofibrils, &c.

These examples may suffice to prove the diversity as well as the intimacy of interrelations between physical and chemical factors underlying the establishment of spatial order which we call structure. They are important not only because of their conspicuous contributions to the microscopic features of terminally differentiated cells, but because of their less overt ordering function at all stages of the differentiation process.

VIII. THRESHOLDS AND STATISTICAL ORDER

Problems of differentiation are commonly dealt with in terms that tacitly imply identity among the cells within a given cell population. In reality, no two cells are ever exactly alike, and random variations during ontogeny are apt to magnify rather than reduce initial inequalities. This being the case, a given cell population, subject to an inductive or otherwise differentiating influence, will give a uniform response in all its elements only if that influence exceeds a certain critical magnitude; corresponding to what in neurophysiology is known as a supramaximal stimulus, or in nutrition as superabundant food-supply. If, however, the influence is of lower magnitude, intensity, or duration, only a given fraction of the elements of the population will respond, namely, those of sufficiently low thresholds to be affected by the given dose of action. This is explained by the diagram, Fig. 8. In the left portion the frequency distribution of the elements of a cell population according to thresholds of responsiveness to a given agent is plotted, assuming that they vary at random (normal distribution curve). The abscissae represent stimuli of increasing dosage while the ordinates (upper base line) represent percentages of cells beginning to show effective response at the particular dosage level. The sigmoid curve (lower base line) represents the integral of the distribution curve, hence its ordinates give the total number of elements activated at any given dosage level. As the middle portion of the integral curve is nearly a straight line, the number of activated elements increases within the median range in almost linear proportion to the stimulus. Some proportionality between stimulus and response is thus to be expected on purely statistical grounds.

(In terms of the model, Fig. 5, we might view the threshold condition as the one in which the deflecting factor k can attract a sufficient number of the pinshaped molecules into master positions to give them the edge over the competing triangular type.)

Let us assume now that the dosage of a differentiating influence is of the magnitude indicated by the heavy black line. This corresponds to an ordinate value

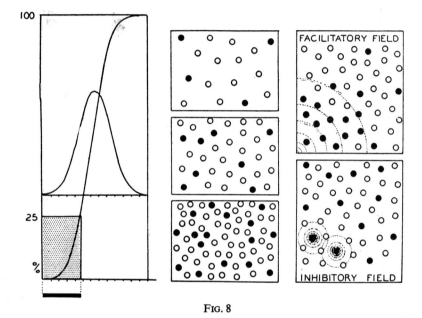

FIG. 8

of 25 per cent. That is to say, the probability is that the cell population will contain, on an average, 25 per cent. cells sufficiently sensitive to respond to this given stimulus condition. Evidently this probability will be the same for any size or density of the population. Sample results are depicted in the middle panels of Fig. 8 for three cell populations differing in densities in ratios of 1 : 2 : 4. Although the number of constituent elements varies, the ratio between responding units (black) and refractory units (white) remains the same. In other words, the relative proportions of the two segments of the population are determined although it is impossible to predict for any given unit whether it will respond or not. Statistical regularities of this kind are doubtlessly involved where the stationary composition of a cell population is to be insured throughout growth, multiplication, and repair.

Similar numerical constancy, however, can also be obtained by an altogether different procedure, namely, by precise cell lineage with differentiating divisions.

Such precision regulation has been suggested, for instance, for the constant ratio between scale-forming cells and ordinary epithelial cover cells in butterfly wings.

Statistical dichotomy of differentiation is further modified in those instances in which elements which have responded to a given stimulus thereby become the seat of secondary actions which either facilitate or inhibit similar reactions by their neighbours. These effects are illustrated in the right-hand part of Fig. 8. The top panel illustrates the case of a responding element spreading an action which lowers the threshold of other elements. As a result, the responding cells will appear grouped in clusters. Conversely, as is indicated in the right bottom panel, a spreading influence which raises thresholds will inhibit a similar cell response within a given radius, entailing thus a higher degree of regularity of distribution.

IX. FIELD EFFECTS

These considerations lead over immediately to a section of great morpho- genetic significance, namely, 'field' responses, in which the fate of a given cell

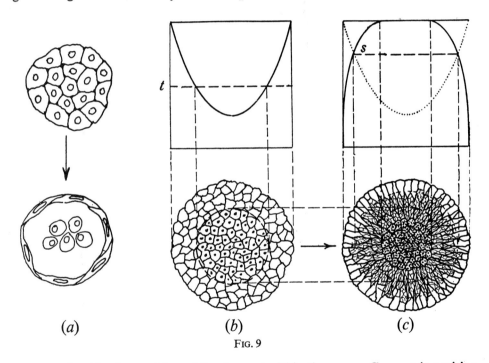

(a)   *(b)*   *(c)*

Fig. 9

is determined by the position of the element within the group. Geometric position evidently signifies tangible physico-chemical constellations at that particular site.

Examples of this type of interdependence are too common to require listing in this place. As one of the simplest cases, let us take the fate of an embryonic blood island in the chick embryo as indicated in Fig. 9a. A group of equipotential cells (top) undergo divergent differentiation (bottom), the outer ones forming a vascular endothelium, and the inner ones free blood cells. The terms 'outer' or 'inner' refer not to properties intrinsic to the component cells but to differentials conferred upon them by virtue of their position within the group. If the original cluster (top) were to be rearranged or bi-sected, or if individual cells simply traded places, each cell would then behave according to its actual, rather than its former, position.

Systems of this kind show the simplest form of field behaviour. Their mode of operation can be conceived of as follows (Fig. 9b, c). Let us assume a simple equipotential cell cluster like the one in Fig. 9a (top), bordering on either cells of some other type or on some other medium. As a result of interaction along the free surface, a certain course of reactions will be activated in the cells that share in the surface. Let us assume that this entails the release from those cells of substances or activities that spread inward with a gradient from their source. In addition, we might assume that the whole cell mass is engaged in activities producing metabolites which can diffuse into the outer medium, hence will tend to be most concentrated in the centre. Additional factors of this kind would be crust-core differentials in accessibility to essential food constituents, oxygen supply, and the like. The resultant gradient of these composite conditions is indicated in the top part of Fig. 9b. Let us now assume that the cells of the group can react to a particular constellation or combination of such factors by a differentiating step according to our model, Fig. 5, and that their thresholds for this reaction lie at the level of the broken line t. Then in the central area, within which the mentioned group factors exceed in intensity the requisite threshold minimum, all cells will be switched into the altered course. Whereas in the preceding section we have considered differentials arising within a homogeneous stimulus field by virtue of random variation among cells, the present case demonstrates the response to a graded field of partly self-engendered stimuli. In the transition zone near the threshold level, the prospective response of a given individual cell will again be unpredictable, but the steeper the gradient, the sharper will be the line of demarcation.

A new chain of responses now having been activated in the central cell group (indicated by dots, Fig. 9b, bottom), let us assume that this activity entails two further events, namely, first, a contraction of the central group (from the broken to the solid contour) and second, the release by them of new reaction products into the vicinity. As a result of the former, the outer cells attached to the shrinking core will be subject to centripetal tension, hence become radially elongated, as shown in Fig. 9c (bottom), which in turn may facilitate the outward diffusion or transmission of the newly produced agent. Its emergent diffusion field is represented by the solid curve in Fig. 9c (top). Let us assume further that the earlier substance gradient has persisted (dotted line) and interacts with the new centrifugal factor so as to form an impervious precipitate which will stop further diffusion, and that this happens in the area where the concentrations of the two agents are about equal, as indicated by the intersection of the two curves at the level s (broken line). As a result, a new zone intermediate between the original core and the new barrier will arise with new influences, activating in the constituent cells the next step of their differentiation repertory. In this manner, complexity within the originally equipotential group increases progressively according to a definite orderly pattern of organization in which 'position effects' substitute for the various extraneous media k, l, &c., of our models.

It is easy to imagine that by the linking of such spherical fields with linear fields and the further introduction of asymmetries, highly complex products can be obtained. It is also to be noted that by combining such a gradient concept of activating factors with the above molecular concept of discrete differential cell response, the conversion of merely quantitative differentials into qualitatively diverse responses finds a ready formal explanation. On the other hand, too many abstractions of this kind have been accepted in the past as 'explanations' prior to and without tangible verifications. They are presented here, therefore, primarily as guides to further analysis and as a hypothetical framework within which proper experiments can be designed. This cautioning note is not to support the scientifically untenable contention, heard on occasions, that a formal operational analysis in dynamic symbols is valueless unless accompanied by a precise identification of the substantial nature of the agents involved. This would be like saying that perfectly valid laws of optics cannot have been developed, as they have, in ignorance of the electromagnetic nature of light. Considering current trends, it would indeed seem more appropriate to caution against the illusion that the mere identification of physico-chemical systems can have much explanatory value unless their formal order of operation in the living system has likewise been revealed. It is of greater pragmatic value to set forth at least formal models of phenomena known to occur in living systems than to ignore or even deny the occurrence of those phenomena just because our present incomplete, oversimplified, and elementary schemes cannot account for them. Field phenomena in supercellular systems are a firm reality to all those observers and analysts of living phenomena who have not deliberately confined themselves to the investigation of elementary and fragmentary processes in which field properties can be legitimately ignored. At the same time, little has been done to instil more concrete content into the various field concepts. The remarks in this last section merely constitute one small and crude effort towards concreteness.

X. CONCLUSIONS

As indicated in the beginning, the study of differentiation stands to make much faster progress if instead of looking for sweeping generalizations or insisting on generalities, which by the very nature of the differentiation process can have little practical meaning, we first disassemble the complex process into its constituent components, many in numbers and of diverse kinds, and then accord each a separate and systematic study in its own right; and if, furthermore, in doing this, we visualize the real objects as concretely as our knowledge and reasoning will permit us, instead of operating solely with formal and symbolic notions as heretofore. The models introduced here for the purpose of concretization are undoubtedly far too simple and incomplete. Yet they may have the merit of pinpointing the targets at which we shoot our questions and direct our experiments on 'differentiation'.

The reader will recognize that many of the current discussions on virus reproduction, plasmagenes, adaptive enzymes, and somatic 'inheritance' in their relations to the differentiation process can be well accommodated in the terms of the concepts here advanced. However, in view of the narrative character of this article, no literature has been quoted. The main earlier publications of the author referred to in the text as precursors of the present discourse are as follows:

Weiss, P. (1939). *Principles of Development.* New York: Henry Holt and Company.
—— (1940). The problem of cell individuality in development. *Amer. Nat.* **74**, 34–36.
—— (1947). The problem of specificity in growth and development. *Yale J. Biol. Med.* **19**, 235–78.
—— (1949). Differential growth. In: *Chemistry and Physiology of Growth*, 135–86. Princeton University Press.
—— (1949). The problem of cellular differentiation. *Proc. 1st Nat. Cancer Conf.* 50–60.
—— (1950). Some perspectives in the field of morphogenesis. *Quart. Rev. Biol.* **25**, 177–98.
—— (1952). Self-regulation of organ growth by its own products. *Science,* **115**, 487–8.

Reprinted from The American Naturalist, Vol. LXXIV, pages 34–46, January–February, 1940.

CHAPTER 7

THE PROBLEM OF CELL INDIVIDUALITY IN DEVELOPMENT

PROFESSOR PAUL WEISS
University of Chicago

Before considering its embryological implications, let us first scrutinize the concept of cell individuality in the developed organism, from which it was originally derived. When Schwann says, "Each cell is within certain limits an individual," his qualification "within certain limits" seems to disavow the doctrinary rigidity which some of his followers have injected into the cell theory when they proclaimed that anything that ever happens in an organism is the resultant of individual cell activities. The severe and scornful criticism which these extreme "elementarians" had to take from the opposing camp of extreme "wholists" who wanted the attribute of individuality strictly reserved to the organism as a whole, is too well known to need further comment here. Much of this fight was carried out on philosophical grounds rather than on the factual grounds of observation and evidence. Many arguments advanced for or against the universality of the cell concept were merely rationalizations of beliefs of their authors that an organism ought or ought not to consist of discrete elements.

It is doubtful, for instance, whether the question of contiguity *versus* continuity among the elements of the nervous system would ever have become such a perpetual issue but for the fact that one group needed membranes and the other group "through" connections to explain nervous function as they saw it. Similar preconceptions were allowed to intrude into the problem of plasmodesms, that is, protoplasmic connections between cells. Those who think of protoplasmic communication and control as transportative in nature would rather have them, while transmissive theories can very well do without them. The observed facts themselves conspired with speculations on both sides by affording the expedient of artifacts caused by histological treatment: If a preparation shows cell bridges, one can not always be sure that they might not be coagulated intercellular matter; and if a preparation shows no cell bridges, one can never be sure that they might not have been present in the living but ruptured by violent contraction during fixation.

As these few examples indicate, there has been a definite tendency to rate cellular individuality by the criterion of morphological discreteness. Insularity of cell behavior seemed to presuppose completely closed boundaries as of an island, and the evidence for this was sought in uninterrupted microscopic delineation. Obviously such an attitude is unwarranted in view of present-day biological knowledge which has stripped the microscopically visible of its former prerogative as vital standard and has pointed us more and more toward the ultramicroscopical and molecular realm. A submicroscopical boundary along which colloidal and other physico-chemical properties of the protoplasm change is no less real and physiologically effective, even though the microscope fails to reveal its existence. Nevertheless, some of the old practices are still with us, and to this day much of the discussion of cell individuality continues to revolve around microscopical arguments.

From this angle, however, the question is unsolvable. There is as much evidence for the existence of discrete, well-demarcated cell individuals as there is for that of plasmodia, large protoplasmic continua containing numerous nuclei but no visible cell limits to fence off nuclear domains. Free migratory cells are clearly unicellular individuals; but the heart muscle is clearly a syncytium; so is the blastoderm of the insect egg—to mention only a few prototypes. The occurrence of both protoplasmic continuity and protoplasmic fragmentation has been amply demonstrated. But more than that: it has been demonstrated that either condition can change into the other. Morphological delineation has thus turned out to be a rather inconstant character and by no means a true test of cellular individuality. Cell bridges can break and reform; symplasms can divide up into cells; and cells can merge again into syncytia. Let us choose three examples to illustrate these facts, one from the lowest and the other two from the highest group of animals.

(1) In continuation of earlier studies by Wilson, Galtsoff and others, Brondsted has recently made a thorough reinvestigation of the problem of cellular individuality in the reconstitution and germination of sponges. His observations leave no doubt that cell discreteness is of a transitory character, which comes and goes according to circumstances. Repeatedly, cells merge into large plasmodial masses, thereby losing their outlines, and later emerge again as individualized, well-circumscribed units. Free circulation of granules and other inclusions between the formerly isolated cell body and the ground substance into which it has opened, attests to the disappearance of morphological cell boundaries in the act of fusion.

(2) The syncytial character of the mesenchyme of vertebrates has been asserted by Rohde, Hueck, Studnička and others; again it has been recorded that under certain conditions nucleated parcels of protoplasm emancipate themselves from the syncytial continuum as mobile cells which can probably later become reincorporated in the common plasmodium (v. Möllendorff).

(3) Perhaps in no other tissue has the dogma of the morphological discreteness of the cellular individual been so vigorously defended as in the nervous system. As for the embryonic origin of the nerve fibers, the neurone doctrine has come out victorious, inasmuch as the neurite of each nerve fiber has been found definitely to be the product of a single discrete nerve cell. However, evidence of secondary protoplasmic anastomoses of the individual units keeps steadily accumulating. To the best of our histological knowledge, particularly according to Boeke, the terminal branches of a motor nerve fiber pass into the protoplasm of the muscle fiber without morphological interruption. Similarly, the connection between a cutaneous

II. On Differentiation   179

nerve fiber and its sensory end organ has been described and depicted as intraprotoplasmatic. Anastomoses among nerve fibers were observed in the living
object by Speidel and in tissue culture by Guiseppe Levi,
who otherwise is one of the staunchest supporters of the
individuality of the neurone. In nerve regeneration the
newly outgrowing nerve fibers merge with the protoplasmic syncytium of the so-called cords of Büngner and
only later become set off from the matrix and from one
another by insulating sheaths.

Apparently, cell contour is a much more variable character than one would have anticipated. It can fade and reappear. But does this upset the concept of cell individuality? I do not think it does. In the studies on sponges
quoted above, Brondsted has reported that even after
merging into plasmodia, the different types of constituent
cells can still be distinguished by differences of their nuclei
and perikarya, and when a cell becomes released again, it
behaves true to its original kind. Evidently, cellular individuality can survive protoplasmic confluence. We may
adjust our picture of the cell to this situation by de-empha-
sizing the criterion of delineation. The only definitely discrete element in a cell is the nucleus, and since each nucleus
keeps protoplasm within a certain radius under its control,
protoplasmic territories have the value of cells, no matter
whether their boundaries are marked by visible surfaces
or merely by a change of physiological properties along
the border. Any change in the colloidal consistency of the
protoplasm attended by biochemical and bioelectrical differences will necessarily produce a definite orientation of
ultramicrons along the boundary and create some sort of
physiological barrier and some degree of physiological
isolation.

Within these limits, the cell is an individual, anatomically and physiologically speaking. But what about the
cell in development? Let me briefly outline the crux of
this problem also.

At the end of development we are confronted with a

unitary organized system, called an "organism," which, at the same time, is a collective of cells. At the beginning of development we find just one primordial cell—the egg. We call a system "organized" when its multiple elements appear in typical diversity, typical spatial distribution and typical temporal order. The elements are subordinated to this order and their freedom is restricted by it; hence, the order is a supra-elemental property of the system. In the developed system, "organism," the cells represent the elements; hence, organization is a supra-cellular property. But the primordium of the organism—the egg—does not consist of cells. Now, there arises a dilemma. Either the egg already possesses supra-cellular organization of the same order as the later body—then it is not just another cell, but an uncellulated organism; or it is merely a cell like others—then it can not be at the same level of organization as the later body. In this case, development would create organization of a higher order. It is to this latter view that the cell theory has committed itself. In the words of Schwann, "the individual cells so operate together in a manner unknown to us as to produce a harmonious whole," and the stress lies on "produce." Cells springing forth from repeated divisions would join hands, as it were, as equal participants in the building up of an organization all of their own making. The organism would be synthesized by progressive integration of cells into higher units, tissues, organs and the body as a whole. Cells would form the organism.

This view has met with vigorous opposition, culminating in a number of pronouncements about the inadequacy of the cell theory of development. In his address at the World's Columbian Exposition, Whitman argued the case in the most trenchant manner. One must realize the philosophical implications of the problem. If organization was to be accepted as something created *de novo* in every ontogeny, some principle had to be invoked which could mold order out of chaos, and the resort to vitalistic agents, such as Semon's "Mneme" and Driesch's "Entelechy,"

II. On Differentiation   181

was a logical outcome. Faced with the alternative, the assumption of some primordial organization inherent to the egg seemed to many a much more palatable solution. Thus, the egg was vested with organizing powers representing the supra-cellular organization of the later organism, and research was directed toward the establishment of external signs of this organization. The egg and the young germ were considered as primarily integrated wholes within which parts gradually arise by individuation. At no time would the cells constitute independent units, but from the very beginning they would be subordinated to the actions of the organism as a whole. The cells would not form the organism, but the organism would break up into cells.

Clearly the two opposing views represented a modern edition applied to organization of the old antithesis: epigenesis *versus* preformation. Epigenesis of organization was the claim of the "egg-equals-cell" theory, while preformation of organization was the tenet of the "egg-equals-organism" doctrine. The latter soon gathered momentum from experimental evidence. Lillie showed that activated eggs of a worm, Chaetopterus, when prevented from cleaving into cells as in ordinary segmentation, still underwent a considerable degree of differentiation, involving development of parts within the protoplasmic continuum of an undivided egg. Localized differentiations of the egg cytoplasm of ctenophores, annelids, molluscs, insects and amphibians became known which imparted definite substantial and dynamic properties upon the cells to which they happened to become apportioned during cleavage. The cellulation of the egg was gradually recognized as a sort of epiphenomenon superimposed upon the differentiation of the germ rather than instrumental in its production. More and more one became impressed by the fact that the organization of the germ as a whole has stability as such, regardless of the extensive fluctuations to which its cells are subjected in nature and experiment. The individual cells began to appear as slaves, rather than bosses, of the organism. The existence of individual cells as units was still acknowledged, but their rôle in embryonic organization was strongly de-emphasized.

As so often happens, however, in the wake of this sound reaction to exaggerated claims of the cell doctrine, an equally intransigent anti-cell doctrine raised its head. It tried to deny cellular individuality altogether and advanced a veritably totalitarian concept of development. Cells were ignored. The mass of the developing organism was considered as clay in the hands of the sculptor, passively submitted to molding forces which neither respect internal boundaries, nor admit of constitutional autonomy of individual units. If to the extremists of one side the individual cell was all and everything, to the advocates of the other extreme the organism as a whole appeared from the beginning in unchallenged control, cells or no cells.

Experimental embryology has, on the whole, steered clear of the two extremes. But it was difficult for the issue to find its proper level so long as one put the problem in terms of an alternative: Is the egg a cell or an organism? Is development epigenetic or preformed? Do the cells establish the properties of the developing organism or does the organism determine the properties of its cells? And so on. As we now see them, these questions are about as pertinent as if one asked: Has the face of the earth developed by volcanism or by erosion? The face of the earth is a highly complex affair, and so is its development. So also is the development of an organism. The time has passed when one could speak of development as if it were a single simple unitary phenomenon, like lightning or crystallization or the casting of a mold. Experimental analysis has revealed that what, in one word, we plainly call "development," is in reality an intricate combination of innumerable component processes, diverse and often disparate in character, which merely simulate oneness in that they all affect the identical material system—germ. Of course, it has been customary all along to single out growth (meaning increase in mass) or differentiation (increase in diversity) or morphogenesis (elaboration of shape) and the like; but there has been a general feeling that all these features are manifestations of a common principle, and that to separate them was pardonable only as an act of mental abstraction. The truth, however, is that they are essentially separate phenomena, and, in fact, each one in itself highly composite. Nuclear division, cell growth, cell division, cell aggregation, movements of cell complexes, differential growth, cytological differentiation, polarity, orientation—these are only a modest selection from the list of component phenomena into which we have learned to decompose development.

The revelation of the multiplicity of developmental processes and mechanisms has been a sad disappointment; for it has removed all hope of a general, comprehensive and universal formula of development. At the same time, it compels us to ask every question which formerly was aimed at development in general, separately for each one of its manifold components. We no longer ask: "Is development epigenetic or preformed?", but focus on a single contributory phase, asking: "How much of it is due to epigenetic and preformed conditions?", only to find that the answer varies with the object. It is this abandonment of the unitarian claim which has rendered us immune to both the strictly elementarian and the strictly totalitarian view, and which has steadied our picture of the relative rôle of cell and organism in development. Instead of a sweeping generalization, we expect a precise description of just how much of a given developmental phenomenon is due to active participation of the cells and how much to effects of supra-cellular order; what does a cell do, and what is being done to it, in a given phase of development? These are questions with which one can deal in matter-of-fact fashion, without even touching the sore spots of principle.

A few specific cases may serve as examples. Let us consider, first, histological differentiation. Do cells produce specific histological characters by intrinsic capacity or through external influences? For instance, is the elaborate conducting and contractile apparatus of a cross-striated muscle fiber developed by virtue of a constitutional property of the myoblast cell or can muscular development perhaps be imposed upon any protoplasmic mass by proper influences from its surroundings, as has been claimed by Carey? Observation and experiment have answered in no uncertain terms: Cellular differentiation is founded on innate properties of the cells themselves.

To prove the point, we remove cells from the community of the organism, thus depriving them of possible outside directives. We choose cell groups of an early germ with no manifest signs of differentiation, explant them into an extraneous medium and watch their fate. Morphogenetic development remains poor. But histologically, the explanted cells differentiate with amazing perfection. As Holtfreter and others have shown, they give rise to typical nerve cells, pigment cells, muscle cells, cartilage cells, notochordal cells, goblet cells, pronephric cells, etc. These productions are absolutely definite and discrete, each cell differentiates distinctly into one type or another, and there are no intergradations, hybrids or blends between the established cell types. Obviously, even very young cells "know" how to make a muscle fiber, a neurone, a chromatophore, etc., and we may conclude that the mechanisms for histological differentiation belong to the preformed endowment of a cell.

The same experiments have revealed, however, that cells are by no means single-tracked from the beginning. We know approximately what is to become of any given cell group of an early germ during normal development. Now, one has often noted that cells, when reared in isolation, can deviate considerably from their normal fate. Presumptive nerve cells, for instance, can become muscle cells or chorda cells, and the like. This means, evidently, that each cell of the early germ possesses a definite repertoire consisting of several discrete differentiation po-

II. On Differentiation   185

tencies. A limited number of clearly circumscribed courses are open to each cell.

In isolation, chance may decide which course is actually followed. But inside the organism, the choice is definite: there, each cell develops in conformity with the character of its surroundings. It becomes a cone or rod when in the retina, a cartilage cell when in the center of a limb bud, and a neurone when in the brain. Intrinsically capable of a variety of performances, the cell receives some definite cue from the locality indicating which trend it is to follow. These cues are decidedly of supracellular origin. Their effects have been beautifully demonstrated by transplantation experiments for which the schools of Spemann and Harrison have become famous. Transplant a young and undifferentiated cell group into the region of the head, and it will form eye or brain; transplant it to the anterior trunk and it will form limb, or further back, kidney; and, transplant it to the rear, and it will form tail—the same cells forming different structures depending on their locations. We may say: "Organizing factors take hold of the cells and direct them to appropriate formations."

We must be careful not to lapse again into the erroneous metaphor of the sculptor molding clay; let us stress, therefore, that no organizing factor has yet been observed that would have made cells assume histological structures strange to their inherited repertoire. This statement is based on crucial evidence obtained from transplantations between different species and orders of animals. In provoking specific histological characters, organizing influences are bound, therefore, to operate through the cells as their executives, and the specific character of the execution is determined by the properties of the reacting cells. To this extent, differentiation is active cell work. But this is not the whole story: The factors which turn a given cell into a definite histological trend do not, at the same time, fix all the particulars of its future course. Take a nerve fiber, for example. The factor which turns an indifferent epithelial cell into a nerve cell does not, at the same time, decide the spot at which the nerve fiber will leave the cell body; and the factor which opens the door for the fiber has no control over the further journey of the outgrowing sprout; and, again, the factors which map this course are different from those which decide where it will terminate. To put it drastically, the nerve fiber is elaborated in assembly plant fashion. In some phases of this sequence the individual nerve fiber plays an active role; in others it behaves purely passively. The original outgrowth of the sprout is free, guided presumably by oriented traffic routes of the surrounding body. But once the free tip has become hitched to a peripheral migratory cell, it is taken in tow and dragged to a destination no longer of its own choosing. It is not at all easy to tell how much of the winding course of a nerve fiber is due to active orientation and how much to passive distortion.

The same holds for cell shape in general. Part of it can be ascribed to autonomous transformations of the individual cell body, the rest to passive deformation caused by pressure, growth and spatial limitation on the part of the cell collective. With cell movements it is the same story: Free cells may aggregate in response to a local stimulus and thus form a crowd whose further growth and movement, as a whole, sweep the participant elements along without leaving them much further individual choice. Similarly, erstwhile free cells which secrete a cementing substance thereby imprison themselves and become subjected to all the dislocations of their common matrix. An opening or canal may be formed either by the active recession of the cells lining the prospective lumen or by the passive destruction of cells with subsequent resorption. When one sees pigment cells arranged in regular geometric tracts, one suspects that they have been forced into this alignment by the topography of their surroundings. But how much the pattern is really of their own making has been shown by Twitty when he interchanged the source of the pigment-forming cells between two species of distinctly different color patterns: the transplanted cells

II. On Differentiation   187

assumed the distribution characteristic of the species of their origin rather than that of the host body. I would venture to say that what the individual cells actually bring into the deal in this instance is a tendency either toward dispersion or toward aggregation; whereas the loci of aggregation in the latter case are presumably a matter not of the cells themselves, but of their matrix, so that the resulting pattern would, again, be of composite origin.

I have deliberately dwelt on these varied examples, in order to make clear that practically every step in development reveals the cell in a double light: partly as an active worker and partly as a passive subordinate to powers which lie entirely outside of its own competence and control, *i.e.*, supra-cellular powers. Now, it is perfectly true that some of these latter result from interactions of cell individuals and are, therefore, of cellular origin. But it is equally true—and the findings of experimental embryology are one rich store of evidence for our assertion—that many of them are supra-cellular from the beginning. They are those organizing conditions through which the fate of the individual cells—undecided, as we have seen, at first—is guided, controlled and progressively fixed. They are definite at a time when the individual cell fate is still indefinite. They impose order upon what otherwise would be an anarchic cell chaos. They are inherent properties of the living system, germ, as a whole, in contradistinction to the inherent properties of its constituent cells of which we have spoken before.

One frequently refers to these organizing entities under the term of "fields." Their existence can be traced back to the egg. In fact, just as there is a continuity of cells from the egg to the organism through successive cell divisions, so there is continuity between the primordial organizing fields present in the undivided egg and the localized fields of the later germ. Primordial fields segregate progressively into more restricted fields, and, furthermore, induce new fields in neighboring areas of the germ. Thus, the organizing principles of a germ have an ontogenetic history of their own which is not cell history. Their possession marks the egg as an entity of the rank of the organism; this, in answer to a question put above. Their development is a matter of the developing system as a continuum, like tensions, currents, potentials, and the like, and they pay no heed to cell boundaries, although sooner or later the intricate interplay between them and the cells sets in, of which we have spoken before.

The existence of these primordial organizing principles in the egg has been firmly established by modern experimental embryology. No pure cell theory derived from the developed organism can embrace them, unless by a vicious circle.

In conclusion, we may say that the cell theory is correct: The egg is a cell and it gives rise to all the successive cell generations which contribute to the organism. But the organismic theory is likewise correct: The egg is also an organism, and it passes its organization on continuously to the germ and the body into which it gradually transforms. Only this dual concept seems to fit the facts, as we see them at present. To be consistent, we should supplement Virchow's well-known tenet of the cell theory: "Omnis cellula e cellula," by its counterpart: "Omnis organisatio ex organisatione." If the former denies spontaneous generation of living matter, the latter denies spontaneous generation of organization. In admitting this, we merely paraphrase what Whitman has called the "continuity of organization." But within these specified limits the cell, even in development, is still, as Schwann has said, an individual.

II. On Differentiation   189

Reprinted from CHEMISTRY AND PHYSIOLOGY OF GROWTH. Editor: A. K. Parpart, Princeton University Press, Princeton, New Jersey, 135-186, 1949.

CHAPTER 8

DIFFERENTIAL GROWTH

BY PAUL WEISS[1]

THE appearance of different properties in initially equivalent cells is called "differentiation." Growth being only one cellular attribute among many, "differential growth" is to be subsumed under the larger topic of "differentiation," and our discussion therefore will have to center on the latter. Since the accent of this book lies more on prospects than on retrospects, I intend to concentrate on the gaps in our concepts and knowledge of differentiation and point to possible ways of filling them rather than dwell on past achievements, which, though impressive in themselves, are dwarfed by the task that remains to be accomplished.

Biology has discovered the strength it can derive from a more rigorous analysis of biological phenomena by the tools and standards of the physical sciences, and if the study of development is to share the new light with its more alert sister branches, such as physiology, genetics and immunology, to name only a few, it will have to submit to the same reorientation that has proved so beneficial with the others. This reorientation is in the direction of greater precision, objectivity, and consistency in the description and interpretation of phenomena, preferably in terms of physical and chemical order.

It implies the adoption not only of the tools of the physical sciences, but above all of their disciplined methodology. Analytical embryology is still full of problems which it can only settle within its own right and by its own techniques. But in doing this it must apply standards, criteria, and methods of procedure as rigorous as those of the physical sciences. It must strive to replace the abstract formalism and verbalism of its groping infancy by a mature realism, in which a term stands as symbol for a known or knowable entity or relation, rather than as a soothing device of convenient elasticity to pretend knowledge where there is none. We must not allow its analytical principles to be contaminated by relapses into animistic mythology, as we are doing when we charge the unresolved residue of meticulous determinations of physico-chemical properties to "organizers," "inductors," and similar ill-descript agents. Labels must no longer be allowed to pass for content, nor generalities for explanations. What we need most in entering

[1] University of Chicago. Original work referred to in this article was aided by the Dr. Wallace C. and Clara A. Abbott Memorial Fund of the University of Chicago.

the era ahead is the will to drive the analysis of developmental phenomena far beyond the mark where we used to stop to coin a term, and the determination to become as exact, specific, objective, and realistic in the formulation of our problems and in the description of our observations and results as has become the rule in the physical sciences. This is what we mean by reorientation. It is just as important as is the gathering of more data. With this in mind, we turn to the question which forms the core of my assignment: What is differentiation?

WHAT IS DIFFERENTIATION?

By the standards just advocated, this question can immediately be recognized as far too indefinite to be answered by anything but another generality. For "differentiation" refers not to a single circumscribed natural phenomenon, but is a summary name for a heterogeneous collection of notions about a variety of events associated with development. Consequently, the various statements and generalizations which have been made in the past concerning "differentiation" and "dedifferentiation" are equally heterogeneous and full of contradictions. Only a few authors have deemed it necessary to be explicit about terms of such general currency. In order not to add to the confusion, I shall begin by developing my own definition. Then whatever I shall have to say about "differentiation" will refer to the phenomenon as here defined, and not to anybody else's "differentiation."

A definition of "differentiation" is partly a matter of objective description, partly of plain logic. We shall deal with the logical portion first, because it has been more persistently ignored.

CHARACTERS VS. PROCESSES

One source of confusion is the tradition of confounding the process of differentiation with its criteria. We usually tell differentiation by its more conspicuous products; e.g. melanin granules in a pigment cell, myofibrils in a muscle fiber, secretion granules in a gland cell. Our criteria are predominantly morphological and we judge mostly by appearances. In the examples just mentioned, our judgment happens to be well founded. It becomes dubious when based merely on superficial differences of shape, arrangement, or size, and it misleads us completely when we try to reverse the argument and take the absence of obvious differences for evidence of lack of differentiation. To appraise a cell correctly we must take into account all of its properties, not just the ones that impress us optically, and to appreciate the development of these properties we must bear in mind that we are dealing with material systems in space and time, in which whatever is present at any one stage is the result of antecedent *processes*. Thus the conventional "characters" by which we distinguish differentiation are in reality only incidental manufacturing samples of the cellular production plant. And just as one kind of industrial plant differs from another not only by its different end products but by the whole machinery and processing method turning out those products, so a differentiated cell of one kind differs from that of another kind not only in its final equipment (pigment, fibrils, secretions, etc.) but in the basic mechanisms through which that equipment has been produced.

Tracing pigment cells, muscle cells, or blood cells back into earlier phases of their development, we find, or postulate, that they contain then, instead of the terminal products of melanin, myosin, or hemoglobin, half-finished "precursors." But if we go back still further, beyond the "precursors of precursors," the future products lose their identity and we are faced with nothing but the plant capable of turning them out, and perhaps some of the ingredients. Now, evidently a plant designed to turn out chiefly melanin must be different from one designed to make contractile myofibrils. We express this fact in the names "melanoblast" and "myoblast" given to those early stages, and we generally concede that they are intrinsically different, although the differences are no longer demonstrable by ordinary microscopic or chemical techniques. By further refinements of technique we can extend the range of our discriminative powers; structurally, by the ultramicroscope, polarization microscope, electron microscope and x-ray diffraction; chemically, by microtests (including cytological stains), spectroscopy, microincineration, etc. These techniques will undoubtedly reveal criteria of differentiation much earlier in the embryonic history than was possible with the cruder classical tools of observation. This is a line of investigation worthy of vigorous prosecution. But we must expect that it too will come to an end short of retracing the whole ontogenetic course of a given cell strain, for we are still essentially concerned with tracing "products" rather than the production processes that lie behind. Once we can detect the "product," most of the story of its production is over and we have certainly missed its essential beginnings. It looks as if the analytical methods at hand are of no avail.

Preformistic theory would deny the validity of our dilemma. It would assert that all distinctions of later stages have existed, though less manifestly, from the beginnings of ontogeny. But experimental embryology has thoroughly discredited this concept by proving that mature cells which differ profoundly in their physico-chemical constitution may be derived from cells that are demonstrably equivalent and, to all practical purposes, identical. When and by what means have their developmental courses then become divergent? This is the crucial question of cytodifferentiation. Fortunately, though analytical techniques fail us, we can answer it by a test that might be called "behavioral." This is where we must draw on logics rather than gadgets.

Suppose we are to compare two closed material systems, whether they be molecules or cells or organisms, about whose character and composition we know nothing. How are we to tell whether they are the same or different? The only way of telling is by watching their behavior. If both behave identically under identical conditions, we decide that they are alike. However, if both, again *under identical conditions, behave differently*, then we must conclude that *they have been intrinsically different* from the very beginning of the test.

Against the cogency of this postulate even the most persistent failures to detect more tangible signs of difference would prevail nothing. Two seeds giving rise to different plants in the same soil, or two eggs giving rise to different animals in the same pond, would have to be acknowledged as intrinsically different even if they were otherwise indistinguishable. Similarly, if two cells under identical conditions behave and respond differently in any respect (e.g. movement, growth rate, chemical activity, sensitivity to radiation or drugs, etc.), then they *are* different, which implies that if they have descended from a common source of identical cells, they must have *become* different, i.e. have gone through a process of differentiation, no matter whether they show it by morphological signs or not. The behavioral test is as compelling as any morphological test, and far more pertinent; for morphological criteria are but residues and convenient indicators of prior activities.

Early differences of this type, which are not immediately recognizable and can only be deduced from later formative activities, are commonly described as differences of "potency." If this were to imply that they are virtual rather than real (in the sense of Driesch), then it would mean reinstating the old "vital spirits" under another name. Actually there can be no doubt that we are dealing with differences of equipment and conditions which are real, yet too subtle to be detectable by present techniques.

In conclusion, we propose to define "differentiation" as the unidirectional changes in the character of cells and cell generations during their life history, transformations by which they become increasingly different from their own former selves, their parent cells, and the cells of other strains which have taken divergent courses. And, as we have already intimated, "character" is to include the totality of reactions of which the cell is capable at a given stage. In any given situation, only a fraction of these possible reactions will be realized, and only a small part of this fraction will become externally recognizable.

MODULATION

We have said that cells that look alike may be intrinsically very different. The opposite is equally true, namely, cells may look very different and yet be essentially alike in character. Let me cite a striking example from my own experience, the vagaries of the adult sheath cell of Schwann of adult mammalian nerve. In the embryonic stage these cells are long spindles. They then envelop the nerve axons as thin protoplasmic cylinders. When axons degenerate, as after nerve section, the sheath cells regain mass, become spindly again, and glide out from the nerve tubes. After nerve regeneration they resume their original sheathing position and corresponding cylindrical shape. But they can vary even much more than just shuttling back and forth between these two states.

Embryonic Schwann cells, reared in tissue culture along the interface of cover glass and fluid medium, tend to transform from the spindly type into a flat round form of the appearance and behavior of macrophages (49). A repetition of the experiments with adult nerve proved that the fully mature Schwann cell is capable of the same transformations and more (58). Plate I illustrates a series of cell forms obtained in vitro from Schwann cells of peripheral nerve.[2] They include tandem strands of filamentous sheath cells, ramified astrocyte-like forms, giant macrophages with ruffled membrane in active phagocytosis, monocyte-like round cells with clear ectoplasm and eccentric nucleus, and foamy round cells whose vitality is proven by their mitotic activity. Each one of these phases occurs under certain definable conditions, among which the fine-structural constitution of the medium seems to play a predominating role. Some of the transformations are easily reversible, others have as yet been observed in one direction only. Since they can be followed directly as they take place in the cultures, valuable

[2] Save for a brief note, the results of these experiments have not yet been published.

insight into the processes underlying the morphological results has been obtained. There was evidence of significant changes in the distribution of the cell content, size of the ectoplasmic border, constitution, elasticity, rigidity, mobility, etc., of the cell surface, average number and shape of pseudopodia, mode and rate of locomotion, viscosity, vacuolization, adhesiveness, orientation of fibrillar plasma constituents (as indicated by silver impregnation), fat production, distribution of nuclear materials, shape and hydration of nucleus and nucleoli, phagocytic power, and other associated properties. In short, these cells, in response to different environments, have undergone rather profound reorganizations of the kind unhesitatingly classified as "differentiation" or "metaplasia" by a nomenclature based on sheer appearances.

But is this type of change directly comparable to the progressive cytodifferentiation during ontogeny of which we have spoken before? It evidently is not, and it is in this place that facts supplement logic.

As we mentioned before, many of the observed changes are reversible if the cell is returned to its prior environment. There are many more cell types of the mature organism which behave similarly in that their behavior and shape change reversibly with their environment. Changes in the nutrient and hormone composition of the internal milieu, sublethal toxic agents, inflammatory processes, seasonal fluctuations, denervation, and in fact, on a more moderate scale, practically all ordinary physiological stimulations, cause a more or less thorough reshuffling of the cell content with consequent change in cell behavior, and often in cell form. The important point to keep in mind is that in all these instances the cell can revert to its original state when the original environmental conditions are restored. Evidently the cell passes through these transmutations with no alteration, either gain or loss, of its basic equipment. At the end of a cyclic change it turns out to be the same cell that it was before, and it is quite beside the point that sometimes the reversal may not occur in the same cell, but in one of its descendants which then prove to be "chips off the old block." There is no reason why the reversible changes observed in our Schwann cells should be regarded in any different light than, let us say, the contractions and expansions of a melanophore in response to cyclic nervous or humoral stimuli, or the change in the secretory state (and associated morphology) of a hormone-sensitive cell in response to varying concentrations of the respective hormone. They all are merely expressions of how a given cell can react to a variety of conditions—its "reaction repertory," based on its material constitution such as it is at the time. The living cell is an unstable system which can have neither existence nor form except in relation to its environment. Few animal cells (e.g. red blood cells) freeze their shape by assuming a rigid envelope. For the rest, shape is variable, merely an index of substance disposition attained in given surroundings and valid only for these particular surroundings.

Obviously, then, the misleading mental picture of absoluteness and static rigidity of cell shape, conjured by the microscopic picture of fixed cells on slides or in textbooks, must be replaced by a dynamic definition based not on the momentary expression of the cell but on all the possible reactions of which it is capable. The variety of expressions that thus can be assumed by the cell on any level of differentiation I have called "modulations" (4, 47). If we may draw a chemical analogy, differentiation would be comparable to the synthesis of a new compound by an irreversible reaction chain, while modulation would correspond to the isomerism and allotropy of intermediate or terminal products. The former involves a change of *composition*, the latter merely of *disposition* in space, of the constituent elements. The implications of this simile for the concept of the cell will be dealt with later.

It can be plainly recognized now how past concepts of differentiation have been vitiated by undue reliance on visual criteria. Cells that looked alike were thought to be essentially of like character, although they often were radically different; and conversely, cells that looked very different were considered as very disparate, although they actually were often of the same kind. The literature of histology, embryology, and pathology abounds with notable examples of victims of this "visual illusion."

DIFFERENTIATION AND DEDIFFERENTIATION

If the test of differentiation, in contradistinction to modulation, is the unidirectional, progressive, and irreversible trend of the former, how safe is the evidence that differentiation in this strict sense ever really occurs? Might not all differentiations be cases of modulation, merely obscured by the difficulty of always finding the right conditions to reverse the fate of the cell?

As long as we confine ourselves to normal development, where different cell behavior is generally associated with different local conditions, the answer remains in doubt. For a crucial test of whether different cells differ in character, we must remove them from their different environments and observe them (*a*) in identical environments, and (*b*) in each other's places. This can be done by the techniques of explantation (tissue culture) and of transplantation, respectively. The answer has been unequivocal.

Cells taken from different tissues of late embryonic or mature organisms and transferred into standard extraneous media offering nutrition, protection, and support lose many of their specialized aspects, i.e. criteria by which we used to tell their "differentiation." Spectacular as their differences may have been during residence in the body, they soon come to look and behave much more alike, in adaptation to the common environment. In the early days of tissue culture this simplification of appearance was actually often interpreted as a sign of "dedifferentiation," in the sense of a return not only to a more primitive but to a veritably primordial state. This was again judgment by appearances with all its pitfalls. Yet later work showed clearly that the explanted cells, in spite of their similar disguise, retain distinguishing features of fundamental nature which they pass on unaltered to their descendant cells for indefinite numbers of generations. As a glance at Plate I will illustrate, Schwann cells and endoneurial connective tissue cells, while assuming both the aspect of common "mesenchym," i.e. spindle shape and amoeboid activity, can be well distinguished by their sizes, nuclear characteristics, staining properties, affiliations, and other criteria revealed only by subtler analysis.

In other words, the parallel modifications which the cell strains undergo on transfer to tissue culture are merely modulations carried out by variously "differentiated" cells that otherwise preserve their basic distinctions. Their distinctiveness persists even during such further modulations in culture as the transformation to the macrophage stage. This transformation has thus far been observed in cultures of various connective tissues (25), Schwann cells (see above), and muscle (7). They all can give rise under similar conditions to a highly phagocytic amoeboid cell with hydrated nucleus, dense entoplasm, and a more hyaline ruffled border, i.e. macrophage-like characters. But in doing so they also retain the signs of their descendancy, so that a "macrophage" resulting from the conversion of an endoneurial cell can usually be clearly distinguished from one that has come from a Schwann cell.

Tissue culture thus does not really abolish character differences established during ontogeny. This conclusion is fully borne out by the critical literature on the subject (4, 40). It is strikingly confirmed by experiments in which cells that had lost their more specialized aspects during intensive proliferation in tissue culture were again given opportunity to produce their specialties, either by suppressing further proliferation in vitro or by transplanting them back into an organism. Under such conditions they were essentially true to the kind from which they stemmed: those derived from pigment epithelium would resume the production of pigment granules (8); those from thyroid, of colloid (13); those from gut, of digestive enzymes (14). Cultured cells returned to an organism, but into atypical locations (42), moreover, showed no tendency to differentiate in conformity with their new environment, proving that the culture period had not restored to them the ability of younger cells to switch into a new local course of differentiation.

All this adds up to the realization that the measure of true differentiation is the absence of true dedifferentiation, if dedifferentiation connotes not the mere assumption of a semblance of primitivity but the actual return to a state of wider potency, i.e. the recuperation of something that had ostensibly been lost. By the testimony of the reported and similar experiments, vertebrate cells are incapable of such true dedifferentiation; hence, for them, true differentiation is the rule.

TISSUE DIFFERENTIATION

All our statements thus far refer to cytodifferentiation. Histodifferentiation and organ differentiation require some additional comment. Tissues and organs consist of cells and matrix in definite mutual relations and groupings. What happens during the differentiation of a tissue is in large measure a direct reflection of the differentiation of its constituent cells. Their order of proliferation, their movements, transformations, affinities, secretions, etc., and the spatial arrangements and restraining conditions resulting from group occupancy of a common space determine, for the main part, the character and organization of the tissue or organ. By their collective activity the cells set up conditions for one another that would not develop if they remained single and independent. For this reason, what is commonly called "histodifferentiation" implies an added element of order and novelty over the mere "cytodifferentiation" of the unit cells. This is particularly striking in the formation of composite organs where the combination of cells from diverse sources leads to novel results which neither of the contributing components could have produced by itself. In addition to material contributions, there are dynamic influences of mechanical, chemical, and electrical nature acting from one tissue or organ upon another.

But on close inspection there remains no property of a tissue that could not be ultimately traced to the activity of either its own or other cells. It is by virtue of properties acquired in cytodifferentiation that cells can combine, interact, and arrange themselves in certain specific ways, conditions permitting. "Histodifferentiation" implies merely the existence of conditions under which cytodifferentiated cells can realize their individual and group faculties. This helps to clarify the meaning of the term "tissue dedifferentiation." Evidently cells may lose their typical associations and orderly arrangements just as we have seen before that they may lose internal differentiation products, without giving up their essential character. Destruction of intercellular matrix, for instance, can set cells free, and such cells may resume motility, scatter, and modulate in various ways. Their community has been dissolved, but they themselves have retained their identity. To this extent, what is called "tissue dedifferentiation" is merely tissue disintegration. Since mobilization of cells is frequently attended by the dissolution of internal differentiation products (e.g. myofibrils), the superficial picture becomes one of general simplification and regression. There is no harm in continuing to speak of tissue dedifferentiation, so long as one bears in mind that it need involve no true dedifferentiation of cells.

Which Cell Parts Differentiate?

A cell is a highly complex and heterogeneous system. Then where specifically do the changes occur that constitute differentiation? In the cytoplasm, in the nucleus, or in both? Or rather, which ones of their subdivisions undergo change, and which do not?

It is customary in speaking of cytodifferentiation to emphasize mainly the concomitant profound changes in the content of the cytoplasm. But the nucleus likewise undergoes patent modifications, which become evident in the divergence of nuclear characters and behavior among different cell types. Some of these nuclear distinctions can be deduced from differential reactions to histological stains, which in essence constitute microchemical tests. In special instances even more direct signs of nuclear differentiation have been observed, as in the extrusion of formed secretion bodies from the nuclei of certain nerve cells (30), or in the transfer of the major mineral content from nucleus to cytoplasm during the maturation of neuroblasts (36). On the other hand, it is a basic tenet of genetics that the chromosomes, or at least those parts of them that constitute the genes, remain unaltered in the process of somatic differentiation. Accordingly each nucleus contains both "differentiated" and "undifferentiated" portions.

The question arises whether the cytoplasm might not be similarly subdivided into a differentiated fraction and an undifferentiated residue. Such is generally conceded to be the case for the cytoplasm in the "germ track," i.e. of those cells that lie in direct ascendancy of the primordial germ cells. The same must be assumed for those "totipotent" somatic cells of the lower invertebrates which can give rise to a new individual by budding, regeneration, or other forms of asexual reproduction. The specialized somatic cells of higher forms, on the other hand, have never given any evidence of such totipotence. This could be interpreted to indicate that *all* of their cytoplasm has been diverted into a specialized course, has become liver-specific for the liver cell, kidney-specific for the kidney cell, and so forth. Yet again, there is undeniably some stock of cytoplasmic equipment that practically all cells have in common (e.g. respiratory enzymes, mitochondria, centrosomes, spindle fibers, etc.) and which, thus, seems to have escaped the progressive changes of cytodifferentiation.

In short, any attempts at making general statements for the nucleus as a whole or the cytoplasm as a whole end up in contradictions. The more concretely and specifically we try to state our problems, the more clearly we realize the inadequacy and ambiguity of our traditional vocabulary. So long as we address our questions not to a real living cell, but to a vague and abstract symbol, we can expect no more than oracular replies. Those general concepts of cell and cytoplasm with which we conventionally operate, with their sundry textbook attributes of contractility, excitability, adaptability, and so forth, have just about lived out the term of their usefulness. We no longer view contractility and excitability vaguely as "fundamental attributes of protoplasm," but have recognized them as defined properties of circumscribed physicochemical systems, protein chains and surface membranes, respectively. Differentiation deserves a similarly specific treatment. This will be feasible only if we replace the abstract notion of the cell by a much more specific picture. Such a picture is gradually emerging from the combined efforts of cytology, general physiology, and biochemistry. According to this picture, a cell is made up of known or knowable molecular elements, and cellular activities are resolvable into elementary molecular processes.

Study of these elementary processes is well under way and has yielded spectacular results. However, preoccupation with the elements has also at times tended to obscure the fact that in order to constitute a living cell the constituent molecular processes must not just happen, but must take place according to a highly ordered plan of interrelations based on orderly spatial distributions, orderly chronological sequences, and orderly ratios of rates. No model of a cell can be pertinent unless it takes into account both the elementary processes and their organizational frame. It is imperative, therefore, that we match research on the former by an equally vigorous search for the physical basis of the latter. To facilitate this task I have tried to sketch a sort of crude molecular blueprint of the cell, in which *the molecular phenomena are placed in relation to the organization of the cell as a unit.* This picture is of necessity diagrammatic, oversimplified, tentative, and full of blank spaces, but it does serve to crystallize the problems. Most of it has already been presented on an earlier occasion (54).

MOLECULAR ECOLOGY

Science advances not only by discovering new truths but also by eliminating old errors. Most older cell concepts held to the view that cellular organization is based on rigid structural frameworks of microscopic order, but this view has been ruled out by modern cell research. Cytoplasm behaves as a viscous liquid. It can be thoroughly mixed up by stirring or centrifugation, and yet its countless chemical workshops will continue to operate in good order, which implies proper spatial segregation. No orderly microscopic array could have survived in the commingled content; what has survived is a fixed set of conditions—the rule, as it were—according to which the original order can be restored. And where do those conditions reside? Evidently they must be looked for in those parts of the system which can preserve organization in the shuffle. There are two categories of cell constituents that qualify under this title, and they are of quite different order of magnitude, one molecular and the other supramolecular. The former consists of the large organic molecules or molecular aggregates with their specific structure and faculty to perpetuate this structure in the presence of the necessary ingredients and appropriate sources of metabolic energy. The latter consists of the existing surfaces and interfaces of microscopic or submicroscopic order, whose molecular linings are anchored by adsorption or chemical fixation and thus better protected against random disruption.

Organization within any sample space of a cell is then to be regarded as a result of the counterplay between these two poles: the specific faculties for selective combination and segregation of the molecular constituents on the one hand; and the marginal reference frames of surfaces on the other, which, by their power of adsorbing and orienting the mobile elements, force them into a supraelemental collective order. It seems impossible to account for cellular organization otherwise than by such a dual concept. Its essence can be summarized as follows: Manifest cell organization results from the response of organized elements to fields of organized (i.e. non-random) physical and chemical conditions, here tentatively identified with conditions prevailing along interfaces.

The nearest simile to this type of organization is found in the behavior of human or animal populations, which likewise consists of the reaction of organized individuals to an organized environment. In both cases, the order is one of *ecological* conditions. Thus, cell biology, in molecular terms, becomes not merely molecular physics and chemistry of the cellular constituents, but *molecular ecology* (54), a science not of molecular individuals or pairs, but of *molecular populations and their existential conditions*.

Some day we may hope to have a fairly complete list of the chemical compounds present in any given cell, from simple elements to the highly complex organic macromolecules, and thus develop a taxonomic catalogue of molecular species. But just as a museum collection of animals is never lifelike unless it shows them in their natural environment and mutual ecological relations, so this chemical catalogue would not be representative of the cell unless it were supplemented by information on the groupings, behavior, interdependence, and general operating conditions that tie the various molecular species into a durable, and indeed "viable," community.

Biochemistry is producing an ever-growing wealth of evidence for the complexity of molecular interdependence in living systems. Even very elementary physiological processes, as for instance the energy transfer in respiration or photosynthesis, involve a great number of coordinated steps, each of which requires the presence of a definite set of properly dosed reactants, mediators (enzymes), and physical conditions (pH, temperature, pressure). Undisturbed operation of these energy-yielding systems, in turn, is a prerequisite for the synthesis, among others, of the complex protein molecules, some of which, as enzymes, are again fed back as indispensable links into the energy-producing systems. Thus neither system can maintain itself without the other. Protein synthesis itself proceeds in series of steps, each of which can occur only under specified conditions which will vary for different compounds. The various chemical systems of the cell thus form "symbiotic" interdependences, and the integrity of the cellular system as a whole is contingent on their coexistence and cooperation. For each particular chemical system we can thus postulate a definite set of "existential and operational prerequisites," that is, conditions indispensable for the synthesis, preservation, and operation of specific organized products. The character of a given animal population is determined by the physical environment (climate, physiographic factors, etc.) and by the organic environment (food, competition among species, biocoenosis, etc.) in that particular area. Similarly, the character of the molecular population occupying a given site in a cell will depend on the physical conditions (electric potentials, surface tensions, etc.) and the chemical interactions among the various molecular species in that locality.

The ecological simile is primarily a convenient device to describe the behavior of the molecular populations of cells. It permits us, for instance, to answer in principle the question of how the morphological segregation and orderly distribution of different chemical systems in the cell is brought about, preserved, and restored after disturbance, despite the absence of static internal frameworks. If each complex biochemical system requires highly specific conditions for its formation and operation, then obviously the probability of finding that kind of system will be high at sites where those conditions are satisfied, and low at others. And after commingling, the content will again sort itself out according to the frame of marginal conditions, and if the latter have remained undisturbed, the re-sorting will likewise follow the old order. We will thus find preferential sites for the synthesis of certain compounds, namely, sites where the necessary ingredients, templates, energy sources, and physical factors for that particular synthesis coexist. There must also be preferential sites for the settlement of the finished products; for as synthesis continues, the accumulating products are driven further and further from their production centers and thus brought under conditions not necessarily compatible with their continued existence. I have suggested on a previous occasion (54) that such preferential settlement might be a matter of the configuration of the molecule, as a stable layer of molecules with specifically shaped ends firmly adsorbed to an interface could readily trap roaming molecules of complementarily fitting configuration.[3]

[3] This hypothesis leans on the immunochemical concepts of Pauling (32).

II. On Differentiation

The concept of molecular ecology gives a more truly representative picture of cell life than the older morphological concepts. It gives expression to the fact that what is "determined" in a given cell is only the general statistical norm of the proportions, arrangement, and distribution of its constituent elements, but not an absolutely fixed and stereotyped pattern exactly repeated in each individual cell specimen. No two cells are ever precisely congruous. It is in the very nature of the cell that we can predict no more than the degree of *probability* of the occurrence of a particular process in a given place, and this only with reference to what is happening in the rest of the cell.

In animal ecology the sizes of the mobile units vary greatly. Some animals move independently, others in flocks, and still others as heterogeneous groups. Similarly, in the molecular populations of the cell some molecules move singly, others combine into larger bodies, submicroscopic or microscopic particulates, and still others form mixed aggregates, with members of different species compounded in definite proportions (e.g. lipoprotein complexes, coacervates, etc.). The formation of any such larger unit introduces a new element of complexity (new specific surface, new conditions for adsorption, restraints of mobility, etc.), and if one thinks the matter through to its logical conclusion, one realizes that cytodifferentiation effects its manifest order through a series of steps of progressive complication, in which the basic determinants are the properties and mutual affinities of the different molecular species, and a frame of "conditions" prevailing in their common space, favoring the various species differentially. This frame or ground plan, in relation to which the mobile elements will become ordered and sorted, must evidently possess a greater measure of physical stability than the rest of the system. In a relatively fluid system, such as a cell, the entities to answer this demand are the surfaces, in which molecular mobility is restrained. Surfaces thus assume foremost rank as ecological niches for the assembly and segregation of different segments of the molecular population of the cell. Their mode of operation can be briefly sketched as follows.

SURFACE ORGANIZATION

Let us consider the interfaces which set off a cell from its surroundings, or a cytoplasmic inclusion from the matrix, or a chromosome from the nuclear medium. Heterogeneous systems meeting along a common border immediately affect each other across the border. This is due to the fact that the molecular population of one system begins to react differentially to the conditions introduced by the new contact system (electric charges, surface tensions, combinations and reactions between border molecules). The resulting interactions lead to (1) the segregation of a selected fraction of the molecular population into a surface layer, and (2) the immobilization, hence, solidification, of this portion. The mechanisms involved will vary, but for our present purpose they can simply be lumped under a common term as *"selective adsorption."* Their common denominator is the fact that a given interface of given constitution will intercept and retain certain member species of a mixed molecular population more easily and more strongly than others. Consequently, the favored species will become concentrated at the surface and the less favored ones correspondingly diluted. A change in the outer conditions may force the replacement of the former border population by new and more appropriate species from the interior. A complex system of this kind will have a labile phase, during which the molecular surface grouping responds to changes in the physical state and chemical composition of the adjacent outer system, and a secondary stable phase, which follows the consolidation of the surface due to chemical bonding, gelation, precipitation, etc. If the outer conditions are not uniform over the entire surface, but vary locally, the adsorbed molecular populations will be correspondingly variegated. The same cell can thus develop different types of surface zones on different sides exposed to different environments. Moreover, each of the manifold types of interfaces inside the cell will also acquire its own appropriate type of covering. The segregative power of specific surfaces can thus be readily understood.

In becoming adsorbed to an interface, polar molecules assume a common orientation. Their exposed ends thus constitute a new free surface to which further molecular species of fitting configuration may be built on in a second layer. This, in turn, may serve for the deposition of a third group, and so layer upon layer may be stacked up in a specific sequence, the molecules of each interlocking with those of the next.[4] Although factual information is scarce, it might be surmised that the apposition of new layers in living systems is governed by highly specific relations among the combining molecules, perhaps due to the steric fitting, key-lock fashion, of the specifically shaped ends of the conjugat-

---

[4] Such stacking up of molecular films has actually been demonstrated by Langmuir and Blodgett (23), and although their observations were made on rather simple models, evidence of a similar molecular lamination at the surface has been found in at least certain types of cells (10, 34).

ing molecules (54). As complex macromolecules will mostly have different configurations at opposite ends, successive layers will consist of different molecular species. Here we have, therefore, a process that, once it has started from a given surface, could lead to a progressive segregation with stratification of molecular species which were originally intermingled. In this process molecular orientation is of paramount importance, since it exposes the maximum number of specific receptive ends to the interior. This consideration gains added significance in the case of those polar molecules that have enzymatic properties, as parallel orientation and packing might increase their effectiveness by turning all active groups toward the substrate. Then selective surface adsorption would become a prime factor not only in the segregation of existing molecular species, but also in the creation of new ones. Moreover, the molecular fringe which settles along the surface acquires master control over substance transfer across the cell boundary in that it can selectively facilitate or prevent the exchange of materials between the areas it divides.

The extent to which a surface layer grows inward by apposition, will vary with the circumstances. For example, in some cells the outermost layers form a thick and distinctive coat (myelin sheath of nerve fiber, envelope of red blood cell, vitelline membrane of egg), while in others they remain an inconspicuous "plasma membrane." The subjacent layers (crust of egg, ectoplasms) likewise vary in thickness as well as in sharpness of demarcation. It must be remembered, however, that stratified molecular segregation need not be so massive as to become microscopically discernible.

The picture of surface organization here presented is incomplete and greatly oversimplified. For instance the specific key molecules, which were given such prominence, obviously lie interspersed with many more, smaller, simpler and trivial molecules of less specific behavior, a mosaic feature which we have not duly taken into account. It has been suggested on the basis of studies on drug action, immunological reactions, and radiation effects, that target points of specific response are present in isolated patches occupying only a small fraction of the cell surface. A similar discontinuity characterizes the structure of the chromosomes. Accordingly, many of the specific effects assigned above to surfaces in general will actually be confined to certain portions of the total surface only. An extension of this thought leads to the realization that complex molecular aggregates (particulates) may be subject to specific adsorption and fixation, with one of their constituent key patches effecting the specific attachment, and the rest remaining uninvolved. But to discuss these and other qualifications in detail, would exceed the scope of this paper.

DIFFERENTIATION IN MOLECULAR TERMS

We return from this excursion into the molecular realm better equipped to phrase questions about differentiation and growth in tangible form. We can now ask, for instance, whether differentiation really signifies a change in the composition of the molecular population of the cell or merely a change in the distribution of preexisting molecular species. The latter view has been implicit in those embryological and pathological theories which hold that the various cells of a given organism are all composed of essentially the same basic and immutable "protoplasm," and that their diverse appearances and performances are merely the overt responses to a variety of external situations or "stimuli." A given response display (called "differentiation") is thought to last only as long as the given stimulus situation lasts. Change the latter and the display will change accordingly ("dedifferentiate" and "redifferentiate") in a new direction.[5] Yet, when we view the facts from the molecular level, there seems to be no doubt that cytodifferentiation does not leave the chemical composition of "protoplasm" basically unaltered.

The ubiquitous species of small and simple molecules going in and out of cells remain about the same at all stages except for differences in concentration. Some species of large and complex molecules, including many of the proteins, are also present throughout development (5, 28). All these comprise what we have called above the stock in common to all cells. But in addition to these, differentiation brings with it, or virtually consists of, the continuous elaboration of molecular novelties which have not been present from the beginning. Thus, a liver cell, a nerve cell, a pigment cell, a mucus cell, a thyroid cell, etc., each produces its own chemical specialties. Can this really be taken as evidence that the respective "protoplasms" are chemically different? Might it not be that all mature cells have the synthetic faculty for the whole array of cell products known to the body, and that a given cell is merely prevented by local external conditions from materializing any but the appropriate one? This contention is contradicted by the experimental evidence quoted above, showing that cells transferred into a different environment

[5] A certain partiality toward such a view can be detected, for example, in the earlier writings of Child, and there have appeared several more modern, if less thoughtful, versions of the same theme.

(explantation or transplantation), though they may cease to produce their customary products, never reacquire the faculty to give rise to any but the products of their original line, when given another chance.

Thus the conclusion is inescapable that the various cell types mentioned do not just temporarily exhibit different production processes, but have themselves become chemically converted into different production plants. That is, each cell type develops, in addition to the molecular species which it contains in common with all or several other cell types, its own peculiar and distinctive array of molecular species. These are the ones that account for the differential behavior of the various protoplasms; they are the true objects of differentiation. One is tempted to identify them with specific proteins, but factual knowledge is still far too inadequate for any generalization. We need much more precise information on the time and mode of appearance of various cytochemical constituents in the ontogeny of various cell strains. Such an inventory is bound to contain the answers to numerous questions about which we can as yet do no more than speculate.

In contrast to differentiation, cell modulation implies no essential change in the composition of the molecular population, but merely a regrouping of the existing species. According to our previous discussion, any change in the environment of the cell that alters the conditions in the surface could cause a reshuffling of the content and replacement of the surface population with marked changes in the behavior and morphology of the cell, but without necessarily affecting its basic composition. However, whether such a modulation is fully reversible, i.e. whether or not the reshuffled population can return once more to its initial distribution when the original environmental conditions are restored, depends on the degree of consolidation the surface has undergone in the interim.

These considerations suggest the possibility that all differentiations start out as modulation, as mere molecular regrouping, in response to changes in the cellular environment. Let us consider an embryonic cell shifting from a site A to another site B, or, what amounts to the same, a change in the environment of a stationary cell from a condition A to a condition B. Exposure to the new contact conditions brings molecules with affinities to B to the surface. Now, suppose this molecular coat initiates a chain of chemical reactions, through which the composition of the interior is materially changed. This implies the appearance of novel molecular types as well as the disappearance of some of the existing species as they become converted into new ones. Such change cannot take place all at once throughout the cell. We must assume that it starts from a localized area (in our concept, the surface) and spreads at a measurable rate. Therefore if the cell were returned to its original condition A during the early phases of this process, it would still contain sufficient amounts of all the species it had before to resume its original state. To the observer this would appear as reversible modulation. But continued residence under condition B would permit continued progress of the changes indicated, until eventually the cell will have lost a major segment of its earlier constituents. It would thus have undergone an irreversible change of its constitution, i.e. true differentiation. The lability of differentiation in its incipient stages is a matter of record.

If a cell that has assumed character B then moves on to a site C (or if its neighbor cells or the intercellular milieu changes to a condition C), its surface population will again become reconditioned and then initiate a new chain of processes, marking a further step of differentiation, the earliest phase of which would again be labile. A cell transferred from A to C directly may or may not respond in the same fashion as when coming by way of B, depending on how decisively its molecular population has been altered during the intermediary stay at B.

We are portraying "differentiation" as a chain of events consisting of alternate physical regrouping and chemical alteration of the molecular populations, the latter phenomenon involving the emergence of novel species of compounds. It can be seen that according to this concept the potentialities for differentiation are strictly limited by the initial chemical endowment of the cell and that they become further restricted as the cell passes through the various phases A, B, C. What we conventionally call "loss of potency" is therefore merely the counterpart of the positive increase in definite chemical specialization incurred by the "differentiating" segments of the molecular population. We need assume no separate inhibitory agents.

This concept contains also, in principle, the answer to the problem of divergent cytodifferentiation, that is, of how two initially identical cells may become the source of two qualitatively different strains. Let us assume that of two cells with identical molecular populations, including key species $\alpha, \beta, \gamma, \delta, \epsilon$, etc., one is exposed to an environment E, and the other to a different environment F. Given some agitation and sufficient mobility of the cell contents, we may further assume that in each cell a selected fraction of key species will be concentrated at the surface, namely, those species which best conform to the adjacent medium; let

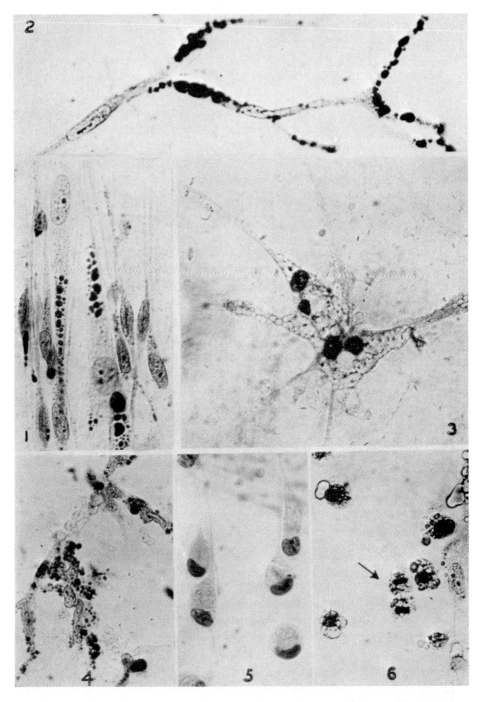

PLATE I. Transformations of Schwann cells from peripheral nerves of adult rats in tissue culture (modified from Weiss and Wang, 1945). All figures at identical magnification (× 710). *Panel 1* shows the original slender spindle shape of Schwann cells freshly emigrating from nerve. The picture also contains two large endoneurial fibroblasts (large, oval, lightly-stained nuclei; dark, sudan-stained fat droplets in cytoplasm). *Panel 2* shows beginning ramification of Schwann cells in plasma clot (note rod-shaped nuclei and fat droplets). *Panel 3* shows giant multinucleate "macrophage" developed through transformation and coalescence of a group of Schwann cells at interphase between cover glass and capillary exudate below plasma clot. *Panel 4* shows extensively arborized Schwann cells with polymorphic nuclei at fringe of culture (note active extrusion of fat bodies). *Panel 5* shows transformation stages from spindle type through monopolar "flask" shape to spherical "monocytoid" form with sickle-shaped eccentric nucleus and distinct zone of hyaline exoplasm. *Panel 6* shows small macrophages formed from Schwann cells in liquid areas of medium. (Note mitosis in telophase indicated by arrow.)

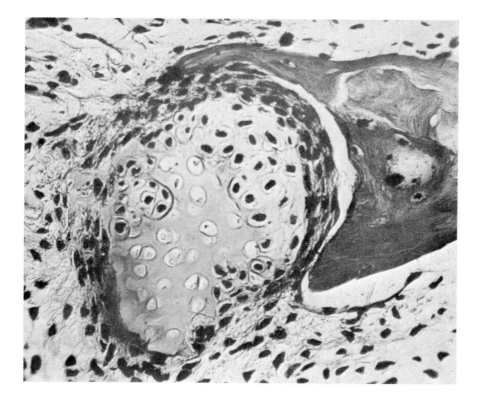

PLATE II. Induction of cartilage formation by dead cartilage graft in Amblystoma tigrinum ($\times$ 185). A metatarsal bone, quick-frozen in isopentane at $-159°$C, dehydrated in vacuo at $-40°$C, rehydrated in Ringer's solution, and transplanted into the tarsus of another premetamorphic specimen of the same species, has healed in; new cartilage and a new perichondrium have formed along most of the surface of the epiphysis of the graft, which has become detached from the bony shaft, seen in the right half of the picture. The new cartilage can be clearly distinguished by its darkly stained nuclei from the dead graft with unstained ghosts of dead nuclei in its capsules.

II. ON DIFFERENTIATION

us say, compounds ϵ in the cell exposed to the E surface, and compounds ϕ in that at the F surface, where ϵ and ϕ may be entirely unrelated. According to the preceding section, such adsorbed and oriented border species can exert key effects in turning the chemical reactions, segregations, and syntheses among the rest of the population into a definite course. Consequently the chemical fates of the two cells, coated by qualitatively different key species, will likewise become qualitatively different. Thus a difference which started out as merely one of distribution gradually develops into one of composition and character. This would be the fundamental step in the dichotomy of cell fates. This concept remains, for the present, hypothetical. But being more concrete and specific than the conventional verbal concepts, it has the advantage over the latter of being amenable to experimental verification. Tentatively, it fits the known facts of development very satisfactorily, as will be shown in the following.

On an earlier occasion (47) I outlined three basic principles of cytodifferentiation, which I called (1) discreteness, (2) exclusivity, and (3) genetic limitation. Any valid concept of differentiation must be consistent with these principles.

The principle of discreteness refers to the fact that "differentiation produces a definite number of discrete, distinct, discontinuous and more or less sharply delimited cell types which are not connected by intergradations." The principle of exclusivity states that once one type of differentiation has taken hold of a cell, no fraction of such a cell can ever become engaged concurrently in any other type of differentiation; the cell in differentiation acts as an entity in all-or-none fashion. For instance, there is a sharp dividing line at the rim of the optic cup separating optic retina from tapetum; one cell is a typical retinal cell and the next cell is just as typical a pigmented tapetum cell. Although we know from experimental evidence that originally both have the endowment to transform into either type, they never form a half-way blend. Examples of this kind could be multiplied at will. It is evident that these principles, derived from empirical facts, can be logically deduced from the postulated discreteness of the molecular key species α, β, γ, δ, ϵ, and their mutual exclusiveness in the competition for surface positions. Thus, a given cell must follow either the α-determined or the β-determined or the γ-determined, etc., course, and follow it wholly and with no compromises.

The third principle, genetic limitation, stating that each cytodifferentiation occurs according to the special methods characteristic of the parent species, reflects the fact that the numbers and kinds of molecular key species available for differentiation are limited by the inherited, hence, genetically controlled, initial endowment. It would seem premature, however, to speculate about the relation between our postulated key species and genes.

Phenomena of dominance and recessivity of cellular activities can likewise be explained in terms of our hypothesis. Any molecular species assuming a monopolistic surface position thereby gains "dominance" over competitively inferior species which, being barred from the surface, are in no position to express themselves effectively.

INDUCTION

After this consideration of cellular responses in differentiation, some comment should be given to the factors calling them forth in the organized play of development. In view of the fact that the problems of "organization" are discussed by Dr. Nicholas in the next chapter, we shall forego any but the most cursory treatment of the subject here.

According to the testimony of experimental embryology, development proceeds as follows. Cleavage subdivides the egg into blastomeres, each of which receives in addition to a nucleus a certain fraction of the original cytoplasmic system. While the nuclei are essentially equivalent, the cytoplasmic allotments vary, depending on the degree of specific molecular segregation attained in the undivided egg. The role of the most nearly stabilized part of the egg, namely its surface, must be especially taken into account.[6] Different blastomeres inherit different parts of the original egg surface. This initial differential among egg parts is then further elaborated and amplified by the subsequent reactive transformations which the various cleaved-off cells undergo in response to the organized conditions of the system which they constitute.

Descriptively, these conditions have been referred to as "fields"; but we shall not concern ourselves here with their character and operation. We need only keep in mind that in the early phases of development any area of the cleaved egg is endowed with dual capacities. On the one hand, it is composed of a large number of individual cells which in many cases, according to experimental evidence, are practically equivalent ("equipotential") and hence interchangeable as far as their response faculties are concerned. On the other hand, the totality of these cells, as a collec-

[6] The organizational role of the surface of the egg has been emphasized by several embryologists (9, 21, 33).

tive unit, exercises specific functions not exhibited by the individual elements as such. To mention a familiar though crude example, the differential exposure to environmental factors of cells in the core of a given group, as contrasted with those near the surface, is a property of the collective not owned by its members.

Cell groups under the influence of such collective factors, but prior to the appearance of a manifest response on their part, are usually referred to as "determined." There follow material transformations of the kind discussed in the preceding chapter, introducing novel relations among parts. Stepwise, in assembly-line fashion, the diversity of activities increases. As initial time differences become accentuated, some parts of the germ soon outstrip others in their tempo of differentiation. From then on, influences of the more advanced ("determined") upon the less advanced ("undetermined") parts become increasingly prominent. These influences find essentially two forms of expression: (1) influences spreading within a primary cell continuum, and (2) influences exerted from one cell continuum upon another contiguous one, usually after secondary junction.

Influences of this type are customarily referred to as "inductions." The term is purely descriptive and covers a considerable variety of heterogeneous phenomena which have often no more in common than the label. There is not the least justification for the supposition that a single operative mechanism underlies them all, much less so a single chemical entity. Much more pertinent than the illusory search for a master solution to the "induction" problem is an objective analysis of each and every one of the various concrete instances of induction that have thus far been recognized. Therefore, instead of discussing induction in general, let us concentrate on two specific examples, one for each of the categories mentioned above.

The first example concerns what might be called "homoco-induction," that is, the progressive recruiting of cells for a given type of differentiation, spreading from a focal area like an infectious wave (e.g. the antero-posterior progression of myotome formation). One gets the impression that cells which have attained a certain differentiated character can communicate their state to their neighbors, which then pass it on further, and so on down the line. The experimental test lies in transplanting a differentiated fragment amidst less far advanced cells (37). In many embryonic operations of this type, completion of the fragmentary graft structures by recruitment from surrounding cell sources has been observed, but in no case has the precise mechanism been revealed.

Technically it might be easier to start with post-embryonic stages, particularly regeneration. For instance, the fact that transplants of mature cartilage may induce new cartilage formation lends itself to a much more detailed analysis than has as yet been undertaken. Nageotte (27) reported that the implantation of alcohol-fixed cartilage into the rabbit's ear stimulates the formation of new cartilage in the adjacent connective tissue. In preliminary experiments (50) in which I transplanted pieces of salamander skeleton which had been devitalized by quick-freezing and drying, I found that cells of the larval host built a new shell of cartilage onto the dead model (Plate II). Since new cartilage has formed only in direct continuity with the graft, we must conclude that the influence in question is transmitted only by contact and not by diffusion. It seems that objects of this kind would be very favorable for a detailed physical and chemical analysis of communicable differentiation.

In terms of a molecular population concept, this type of inductive effect could be described as follows. The exposed surface of the cartilage contains certain cartilage-specific key compounds. Any potentially chondrogenic cell would contain similar key species in its interior. On making contact with cartilage, these key species would become concentrated and oriented along the surface by virtue of their affinity to the cartilage. After saturating the surface, they would then direct the subsequent chemical transformations in the cell interior into a cartilage-specific course. Thereupon the cell can, in its turn, act on the next one, and so forth. In the case of cartilage, the selective effect would be exerted by the secreted ground substance rather than the cell body itself, but in other instances it would pass directly from cell to cell. This concept has been developed in greater detail in an earlier publication (54). Its main thesis is that adjacent cells influence each other's development by means of selective attraction among specifically configured molecules, key-lock fashion. An alternative assumption would be an actual transfer of specific master units from cell to cell, in the manner of a spreading infection. Present evidence is insufficient for a final decision.

The second category to be illustrated here is "heteroinduction," in which one tissue induces another tissue, with which it has secondarily come in contact, to differentiate in a direction different from that of the inducing part. Inductions of neural plate by chorda-mesoderm and of lens by retina are familiar examples. According to our hypothesis the mechanism might be as follows. The cells of the inductive layer have a certain specific surface organization, that is, they possess already a surface layer of specific key molecules in relatively stable configuration and orientation. Corresponding molecules of complementary configuration would be present in the ectoderm cells; but during the earlier labile phase of these cells, these would still lie scattered in the interior, mixed with an assortment of other species. As the organized substratum then doubles under the ectoderm, its surface molecules would attract their molecular counterparts in the ectoderm cells into the contact surface. From their oriented surface positions, these molecular layers in the basal face of the ectoderm cells would then control the further chemical developments in their cell bodies, as outlined above. It can readily be seen that the type of response to be obtained from such an inductive action depends both on the character of the substratum and the types of molecular species present in the reacting cells at the time of exposure. The progressive transformation of the molecular equipment of the cell explains why there is often a limited period of "competence" (43) during which cells can appropriately conform to a given inductive stimulus.

In contrast to previous hypotheses, which for the most part have ascribed the inductive effect to diffusible "evocator" substances, the theory here advanced presumes a definite spatial organization of the molecular population as essential for inductive success. Induction of this type could be transmitted only by direct contact. This is in agreement with experience. Lens induction requires intimate adhesion between retina and epidermis. Neural plate arises in direct contact with the chorda plate. Many more examples could be cited that would seem to discount the diffusible nature of the inductive agent. One might object that neural inductions have been obtained in explants floating in body fluids, but it is conceivable that in these cases the active role is played by a protein film adsorbed to existing surfaces to which the explant has temporarily adhered, or a similar film settling along the outer surface of the cells. Since there is a direct relation between the presence of disintegrating cells and the occurrence of neural inductions (18), it is plausible to assume that some neural key species, set free from the destroyed cells, collect and become oriented along the outside of the surviving cells and thereby become an inductive film in the sense discussed under homoeoinduction.

These comments are conjectural. There are not enough facts known on which to decide the issue of contact versus distance action conclusively. It stands to reason that a precise and objective description of the cytological changes attending induction, together with experimental tests of their relevance for the process, could provide some of the wanting clues. A suggestive instance is the following. In observing tissues in inductive interaction I have often been struck by the remarkable correspondence of their cell and nuclear orientation just during the crucial phase. This phenomenon is very marked, for instance, in the formation of the lens. When the eye cup makes contact with the overlying epidermis, the irregular cells of the latter, within an area exactly coextensive with the area of contact, turn into a prevalent radial orientation, each elongate nucleus lining up in the direction of the axis of the opposite retinal cell. Precisely these oriented cells will later constitute the lens placode. The orientation effect is transitory, being obliterated by the subsequent transformations of the placode. While experimental tests remain to be applied, the cell-sharp coincidence between area of contact, orientation, and lens determination makes it almost certain that the observed cell changes are directly related to the primary inductive effect. Evidently, what happens is that the content of the epidermal cells undergoes a thorough reshuffling, with the underlying retinal cells acting as attractive foci. In short, the elongation and orientation of cell structure would be visible expressions of the oriented shifts of the molecular population, which would have to be expected from our theory. A systematic investigation of this orientation phenomenon is under way.

The foregoing comments on induction were presented mainly as illustrations of how problems of development gain in sharpness when resolved into terms of molecular behavior. The special solutions suggested should be considered merely as a first crude and wholly tentative effort to replace the vague, noncommittal, and unrealistic notion of "inductor substances" by a specific, concrete, and testable concept. The limitations of this concept are self-evident. Therefore our insistence lies more on the merits of the particular manner of viewing the phenomena than on the tentative views presented.

As for the limitations, one must not lose sight of the fact that the concept pertains only to the instrumentalities of biological processes and not to the overall order which these processes follow and which we usually refer to as organization. It can explain the manifest differentiation of a given molecular population, provided only that there is a regular frame of organized surface conditions to start with. The organization of the cell content is thus referred back to a prior topographical organization of the various surfaces of cell, nucleus, chromosomes, etc., and the principle, *omnis organisatio ex organisatione* (48), still holds. The "field" character (47) of the organizing conditions in development, which cannot be reduced to molecular terms, has not been touched at all in our discussion.

II. On Differentiation   217

Another point is that our concept is perhaps too exclusive in assigning the major ordering role to the surfaces. Some faculty of self-sorting of molecular mixtures (as in the formation of tactoids [3]) will perhaps have to be brought into the picture to make it complete. Also our assumption that selective transmission effects and affinities are essentially explicable in terms of steric conformance of specifically shaped molecules remains to be verified. It is still possible that ultimately we may have to take recourse to some other sort of "resonance" mechanism, "resonance-like" interrelations among tissues being clearly recognizable on the biological level (46).

GROWTH

The problem of growth likewise resolves itself now into a number of tangible issues. The definition of growth gains precision. Growth is often defined as any increase in the mass of an organic system. Since such an increase may occur by the mere intussusception of water—and even a dead seed can swell—this popular definition is scientifically of no value. I propose to define growth in a more restricted sense, as the increase in that part of the molecular population of an organic system which is synthesized within that system. The other part of the population, consisting of inorganic substances like water and salts, passing in and out of the system, is left out of consideration, and so are the smaller organic compounds that enter as ready-made building blocks, e.g. sugars and amino acids, until they have become assimilated and thereby lost their identity. Since, as far as we know, the synthesis of the complex high molecular systems in a cell occurs only in the presence and with the intervention of other similarly complex and cell-generated systems, after their own image, growth in our definition is essentially synonymous with reproduction. In population terminology, we might say that immigration of molecules into the cell space as such does not constitute growth, but their assimilation into complex systems after the model of preexisting indigenous systems does.

We can thus divide the cell content roughly into two categories: systems capable of turning out more of their own kind, and others devoid of that faculty—reproductive and non-reproductive ones. The non-reproductive group includes all the elementary ingredients for the higher compounds, as well as the sterile terminal cell products—membranes, fibers, secretions, pigments, etc. In thinking about these matters it is rather distressing to note that the proteins about which we have the most direct information, such as collagen, keratin, melanin, myosin, pepsin, etc., belong in the latter class and hence can divulge nothing about the reproductive systems, which are still very much of a mystery. At any rate the one proposed enucleates the core of the growth problem. We shall say of a pigment cell that it grows not when it synthesizes more melanin granules, but only when it synthesizes more of that peculiar part of its protoplasm which, other conditions favorable, synthesize pigments. Similarly, growth of a myoblast is not the synthesis of more myosin and actin, but of more of the basic muscle protoplasm possessing, among others, the faculty to synthesize myosin and actin; and so forth. For the sake of simplicity let us ignore for the moment the fact that "more" refers not to absolute production, but to the net excess of production over destruction in opposite metabolic phases. It would also be premature to regard the "basic" systems in question as nothing but enzyme mixtures, although enzymatic activity is one of their prime characteristics.

The fundamental feature of growth, therefore, is the capacity of a reproductive system to procreate more systems with similar reproductive faculty. In comparison with this fundamental phenomenon of multiplication of the molecular key population, the often superimposed phenomenon of cell division is as incidental as the setting up of new administrative districts in a growing country. The relation between cell division and growth is by no means as close as the widespread habit of treating them interchangeably would make it appear. During cleavage, an egg undergoes repeated cell divisions with practically no growth at all, while conversely, certain cells, such as mature neurons or the single-celled kidneys of some nematodes, grow to comparatively enormous sizes without division of either cytoplasm or nucleus.

The growth process is being explored from many directions and with increased momentum. Pertinent information comes now from genetics, now from virus research, now from embryology. Views begin to converge, and concepts, to become unified. Yet the more diversified the contributions, the more we must guard against the obvious danger of confusing a common label with a common object. If growth means cell division to one, increase in weight to another, elongation to a third, metabolism to a fourth, protein synthesis to a fifth, their mutual communication on "growth" is apt to become deceptive unless they are fully explicit. I want to make it clear, therefore, that I shall use the term strictly as defined above, that is, as the multiplication of that part of the molecular population capable of further continued reproduction, irrespective of whether or not accompanied by cell division.

II. On Differentiation

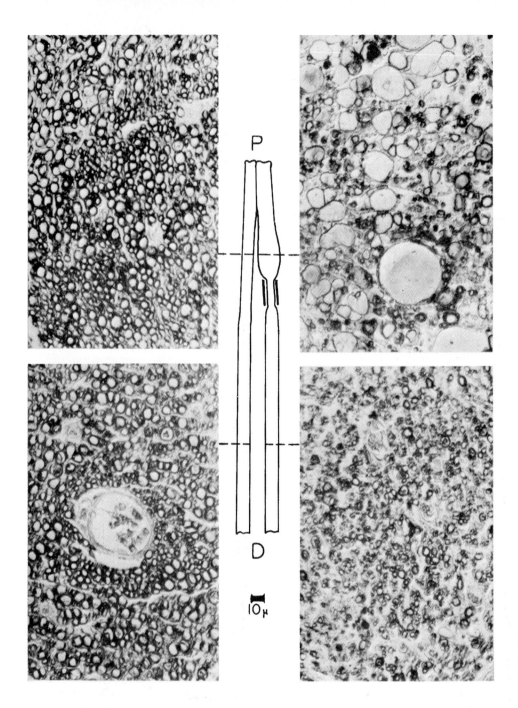

PLATE III. Unequal growth of constricted nerve fibers. Cross sections through a constricted nerve (right) and neighboring control nerve (left) at levels indicated in center diagram, all at identical magnifications ($\times$ 425). (Rat, 35 weeks after operation; P, proximal; D, distal). The same nerve fibers which are greatly shrunken distal to the constriction (lower right) have become blown up to giant dimensions just proximal to the "bottleneck" (top right); some fibers have become as wide as a small artery (compare the artery in the center of the distal control nerve at the lower left).

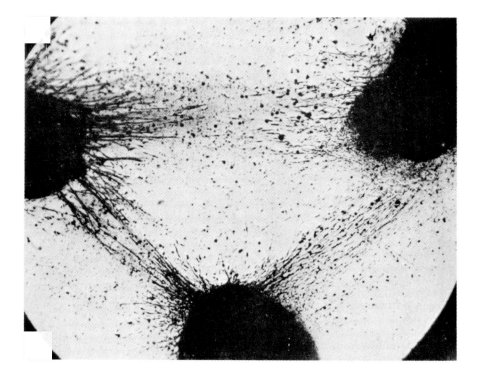

PLATE IV. Oriented cell growth between three explants cultivated in a thin film of coagulated blood plasma. Silver impregnation makes the condensations of oriented fibrin between the centers (compare Figure 3) recognizable as darker streaks in the medium; the cells can be seen to follow these dark bands.

With that definition in mind we can now ask a number of specific questions pertaining to the mode, the rate, and the site of growth. First, do cells ever stop growing? Superficially, cells fall into two categories, those that keep on proliferating, and those that do not. The ratio of the former declines steadily during ontogeny. Yet even in the mature body proliferation is still very active in many tissues, such as skin, blood-forming organs, many glands, sex cords, and others. Other cell groups become stationary. Since incremental growth can be observed only if there is a net surplus of production over breakdown, it is quite conceivable that the stationary cells are likewise engaged in growth, but at a rate just sufficient to balance the metabolic losses. In this case, protoplasm would be continually regrown in all cells.

What evidence is there for this view? Our habit of lumping all synthetic processes in one class and all degradation processes in another tends to make incremental growth appear simply as a shift in the balance toward the former; in other words, simply as more of the same kind of activity that is going on in any metabolizing cell. But if we distinguish, as we must, between different levels and types of synthesis, the problem loses its aspect of simplicity. To crystallize the issue let us focus on the most complex molecular systems, the proteins. We know from isotope tracer work (35) that proteins are in a state of continuous renovation. Now, this may mean either that the individual molecule merely interchanges constituent parts with its environment while preserving its identity and individuality, or that the molecule breaks down completely while another whole one, freshly delivered from the reproductive system, appears in its stead. The difference is as between keeping an automobile in repair by replacing the worn-out parts one by one, or scrapping it and replacing it by a new one. If the latter interpretation is correct, protein replacement and reproduction, i.e. true growth, would definitely be two different processes. To the best of my knowledge, the matter has not been finally settled.

Let us then confine ourselves to the cells which are known to grow perpetually (basal skin, lymph, intestine, etc.). Each of these cell types is known to produce each more of its own kind. What does this imply at the molecular level? Again we are faced with alternatives that we can only present, without deciding between them. The question is whether the basic systems of reproduction are the same for all cell strains or whether they change in character with progressive differentiation. Perhaps all cell strains retain a certain germinal core population which remains exempted from differentiation and constitutes the sole source of new protoplasmic compounds; in that case the synthesis of each specific compound in skin, blood, gut, etc. would have to recapitulate the whole course of the ontogenetic differentiation of that cell strain. The new units would be turned out in a primordial mold common to all cells and then gradually pass through a series of transformations after the pattern of the already present skin-, blood-, gut-specific units, as the case may be. The other possibility is that differentiation gradually changes the reproductive apparatus of the various cell strains itself so that the skin-, blood-, gut-specific systems would each be enabled to reproduce themselves directly. If we were to postulate that the genes represent the only self-reproducing systems, and assume, on the basis of genetic evidence, that genes do not change in character during ontogeny (60), the former alternative would become axiomatic. Some new experimental contributions bearing on this problem will be reported later.

Growth Rate

The fact that we do not know just what fraction of the molecular population of a cell constitutes the actively reproductive growth apparatus makes it impossible to determine "true" growth rates, i.e. the rate at which the reproductive fraction procreates more of its kind[7]. This places serious limitations on our interpretation of relative growth, for in all our measurements of growth the reproductive and non-reproductive fractions of the cell are lumped. As long as their ratio remains constant the error in estimating true growth from measurements of total growth is constant and can be discounted for most purposes. For example, if identical cells, kept in different conditions (temperature, nutrients, etc.), grow at different rates, the difference can be properly ascribed to effects on the true growth process. However, if cells of different character grow unequally, even under identical external conditions, this does not necessarily reflect differences of their true growth rates; it may merely be an expression of the fact that different fractions of their total mass are in reproductive activity.

If we designate the reproductive fraction of the population as R and the nonreproductive, hence, constant fraction as D, the true growth rate would be $V_r = \dfrac{dR}{Rdt}$ while the measured growth rate would be

[7] True growth rate could theoretically be expressed by the time required for a unit quantity of the reproductive system to increase to two units.

$$V_m = \frac{d(R+D)}{(R+D)dt} = \frac{R}{R+D} \cdot V_r. \text{ Since } \frac{R}{R+D} < 1, \text{ the measured}$$

growth rate will be smaller than the true growth rate by an amount varying inversely with the unknown constant D. Consequently, of two cells with different proportions of R and D but identical fundamental growth rates of the portions R, the one with the larger D will appear to grow more slowly. The fraction D is itself heterogeneous, including water, inorganic compounds, and all organic compounds that have either not yet been incorporated in the reproductive systems or have become detached from them as metabolites and differentiation products. Accordingly the ratio between D and R will be significantly different in different cell types, and hence different actual growth rates must be expected simply as a corollary of differentiation. Whether there is also a concurrent change in basic growth rate remains undecided.

Moreover, the ratio R/D does not remain constant for a given cell strain over its whole ontogenetic history. As differentiation proceeds, the proportion of "differentiated" terminal products, making up a large fraction of D, increases progressively at the expense of R. This increasing drain on the reproductive fraction of the cell is reflected in the well-known decline of measured overall growth rate with age. After our earlier discussion of differentiation it seems hardly necessary to reiterate that the D-fraction in question includes not merely the formed cell products (pigments, secretions, fibrils, etc.) but also that part of their formative apparatus that is not self-reproductive. Even after the loss of the formed products, as in tissue culture, cells of different strains of differentiation exhibit different growth rates (31).

Comparative evaluation of differential growth rates among different cell types is further complicated by the fact that in some cells the formed differentiation products stay within the cell and are therefore included in all measurements, while in other cell types they are extruded and lost. For instance myofibrils, neurofibrils, most pigment granules, membranes, and the like stay with the protoplasms from which they arose, while formed secretions, intercellular ground substances, and the like do not. Comparisons based on visible increments are therefore quite deceptive.

Not until most of these matters have received much more precise formulation and investigation will it be possible to develop a rational theory of differential growth. Attempts have been made to account for differential growth rates by different partition coefficients in the appropriation of nutrients, by competition for "growth substrate," by differences in the efficiency of utilization of foodstuffs, and by differential growth stimulators and growth inhibitors.[8] Plainly, many of these theories miss the point in that they try to explain differences which only appear to be differences of basic growth but in reality are natural results of differentiation. Further compelling evidence on this point will be presented below, where we shall deal with tissue growth.

THE SITE OF GROWTH

The distinction between a growing (R-) and non-growing (D-) fraction in protoplasm raises the question of the sites of growth. Evidently, the R- and D-fractions might be either segregated or interspersed. They might be mingled as mosaics of molecular, submicroscopic, or microscopic .(particulate) dimensions; they might be uniformly dispersed or each concentrated in certain localities, which might or might not correspond to the nucleus, the chromosomes, nucleolus, cell surface, or other topographical features.

Some unexpected light has been shed on these problems from recent observations on nerve growth, which I shall briefly relate. A nerve fiber is the long cytoplasmic extension of a nucleated cell body, surrounded by sheath cells. The fiber grows enormously in length and width during ontogeny, and even more dramatically during regeneration. When a fiber is cut, the distal stump, severed from its nucleated base, perishes, while the proximal stump, still connected with its central cell, grows out anew. The rapid elongation of the regenerating fiber is a process of protoplasmic movement rather than of growth in the proper sense (51). But true growth also supervenes and becomes most conspicuous in the increase in width of the fiber (from less than 1 micron to often more than 10 micra, i.e. more than a hundred times in mass). Since the fiber is surrounded by accessory sheath cells and well supplied with blood, one could well imagine that its growth is sustained from local sources. The experimental evidence, however, is that growth takes place exclusively in the nucleated central part of the cell and that all the specific cytoplasm for the whole axon is synthesized there. The evidence comes from constriction experiments as follows (55).

A local constriction (by a ring of artery) reduces the diameters of the neurilemmal tubes which contain the nerve fiber. Fibers which have regenerated through such "bottlenecks" remain permanently undersized from that point on distally. In contrast, these same fibers swell to excessive dimensions just proximal to the constriction. Plate III shows

[8] For some recent examples, consult Needham (29) and Spiegelmann (38).

II. ON DIFFERENTIATION　　　　　　　　　　　　　　　　　　　　225

cross sections through an experimental and adjacent control nerve at the indicated levels, located one centrally and the other distally to the bottleneck. A detailed analysis of this phenomenon in a large series of varied experiments has led consistently to the same conclusion, namely, that new axoplasm is constantly produced in the nucleated part of the cell body and that the whole column of axoplasm is in steady slow motion in proximo-distal direction. If this translatory movement is partially throttled by constriction, the supply to more peripheral levels is correspondingly reduced, while the surplus piles up at the entrance to the narrows. If the constriction is later removed, the effect is much as that of opening flood gates; that is, the dammed up material moves downward, its front advancing, according to preliminary determinations, at the order of one millimeter per day.

There is evidence that this proximo-distal growth of the axoplasm occurs not only during regeneration, when it adds directly to the width of the fiber, but actually goes on continuously in all axons, even the mature ones, which have reached stationary size. The central cell body seems to turn out a steady supply of new axoplasm, most of which is then conveyed distad. This raises the puzzling question as to where the material goes, if it no longer adds to the size of the fiber. I submit that it simply serves to replace the protein systems of the cell as they wear off in a steady natural degradation process, and that the nucleated cell body is, in fact, the sole source of such replacement. We know that any isolated fragment of nerve can exhibit the synthetic phases of respiratory metabolism with the aid of the enzyme systems it contains. But in the light of our experiments, it would seem that the enzyme systems themselves cannot be synthesized anywhere except in the central cell body, whence they are then distributed over the whole neuron.

As supporting evidence for this view we can quote the intensive production of nucleoproteins at the nucleus of the nerve cell (20), the identification of nucleoproteins in the axis cylinder (2), and finally what might be taken to be the external sign of their peripheral breakdown, namely the liberation of nitrogen in the form of ammonia from peripheral nerve (15). As a matter of fact, a preliminary estimate of the rate of protein breakdown from the known values of ammonia production leads to a figure which is satisfactorily close to the rate of proximo-distal replacement determined from the damming-and-release experiments reported above. While details of this concept remain to be worked out, it leaves no doubt that the site of growth and replacement of the neuron is strictly confined to the nucleated portion. It is probable, though not proven, that the same would then hold for other cell types.

These facts have an immediate bearing on efforts, currently in the foreground of interest, to bring concepts of genetics and growth in line. Our experiments place the site of growth near the nucleus, not necessarily in the nucleus. We mentioned before that if the reproductive sources should prove to be identical in all types of cells (as is assumed for the chromosomal genes), we would have to assume that their products are instantly recast into specific patterns corresponding to the surrounding differentiated cytoplasm. In our present case the nucleus may continuously reproduce basic protoplasmic systems of the general character dictated by the genic constitution, which, however, under the molding influence of the existing specific neural population to which they are added, would assume specific nerve-character. The alternative would be that the intranuclear extrachromosomal systems themselves have become specifically transformed in the course of differentiation (see p. 144), and in that case the proliferated new compounds (nucleoproteins, etc.) could acquire their full nerve-specific character at their intranuclear site of origin *in statu nascendi*. Our data are noncommittal on this point. They do prove, however, that growth is not a ubiquitous process in cytoplasm and should not be treated at par with the type of synthesis commonly met with in other types of metabolism.

TISSUE GROWTH

Analogous considerations, as applied in the foregoing to cell growth, apply to the growth of tissues and organs. Just as the molecular population of the cell consists of a reproductive and a nonreproductive fraction, so the cellular population of a tissue is composed of proliferative and nonproliferating components. In most organs proliferation becomes gradually confined to certain restricted areas, so-called germinal zones, arranged in layers (e.g. central nervous system, skin), cords (e.g. sex cords) or foci (e.g. buds). Of the newly formed cells, some remain in the germinal zone, while the rest move away from it. The former continue to proliferate, while the latter cease to multiply and usually undergo some terminal specialization.

This dichotomy has sometimes been interpreted as signifying an antagonistic relation between proliferation and differentiation.[9] Statistically speaking, it is correct that as differentiation progresses, proliferation declines. But this does not imply that differentiation in the strict

[9] For a brief summary, see Weiss (47), p. 85.

sense and proliferation are mutually exclusive (11). From our earlier discussion, it should be plain that the proliferating sources of epidermis, bone, blood, muscle, glands, etc. are differentiated to the extent that each can reproduce only its own kind and none of the others. Yet they are still actively multiplying. Only the full transformation into terminal stages seems to be held up in those cells that are kept in mitotic agitation. Proliferation does not interfere with differentiation, but it does impede the elaboration of certain manifest products of differentiation. Conversely, the general impairment of proliferative capacity in cells undergoing terminal specialization can be ascribed to the fact that most of their substance, including mitotic prerequisites, is diverted into the building of differentiation products (e.g. intracellular fiber systems).

In early developmental stages practically all cells proliferate. With progressive differentiation, however, the outlined segregation into reproductive and sterile groups takes place. Since the latter do not further contribute to the increase in mass, the total growth of a tissue or organ will vary directly with the ratio of reproductive to nonreproductive cells. This ratio in turn depends on the geometric configuration of the proliferating zone and the rate at which newly proliferated cells desert it. Further variability is introduced by the fact that most organs receive secondary additions from foreign sources (e.g. blood vessels, nerves, pigment cells), and, conversely, lose some of their own cells by dispersion or destruction. Since each organ has its own rule, it is evident that bulk measurements, which fail to separate the actual sources of the measured materials from the "dead weight," can furnish no valid basis for comparing "differential" growth of parts differing in constitution. If percentage increase is faster in one organ than in another, this could be due to either (a) a larger initial proportion of reproductive cells, or (b) a larger quota of retention of new cells in the germinal zone, or (c) greater infiltration of foreign elements, or (d) less dissipation of cells and cell products, or finally (e) an intrinsically higher rate of reproduction in the respective germinal cells. Only differences not accounted for by (a), (b), (c) or (d) are to be ascribed to (e), that is, to a higher "growth rate."

For an illustration let us choose the growing vertebrate eye. Due to its simple form and accessibility it has been a favorite object for comparative studies on growth, with the increase in dimensions serving as index. How heterogeneous a collection of events this index covers can be readily seen from the following account. The original eye vesicle consists of a certain initial allotment of cells from the embryonic brain wall. At first, all of these cells divide. The growth function at this stage is therefore a volume function. In the cup stage the retina becomes multilayered, with a sharp division into a germinal and a sterile zone. Only the cell layer in contact with the outer surface, corresponding to the ventricular (ependymal) layer of the brain, continues to proliferate, while the cells released into deeper layers differentiate the various retinal strata without further multiplication. The source of growth thus has become reduced to a two-dimensional one, causing a marked decline in the relative growth rate taken over the whole organ (e.g. from measurements of diameter). Later, the cells of the germinal layer themselves cease to proliferate and transform into sensory cells, a process which starts from the center (macula) and spreads rapidly toward the periphery (ciliary zone) of the retina. Eventually, only the cells at the rim retain residual capacity to multiply. Further growth is then essentially by apposition from this rim; that is, the growth source has shrunk from planar to linear extension. Meanwhile some of the neuroblasts, though no longer multiplying, grow in size as they sprout nerve processes, which, grouped into plexiform layers, add to the thickness of the retina. During the later stages a gelatinous secretion, supposed to come from cells of both retina and lens, fills the interior with vitreous humor, thereby progressively distending the eyeball. In addition, blood vessels and other mesenchym penetrate into the eye from the surroundings.

This diversity and complexity of the component processes contributing to eye size makes the search for a single "growth-controlling" principle appear utterly unrealistic, and comparisons between different eyes on the sole basis of size must be fallacious. To exemplify briefly the direction that an analytical approach to "differential growth" would have to take, I shall select a few of the concrete questions raised by this picture of eye growth and point out their general significance.

First, what causes the confinement of proliferation to the single cell layer at the convexity of the retina? The answer is unknown. But a clue lies in the fact that in this and most similar instances the site of proliferation corresponds to some prominent geometrical feature of the system, such as a surface or fold or tip or crest. For the cells concerned this means unique physical conditions (e.g. tensions, pressures, unique exposure to surrounding agencies) not shared by the rest of the tissue. In neural tube and retina, for instance, the direct exposure to the fluid of the central canal and ventricular spaces might be the crucial factor. The fact that injury can activate cells of the interior to take part in proliferation[10] could be explained by the penetration of the central fluid to the deeper layers as a result of the break in the internal limiting membrane. This is purely conjectural, but capable of an experimental test.

Next, what determines the further fate of the cells proliferated in the germinal layer? Of the two daughter cells of a germinal cell, either (a) both stay in the germinal layer, or (b) both move away, or (c) one stays and the other moves off. The same alternatives recur at subsequent divisions. Continuation of procedure (a) would lead to exponential increase of the proliferative source with no cells left for terminal differentiation, while procedure (b) would rapidly deplete the proliferative source. The actual growth of the eye at this phase is therefore determined by the relative incidence of events (a), (b), and (c). But again, factual information on this point is practically wholly lacking. In anuran amphibians the number of dividing cells in the retina has been reported to remain remarkably constant (16), which would mean that all divisions follow scheme (c). But this is definitely not true of the chick, for instance, in which the number of proliferating cells increases with the expansion of the retina.[11] Here the rate of growth is therefore largely dependent on the statistical apportionment of the newly formed cells according to (a), (b), or (c). This in turn seems to resolve itself into a matter of the relative orientation assumed by the daughter cells, either during mitosis or immediately after, tangential orientation leading to (a), radial orientation to (b) or (c). A plausible hypothesis is the following.

It has been demonstrated that dividing cells separate in the direction of maximum cell elongation, which corresponds to the direction of maximum tension. Hertwig's rules of cleavage are an illustration of this principle. Assuming that retinal cells follow the same principle, we may conclude that distribution of type (a) will result whenever the locally prevailing tension is tangential, while types (b) and (c) will result from radial tension or lateral pressure. According to this concept the proliferating layer can expand only as long as the retina is under tangential tension. The source of such force is found in the distension of the eyeball by the growing lens and the vitreous body. Retina, lens, and vitreous body would thereby form a self-regulating system operating as follows.

Accumulation of vitreous fluid would stretch the retina, producing a

---

[10] See for instance the compensatory proliferation of cells throughout the remaining half of an embryonic hindbrain after removal of one half (12).

[11] Recently confirmed by unpublished investigations in this laboratory by Mrs. Eleanor Gould.

prevalence of tangential tensions, which, in turn, would cause a large proportion of newly proliferated cells to stay in the germinal surface according to (a). As these additional cells are intercalated, the lateral stretch is reduced and may eventually even be reversed into pressure if the cells begin to crowd laterally; in that case, radial movement (b) or (c) would result. Meanwhile, however, further secretion of vitreous humor has produced further distension so that more new cells can be accommodated in the surface expanse, and the cycle repeats itself. Because of widely varying local conditions, these events will not cause a regular alternation of phases of tangential and radial divisions as in cleavage, but will merely statistically increase the probability of the occurrence of one or the other. There are some experimental results suggesting such a mechanism of growth regulation. When the vitreous humor of an embryonic chick eye is drained, the collapsed retina remains greatly reduced in size, but increases in thickness (56). The loss of a progressively distending core would properly account for the result. It has also been demonstrated that when incongruous eye-lens combinations are produced by heteroplastic transplantation, a mutual adjustment of size occurs (1, 17). This could be ascribed to the fact that an undersized lens, both because of its smallness and because it will permit vitreous body to flow out, will not be able to exercise the distensive function which promotes retinal growth until it has caught up with the size of the eye and sealed the borders. Reciprocally, an oversized lens will produce greater distension, hence, expansion of the germinal layer, resulting in intensified retinal growth. In a similar capacity, distension of vesicles, glands, ducts, etc. by their own or foreign discharges may be a major factor in regulating the extent of their proliferating layers, and consequently of the amount of growth. Perhaps the excessive size of the gut in amphibians fed on a bulky vegetable diet as compared with meat-fed ones (22) is due to the same principle. One could even envisage that the air in the larval lungs of gill-breathing aquatic amphibians is there for a growth-supporting rather than respiratory purpose.

These examples may suffice to illustrate the dependence of overall "growth rate" on the special mode of growth of each organ, particularly the geometric configuration of its germinal zone.

ORIENTATION OF GROWTH

Growth as such, i.e. the mere increase of the specific molecular population of the cell, is a scalar process. It has no intrinsic direction. Its directiveness is given to it by vector components of the physical framework in which it occurs. Yet growth itself plays a decisive role in creating those physical conditions which, in turn, will orient its further course. A few examples may illustrate this principle of reverberation.

In the preceding section reference was made to the effect of tension on the form of growth in epithelial tissues. Even more spectacular is its effect on the architecture of tissues of mesenchymal origin. Connective tissue, tendon, fascia, cartilage, bone, blood vessels, muscles, all respond to mechanical tensions by orientation of the component cells and intercellular fiber systems along the lines of stress. Experimental analysis of this response in tissue culture showed that the observed cell orientation is a secondary product following a primary orientation of the colloidal matrix in which the cells lie embedded (44). The sequence is as follows: (1) Tension produces alignment of aggregates of the anisodiametric molecules of the matrix into chains. (2) Such chains join to build up submicroscopic fibrils, which eventually may merge into microscopic fibers. (3) Cytoplasmic filaments (filopodia) from cells inhabiting the matrix extend along the interfaces between these solid micellar threads and the interstitial liquid. (4) The rest of the cell body follows the advancing filopodia. In this manner the cell becomes greatly elongated and is made to trace the ultrastructural pattern of its surroundings. Nerve processes advance in the same fashion, except that their cell body remains anchored (45).

In the original tissue culture experiments the matrix consisted of blood plasma, with fibrin constituting the orientable ultrastructural reticulum. Later experiments (53) revealed that even in so-called cultures in liquid medium, cell orientation is brought about by a similar reticulum, namely, coagulable fibrous material exuding from the cells themselves, spreading along the surfaces of the liquid and constituting a trellis ("ground mat") on which the cells then advance. Tensional fibrin organization in blood clots has likewise been disclosed as the mechanism orienting cell growth in wound healing and nerve regeneration in the body (57). In embryonic tissues, other fiber proteins can be expected to operate in like manner. All pertinent observations and experiments made thus far confirm the conclusion that tension, by way of its organizing effect on the colloidal cellular milieu, can determine (a) orientation of cell movement, (b) cell elongation, (c) cell shape, (d) orientation of mitotic spindles, hence, direction of division, (e) rate of movement, (f) rate of multiplication, (g) deposition of fibrous cell products. Figure 1 illustrates diagrammatically how different degrees of stretch applied to a fibrous colloid translate themselves into the morphology of inhabitant, as well as immigrating, cells. Potent as these effects are, they could not be a major factor in development if only tensions of extraneous origin were involved. But tensional stresses also arise within the developing organism as a result of shifts, growth, and chemical activity of its parts.

Figure 1. Diagram illustrating the effect of graded stretching (in the direction of the arrows) of a reticular matrix on the shape of mesenchymal cells.

Let us consider the two main intrinsic sources of morphogenetic tensions—expansion and contraction. Their development is illustrated in Figure 2. On the left side (A_o, B_o) are sectors of a tissue possessing a continuous reticulated matrix and containing a circumscribed area (dark circle) in which some localized developmental process goes on, as follows. The lower half ($B_o \rightarrow B_e$) represents a case in which the marked area expands relative to its surroundings, either by rapid proliferation or by deposition of large amounts of ground substance or, in the case of hollow organs, by distension. This local expansion creates in the surrounding continuum tensions that are oriented circumferentially, declining in strength with increasing distance from the border. The fibrous matrix assumes a corresponding pattern as depicted, and this in turn imparts itself to the cells. Cell movements and cell divisions will follow tangential courses. It is evident that the familiar concentric architecture of the mesenchymal sheaths, coats, tunics, and membranes around both solid organs and hollow ducts and vesicles finds a natural explanation on this basis. Essentially the same effect is observed if the surrounding tissue contracts against a rigid core, as in the formation of foreign-body capsules.

A wholly different pattern results if a tissue contains a focal area that contracts. In this case, illustrated in the top row of diagrams ($A_o \rightarrow A_c$), the matrix adhering to the shrinking area is subject to radial tensions, and the surrounding tissue will thus be gathered into a star-shaped pattern. Cell movements and divisions will be directed at the area, instead of around it as in the preceding example. Many embryonic invaginations (e.g. blastopore, neural tube, nasal pits) seem to take their origin from such localized contractile centers in the surface coat of the germ (24).

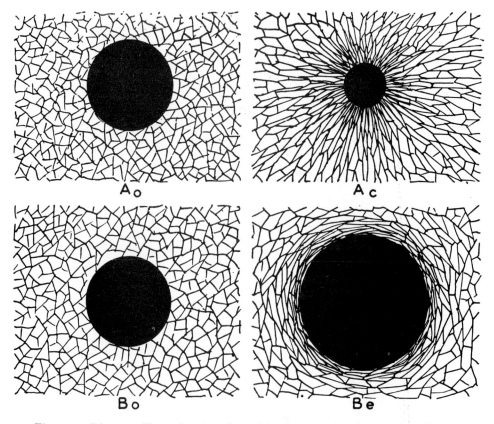

Figure 2. Diagrams illustrating the effect of local contraction (top) and local expansion (bottom) on the structure of a reticular matrix.

It was very illuminating to find that certain biochemical activities attending growth and differentiation may lead to the same type of expansion and contraction patterns (45). The expansion type, for instance, is observed around areas with proteolytic activity. When a local process dissolves part of an erstwhile coherent protein network, the remaining portion, owing to its elastic tension, retracts from the hole and assumes a pattern like that shown in Figure 2 B$_e$, much like a pierced cobweb. A contraction pattern such as in Figure 2 A$_c$, on the other hand, arises around actively proliferating centers through the following chain of events. Cells in proliferation, and apparently only then, liberate agents with strong dehydrating potency. The result is a progressive condensation of the surrounding fibrous phase of the medium. Such local condensation in a rigid tissue frame generates tensions, which in turn orient the condensing system toward the center of shrinkage. The familiar ray-shaped growth of ordinary tissue cultures in blood plasma is due to this effect. Analogous configurations can be recognized in many embryonic formations. By its radial orientation such a contraction pattern can evidently guide peripheral cells toward the center. This is one of the major mechanisms by which cells are drawn toward distant destinations, as if "attracted." Its morphogenetic importance becomes particularly impressive when several such centers act on a common matrix, where they create interconnections of great regularity among the otherwise independent centers. As is readily understood from Figure 3, two

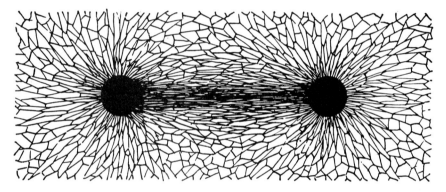

Figure 3. Diagram illustrating the orientation of a reticular matrix along the connecting line between two centers of contraction.

contractile centers produce the greatest amount of shrinkage, hence tension, and thereby orientation of the matrix along their connecting line; this automatically establishes a bridge for cell traffic between the centers. Three centers determine a triangle (Plate IV), and so forth. In principle, these experiments have solved the problem of orientation in cell movement and growth.

In acknowledging mechanical stress as a major morphogenetic factor, two reservations must be borne in mind. Firstly, tension is not the only agent capable of orienting tissue structure. Any other physical force that is capable of affecting the orientation and aggregation of polar molecules (electrostatic fields, electrophoresis, streaming, etc.) may have comparable effects. Secondly, stress is decidedly not a determining factor in cytodifferentiation. Though it seems to be prerequisite for some cell types in order that they may be able to carry out terminal transformations for which they are already predisposed by previous differentiation, it is not the cause of such differentiation. Without stretch, myoblasts cannot transform into muscle fibers (39), but stretch itself neither makes a cell into a myoblast nor ever transforms a non-myoblast into a muscle fiber, as has once been claimed (6).

An interesting combination effect of both contractile tension and proteolysis has recently been observed in the healing of nerve stumps (57). Tension transmitted from stump to stump through an intervening blood clot, intensified by the syneretic contraction of the clot itself, orients the fibrin network in the latter in a prevailingly longitudinal direction. The oriented strands tend to fuse and thus become heavier than their irregularly arranged cross links. Then proteolytic enzymes appear on the scene and rout out the cross links, while leaving the straight linear bundles standing. Thereby preferentially linear order becomes strictly linear.

The main value of these varied results is that they demonstrate in principle how chemical activity (proteolysis, dehydration) can engender spatial order of a high degree, although it has itself no space order. Through its physical and chemical by-products, plain growth thus leads to ordered growth. As the order, arrangement, and consequent exposure of the growing tissue reciprocally affect the condition, manner, and hence, rate of its further growth, we realize how complex and diverse are the factors that contribute to the growth of a given tissue. Some of the factual sequences are summarized in Figure 4. Many more are still unexplored. Carrying this exploration of the "microtechnology" of development, i.e. of "Entwicklungs-mechanik" in its original sense, for-

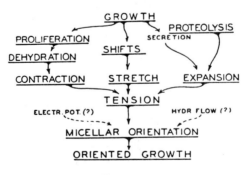

Figure 4.

ward on an intensified scale will go far in rectifying the current rather abstract and often patently unrealistic concepts of "differential growth." But even the existing knowledge, of which we have presented some samples, is sufficiently compelling to discredit the growing fashion of reducing, on paper at least, problems of differential growth to terms of simple chemical reaction rates, with "growth stimulators" and "growth inhibitors" called upon to do the "regulating."

ELABORATION OF GROWTH PATTERNS

According to the preceding sections, differential growth is simply a corollary of differentiation, and differences in growth rates result from a great variety of causes. As such, we have singled out the unequal depletion of reproductive protoplasm in the building of specialized differentiation products in different cell types, the various inequalities arising from shifts, aggregations and dispersals of cell groups in accordance with properties acquired during differentiations, the resultant differences of geometric configuration, and hence, exposure of different parts of a tissue, with the ensuing confinement of proliferative activity to circumscribed zones of varying sizes, the different fates of differentiation products, which are retained in some tissues but extruded in others, and other related disparities contingent upon the divergent courses of differentiation. Such inequalities in the growth pattern can, in turn, secondarily influence the setting for subsequent steps of differentiation, and the resulting changes will further modify the growth pattern. The actual elaboration of the mature body thus appears as a complex sequence of transforming patterns in which the contributions of differentiation and differential growth are intimately interlocked. Basically, however, differential growth is not an agent but merely a product and index of differentiation.

This realization is of the utmost importance when it comes to interpreting the manifold general non-localized influences which modify the growth process during later developmental stages, and which are impressively illustrated by the diversity of phenotypic variants that can be obtained from genetically identical germs: giants, dwarfs, duplications, defects, disproportions, excesses, and suppressions. Since these can often be directly related to aberrations of nutrition, hormones, oxygen supply, physical factors (radiations, pressure), and even climatic changes, all of these agents have at one time or other been claimed as the "dominating" factors in the growth process. We realize now that any factor that has an effect on the physico-chemical setting in which growth occurs may thereby alter the manifest form which the growth process will take. But it is evident that such actions will be in the nature of merely modifying already established patterns rather than instituting wholly new ones. Consequently an agent applied systemically (non-topically) may accentuate or reduce inherent differentials but cannot create differentials not yet in existence. A few illustrations may be in order.

As the developmental realization of the primordial growth patterns of the germ progresses, small initial differences among parts become greatly amplified. Minor deviations from normal become exaggerated. Therefore, if each part were to keep on growing independently of the other parts, the harmony of development would soon be endangered. The setting up of new systemic growth controls coordinating the separate parts thus becomes vital for harmonious development. This is achieved through the humoral, more specially the circulatory, systems for effective and rapid diffusion of substances. But we understand now that a humoral pool could exercise no discriminative control over the various parts supplied by it unless these parts had already previously acquired different response qualities. For instance, the fraction of a common supply that a given part will be able to secure is a function of the accessibility of the part to the circulatory channels; this relation, however, is entirely a matter of previous differentiation and growth of both the part and the vessels. Similarly, the capacity of a part electively to remove, concentrate, utilize, or store certain components from a common pool only reflects affinities of its cells acquired during their prior biochemical specialization; affinity for iodine in the thyroid cell, for vitamin A in the sensory cell of the retina, etc. Or if one cell group responds to a circulating hormone in a manner or degree different from that of another cell group, equally exposed, the two must have been already different constitutionally at the time the hormone appeared on the scene. The case is no different from that of selective drug response of mature cells, which is likewise based on the specific sensitivity of the reacting targets.

Since no nutrient or hormone or other chemical of ubiquitous distribution can make a cell render any contribution for which that cell has not been predisposed by its previous differentiation, none of those agents can be rated as primary causes of differential cell behavior. But they do affect the subsequent elaboration and final proportions of the growth pattern in a variety of ways. A general deficiency of nutrients, oxygen, or specific growth accessories, for instance, hits different growing parts differentially, depending on their respective metabolic rates, margins of tolerance, reserves, substitutive capacities, and the like. Faced with deficiencies, they compete for the available requisites in what Roux has described as the "struggle among parts." One part's loss may either be another part's gain, or may doom other parts, depending on the mode and degree of their mutual dependence. The production of differential growth defects by different vitamin deficiencies (59) may be pointed out as a most instructive example. But the lessons from such deficiencies are not reversible. While increasing the supply of growth essentials from deficient to adequate levels will benefit growth correspondingly, a further increase to levels of overabundance does not keep on enhancing growth indefinitely. A limit will be reached when the production plant works at full capacity, the maximum rate of production being determined by the intrinsic rates of the component processes, which, because of the assembly-line character of growth, are limited by the slowest links in the chains. A fuller discussion of these problems has been ably presented in other chapters of this book.

Of course, since what we call the "normal" level of nutrients, vitamins, hormones, and other growth requisites still falls short of the attainable optimum, the growth potentialities of the organism are "normally" not fully realized. Raising such growth factors to supranormal dosages can therefore occasionally have the spectacular effect of bringing out latent formations which under "normal" circumstances would have remained unrealized. For example, overdosing a female opossum with sex hormones causes the formation of prostate glands (26), a capacity inherent in the urogenital tract but failing to come to expression at "normal" hormone levels.

To the various humoral principles thus far recognized as secondary regulators of growth will perhaps have to be added an even more general category, namely, systems of the antigen-antibody type consisting of organ-specific discharges and their molecular counterparts. The case for this assumption has been presented on an earlier occasion (54), together with supporting experimental evidence. In the most condensed version, it is as follows. The specific key molecules which arise progressively in the course of differentiation (see above, page 153) act as templates ("positives") in the intracellular reproduction of more of their kind. At the same time they cause the formation of molecular species of complementary shape ("negatives"), which may or may not be an integral step in the reproductive process itself (double reversal). Complementary pairs could inactivate each other by conjugation. The amount of free active growth catalysts in a cell would thus vary with the excess of one type over the other. Now, if we assume that the "positives" are retained in the cell (e.g. because of large dimensions), while the "negatives" can escape, circulate, and enter other cells (e.g. because of smaller size), we can see the general outlines of a highly specific and sensitive mechanism of growth regulation, which would coordinate all cells of common character into a single operational system, no matter how much they are scattered. Local reduction (or augmentation) of growth in any part of the system would imply a corresponding decline (or increase) of both "positives" and "negatives" in that locality, and because of their easier diffusibility, of the "negatives" only throughout the system. Since systemic reduction of the "negatives" means an increased ratio of active over conjugated "positives," a compensatory increase of growth throughout the rest of the system will ensue; and conversely, enhancement of growth in one part will automatically entail depression of growth in all other members of the system. Thus the total growth of a given system would be held under joint control of all its components.

That some such mechanisms exist is becoming increasingly evident from a study of compensatory growth reactions. That these mechanisms operate, at least in part, after the fashion of immune reactions is indicated by experiments in which organ-specific antibodies were shown to have selective effects on the growth of the homologous organs (54). But the precise mode of operation here sketched by way of illustration is sheer supposition. Whatever its nature may eventually turn out to be, one must remember that it constitutes merely another contributing factor to the highly intricate system of dependencies and interactions which control the growth process. It is no more the master key to differential growth than are any of the other factors discussed before.

CONCLUSION

What our whole discussion adds up to is that there is no single master clue to the problems of differential growth or of growth in general. The measurable growth of different parts is so intimately dependent on their peculiar configurations and cellular differentiations, which themselves differ sharply and qualitatively, that a purely quantitative comparison of different growth processes is a deliberate abstraction. There are instances where such abstraction has led to the recognition of striking regularities capable of mathematical formulation (41) (e.g. allotropic growth (19)). Whether these regularities are due to the fact that the compared systems resemble each other closely enough in their mode of growth to be essentially commensurable, or to the existence of some undefined principle of organismic nature, exercising overall control over the multiplicity of heterogeneous components involved in growth, can only be decided after a detailed analysis of the underlying phenomena. At any rate, a purely formal treatment of growth, as is often attempted through the interpretation of growth curves, is only a valuable guide to and supplement of, but never a substitute for, a precise analysis of the different forms in which growth manifests itself.

There can be no research on growth as such. We can only study growing objects. And different growing objects follow different methods. The growth of a differentiating metazoan introduces aspects not presented by the growth of a bacterial culture, and the growth of a higher plant has still other peculiarities. As we have pointed out repeatedly, each individual tissue and organ has its peculiar mode of growth. To know growth we must therefore first break down each one of its manifestations into its constituent elementary processes and then study these and describe them in objective terms. This is a long way to go, but there is no short cut.

In our present primitive stage of knowledge impatient attempts to formulate a general and universal theory of growth seem to have little chance of success. Those general theories that are sporadically being advanced are usually "general" only in the sense that they are first derived from some rather special segment of the growth problem and then broadly generalized by proclamation or mere implication. Consequently they may be sound and pertinent within the area of their derivation but gratuitous and invalid in their illegitimate extensions. Oversimplification of facts and overgeneralization in theory are the main causes of the existing discordance among the several contending concepts of growth. Unless we desist from those practices the confusion is bound to grow, in spite of all the excellent experimental work being done in various areas.

Growth is not a simple and unitary phenomenon. Growth is a word, a term, a notion, covering a variety of diverse and complex phenomena. It is not even a scientific term with defined and constant meaning, but a popular label that varies with the accidental traditions, predilections, and purposes of the individual or school using it. It has come to connote all and any of these: reproduction, increase in dimensions, linear increase, gain in weight, gain in organic mass, cell multiplication, mitosis, cell migration, protein synthesis, and perhaps more. It is gravely inconsistent to apply the most exacting standards of precision to our research

II. On Differentiation   241

data and then proceed to mix into their description and interpretation such vague terminology as this. The mixture can be no more precise than its vaguest ingredient. To deal with growth as an entity, which can be "activated," "stimulated," "retarded," or "suppressed," is only part science, and for the other part, fiction. The less we let our work and thoughts be misled by the delusion that "growth" is basically but a simple elementary process, like a "bimolecular reaction," the faster will be our progress toward true insight into the real mechanisms of development. To promote this more factual approach to the problem of growth, I have tried in this sketchy survey to portray the problem in its natural complexity. If the discussion has helped to make clear what the problems are, it will have served its purpose, even if some of the special interpretations and hypotheses presented may not stand the test of time.

REFERENCES

1. Ballard, William W. Mutual size regulation between eyeball and lens in Amblystoma, studied by means of heteroplastic transplantation. *J. Exp. Zool., 81,* 261, 1939.
2. Bear, Richard S., Francis O. Schmitt, and John Z. Young. Investigations on the protein constituents of nerve axoplasm. *Proc. Roy. Soc. London,* Ser. B, *123,* 520, 1937.
3. Bernal, J. D. Structural units in cellular physiology, in *The Cell and Protoplasm.* Publ. Am. Assoc. Adv. Sci. No. 14, 199, 1940.
4. Bloom, William. Cellular differentiation and tissue culture. *Physiol. Rev., 17,* 589, 1937.
5. Brachet, Jean. Embryologie Chimique. Masson & Cie., Paris, 1944.
6. Carey, Eben J. Direct observations on the transformation of the mesenchyme in the thigh of the pig embryo (Sus scrofa), with especial reference to the genesis of the thigh muscles, of the knee- and hip-joints, and of the primary bone of the femur. *J. Morph., 37,* 1, 1922.
7. Chèvremont, M., and S. Chèvremont-Comhaire. Recherches sur le déterminisme de la transformation histiocytaire. *Acta Anat., 1,* 95, 1945.
8. Doljanski, L. Sur le rapport entre la prolifération et l'activité pigmentogène dans les cultures d'épithélium de l'iris. *Compt. Rend. Soc. Biol., 105,* 343, 1930.
9. Dalcq, Albert. *L'Oeuf et son Dynamisme Organisateur.* Albin Michel, Paris, 1941.
10. Danielli, J. F. The cell surface and cell physiology, in Bourne's "Cytology and Cell Physiology." Clarendon Press, Oxford, 1942.
11. Dawson, Alden B. Cell division in relation to differentiation. *Growth* (Suppl., Second Symposium), 91, 1940.

12. Detwiler, S. R. Restitution of the medulla following unilateral excision in the embryo. *J. Exp. Zool., 96*, 129, 1944.

13. Ebeling, A. H. A pure strain of thyroid cells and its characteristics. *J. Exp. Med., 41*, 337, 1925.

14. Fischer, Albert. *Tissue Culture*. Levin and Munksgaard, Copenhagen, 1925.

15. Gerard, R. W. Nerve metabolism. *Physiol. Rev., 12*, 469, 1932.

16. Glücksmann, A. Development and differentiation of the tadpole eye. *Brit. J. Ophthal., 24*, 153, 1940.

17. Harrison, R. G. Correlation in the development and growth of the eye studied by means of heteroplastic transplantation. *Arch. f. Entw. -Mech., 120*, 1, 1929.

18. Holtfreter, Johannes. Neural differentiation of ectoderm through exposure to saline solution. *J. Exp. Zool., 95*, 307, 1944.

19. Huxley, Julian S. *Problems of Relative Growth*. Methuen & Co., London, 1932.

20. Hydén, Holger. Protein metabolism in the nerve cell during growth and function. *Acta Physiol. Scand., 6* (Suppl. XVII), 1, 1943.

21. Just, E. E. *The Biology of the Cell Surface*. P. Blakiston's Son & Co., Philadelphia, 1939.

22. Klatt, B. Fütterungsversuche an Tritonen. I. Allgemeines. Wachstumsverhältnisse. Darmlänge. *Arch. f. Entw. -Mech., 107*, 314, 1926.

23. Langmuir, I. Molecular layers. Pilgrim Trust Lecture, *Proc. Roy. Soc. London*, Ser. A, *170*, 1, 1939.

24. Lewis, W. H. The superficial gel layer and its role in development. *Biol. Bull., 87*, 154, 1944.

25. Möllendorff, M. von. Über Histiozytenbildung aus Fibrocytenreinkulturen des erwachsenen Kaninchens nach leichten chronischen Reizungen. *Zeit. Zellforsch., 12*, 559, 1931.

26. Moore, Carl R. Prostate gland induction in the female opossum by hormones, and the capacity of the gland for development. *Am. J. Anat., 76*, 1, 1945.

27. Nageotte, J. *L'Organisation de la Matière dans ses Rapports avec la Vie*. Alcan, Paris, 1922.

28. Needham, Joseph. *Chemical Embryology*. Cambridge University Press, 1931.

29. Needham, Joseph. Chemical heterogony and the ground-plan of animal growth. *Biol. Rev., 9*, 79, 1934.

30. Palay, Sanford L. Neurosecretion. V. The origin of neurosecretory granules from the nuclei of nerve cells in fishes. *J. Comp. Neur., 79*, 247, 1943.

31. Parker, Raymond C. Physiologische Eigenschaften mesenchymaler Zellen in vitro. *Arch. f. Exp. Zellforsch., 8*, 340, 1929.

32. Pauling, Linus. Molecular structure and intermolecular forces, in Landsteiner's *Specificity of Serological Reactions*. Harvard University Press, 1946.

33. Runnström, John, and Ludwik Monné. On some properties of the surface-layers of immature and mature sea-urchin eggs, especially the changes accompanying nuclear and cytoplasmic maturation. *Arkiv für Zoologi, 36,* 1, 1946.

34. Schmitt, Francis O., R. S. Bear, and K. J. Palmer. X-ray diffraction studies on the structure of the nerve myelin sheath. *J. Cell. and Comp. Physiol., 18,* 31, 1941.

35. Schoenheimer, R. *The Dynamic State of Body Constituents.* Harvard University Press, 1942.

36. Scott, G. H. Mineral distribution in some nerve cells and fibres. *Proc. Soc. Exp. Biol. Med., 44,* 397, 1940.

37. Spemann, Hans. *Embryonic Development and Induction.* Yale University Press, New Haven, 1938.

38. Spiegelmann, S. Physiological competition as a regulatory mechanism in morphogenesis. *Quart. Rev. Biol., 20,* 121, 1945.

39. Stillwell, E. Frances. Cytological study of chick heart muscle in tissue cultures. *Arch. f. Exp. Zellforsch., 21,* 446, 1938.

40. Thomas, J. André. Des épithéliums en culture et dans l'organisme. *Arch. f. Exp. Zellforsch., 22,* 15, 1939.

41. Thompson, D'Arcy W. *On Growth and Form.* Cambridge University Press, 1942.

42. Törö, E. Über Einpflanzung von Gewebekulturen. *Arch. f. Zellforsch., 15,* 312, 1934.

43. Waddington, C. H. *Organisers and Genes.* Cambridge University Press, 1940.

44. Weiss, Paul. Functional adaptation and the role of ground substances in development. *Am. Nat., 67,* 322, 1933.

45. Weiss, Paul. In vitro experiments on the factors determining the course of the outgrowing nerve fiber. *J. Exp. Zool., 68,* 393, 1934.

46. Weiss, Paul. Selectivity controlling the central-peripheral relations in the nervous system. *Biol. Rev., 11,* 494, 1936.

47. Weiss, Paul. *Principles of Development.* Henry Holt & Co., New York, 1939.

48. Weiss, Paul. The problem of cell individuality in development. *Am. Nat., 74,* 34, 1940.

49. Weiss, Paul. In vitro transformation of spindle cells of neural origin into macrophages. *Anat. Rec., 88,* 205, 1944.

50. Weiss, Paul. The morphogenetic properties of frozen-dried tissues. *Anat. Rec., 88* (Suppl.), 48, 1944.

51. Weiss, Paul. The technology of nerve regeneration: A review. Sutureless tubulation and related methods of nerve repair. *J. Neurosurg., 1,* 400, 1944.

52. Weiss, Paul. Evidence of perpetual proximo-distal growth of nerve fibers. *Biol. Bull., 87,* 160, 1944.

53. Weiss, Paul. Experiments on cell and axon orientation in vitro: the role of colloidal exudates in tissue organization. *J. Exp. Zool., 100,* 353, 1945.

54. Weiss, Paul. The problem of specificity in growth and development. *Yale J. Biol. Med., 19*, 235, 1947.

55. Weiss, Paul, and Helen B. Hiscoe. Experiments on the mechanism of nerve growth. *J. Exp. Zool., 107*, 315, 1948.

56. Weiss, Paul, and R. Amprino. The effect of mechanical stress on the differentiation of scleral cartilage in vitro and in the embryo. *Growth, 4*, 245, 1940.

57. Weiss, Paul, and A. Cecil Taylor. Histomechanical analysis of nerve reunion in the rat after tubular splicing. *Arch. Surg., 47*, 419, 1943.

58. Weiss, Paul, and Hsi Wang. Transformation of adult Schwann cells into macrophages. *Proc. Soc. Exp. Biol. Med., 58*, 273, 1945.

59. Wolbach, S. B., and O. A. Bessey. Tissue changes in vitamin deficiencies. *Physiol. Rev., 22*, 233, 1942.

60. Wright, Sewall. Physiological aspects of genetics. *Ann. Rev. Physiol., 7*, 75, 1945.

II. ON DIFFERENTIATION   245

Reprinted from the *Journal of Cellular and Comparative Physiology*
Vol. 43, Supplement 1, May 1954.

CHAPTER 9

SUMMARIZING REMARKS

PAUL WEISS

Department of Zoology, University of Chicago
Chicago

If I may presume to talk to the non-embryologists and those who are not normally conversant with the developmental process, I shall address myself mainly to them.

In the first place, embryogenesis is not something which proceeds in a few discontinuous, nicely, neatly marked out phases. It is something which starts off with the whole plasma and the genetic system in it in a continuous series of transformations, a large number of separate chains of events in different parts of the organism, whose paths soon diverge but remain, as Dr. Willier has shown here, interrelated, in increasing complexity. If this is kept in mind, there will be no danger of misinterpreting such terms as "critical phase" or "sensitive periods" as if they were sharply demarcated phases. An objective view will be taken of the system at any one moment as it is at that time, with the idea of determining what a particular agent does at a particular time to a particular system, considering that the system prior to that time has already undergone a long series of transformations. The genes are no longer in the same setting of protoplasm that was present originally. The system has been transformed, and it must not be expected to be the same ever again.

It should be remembered that an agent may undergo a number of modifications in its course through the organism in the blood, or in the uterus if the object is a mammal; in the membranes if it is a chick; at the border of the cell, etc. It is not even known whether it gets into the cell. Therefore, it should be borne in mind that when different agents are employed they will neither reach the target all at the same moment, nor will all of them perhaps ever get to the target in the forms in which they were introduced.

In speaking of the reaction of the tissue, which is used as a final criterion, it should be remembered that this reaction consists of chains of physical-chemical processes in time, that is, kinetics. It takes time to get a tangible, recordable result; and that time, depending on the rate of the process, may be of the order of the second, the minute, the hour, the month, or the year. It is, therefore, impossible to tell from a particular external criterion or signal which has been chosen arbitrarily just when the action that later produces it set in. It is known only that the agent must have struck some time before its effect appears. The change is in the system, but its detection depends entirely on the observers, not on the system itself. Things which are properties of the human powers of observation, and their limitations, should not be projected back into the system. Finally, the developing organic system must not be considered as a microprecision clockwork in which any two individuals or any two systems can possibly ever behave in exactly and precisely the same way. This is what the geneticists describe as variability. The same thinking should be transferred to the group, to the population of cells being studied. They come from a common origin, but do not remain identical. They are subject to different accidents in their individual histories, and they eventually constitute a population, which in all respects, presumably conforms to a normal distribution curve. In such a cellular population, the ordinates being frequencies and the abscissas being thresholds of response, obviously, if an agent of a given dosage, intensity, or duration is applied all the cells up to a certain point on the abscissa are activated. An increase in the dosage will activate the cells up to the next point. It is the property of the integral of this normal curve that it is a nearly straight line for the median in about 85% of its course.

Therefore a linear increase with dosage of any such effect is really suggestive evidence of a random population of cells, where increasing numbers of singly dispersed cells are hit as the agent is increased.

II. On Differentiation   247

If this were the case in radiation — it definitely can be shown in embryology — or with chemical agents, then the question arises: How can solid patches of cellular destruction — a multicellular mosaic — result from chemical or radiation agents?

Apparently, if an affected single cell in turn starts to emanate secondary effects around it, perhaps due to the release of chemicals from the necrosis of those primary target cells, those secondary effects will mark the spot, amplify it, so to speak.

Quite a number of the phenomena that have been described in the literature might be explained on this assumption. This type of thinking should be applied when dealing with tissues — statistical responses, secondarily modified by chain reactions, so to speak, of releases within the tissues.

Now with these few general comments, I shall return finally to some positive statements. Obviously, mitosis cannot be the sole factor affected by these agents. Whether chemical or radiation induced, their effect on synthesis of cytoplasm cannot be the sole trouble. Neither can the separation of the chromosomes be the sole defect, because some of these effects — radiation, for instance — do not show until very late, after a long lag period when all the cells which divide are well beyond the duration phase of the radiation.

Hence, perhaps for the time being, no single agent should be postulated, no royal road or key answer to the questions which have been posed. As can be seen, little is as yet known about the embryo, or about development; what really goes on inside is still shrouded in comparative mystery.

III. ON GROWTH AND AGING

Reprint from
THE HYPOPHYSEAL GROWTH HORMONE, NATURE AND
ACTIONS, *Editors:* R. W. Smith, Jr., O. H. Gaebler, and C. N. H. Long
— THE BLAKISTON DIVISION, McGraw-Hill Book Company, Inc.

CHAPTER 10

What is Growth?*

Paul Weiss
Rockefeller Institute for Medical Research, New York

The invitation to give an introductory address to this meeting is a distinct honor; it also carries a mandate to set the phenomenon of growth in proper perspective before discussing "growth" hormone. Unfortunately, "growth" itself has received much less critical attention than have the agents for which it serves as indicator and assay. To put it bluntly, "growth" is a term as vague, ambiguous and fuzzy as everyday language has ever produced. Adopted into scientific language without precise and consistent meaning, it may be passable for crude description, but is ill-fitted to analytical application. Hence, if you ask: "Just what is Growth?", the correct answer is: "A word that covers, like a blanket, a multitude of various things and meanings." To know "growth" for what it really is, rather than what we are wont to call it, we must remove that blanket and uncover the underlying facts it has concealed. This I propose to do in rudimentary sample form as time permits. A close look at the facts will do far more for clarification than would a host of academic definitions and circumlocutions.

Our notions about growth have been shaped more by usage than by incisive study; they form a sort of scientific folklore. As a result, we find that various groups, while they all just plainly speak of "growth," do not all mean and talk about the same thing. Thus growth has come to connote any and all of these: reproduction, increase in dimensions, linear increase, gain in weight, gain in organic mass, cell multiplication, mitosis, cell migration, protein synthesis, and perhaps more. It would seem inconsistent to apply the most exacting standards of precision to our research data and then proceed to mix into their description and interpretation such vague ter-

---

* Research supported by grants-in-aid from the American Cancer Society upon recommendation of the Committee on Growth of the National Research Council, and from the National Institutes of Health, Public Health Service.

III. On Growth and Aging   251

minology as this. The mixture can be no more precise than its vaguest ingredient.[1]

Then, what is wrong? Why such diversity of views and versions? The reasons lie in our unfounded expectation that growth is a single, simple, measurable entity. In this conviction, each of us has tended to deal with his own limited aspect of the problem as if it were a representative sample of the total perspective. Yet, far from being a single, simple and unitary phenomenon, growth is conglomerate, complex and intricate, and this is why it defies formulation in simple terms. What usually deceives us is the simplicity of our tools and terms of measurement, which all too easily produce the illusion of similar simplicity of the measured systems.

Just bear in mind how we get to know about growth: by taking measurements at different times, comparing them and noting a net gain—of size or mass or numbers. These serial measurements then define a growth curve, as descriptive of the particular system as, let us say, a fingerprint—and equally empirical. This is the blanket under which a host of disparate events lie hidden; events, moreover, of opposite signs, some adding to, others subtracting from, the measured body. Since they are not all of one kind and their shares are unequal, growth can be recorded, but it cannot be understood, without identifying these tributaries and determining their respective contributions.

Let me phrase this in terms of an analogy. The body is a community of cells; each cell a community of smaller particles; and each particle an assemblage of molecular species. Thus, the proper analogue of biological growth is the growth of a human community; for example, of a city. Here we rate as growth, for instance, any increase in population over a given interval. But a simple tally will not tell us how the increase has come about. It takes census data to give a more detailed accounting. They reveal that additions come from two different sources: reproduction from within, and immigration from without; losses, likewise, from death as well as emigration. The results would be altogether different if instead of just counting noses, we chose to include in our considerations the physical wealth of the community, that is, the net gain in goods and estates produced by the members of the population. To understand its sources would require running inventories of raw materials, production, conversion, consumption, imports, exports, storage and wastage. Moreover, in either reckoning, the data can have meaning only in reference to fixed boundaries which divide what we count as "within" from that which we count as "without."

Now, as we apply this simile to biological growth, the whole indefiniteness of our customary position becomes obvious. First, let us consider the matter on the tissue level. Suppose we note an increase in the number of cells of a given organ. What does this really tell us? As in the human population, some cells have reproduced, others have immigrated, still others have been lost by shedding or disintegration, the proportions and rates of these component events varying from tissue to tissue. The final tally—no more than a crude balance sheet—discloses none of these details. According to Hamburger and Levi-Montalcini,[2] for instance, abnormal enlargements in the early central nervous system, formerly ascribed simply to "hyperplasia," that is, overproduction, are partly due to the fact that fewer cells degenerate, rather than that more are being proliferated, and partly to the fact that the cell group being counted has received additions from an indifferent pool outside the counted area. Another shortcoming of plain cell counts is that they ignore all growth of individual cells (e.g., hypertrophy) not followed by division.

If, then, we turn from cell counts to over-all dimensions or total mass, we are on even weaker ground. In terms of our community analogy, we first have to agree as to what is, and what is not, real property of the system we measure, or what has been acquired and what discarded during the measured period. Food in the alimentary tract is still distinctly out-of-bounds; even if stored for weeks, as in a hamster's pouch. But what of this mass once it has passed into the blood and lymph stream? Though strictly on the inside, it still has not become converted into substance of the body proper. Then, what about the food stuffs stored in modified form, for instance, as glycogen or fat in liver or fat bodies? Their fluctuations up and down are not conventionally considered growth and degrowth. Why? Because we sense that growth connotes some *permanent* addition, and merely temporary physiological variations do not qualify under this title.

Then, what about the wastes not yet eliminated? And the products manufactured by our organs? Take hair or nails or even red cells—terminal products destined to be shed or otherwise eliminated. In counting bodily productions, is it fair to include just those fractions which happen to be present on the measured body when we take our measurements, and leave out all the unknown mass that has been similarly produced in the interim but irretrievably lost? Evidently, we ought to be consistent and either count it all in or all out, neither of which is practicable. We certainly would not collect secretions, such as slime, urine, sweat and sebum, over a measured period and add them to the growth record. Yet, we do customarily include the bulk of cartilage and bone and other connective tissues, which consists of residues of cellular secretions, just like those other ones, but incidentally deposited, instead of extruded, hence accruing to the measured mass. Thus what we measure, is related not so much to the process of production as to the accident of the disposal of the products. If they persist, we count them; if they drop out, we miss them.

The arbitrariness attached to our measurements is about the same, whether we use total mass, dry weight, nitrogen content, volume, length, or what not, as reference system. It is even worse when we turn from the body to its component cells. The cell is bounded by a surface, and we are in the habit of ascribing any increase in the volume thus enclosed to "growth." But

III. On Growth and Aging 253

water, electrolytes, food stuffs and wastes pass in and out across that boundary without revealing to the observer just when they lose their original small-molecular identity to become merged with the complex specific compounds of the cell, or when they emerge again from decomposition of the latter; we thus cannot determine what fraction of a given increase to allocate to protoplasmic synthesis proper, and what fraction to substances that just reside within the confines of the cell but, strictly speaking, belong to it either not yet or no longer, comparable to food and wastes in the alimentary tract. Then how should we consider the formed cell inclusions which vary according to each type of cell? Obviously, the production of myofibrils, which accumulate within the muscle cell, thus adding to its mass, is not fundamentally different from the production of collagen fibrils, which leave their cell of origin—the fibroblast—hence, do not enlarge it. Conversely, a fat cell inflated by unextruded fat of its own production cannot be compared to a macrophage similarly dilated by fat from stuff it has engorged from the outside. Evidently, the attainment of equal or unequal sizes, as such, can give us little information about the identity or difference of the underlying causes of the enlargement.

It is this habit of confining ourselves deliberately to a single parameter of a highly complex system—for instance, mass—and then measuring all changes on this single scale only, which tempts us, as I said before, to confound the arbitrary simplicity of our method with an inherent simplicity of the object itself. We are fundamentally in error if we prorate the average over-all changes in a complex system evenly over all parts of the system, as if all of them took equal shares in the result. In doing so, we commit an act of "mental homogenization," as it were, obliterating the organized complexity of the heterogenous events which are the very center of our interest.

To this we usually add a second and similarly unfounded simplification whenever we consider growth rates. We measure two comparable parameters—let us say, weight or length—at the beginning and end of a convenient period, divide the difference by the time elapsed, and call the quotient "growth rate." Empirically, this gives useful descriptive data. But before using these for analytical or comparative purposes we should realize that we have again "homogenized"; in this case, the time interval. We have tacitly assumed that during this interval the bracketed change has been steady and continuous. Yet, if the change occurred unevenly, in spurts separated by phases of quiescence, any inference as to the kinetics of the system would be misleading. Two different systems can achieve the same amount of growth increment in precisely the same period by wholly different means: the one by growing faster for shorter spells; the other growing more slowly, but with fewer interruptions or shorter lags.

But this example also contains the clue to overcoming our predicament. You note that the two systems just mentioned would become readily distinguishable if they were sampled at shorter intervals; that is, by letting factual information replace supposition and extrapolation. And this is a lesson which cannot be overstressed. Interpretations and comparisons in matters of growth will remain of little value and validity if they are based on over-all generalizations instead of such detailed and painstaking sorting and recording of the component events as is practically feasible. Tally must give way to detailed census. There are insuperable limitations set to this census by our ignorance of cell life and the inadequate resolving power of our tools; but it would be unpardonable not to carry the analysis at least as far as the objects and techniques would permit.

Therefore, turning to practical correctives, let me indicate how a first breakdown of the problem could be attempted. Growth is the surplus accruing in the balance sheet of a complex account. Our task is to itemize the accounting. Here are some major items. In listing them, I shall artificially separate them as if they were consecutive steps, while in reality, you understand, they proceed more or less concurrently (Fig. 1).

We follow a system through a period of growth from A to G. The system may be a cell, an organ, an individual. Now, in the first place, we must distinguish within each such system between two major fractions—a reproductive and a non-reproductive one; the former capable of giving rise to more of its own kind—more protoplasm in a cell, or more cells in a tissue—the latter merely dead bulk in point of growth—fibers, granules and all sorts of functional equipment in a cell; or fully differentiated, non-proliferating

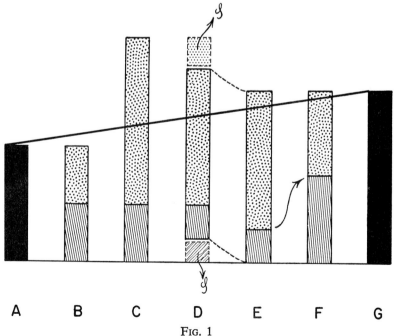

A   B   C   D   E   F   G

Fig. 1

III. On Growth and Aging

cells in a tissue. Their proportions vary according to cell type, species, age and activity. In our example, about 50% are assumed to be reproductive, as shown (stippled) in B. During the recorded growth period this mass will increase to the dimensions shown in C. At the same time, however, there will occur some losses from degradation, lysis, shedding, extrusion, emigration, and so forth. These will reduce both fractions by the amounts indicated in D. The resulting state E has larger mass than the initial A. You note, however, that the proportion between the reproductive and non-reproductive fraction comes out markedly altered. In most cases, this is then rectified by the conversion of freshly reproduced protoplasm into sterile differentiated products, as indicated in F. As you can readily see, this conversion, which drains the reproductive potential of the system, is of crucial influence on what we externally discern as "growth rate." In the end, we find our system in the enlarged state G. The point I want to stress with this diagram is that even assuming the simplest case, namely, proportionate growth, the result cannot be interpreted to mean that all parts have taken part in growth, let alone grown at equal rates or taken equal shares in the composite effect. And nothing short of detailed study will tell the real story.

    This confronts us with a major task, which we mostly dodge rather than face: dodge by attaching real biological meaning to such purely descriptive terms as "growth stimulation" or "depression"; rather than face by finding out just *why* a given system, subject to certain agents, turns out to be either

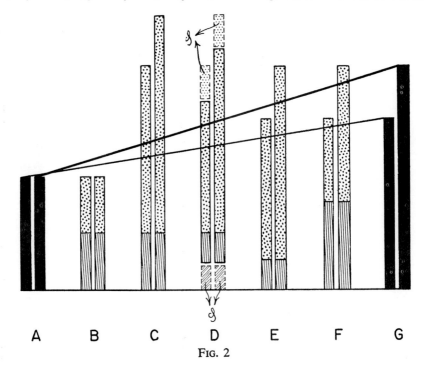

A   B   C   D   E   F   G

Fig. 2

larger or smaller than a chosen reference standard. For this is all we usually determine; and yet, we imply far more. Just let us take for once a closer look. Let us return to the diagram of Figure 1 and assume that it represents a "normal" reference condition. Suppose we now doctor an identical system with an agent, let us say "growth hormone," that entails an even greater increase during the observed period (Figs. 2, 3, 4). We sum up our observation by stating that growth has been "stimulated," like an invested principle that suddenly yields higher returns because the interest rate has gone up. Yet, greater gain may come not only from faster earning through stepped-up production, but also from reduced consumption, or even from diverting less of the working capital to non-productive uses. These various possible modes of action of our agent are illustrated in the remaining diagrams, in which the left one of each pair of bars always gives the control data from Figure 1, while the right one represents the experimental case.

In Figure 2, we actually let the velocity or intensity of reproduction be increased (from B to C). This alone naturally leads to faster growth; but at the same time it also modifies the distribution of reproductive and sterile mass (compare the two bars in F) which, if uncorrected, would distort the whole growth pattern. So, even this seemingly simple change is not so simple as it seems.

The second alternative is shown in Figure 3. Here we assume that our so-called "stimulating" agent has simply reactivated part of the normally quiescent fraction to become reproductive (in B), thus recruiting a larger

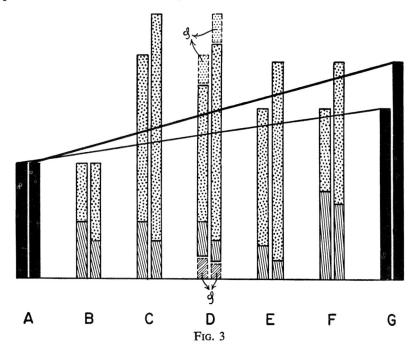

A   B   C   D   E   F   G

Fig. 3

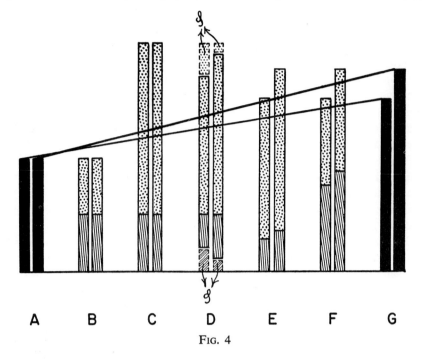

A   B   C   D   E   F   G

Fig. 4

source for active growth, without however, stepping up the growth rate of the elements at all. The target process here would be the conversion of reproductive mass to specialized products. Reduce or retard this conversion, and you have the picture of "growth stimulation."

Lastly, the same picture would be obtained (Fig. 4) if merely the debit side of the balance sheet of growth is reduced (in D), that is, when there is less than the normal amount of destruction, consumption and dissipation.

There are many further variants of this complex account, but these few here will have served our purpose: to show that the same net increase can be brought about by a variety of unrelated means; hence, that comparing different systems or different agents on the sheer basis of net increases observed may lead, or rather mislead, to quite gratuitous interpretations. Perhaps in such critical illumination, the term "growth" hormone may eventually turn out to have been overly suggestive.

But to leave this critical and somewhat defeatist note, what is there in the way of positive information that could give sharper focus to the growth problem? However fragmentary this information still is, it begins to piece together. The tentative result may be summarized in the following seven theses:

(1) *The general common denominator shared by all organic growth is protoplasmic reproduction,* which involves the replication of those high-molecular systems in each cell that are characteristic of the particular cell

type and are compounded only inside of cells of that kind. It is important, however, to make a clear distinction between primary protoplasmic reproduction and secondary elaboration of protoplasmic products. All such cell products, including the fibers and ground substances, whose deposits form the bulk of most bodies, are derived from protoplasm secondarily, either by direct transformation or by synthesis ·with the aid of enzymes which in turn have originated in the process of protoplasm synthesis. The current shorthand habit of equating protoplasmic reproduction with protein synthesis, apart from by-passing the non-proteinaceous constituents of protoplasm, fails to take this fundamental distinction into account. For instance, while such proteins as collagen or melanin cannot themselves give rise to more collagen or melanin under any known conditions, the protoplasms of the fibroblast or melanoblast in their growth evidently multiply the chemical machineries that can synthesize more collagen or melanin. Thus, only a fraction of the cellular protein is actually engaged in protoplasmic reproduction, the rest is sterile; but just what those fractions are, remains to be determined. In the following I shall use the term "growth" in the restricted sense of protoplasmic reproduction.

(2) *Growth in the indicated sense has its sources not uniformly distributed throughout the cell, but is of localized origin, centering on the nuclear territory.* The nuclear source of protoplasmic reproduction has been deduced from cytochemical studies[3] and demonstrated by our own experimental studies on neurons.[4] The nucleated part of the nerve cell body issues continuously a fresh supply of neuroplasm, which, wicklike, moves peripherad in the nerve fiber to replenish basic protoplasmic systems at the rate at which they are breaking down by ordinary wear and tear. Evidently, although any reasonably large fragment of a nerve fiber can carry out complex metabolic functions and syntheses by virtue of its enzyme content, the enzyme systems themselves cannot be reproduced ubiquitously but must be furnished from the nuclear supply center. Although even the mature neuron maintains itself in a state of perpetual growth renewal, not all somatic cells can be assumed to retain this faculty. Those that have lost it are destined to die precociously and they can be replaced only from those reserves that still have it. In the epidermis, for instance, the squamous outer cells are in the former state, the basal cells in the latter. A single neuron could be viewed as comparable to a vertical column of epidermal cells (the perikaryon corresponding to the basal cell), in which cell boundaries had become obliterated.

(3) *The nuclear growth process is intimately related to, perhaps engendered by, the multiplication of chromosomal genes.* As a result, cell growth is closely associated with nuclear growth. Accordingly, haploid nuclei and cells, with only half the normal complement of chromosomes, are of half the normal size, whereas polyploid nuclei and cells, with multiple chromosome sets, acquire correspondingly excessive sizes. Parenthetically,

III. ON GROWTH AND AGING

there is by no means a binding relation between cell size on the one hand, and organ and body size, on the other; while haploid organisms are dwarfs because they are made of dwarf cells in normal number, polyploid organs and animals often remain of normal dimensions, as the giant size of their cells is offset by the use of smaller numbers.[5] Although true cell growth is reflected in nuclear increase, the reverse is not always true, and nuclear enlargement as such is not sufficient evidence of true growth, but may be simply a sign of functional hyperactivity. The two types of increase can sometimes be told apart cytochemically, as true growth is accompanied by augmentation of nuclear DNA, whereas hyperactivity is not.[6] DNA seems to assume ever increasing significance as index of true growth.[7]

(4) *Despite the fact that the genic equipment, according to current thought, is essentially the same in all somatic cells, mode and rate of growth of each cell type are specific for that type;* that is, neuroplasm begets more neuroplasm, myoplasm more myoplasm, thyroplasm more thyroplasm, and so forth. To deal with this matter, which touches on the complex problem of differentiation, is beyond our scope here. We mention it merely because it reveals that genic replication, while seemingly the initiating step in the growth process, is followed by chains of events the nature of which is determined by the specific constitution of the (extragenic) rest of the cell; and in this regard, the different cell types differ crucially. Besides certain basic prerequisites in common to all of them, each type has its private needs; for instance, in tissue culture, one tissue will grow better in one kind of medium, while another tissue thrives better in another kind, usually without our knowing just why. Understandably, the more conspicuous common factors necessary for all growth have received more attention. Those basic to the maintenance of life in all cells, whether growing or non-growing, have no particular relevance to the growth problem as such. But up and above these common prerequisites for life are the special needs of growing, versus non-growing, systems, and the differential needs of different growing systems according to their kinds. These include specific building blocks, factors to insure the proper physical framework for growth, and perhaps separate energy resources and enzyme systems to support growth, in contradistinction to stationary maintenance. Signs are increasing that the growth process is not a mere quantitative shift of the normal steady state between anabolism and katabolism in favor of the former, but a process of different character sustained by auxiliary biochemical systems dormant during states of sheer maintenance. At least, this is a conclusion suggested by metabolic studies on forms in which periods of maintenance and growth are clearly separated, as in some insects.[8]

In view of the qualitative and quantitative differences between the growth requirements of different tissues and organs, we must concede to each one its characteristic chemical kinetics. And this explains the following familiar experience.

(5) *Given optimal conditions, superabundance of all prerequisites and freedom from active inhibitions, each cell strain, tissue and organ grows at its own characteristic rate.* This rate represents the maximum output of which the particular system is capable (at a given temperature). It is a "ceiling" rate and varies according to species, genetic constitution, kind of tissue and organ, state of differentiation, and perhaps age. It manifests itself, for instance, in the fact that transplants between different species or breeds of different growth rates, if successful at all, continue to grow according to their native growth patterns.[9]

Now, under normal conditions in the body, tissues do not grow at ceiling rates. This shows that conditions are not usually optimal for total output; either because of bottlenecks in the supply of some essential elements, which may be a matter of timing or of competition for limited supplies, or because of suboptimal physical conditions, or lastly, because of the presence of factors that actively repress or retard one step or another in the reaction chain. Now, as you note, any of these situations, by holding growth to a level below its potential ceiling, gives the external appearance of an "inhibitory" influence, and is likely to be so labelled. Yet, between them, they have no more in common than the various agents that may cause an automobile to stall. Reversing the argument, any lessening of the effectiveness of any one of them may enable the system to come closer to its optimal growth output, and if so, will give us the impression of stimulation; a more correct term would be "desinhibition." Some recent work of ours strongly supports this view.[10] In the epidermis of starved amphibians, cell growth followed by mitosis can be evoked in two different ways: (a) throughout the skin by sudden feeding; or (b) in a localized spot by fragments of certain organs inserted under the skin. The magnitude of the unit response is the same for (a) or (b) alone and for (a) and (b) combined. This demonstrates the existence of a "ceiling." On the other hand, when the same agents are applied at any stage at which growth is submaximal, a new peak of growth activity appears which again has the height of the "ceiling." This evidently proves that these agents have acted not by boosting the growth process by a given amount, as would be the connotation of "stimulation," but by releasing the system from certain depressive effects that had previously been in operation.

Obviously, factors restraining growth by holding any of its contributory steps to a suboptimal level belong to a great variety of categories and need have nothing in common except their eventual effect on the growth result. Similarly then, the agents which offset these restraining factors, thereby reversing the net effect on growth, are likewise manifold and varied. One might thus question whether the search for over-all "growth stimulators" is at all realistic and promising.

To judge from virus reproduction, the multiplication of basic protoplasmic units presumably takes up no more than a very minor fraction of the time needed for a given growth interval, let us say, the interphase between two somatic mitoses, where according to the DNA index the crucial events seem to be crowded toward the end.[11] The rest of the time is evidently used for processes preparing this crucial step, as well as for the subsequent elaboration of differentiated products. Since these vary from one cell type to another, any observable shortening of interphase—appearing as "growth stimulation"—is due more probably to the speeding up of some of the prefatory or consecutive steps, rather than of the crucial act, of genic reproduction.

(6) *Growth and cell division are often, but not necessarily, coupled.* Present evidence indicates that in this correlation cell growth is the primary event, with mitosis then supervening facultatively. In such cases, and with due caution, mitosis may be used as index of preceding growth. Even then, comparisons are difficult for in all cases in which only a fraction of the cell population takes part in proliferation, a rise of the mitotic index need not signify either a shortening of interphase or a protraction of the mitotic act, but simply the mitotic involvement of a larger portion of the population. Again, speaking of "growth stimulation" would add little to our insight.

(7) *As complex as the growth process itself, are the means that keep the various steps of the process, as well as the various growth centers among one another, in mutual harmony.* It cannot be stressed too strongly that growth is regulated by a great multiplicity of factors of chemical and structural nature, no single one of master rank.[1] One potent growth-regulating principle, which we discovered relatively recently, is a "feed-back" equilibration between a growing organ on the one hand, and organ-specific discharges of its own production that restrain the growth of all cells of the homologous type, on the other. Each cell type may be assumed to give off substances specific to its kind which inhibit the multiplication of any protoplasm of the same type in proportion to their concentration in the extracellular environment. As this concentration increases with the growing mass of an organ, growth of each cell type will become self-limiting regardless of how widely dispersed the total mass is throughout the body. This principle explains the "compensatory" growth reactions after partial removal or injury of a given organ, as well as the observation that the presence of crushed cells of a given type in the blood stream (by injection) or in a culture medium enhances the growth of homologous cells; for a more detailed review of the evidence.[12] While this principle operates among homologous members of a cell and organ population, growth effects between heterogenous types are a major function of "hormones."

* * *

Since growth processes are so diverse and composite, it would be a miracle if any one hormone were to act on all of them in such fashion as to net always a positive balance, which would make it a veritable "growth hormone." Any hormone affecting any one of the innumerable component steps of growth, is in a sense a "growth hormone," although nothing seems to be gained by naming it so. The obligation remains to find out precisely where and how it acts. This requires more than just measuring over-all changes of size or bulk or composition.

To illustrate the danger of shorthand explanations of the relation of hormones to growth let me conclude with an instructive example from our own laboratory experience. Amphibian metamorphosis from larva to adult is, as you know, dependent upon thyroid secretion. The hormone activates a pattern of proliferation and involution processes, and all organs undergo profound reorganization, including the brain. In this process, a peculiar pair of giant hindbrain cells concerned with larval swimming, so-called Mauthner's cells, atrophy, while other brain parts grow. This could have been ascribed to the loss of their functional terminations, but our experiments[13] proved otherwise. We implanted fragments of thyroid or thyroxin-soaked agar above the 4th ventricle so as to allow the hormone to diffuse into the brain wall; according to earlier results,[14] under such conditions, a circumscribed metamorphosis of the tissue complex within the diffusion field could be expected. This actually occurred, with the following results on brain growth: while all other ganglion cells of the area grew conspicuously, that one pair of cells in their midst which was destined to regress—Mauthner's cells—did not grow, but shrank (Fig. 5). We could show that this reverse behavior had nothing to do with peculiarities of size or position, but was simply an expression of a different biochemical constitution of these cells which predisposed them to react to thyroid hormone with a growth response of opposite sign from that of the other cells of the group. In conventional terminology, one would say that the same hormone in the same dosage at the same time can induce either hypertrophy or atrophy, depending on the kind of cell it strikes. If this holds true for cells so closely related as the different neurons, how much more generally will it apply to more widely

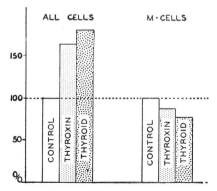

Fig. 5

diversified cell populations. It seems hardly necessary to labor the cautionary lesson contained in this example.

To sum up, what we measure as growth is the resultant of a heterogeneous array of processes of most diverse kinds, and there is no way of telling in advance, without careful analysis, just what component may be affected by any particular agent in any particular tissue in any particular species, and how. As long as we keep this in mind, we are on safe ground even if, for convenience, we resort to such shorthand reference terms as the one heading this conference. Even if the "growth hormone" should turn out not to be just what its name implies, I doubt whether this would in the least detract from the superb factual achievements in its study of which we are to hear an up-to-date account; just as I doubt that progress in X-rays would have been delayed if they had instead been named "death rays" because undeniably they sometimes kill. I also have little doubt that those with firsthand experience in this field are essentially aware of what the real situation is, and how complex it is. Yet, there are also those whom the simple label "growth" might delude into holding complacent, oversimplified and unrealistic notions of its content. My comments were intended to restore the complex problem to plain view—not for discouragement, but simply for clarification of the problems so that research may orient itself toward the real thing, instead of to a verbal symbol. As for my final answer to "What is Growth?", I am tempted to dodge by saying: "Let us go back to work and find out more about it and not pretend we know."

References

1. Weiss, P.: *Chemistry and Physiology of Growth*. Ed. A. K. Parpart. Princeton University Press, 1949, 135.
2. Hamburger, V., and R. Levi-Montalcini: *Genetic Neurology*. Ed. P. Weiss. University of Chicago Press, 1950, 128.
3. Caspersson, T. O.: *Cell Growth and Cell Function*, New York, W. W. Norton & Co., Inc., 1950, 185.
4. Weiss, P., and H. B. Hiscoe: *J. Exp. Zool.* **107**:315 (1948).
5. Fankhauser, G.: *Quart. Rev. Biol.* **20**:20 (1945).
6. Schrader, F., and C. Leuchtenberger: *Exp. Cell Research* **1**:421 (1950).
7. Swift, H. H.: *International Review of Cytology*. Eds. Bourne, G. H., and J. F. Danielli, New York, Academic Press, Inc. **2**:1 (1953).
8. Schneiderman, H. A., and C. M. Williams: *Biol. Bull.* **105**:320 (1953).
9. Harrison, R. G.: *Harvey Lectures, 1933–34* **29**:116 (1935).
10. Weiss, P., and J. H. Overton: *Excerpta Medica* **8**:424 (1954); Overton, J. H.: (in press).
11. Walker, P. M. B.: *J. Exp. Biol.* **31**:8 (1954).
12. Weiss, P.: *Growth Symposium* (in press).
13. Weiss, P., and F. Rossetti: *Proc. Nat. Acad. Sci.* **37**:540 (1951).
14. Kollros, J. J.: *Physiol. Zool.* **16**:269 (1943).

Reprinted from *The Yale Journal of Biology and Medicine*
Vol. 19, No. 3, January, 1947

CHAPTER 11

THE PROBLEM OF SPECIFICITY IN GROWTH AND DEVELOPMENT*

PAUL WEISS

Introduction: Biological Specificity

The frequency with which such terms as specificity, selectivity, conformity, correspondence, etc., appear in biological literature is ample proof that they denote a universal and fundamental trait, running like a common theme through all manifestations of life. Yet, they are used with so many different shades of meaning and degrees of precision that it is impossible to tell whether the various phenomena to which they are applied bear a purely formal resemblance to each other or whether there is essentially a single principle in back of them all. A random list of examples will illustrate the case. We describe as "specific" the absorption by certain compounds of certain wave lengths of light; the relation between enzymes and their substrata; the matching between egg and sperm; the action of a hormone on its end organ; the effect of genes on characters of development; the association between a parasite and its host; the immunological response to a foreign protein; the adequate response of our nervous system to a given stimulus; the acts of recognition and evaluation, which characterize our highest mental functions. What do these various "specificities" have in common? Are they merely superficial parallels, or does one or the other of them perhaps contain the key to the rest so that specificity in all manifestations of life could be resolved to a single operative principle?

It may be too early to attempt an answer to this question, but it does not seem too early to ask it. Therefore, let us take a closer look at relations and activities in growth and development which we commonly describe as "specific," and examine to what extent their specific character might be explicable in terms of better known and better understood specificities at other biological, or preferably simpler physical and chemical, levels. In particular, let us explore the

---

* From the Department of Zoology, University of Chicago. The present paper is an expanded version of an address given at a symposium on "Specificity" of the Society for the Study of Development and Growth held on 24 July 1945

pertinence of the model of serological specificities as a model of developmental processes, inasmuch as recent studies in immunochemistry have brought those specificities within our grasp.[16] Perhaps, the study of growth and development could then profit from this faster advance of one of its biological sister lines.

But let us first clarify what biologists mean when they speak of "specificity." In its common connotation, the term refers to that relation between two systems which enables members of one system to exert a *discriminative* effect upon certain members only of the other; it implies *selectivity* of action and reaction even in the absence of separate channels from the acting to the reacting members. A chemical that bathes all tissues, but affects some of them with disproportionately greater potency than others, will be considered as specific for the affected tissues in that sense. By definition, selectivity is the faculty of a process or of a substance to activate, to alter the state of, or to combine with, certain elements in preference to, and to the exclusion of, other elements of the same system. The basic criterion of selectivity, therefore, is the correspondence and mutual fitting between two properties. Primarily, the term specificity applies to this *correspondence*, and to neither of the interacting systems as such. By custom, however, it has acquired a secondary meaning signifying those properties of each system which make selectivity of interaction possible.

Resonance is one simple model of selectivity. Here the specificity is based on time characteristics. The example of fitting keys and locks illustrates specificity of relations based on spatial correspondence. And if we analyze all conceivable types of specificity, it would seem that all can be resolved into characteristic patterns of time or space. Selectivity shows different degrees of sharpness, the intensity of the response falling off more or less steeply from a peak, which marks the point of best correspondence. In these days of radio communication, we need hardly stress the fact that selectivity is a matter of degree. It is important, however, to point out that the degree of selectivity need not be entirely a fixed constitutional property but can often be sharpened as a result of adaptation of the responding system to repeated or lasting exposure to a stimulus of constant configuration.

Perfunctory and incomplete as this definition of specificity is, it will do for the purpose of our further discussion. In the indicated sense, specificity is perhaps the most fundamental attribute of life processes, from the synthesis of the building stones of protoplasm to the orderly performance of our mind, and its elucidation must remain one of the prime concerns of biology. Let us now turn to our object proper—development.

Developmental Kinetics

Brought to its simplest formula, development consists of three types of events: (1) *Growth*: the reproduction of certain basic compounds by synthesis from simpler elements, duplicating the patterns of existing compounds. (2) *Differentiation*: the gradual elaboration of new chemical systems and compounds not previously present as such, presumably by gradual transformation of the patterns according to which synthesis of the protoplasmic compounds occurs. This transformation of basic protoplasm takes divergent courses in different cell strains, producing lines which become increasingly dissimilar in their biochemical and morphological constitution as development proceeds. (3) *Localization*: the sorting and segregation of biochemically different units into definite locations. The field which has for its object the study of these processes might appropriately be called "developmental kinetics."

As the various biochemically differentiated units aggregate in different predictable locations, associate with their own kind and with certain other units, or relinquish their positions and disperse, all depending on whether and how they fit into each particular site, they furnish us with paradigms of selective behavior, and we may use these phenomena as our point of departure. When we speak of shifts and redistribution of units, we refer to the materials of the undivided egg prior to segmentation, as well as to the movements of individual cells or whole cell layers later in the cellulated germ. The accumulation of certain visibly distinct substances in a given sector of an egg and the shift of germ layers during gastrulation have many basic features in common. Both involve convections by which the units concerned are brought into novel combinations with other units and under new local conditions. Whether such new combinations will be lasting depends on the nature of the components. Since cells are in constant activity and activities of adjacent tissues may differ in kind, only those can remain united whose metabolic activities, chemical requirements, chemical discharges, electrical properties, growth changes, surface properties, etc. are mutually compatible.

As we shall explain below, this compatibility is not merely a matter of communal tolerance, but implies active "affinities" among the partners that are to form durable biological unions. Intracellular streaming, cell locomotion, and the shifting of cell masses may be gross mechanical events, but the forces that tie a part to its final location seem to be subtle and specific. True, in the mature organism the attachment between neighboring tissues is secured by various encasing and cementing structures, such as basement membranes, connective tissue fibers, and the like. However, prior to the development of these accessories, tissues must rely for whatever hold they exert upon each other on forces residing directly in their naked contact surfaces. It is to the exposed cell surfaces then that we must look for the revelation of the factors which make or break specific cellular associations.

The phenomena of developmental kinetics thus present us with the following questions. Why do particles, cells, and cell layers shift during development? Why do they cease to move once they have reached certain localities or have combined with certain other groups coming from other directions? What determines their course and how do they get to their proper destinations?

Embryology furnishes numerous striking examples of shifts of tissues relative to each other, moving either as compact masses or as groups of individual cells. Indeed, the processes molding the early embryo after cleavage are predominantly in the nature of translocations rather than growth. Practically the whole germ is on the move. Later, after the basic form has become fixed, mobility is restricted to certain cell types which move within the now consolidated frame. The neural crest,[8] for instance, spreads into the interstices of the embryonic body, laying down different cell types at different stations: ganglion cells along the vertebral column, sheath cells along the nerve fibers, pigment cells along predetermined lines in the integument and its derivatives, and, at least in Amphibians, cartilage for certain elements of the head skeleton. There is circumstantial evidence that these various cell types are already different in character when they leave their common sites of origin. What, then, guides each to its proper final destination? Nerve fibers offer another example.[38] By outgrowth, which is primarily a matter of the movement of their free tips, motor and sensory fibers span considerable spaces, after which each type forms exclusive connections with the peripheral tissues appropriate to its own kind, motor fibers with muscle, sensory fibers with skin organs. How do they get where they belong? Germ cells,[45] at least in some forms, have been claimed to originate in embryonic areas far distant from the gonad and to immigrate only secondarily into the latter, presumably conveyed part-way by the blood stream. The oriented migrations of lateral line organs, muscle buds, various ducts, and capillary sprouts offer further examples of the same kind.

One notes a formal resemblance between these developmental processes and the behavior of some parasites, which enter their host at a specified point and end up at a predictable destination. The comparison may be valid even for parasites with well-differentiated nervous systems, reacting to an orderly sequence of sensory cues furnished by landmarks of the host body; for the properties which endow a sensory cell or a nerve cell with discriminatory ability may yet turn out to be of the same nature as those that permit a cell to "sense" its way through the body.

As to the question of how the migratory cells of the embryo come to gather in certain specified locations, it has long been considered good form in biology to answer by circumlocution, stating that the cells get to their proper location by "attraction," "tropisms," a "sense of direction," and so forth. For descriptive classification, a listing of various types of tropisms may be useful, but it serves no analytical purpose. The term "neurotropism," for instance, in use for nearly half a century, has not only explained nothing but actually delayed real insight into the factors orienting the course of a nerve fiber. Even less realistic is the anthropomorphic and animistic terminology, in which cells are described as acting personalities, making decisions, choosing courses, and quite generally "doing" things. The main objection to symbolic expressions of this kind comes from the fact that instead of formulating the problems, they merely label them. We may not be able to dispense with such descriptive terms for some time to come, but we must guard against giving them any explanatory value. We must treat cells as physical systems in space and time, endowed with definable properties which are subject to the limitations of all physical bodies and their laws of behavior. If a cell gets from one place to another, it can do so only in strict compliance with the physical realities prevailing along its course.

Realistically speaking, then, a cell can have no "sense of direction" unless there is a physical directive agent (intensity gradient, oriented guide structure, or the like) operating right at the spot where the cell is. No agent can affect a cell from a distance other than through the intervening medium, and the physico-chemical constitution and behavior of the latter will determine if and when and from what direction and in what amount an orienting agent will arrive at the cell. What counts, is not the nature, orientation, or concentration of the agent at its source, but only the nature, orientation, or concentration in which it is present in the differential of space immediately adjoining the cell. This commonplace statement is called for by the tendency of some biologists to treat the attractions of cells "toward" distant destinations as if the intervening space were a vacuum, fully transparent, permeable, and non-corporeal.

Just how, in concrete language, does a cell get from its source to its destination? There are three possible answers to this question. They refer to three principles which we shall call (1) *selective conduction*, (2) *selective fixation*, and (3) *selective elimination*. A brief explanation of these follows.

Selective Cell Association

Selective conduction. The locomotor mechanism of tissue cells is still rather poorly understood. Lacking such special locomotor organs as cilia, flagella, and the like, cells move either by rhythmic deformations of their bodies, protrusion of pseudopodia, or a form of gliding. Cell sheets advance actively either by the locomotor activity of the cells along their free edges or by changes in the shape of all their cells, often due to differential contraction or expansion of their surfaces. In general terms, cells move whenever an inner disequilibrium creates pressures and tensions which are unevenly distributed over the surface and yield a resultant in some one direction. If this resultant direction changes at random from instant to instant, the cell will likewise shift at random, in a sort of large-scale Brownian movement, about a stationary center. Any progressive cell advance, on the other hand, implies the presence of constant polarizing forces which make the locomotor pressures and tensions yield resultants in a prevailing direction.*

Accordingly, no cell can make continuous progress in an isotropic field of force. On the other hand, any anisotropy of the surrounding field capable of affecting the pressure-tension configuration of the cell, will thereby have an orienting effect on locomotion. Among

---

* For recent descriptions of cell locomotion, see Lewis[18] and Holtfreter.[18]

the agents potentially capable of such effects are elastic tensions, pressure, interfacial tension, flow, gravity, electric potentials, electric currents, and radiation, as well as steady gradients in the concentration of any chemicals that affect the physical properties of the cell surface. In some cases, the external polarizing force may provide both the drive and the guide for the movement; in other cases, it determines only the direction of the movement, while the driving power is furnished by the metabolic energy of the cell itself. In either case, the orientation is of external origin. Spindle type cells (fibroblasts, mesenchyme, Schwann cells) and nerve processes, for instance, are oriented by interfacial tension along oriented fibrous structures, which, in turn, are but orderly arrays of groups of linear molecules.[40, 41] The locomotor mechanisms of spindle cells and amœboid cells (e. g., lymphocytes) differ somewhat, but even the latter seem to require contact substrata for continued advance.

"Contact guidance" has been shown to be a necessary condition for directive cell movement. If cells are confronted with guide structures that are all aligned in a common direction, the cells are forced into a single course.[41] In this case, contact guidance is not merely a necessary, but it is a sufficient condition for orientation. Yet, if the medium contains multiple intersecting guide structures, the problem becomes equivocal. Which one among the several available and structurally equivalent pathways is a cell to follow in a given instance? Since they seem to be able to "choose" the right track, we must postulate that the contact substrata have different specific properties that act as cues: one particular type of surface would be uniquely suitable for the application of one particular cell type, and another type of surface for another cell type. I have called such a mechanism, which is based on the specific matching between the cells to be guided and their prospective guide structures, "*selective contact guidance.*"[38] Its most plausible explanation would be that temporary linkages are formed between specific molecular groups in the cell surface and complementary groups in the guide structure. The guide structure may be situated in the surface of another cell or in the intercellular matrix.

Selective fixation. This term refers to a mechanism of the following sort. Cells of a given kind spread from their source at random, but certain areas of the body are so constituted as to arrest and hold cells of that particular type when they happen to get there. This local trapping, which might be a matter of mere immobilization or of true attachment, presupposes again highly discriminative powers on the part of both the cell and its prospective "trap." One could envisage a hypothetical trapping mechanism as consisting of the establishment of firm linkages across the contact surfaces between molecular groups of high affinity.

Selective elimination. This process is the reverse of the one just outlined. As in the former, the cells concerned are at first distributed rather ubiquitously throughout the body, but are then actively destroyed in certain regions while being spared in others. The final distribution would not be a matter of selective affinity between the cell and its permanent site, but rather of some disaffinity between it and all other sites.

It seems that these three principles embrace all the conceivable mechanisms by which localized aggregations of cells originating from distant sources could be effected. In a last analysis, they resolve themselves to a single common principle, namely, the existence of specific contact relationships between the various units of the organism, according to which adjacent units may either be bound to each other, or not bound, or even actively separated. In the case of selective conduction, the ties are those between the cell and its guide structure and are transitory. In selective fixation, they are more durable, anchoring the cell to its surroundings, pending reinforcement or replacement by secondary cementing agents. Being contact relationships, they can easily be conceived of as products of intermolecular forces, and their specificity as the result of steric conformances, that is, fittingly interlocking configurations of the molecular species to either side of the surface of contact.* These relations will have to be viewed as dynamic rather than static, and as statistical rather than rigidly fixed; that is to say, as the bonds in question are presumably incessantly made and broken, the rate and frequency of these events are as instrumental in determining the degree of specificity attained as are the nature and arrangement of the molecular groups involved.

Before going into these matters more fully, however, we wish to strengthen the biological evidence on which this theory of contact specificity rests.

---

* We are adopting here essentially Pauling's[24] concept, according to which the strength of intermolecular bonds varies with the degree of correspondence in the shape of the interacting molecules.

Tissue Affinities

In the healing of complex wounds, in which several tissues are involved, it has been repeatedly observed that each tissue component tends to fuse with its own kind. What has not been sufficiently stressed is that after such fusion has occurred, the tissue components automatically cease to expand further. This phenomenon is strictly comparable to the stoppage of the migration of units in the embryo when they meet their own or some properly matching kind. If we try to reduce all observations of this sort to a common denominator, we are led to the following thesis.

Any given cell type remains stationary only as long as certain very specific contact conditions peculiar to its own kind prevail along its exposed surfaces. If these conditions are not fully satisfied, the unit will move and continue to move until it finds itself again either in the original or in an equivalent situation. To use a simile which may prove pertinent, each unit would possess many specifically arranged "valencies," and only if all of them are completely "saturated" by properly matching "valencies" of the surroundings will the unit be immobilized. Any partial unsaturation, on the other hand, would mean instability, and hence, result in mobilization. It is obvious that such "unsaturation" could arise either from without or from within the unit; from without, if the unit is deprived of its matched environment by mechanical lesions or other alterations; from within, if the character or state of the unit itself changes, as happens during ontogenetic differentiation or in pathological states.

Let us now illustrate the operation of this principle on some concrete examples. These can be grouped into three classes, depending on whether we focus on the selective combination of units with their own kind (*"homonomic"* affinity) or with some other matching kind (*"complementary"* affinity), or on the *active detachment* from other units.

The behavior of epithelial sheets may serve as prototype of homonomic affinity. Epithelia with free borders rarely remain quiescent. They expand until edge meets edge and the system becomes closed up in itself. Epithelial coats and linings, therefore, always tend to restore their own continuity and tolerate no gaps. Inflicting a hole deprives the epithelial cells along its border of some of their natural surface contacts, and thus creates the "unsaturated" state which presently sets them in motion. If they remain attached to the rest of the epithelium, they drag it along. The movement is sometimes directed, at other times rather random. In either case, it ceases only after the gap has been covered and there is no longer any free edge. Observations on explants and transplants indicate that in the presence of several kinds of roaming epithelia, reunion mostly occurs selectively, each fusing preferentially with its own kind and by-passing other kinds. Mixed epithelial mosaics tend to break up into their constituents, each type forming a separate coat or cyst. Exceptions are noted in those cases in which two different epithelia are of the kind that normally have a common border, but whether this is an expression of complementary affinity or of purely mechanical welding remains to be seen.

Further examples of homonomic affinity may be cited from such widely different fields as the development of vascular anastomoses, the growth of Schwann cords in nerve regeneration, the formation of nerve trunks, the healing of transplants, and the reaggregation of dissociated sponges.

Capillary networks arise from the fusion of advancing capillary sprouts. Evidently, the blind processes keep growing until they meet other similar processes, with which they then merge.[4] The fact that they anastomose only with members of their own kind, to the exclusion of all the other cell types they pass on their way, is proof of the distinctive constitution of their surfaces, which alone makes such selective recognition possible.

The transection of a nerve trunk is followed by the emigration and growth of masses of sheath cells (Schwann cells) from both ends. If the gap between the ends is occupied by a diffuse scar, the Schwann cell masses expand in it profusely in all directions and form a tumor-like glioma. If, however, the nerve ends are bridged by a trellis of parallel fibrin fibers, growth is checked as soon as the cell strands advancing from opposite ends have merged into continuous bands.[43] Evidently, the difference in the two cases is this. An irregular scar provides the cells with a diffuse net of pathways, along which the chances for any two cells to meet head-on are relatively low, and consequently, growth continues unchecked; while the presence of oriented guide fibers across the gap necessarily leads the cell cords from opposite ends to advance straight towards each other and eventually to meet, whereupon they come to rest. Since the same Schwann cords are not arrested by their frequent encounters with other cell types, such as endoneurial cells, fibroblasts, and macrophages, it is clear that the effect is highly specific.

Nerve fibers show a marked tendency to associate according to their specific character. Not much is known about this except what can be inferred from the fact that peripheral nerve trunks and central fiber tracts do not contain different kinds of fibers (motor, sensory, sympathetic, etc.) in indiscriminate dispersion, but grouped into relatively homogeneous bundles. In view of the fact that each bundle or tract is made up of fibers of widely different ages, the homogeneity of their content indicates that fibers of a given kind growing out at a later stage have gathered preferentially around older fibers of their own kind. This concept of "selective fasciculation"[38] finds some more direct support in certain experiments with an easily distinguishable nerve cell type (Mauthner's cell): supernumerary Mauthner's fibers developing from grafted brain parts tend to follow the normal Mauthner's fibers of the host if they happen to make contact with them.[23]

These examples find their simplest counterpart in the behavior of sponges dissociated into small fragments. It has long been known that such fragments become highly mobile and upon encounter merge into larger bodies. In this reorganization process, the various cell groups become sorted according to their original characters, partly by selective association during the merger, partly by later segregation and regrouping.[3] Though there is again no evidence of attraction among homologous elements, order is restored by virtue of the fact that those elements that happen to come in contact will join more readily and more firmly if they are of the same kind than if they belong to unrelated kinds.

When there is selective association, temporary or permanent, between units of two different kinds, we may call this "complementary affinity." It implies a sort of "plurivalency" on the part of the units concerned. Examples of this principle are found in many cases of "selective contact guidance." For instance, the lateral line of amphibians develops by the tail-ward migration of a streak of cells along several predetermined routes under the epidermis. It has been shown experimentally that any change in the orientation of those parts of the body through which the lateral line is to travel, causes a corresponding change of the course of the line.[11] We must assume, therefore, that the bed for its growth is marked out by specific characters of the underlying local tissue which the tip of the outgrowing line follows.* The nature of these cues is unknown, but

---

* Experimental evidence[30] indicates that the relevant conduction pathway is furnished by the mesodermal structures rather than by the epidermis.

it would seem simplest to envisage them again as distinctive chemical characters of the contact surfaces. A purely mechanical concept seems inadequate to explain the facts.

A similar contact affinity must be assumed as guiding the growth of the Wolffian duct to the cloaca. When the posterior trunk of an embryo is experimentally rotated against the anterior trunk, the duct, on reaching the dislocated portion, deviates from its original course and often turns into the rotated position.[14] This indicates that its channel must have been marked out by local characteristics of the surrounding tissues. The fact that the blind end of the duct finally breaks through into the cloaca, in turn, is indicative of some complementary affinity between duct and cloacal wall, while the merging of the segmental mesonephric tubules with the pronephric and later Wolffian duct, is presumably to be classed as "homonomic affinity."

It was mentioned before that sensory and motor nerve fibers tend to group themselves according to type, as older fibers serve as guides to younger ones of the same category. However, even the early pioneering fibers take already divergent courses, depending on whether they are sensory or motor,[31] a fact which can only be explained by some "complementary affinity" between the respective fiber tips and the preneural pathways over which they travel. There seem to be separate pathways in the mesenchyme for motor and for sensory fibers, each of them impregnated with a specific character permitting just the nerve fibers of the corresponding type to follow it. At the end of their pathways, the nerve fibers enter another phase of selective behavior as they connect with their respective end organs. It has been shown, for instance, that sensory fibers which have been diverted into the path of motor fibers and thus forced to terminate on muscle fibers never form transmissive connections with the latter,[10, 42] evidently because of some incompatibility of the respective protoplasms. Thus, sensory fibers and motor fibers are not only constitutionally different themselves, but their differential is matched by corresponding differentials in the embryonic pathway structures and again among the terminal tissues. This mechanism insures not only that nerve fibers are generally guided to their proper terminal recipients, skin and muscles, but actually effect connections in the proper combinations. As will be outlined below, specific interactions with the periphery continue even beyond this stage.

Generalizing these experiences, it would seem reasonable to assume that all associations between tissues of different character,

whether temporary or permanent, are essentially a matter of reciprocal bonds. The development of composite organs by the joining of
two or more contributions from separate sources is particularly suggestive; see, for example, the combination of the infundibulum with
Rathke's pouch to form the pituitary body, the association between
nerve fibers and sheath cells, adrenal medulla and cortex, liver cords
and capillaries. These combinations are so unique in each case that
it would be difficult to account for them otherwise than by some
specific reciprocal bonding between the combining elements. It is
only later in development that the mechanical frameworks of connective tissue fibers and tunics come in and consolidate the existing
primary unions.

Just as "homonomic" and "complementary affinities" must be
postulated to account for the selective association among parts, so
some active mechanism is called for to explain the separation among
formerly contiguous parts, which is a common embryological phenomenon. In the reassembling of dissociated sponges (see above),
one observes not only positive affinity among similar units, but also
an active separation among dissimilar ones. Likewise, in embryonic
development, certain cell groups regularly leave their former positions, as if forced out.

For instance, the primary mesenchyme of the echinoderm
gastrula leaves the entoderm plate for the blastocoele. The neural
crest abandons its position along the margins of the neural plate and
migrates into the body spaces. The lens and other placodal derivatives become pinched off from the epidermis. The mesenchyme of
the limb bud leaves its somatopleural source, the anterior hypophysis
separates from the oral ectoderm, and so forth. Again, while some
of these effects might be of purely mechanical character, due to
differential retraction, dissolution of connections (fibers or membranes) by proteolysis, or extrusion as a result of crowding, there
remain many more instances that cannot be adequately explained on
such a simple basis, particularly those in which cells leave their
former associates individually and sort themselves out from the rest.
In these instances, we may conclude that former bonds among cells
have been selectively severed. This will happen whenever a cell
type, in consequence of its progressive differentiation, has become so
modified in its surface composition that it no longer conforms to the
surfaces of surrounding cells of other types which have not differentiated in the same direction.

III. ON GROWTH AND AGING

Our discussion up to this point has centered on the demonstration of the fact that living parts engage in, and maintain, their mutual relations primarily by forces resulting from varying degrees of "affinities" between contiguous elements. As indicated in the introduction, specificity is subject to gradations, though, for the time being, the classification rests entirely on crude criteria of biological behavior. Judging from their conduct, most tissues show the greatest affinity to their own kind; that is, "homonomic affinity" dominates over "complementary affinities." Among the latter, there are gradations from those with sharp selectivity ("univalent") among given pairs, which might be termed "conjugated," down to the less restrictive specificities ("plurivalent") permitting multiple combinations to be formed, and in each class there are evidently variations of intensity.* Tissues occurring almost ubiquitously, such as blood vessels and common connective tissue, which prove to be acceptable to a wide variety of tissues other than their own, must be regarded as either endowed with multiple selectivity, their surface being "complementary" to some character shared by all those other tissues, or as altogether non-selective. In view of the notable failure of blood vessels to penetrate into intact cornea, epidermis, and cartilage, one would favor the alternative of multiple selectivity over that of non-selectivity. Following this general scheme of evaluation, it should be possible eventually to develop a systematic list of the various components of the body in which each one would be assigned definite affinities, single or multiple and of different valencies.

The concept here proposed, if valid, immediately raises two further fundamental questions: (1) How do the various "affinities" arise ontogenetically? (2) What is their nature, and can they be expressed in terms of known properties of a simpler order? Let us turn to the first point first.

Ontogeny of Specificity

Biochemical differentials between organs, individuals, races, species, and higher taxonomic categories can be tested by serological reactions in vitro or by the biological reaction to tissue grafts.[21] The

---

* It is an oversimplification to treat the cell as a unit in matters of affinity, as different faces of the same cell may exhibit different affinities (e.g., the basal, lateral, and apical faces of simple epithelia). The conditions making this situation possible will be explained below.

claim that antigenic specificity increases with the progress of development has not been substantiated by recent immunological studies.[6] On the other hand, there is abundant evidence to show that incompatibility reactions between grafts and foreign hosts increase with age. Combinations that are tolerated when host and graft are young may later be dissolved as a result of progressive biochemical divergence of the components. A similar gradual estrangement has been shown by Holtfreter[13] to occur between different tissues of the same species as a corollary of their biochemical differentiation. The experiments in point consisted of grafting together in arbitrary combinations different portions of amphibian germs.

When fragments of ectoderm and entoderm from blastulae or early gastrulae are combined and explanted in vitro, they merge at first into a common mass, but a few days later begin to separate into their ectodermal and entodermal constituents. Evidently, the divergent differentiation of ectoderm and entoderm gradually produces discrepancies which make impossible further intimate association between them. *Pari passu,* fragments of pure ectoderm and pure entoderm develop increasing resistance to combining with each other, until after four days they can no longer be made to coalesce at all. Only the oral and branchial portions of the entoderm, which in the normal organism remain permanently connected to ectoderm, continue to fuse with it in vitro, too. This, incidentally, gives further support to our view that the various parts of the organism are joined not simply as a result of mechanical accidents, but by virtue of specific fitting linkages between contiguous parts.

In contrast to the sharp antagonism that develops between ectoderm and entoderm, mesoderm shows positive affinities to both of them, and consequently can act as intermediary in cementing ectodermal and entodermal components into common compounds. It is noteworthy that this ambivalence of mesoderm is preceded by a brief phase during which it resists fusion with ectoderm. This coincides with the period in which the execution of gastrulation makes it necessary for the presumptive mesoderm to shift independently. Thus, affinities not only vary with the progress of differentiation, and may even change their sign, but are pre-adapted, stage for stage, to the needs of ontogeny. Indeed, they may turn out to be major means of insuring the proper course of ontogeny.

As differentiation continues within the germ layers, so does the development of further specificities of association. For instance, isolated entodermal fragments containing material for the formation of liver and gut later segregate neatly into these two components. Similarly, in the ectoderm, the neural portions detach themselves gradually from the epidermal portions; neural crest separates itself from neural tube; eye from brain; and so forth. In all these experiments, the progressive self-sorting of tissues according to their developing idiosyncrasies has occurred under very unnatural conditions, and the methods by which segregation was effected often differed radically from those of normal embryogenesis. The only common denominator was the net result, namely, that they did become separated. Many more similar examples could be cited from Holtfreter's work. They all show clearly the progressive development of "affinities" and "disaffinities" within the differentiating germ.

These results are fully consistent with our concept, that cells, in the course of their ontogenetic biochemical specialization, assume characters which predispose them to make or break connections with other cells; that these properties are instrumental in guiding mobile cells and cell groups and in determining their associations with others or dissociations from others; and that cells will come to rest only ·after they have become contiguous with others of matching properties, but are mobilized again whenever this correspondence is disturbed by changes from within or without. Cytological differentiation consists of progressive changes in the composition and constitution of the cell. While some of these changes become conspicuous as structural "differentiation products" or as changes in cell behavior, others do not reveal themselves to direct observation. The acquisition of those specific surface configurations on which cellular affinities are based evidently belongs in the latter class. Identical courses of differentiation in a group of cells lead to "homonomic" affinity. Cells whose differentiations take divergent courses, may or may not show affinities to each other, depending on whether or not their differentiations have produced concordant conditions. If surface conditions remain (or become) concordant, then the cell types concerned retain (or acquire) "complementary" affinities. If differentiation results in incongruous surface conditions, the affected cells will fail to combine, or if combined, will become detached.

Surface specificity, thus, is to be viewed not as an independent character, but simply as an outward expression of the specific character acquired by the whole cell during its differentiation. As such, it must be subject to the same developmental rules and limitations as differentiation in general. This is important in connection with the problem of whether all the specific characters of a given cell are determined by its constitution or whether some may be imparted to the cell by its environment. We know from experimental evidence that most of the characters which a cell differentiates are released from an inherent response repertory, with which each cell is genetically endowed,[36] but we also know from the process of antibody formation to introduced foreign antigens that cells can be made to acquire new specific properties not originally contained in their native endowment.

As for embryological differentiation, we know of at least one example in which specific characters are not evoked, but actually impressed by one cell type upon another. This occurs in the "specification" of neurons by their terminal organs. An extensive series of experiments has revealed that each muscle represents a constitutional entity, possessing specific properties that distinguish it from any other (non-homologous) muscle.[35] By virtue of this specificity, which is presumably biochemical, each muscle secondarily modifies its young motor neurons in such a manner as to confer upon them its own precise specificity.* Each muscle thereby "tunes" itself in on the action systems operating in the nerve centers. Sensory end organs exert a similar "specifying" effect on the afferent nerve fibers.[29, 39] The responses are so strictly selective for a given individual muscle or sensory ending that we must assume there to be in operation as many discrete specificities as there are individual muscle specimens and distinct sensory organs. It has been previously suggested[35] that this "specification" process may proceed farther into the centers, the ultimate neurons passing their acquired specificity on to the penultimate ones, and so forth, but the extent of this chain process is still wholly conjectural. It has also been concluded from the slowness with which the "specification" process spreads, that it may be in the nature of an antigenic reaction, the individually distinguishing protein group of each terminal organ impressing its configuration on receptive protein groups in the nerve fiber, where it would be passed on in similar manner down the line.[37] Moreover, one could speculate that specific conformances of this nature along synaptic interfaces

---

* This specification of neurons by their muscles has been called "modulation"[35, 37] but we refrain from using the term here to avoid confusion with another type of "modulation" which will be discussed below.

may determine the course of impulse transmission, so that the specificities of developmental and mental processes may yet rest on common grounds. At present, the experimental evidence is confined to the described phenomenon of peripheral specification. Whatever its nature, it is clear that it involves the imposition of specific characters from one protoplasmic system upon another, even though only in addition to, and on top of, other specific characters already previously acquired by ordinary divergent differentiation. Whether this "infective" type of specification is peculiar to the nervous system remains to be seen.

Molecular Ecology

To speak symbolically of "affinities," is merely to outline the problem, not to attack it. It remains to resolve the described biological phenomena into known phenomena of physical and chemical order. How such resolution could be envisaged will be indicated in the following. It will be essentially an elaboration of an earlier similar attempt to interpret cellular affinities in terms of molecular structure and organization.[38]

By way of preparation, it seems appropriate to transcribe the symbolic concepts of "cell" and "protoplasm" into terms of molecular phenomena. This transcription has a purely pragmatic purpose, namely, to create a more workable model of the cell. Its utility will soon become evident. It has led me to introduce a concept of the cell which can best be characterized as "Molecular Ecology." That is, a cell is to be viewed as an organized mixed population of molecules and molecular groups of the following properties and behavior.

(1) Each population is made up of molecular species of very different composition, sizes, densities, rank, and stability, from trivial inorganic compounds to the huge and highly organized protein systems. Some segments of these populations occur in relatively constant "symbiontic" groupings, often of a limited size range; these form the various particulates of the cell content.

(2) It is one of the fundamental characteristics of cellular organization that the various species constituting the population are not self-sufficient, but depend in various degrees upon other members of the population as well as upon the physical conditions prevailing in the space they occupy. Survival and orderly function of the total

population are predicated on the presence of all essential members in definite concentrations, combinations, and distributions.

(3)   In view of this intricate interdependence, given molecular species can exist and given interactions between species can occur only within a certain limited range of conditions specific for each kind. We might call these conditions the "existential and operational pre-requisites" for each molecular species or group.   The probability of members of a given species to persist, hence to be found, in any but the appropriate setting, would be extremely low.

(4)   If the specific existential and operational prerequisites for the various molecular species and groups differ at different sites of the cell, different species will automatically become segregated into their appropriate ecological environments. As a result, even a wholly indiscriminate mixture can become sorted out into a definite space pattern.   Certain species will assemble in relatively stable combina-tions, like biotic groups, while others, mutually incompatible, will separate.*

(5)   While the conditions and forces which determine the molecular regrouping are of the most diverse sorts—electric charges, surface tensions, coacervation, solubility, chemical affinities, adsorp-tion, enzyme-substrate relations, mobility, elasticity, etc.—their resultant in each case is of such character as to insure relative stability of composition, density, and localization of the given group of species. As they combine, larger units of supramolecular, submicro-scopic, and finally, of microscopic order arise, each durable or "viable" only in a particular typical constellation of conditions.

(6)   Organization in space of the content of the cell, and of any of its constituent particulate elements as well, therefore, presupposes a primordial system of spatially organized "conditions" to set the frame for the later differential settlement of different members of the dispersed molecular populations. Such conditions can presumably only exist in systems with stability like solids. Systems answer-ing this demand are presented by all surfaces and interfaces in the cell, which include the interfaces between one cell and another,

---

* We are omitting here from consideration the fact that many large organic molecules, such as the native proteins, seem to undergo constant metabolic renova-tion, exchanging constituents with their environment, but preserving their identity.** In terms of our analogy, this is the counterpart of the turnover of cells within the individual members of an animal population.

between cell and medium, nucleus and cytoplasm, nucleolus and nuclear sap, chromosomes and nuclear matrix, chromatic and achromatic substance, as well as between all other formed cell components and the interstitial fluid.

(7) A given surface area of given constitution will therefore favor the adsorption of a given assortment of molecular species, which will thus concentrate in that area and thereby crowd out other species not equally fit to occupy that particular zone. In this manner, the various surfaces will gradually become settled by mosaics of "frontier populations" recruited from the subjacent territories.*

(8) Owing to their frontier position, these surface populations acquire a unique rôle in determining the subsequent course of events in the interior. Without necessarily being morphologically distinct, they assume the functional properties of membranes. That is, they control the selective transfer of substances and energy between the molecular realms they divide.

(9) Polar molecules (e. g., the biologically prominent lipoproteins), in becoming fixed to an interface, are forced into a definite orientation relative to that interface, and hence, relative to one another.[17, 27] This orderly array makes it possible for the resulting polarized layer to serve now, in its turn, as a new surface along which further molecular layers from the interior can become fixed, with the selection depending on the physical and chemical properties of the free ends of the righted molecules of the first layer. Thus, a stacking up process is initiated through which organization can be gradually extended into the interior, creating an increasing diversity of conditions as it proceeds.

(10) If the conditions along an interface change in such a manner that the new conditions are no longer compatible with the continued existence of the old frontier population, the latter will be crowded out by a new assortment of species better fitted to the new situation. As this new frontier population settles in the controlling master position, it sets a new master pattern for the events in the interior, causing the further fate of the cell to take a radically different turn. Different contact surfaces can thus entail qualitative

---

* Again for the sake of simplicity, we are ignoring here the fact that specific local conditions favor not only the adsorption of certain existing molecular species, but synthesis of new species as well. This point will be more fully discussed in a later section.

changes in the cell by bringing different segments of the molecular population into the controlling surface positions.

The concept formulated in these ten points takes into account the growing realization that the structural and working order of the cell is based not on the presence of a fixed mechanical framework pervading it—abundantly disproved by the facts—but on a regular distribution in space of the various intracellular processes: a dynamic rather than static skeleton, maintained by metabolic energy and determined in its characteristics by some definite geometrical order in the field of its operation. This order we conclude to be an order of "conditions," going back in last analysis to the typical organization of surfaces—"organization" in this sense referring to the particular non-random distribution of physical and chemical properties (see later). Pending evidence to the contrary, it is also possible to view the organization of genes as residing in their surface properties. In other words, the organization pattern of many, and perhaps all, living systems can be derived from a two-dimensional ground plan to which the third dimension is secondarily added by the selective stacking-up of various polar compounds in consecutive layers.

This concept likewise makes allowance for the statistical variability of many cellular and developmental events. Only the frame in which these events occur is relatively invariant. The highest degree of invariance has been assigned to the gene with a molecular population of remarkable constancy in composition and arrangement; mutations being explicable either by the loss of a given member species of the population or by its mere expulsion from a controlling surface position. However, the microdeterminism of chromosome structure is not matched by an equally rigid determinism of cellular characters. Only "norms" of development and statistical probabilities of the occurrence of given characters, rather than stereotyped constancies of details, are predictably determined. This is precisely what the "molecular ecology" concept of the cell implies in its emphasis on local "conditions," meaning simply *probabilities* that certain processes rather than others will occur with varying degrees of definiteness.

Contact Relations in Molecular Terms

Let us now consider, in terms of this concept, what will happen when two such systems, e. g., two cells, to be named A and B, come together in a common boundary. The surface of each will become a conditioning factor for the other. If the surface constitution of A satisfies the existential prerequisites of the adjacent frontier population of B, the latter will remain unaltered. However, if the A-surface introduces conditions incompatible with the B-surface population, the latter, if sufficiently mobile, may retreat from the surface to be replaced from the interior by another group of species better adapted to the new contact area.

It must be remembered here that stationary frontier populations tend to become rapidly congealed by the recruitment of additional layers, very probably accompanied by the cross-linking of fibrous molecules into fabrics, which with the incorporation of water compose the gel crust described for so many cell forms. Molecular mobility in these gel layers is greatly reduced. Whether and how fast a cell will respond to a new surface by regrouping will, therefore, depend on the condition of its crust. A cell in motion (locomotion or mitosis) or immediately after settling will be more responsive than will one that has been stationary for some time. On the other hand, there are indications that the very presence of an incompatible surface condition may lead to a solvatization and mobilization of the crust, thus restoring the freedom of movement necessary for molecular regrouping.

Now, just what are the surface conditions to which the molecular populations will react? As stated above, they are of complex character, and in part merely a combination of electric charges and surface tensions of the sort that is exemplified by the adsorption and concentration of detergents along inorganic interfaces. However, forces of this description seem too general to account for the high degree of specificity in intercellular relations illustrated in the earlier part of this article. Unless we want to invoke entirely unknown principles, no other explanation of such specificity seems at hand than one based on a concept of interlocking molecular configurations. This concept, traceable to Ehrlich, and culminating in the recent work of Pauling, maintains that the specificity of intermolecular relations is based on "steric conformance," i. e., corresponding or complementary spatial configurations between molecules, or certain exposed atomic groups of them, enabling them to conjugate in key-lock fashion. The theory is that such structural fitting allows the fitting particles to come within range of strong binding forces. A second and equally important point of the theory is the thesis that master molecules of specific configuration may serve as templates or models which would force other surrounding molecules to assume a complementary configuration, in mould-cast fashion.

Suggestive evidence for this concept can be derived from studies on antigen-antibody systems, enzyme-substrate systems, hormone-effector cell relations, and drug action; in other words, from a sufficient variety of biological phenomena to suggest fundamental validity. On the strength of this evidence, an extension of the concept to problems of growth, differentiation, and tissue behavior becomes a legitimate task.

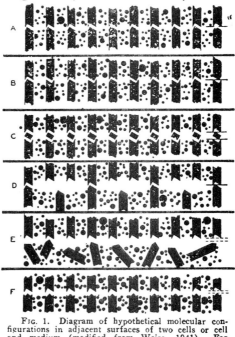

Figure 1 is the slightly modified reproduction of a diagram used previously to explain selective adhesion and non-adhesion among cells in terms of molecular configurations along the contact surfaces. The key molecules, which numerically perhaps constitute only a small fraction of the surface population, are symbolized as bars with char-acteristically shaped ends.

FIG. 1. Diagram of hypothetical molecular configurations in adjacent surfaces of two cells or cell and medium (modified from Weiss, 1941). For further explanation, see text.

The assumption is that two complementarily shaped molecules meeting in proper orientation will become linked by intermolecular forces, which thus become forces of attachment between the two contiguous systems. Properties of the sort required in our model are commonly associated with proteins, or combinations of proteins with lipids and other substances.

Let us now consider the implications of our hypothesis in greater detail. Each horizontal row, from A to F, symbolizes one hypothetical molecular state along an interface either between two cells or between a cell and its medium. In diagram A, for instance, the upper row (notched ends) represents the critical surface molecules of one cell, the lower row their complementary counterparts in the other cell; the white zigzagging band between them is the cell boundary. In this instance, the linkage occurs between members of two different populations, illustrating "complementary affinity" as outlined in an earlier section.

For simplicity, the molecules are shown evenly spaced. This could be true only if the cell surface were a highly organized lattice with the key units disposed in regular two-dimensional periodicity, somewhat along the ideas of Wrinch.[46]   Otherwise, we would have to give the diagram a statistical interpretation, in which the even spacing would merely indicate the average density of the surface population. Probability of encounter between complementary units, rather than absolute spatial congruity between the two complementary populations, would be the determining factor. This will be further explained in diagram D.

Diagram B illustrates the linking between two identical systems —"homonomic affinity" in our terminology—on the assumption that they both contain units of complementary configuration and that both types have identical chances to settle in the surface. The question as to why these molecules should combine with their counterparts across the boundary, rather than with their neighbors, could be answered by reference to the orienting effect of the surface, which would greatly enhance the chances of the former over the latter. As we shall mention below, there is some evidence for the coexistence of complementary proteins in cells, but in the few instances where their position could be ascertained, they were found to be in different locations, one near the surface and the other in the interior. If this condition were to hold generally, it would, of course, invalidate scheme B.

An alternative explanation of "homonomic affinity" is illustrated in scheme C. Here an intermediary substance of complementary configuration is assumed to act as the link between identically shaped molecules. According to Schmitt,[25] histones might play such a cementing rôle in the formation of epithelia. The cementing substances could still be products of the very cells they unite. That is, the cells would again be endowed with complementary units, but the one type of smaller size would be exuded to act as intercellular cement, while the other and larger one would be retained. Evidently, this concept would be wholly compatible with the observed separation of complementary species into surface and subsurface sites, respectively.

Scheme D is intended to show how affinity can vary in strength. The key molecules are assumed to be present in the two surface populations in different concentrations (in the given instance, in a ratio of $3^2 : 2^2$, i. e., $9 : 4$). This means fewer points of coincidence between complementary units, if one thinks in terms of regular surface fabrics; or a lowered probability of encounter, in a statistical concept. In other words, the "bonding" between the two surfaces is weaker than in scheme A, where concentrations are equal.

Scheme E illustrates the effect of disorientation. It is evident that even in a completely random array some complementary molecules would happen to come to lie in interlocking positions. However, the incidence of this occurrence would be very low as compared with the orderly array produced by adsorption to a surface, which turns the key molecules with their receptive groups all in the same direction. Surface orientation thus becomes an indispensable prerequisite for interlocking on an effective scale. This being the case, disorganization of the surface as a result of some change in the condition of the cell would automatically lead to detachment from its neighbors. In none of these speculations does the surface population have to be visualized as static, so long as its *average* composition remains unaltered.

In scheme F, detachment (or non-attachment) is due to lack of correspondence of shape between the key members of the two frontier populations. The molecules facing each other do not interlock. This condition will often arise secondarily as a result of cellular differentiation, which implies profound transformations in the character of the molecular populations. Differentiation gives rise to new species of molecules and presumably also modifies many of the existing ones. These changes will necessarily be reflected in the composition and distribution of the surface populations. Whether or not two systems, which differentiate in contiguity, remain attached or become separated, will, therefore, depend on whether their respective frontier populations undergo parallel or divergent changes. In the latter case, incongruities as depicted in diagram F, will arise. This explains in principle the observations of Holtfreter and others, quoted earlier, that tissues which at one time are closely combined may later separate, as differentiation progresses.

In addition to the permanent incompatibilities resulting from divergent differentiation, the scheme provides an explanation for temporary and reversible separations such as sometimes occur between cells of the same character in response to certain stimuli. If the surface is altered in such a way that it ceases to offer optimal existential conditions to the former frontier population, other molecular species will emerge from the interior and occupy the modified surface. The new settlers may be wholly unrelated to the populations across the border (as in diagram F), and consequently, the old links will be severed and will not be reformed until the old surface condition has been restored. In the meantime, the new frontier population, in its master position, may initiate rather profound and conspicuous changes in the appearance and behavior of the particular cell, but these changes, as one can readily see, are primarily changes in the distribution of existing compounds, and not in basic composition. They represent the transient processes which I have designated as *"modulations,"*[36] in contradistinction to the progressive and irreversible transformations of character and composition to which the term "differentiations" had better be reserved.

It would exceed the scope of this paper to illustrate in greater detail how a wealth of familiar phenomena of development and pathology can be interpreted in terms of this concept.* For the time being, we merely want to point out that it can satisfactorily account for all the phenomena of selective affinity and disaffinity dealt with in the earlier sections. The question, therefore, is no longer whether the concept of "steric conformances" as mechanisms of intercellular relations is fruitful as a working hypothesis—for this seems to have been answered in the affirmative—but whether it can be verified. We have borrowed the concept of "complementariness" of configuration from immunology, where antigen-antibody linkage has been tentatively explained on the basis of steric conformance. We are now to explore how pertinent this model is.

---

* It is evident, for instance, why different faces of the same cell can behave differently, and even show different affinities, as mentioned before: Surface areas exposed to different conditions (either neighboring cells or ambient medium) will become occupied by different segments of the molecular population. The marked polar architecture of epithelial cells is a notable example. In one given type, for instance, compounds mediating attachment to the basement membrane will be drawn toward the base, compounds, say with proteolytic potency, toward the apical face, and compounds for "homonomic" adhesion with like cells, to the sides. It can easily be understood that changes in the physico-chemical conditions along either the basal or apical surfaces could destroy this polar organization of the cell and turn it into a "pathological" course.

Immunological Models

In terms of "molecular ecology," the cell can not be considered as an antigenic unit. It contains numerous and diverse molecular species, which if specific steric properties are a prerequisite of antigenic action, represent a wide variety of antigenic agents. To a certain extent, it has been possible to demonstrate this fact by fractionating the cell content and testing the antigenicity of the various fractions separately. Head and tail fractions of spermatozoa, which can be conveniently separated, provoke each a corresponding antibody.[12] Considering the great limitations of techniques of fractionation, it is reasonable to assume that any given cell harbors an infinitely greater variety of specifically configured proteins than we can reveal by present immunological techniques.

Next to the existence of such specifically shaped units, our concept postulates that those occupying surface positions can act as links between the cell and its surrounding structures. The clumping of scattered cells and blood corpuscles by agglutinins or precipitins is evidence that comparatively large bodies can become affixed to one another by intermediary molecules of fitting complementary configuration. Obviously, there is thus no fundamental difficulty in envisaging bonds between cells in general as effected by a similar principle. The factual basis for such a view, however, is still extremely meager. There is perhaps only a single well-attested case known that could be quoted as supporting evidence. This is the combination of egg and sperm in fertilization, which could be classed under our general heading of "complementary affinity," as it involves the permanent selective union between two cell types of different origin and character.*

The first one to call attention to the similarity between fertilization and immunological phenomena was F. R. Lillie[19] in his classical experiments on "fertilizin." This work has later been carried on by Tyler,[32] and according to his version, provides definite evidence for

---

* The fact that egg and spermatozoon fuse after combining while somatic cells in "complementary" combinations usually retain their identity, is of minor importance. There are somatic cells which behave similar to germ cells; for instance, those that merge into syncytia, or those that penetrate into neighboring cells (e.g., the nurse cells of some oocytes; the melanoblasts that feed pigment into epidermis cells of feathers). Evidently, it requires special mechanisms to cause two cells to coalesce after they have become joined.

the contention that egg and sperm hook on to each other, as it were, by the interlocking of surface substances of complementary configuration, acting precisely like antigen-antibody systems. "Fertilizin" of the egg surface combines with "antifertilizin" of the spermatozoon, and the experimental evidence available strongly supports the view that either of these substances closely resembles the true antibodies to the other, as obtained by immunological procedures. Moreover, the interior of the egg contains a substance which behaves just as "antifertilizin" does and must, therefore, be regarded as complementary to the surface "fertilizin." At least for the egg, the simultaneous presence of complementary substances in the same cell, but at different sites, has thus been made highly probable, and Tyler has correctly appraised the possible general significance of this fact. It is immaterial for the general concept whether the key compounds are still incorporated in the cell surfaces proper, when they act, or are segregated in a distinct outer coat, as is the case with fertilizin.

The egg-sperm relation thus illustrates precisely the type of relation our concept has postulated as controlling compatibility and incompatibility or affinity and disaffinity between tissues in general, and since the former relation has proved to be reducible to simple terms of immunochemistry, particularly interlocking molecular configurations, there is good reason to suspect that the latter relation may yield to the same interpretation. Although the proof remains to be produced, the expectation is logically well founded, and attempts at verification hold some promise of success. Should they prove unsuccessful the problem of "specificity" of tissue relations would revert to the descriptive stage, in which it is now, and a fresh solution would have to be attempted in some other, as yet unforeseeable, direction.

Differentiation in Molecular Terms

It is of interest now to examine how the basic problems of ontogeny present themselves in the light of the concepts developed in this paper. While the bulk of this task must be reserved for a future occasion, a few pertinent comments and experimental data that fit into the present context may be advanced here.

Ontogenic differentiation is characterized by the fact that cells of demonstrably equivalent constitution turn into strains which become increasingly diverse.[36] There is overwhelming evidence that in this process, content and character of the cell undergo irreversible trans-formations.* At the same time, it has become clear that a cell at any given stage of differentiation can assume a variety of morphological and physiological expressions, which are commutable; these fluctu-ating states have been designated as "modulations,"[36] in contradis-tinction to the unidirectional differentiations proper. Translated into terms of molecular ecology, "modulation" would consist of the mere regrouping of the existing molecular key species without basic change in their character, while "differentiation" would involve a change in the composition of the population, with the appearance of new key species and the loss of old ones.

Some examples of "modulation" in mature cells are: the cyclic changes which hormone-responsive cells undergo in accordance with cycles in the hormone concentration in the blood[22]; the reversible morphological changes in lateral line sense organs which accompany the transition of certain newts from aquatic to terrestrial, and back to aquatic life[7]; the conversion of fixed histiocytes to macrophages and the resettlement of the latter; the temporary conversion of osteoblasts to fibroblasts[2]; and many similar cases. In all of these instances, the change in the cell occurs in response to a definite change of the cellular milieu. According to our concept, this would happen whenever that part of the medium which is in immediate contact with the cell surface becomes so altered in its composition or physical properties that it no longer satisfies the needs of the old frontier population of the cell, and the latter retreats and gives way to some other species better adapted to the new conditions. These changes initiated from the surface can then produce a thorough reshuffling of the cell content, in the course of which formerly inactive species may assume prominent functions, and formerly active ones go into eclipse, with the result that the whole cell changes con-spicuously in appearance and behavior.

---

* How far this process goes in any given cell is an empirical question, which must be explored separately for each particular cell type of each particular group of animals; moreover, it cannot be properly answered unless cytoplasm, cytoplasmic products, nucleus, and chromosomes are considered separately. The criterion of differentiation is the irrevocable restriction of potency. The test of this is not whether or not a given cell can lose some of its specialized aspects, but whether in doing so, it regains the capacity to redifferentiate in other, new directions. In all higher forms, perhaps all cœlomates, irreversible differentiation of various degrees is the rule in somatic cells. In cœlenterates, on the contrary, true differentiations seem to be rather the exception.

Differentiation does not appear to differ at first from modulation, and only by the criterion of reversibility can we distinguish them. This suggests the possibility of deriving one from the other. Each embryonic cell possesses an inherent endowment of molecular key species, whose properties, specific requirements, and possibilities of interacting narrow the capacity for future differentiation to a limited number of possible courses. This is the material basis of the embryological term "potency." The actual realization of differentiation requires (a) some factor that initiates one particular among the several potentially feasible courses, e. g., makes the pluripotent cell turn into a neuroblast or spongioblast or myoblast or chondroblast or melanoblast, etc.; and (b) the proper physical and chemical setting for the realization of the selected course, e. g., conditions that permit neuroblasts to transform into nerve cells, myoblasts to produce contractile fibrils, melanoblasts to produce pigment, etc. Now, factor (a) could be assumed to consist of some condition that attracts specifically one particular segment of the mixed molecular population to the surface and fixes it there in a master position. Once settled, these oriented surface molecules would specifically control intake and output between the cell and its environment, and would act as a ground plan for the progressive segregation of further species from the original intracellular pool. They not only would furnish a structural frame to which other species of molecules could be built on in a series of steps leading to a fabric of increasing complexity, but they could also catalyze specific reactions among the other species and thus take the lead in changing the chemical composition of the basic constituents of the population. Evidently, once this change in composition has been effected, reversible modulation has turned into irreversible differentiation. What has started as a mere redistribution and relocation of the cell content has ended in a change in character. Tentatively, all differentiations may therefore be considered to have started as modulations—a view fully in accord with the results of Experimental Embryology which have proved the reversible character of the early steps of differentiation.

Divergent differentiation among initially equivalent cells is to be explained by the fact that these cells were exposed to critically different surface conditions. As outlined earlier, the surfaces of such cells will then become occupied by wholly different segments of the molecular population, and in consequence, subsequent chemical events will take quite disparate courses. Experimental Embryology teaches that the "position" of a cell determines its fate in differentiation. Since, physically speaking, "position" can only mean exposure to certain physical and chemical conditions prevailing at that particular site, our concept proves again in harmony with experience. Some sample applications may be briefly outlined.

Following the classical studies of Spemann, "inductive" effects exerted by one embryonic tissue on an adjacent one have been extensively explored. To quote the most common examples, we refer to the hetero-induction of neural formations in ectoderm by underlying mesoderm,[28] the induction of a lens in epidermis by the underlying eye cup,[28] the induction of feathers in the epidermis by an underlying dermal papilla.[20] It is doubtful that all influences described as "inductions" operate through a common mechanism. Yet, if we confine ourselves, for the present, to "hetero-inductions" of the type just mentioned, the following points may be considered as fairly well established. (1) The "inductive" effect is in the nature of an evocation,[33] rather than an imposition; that is, it merely calls forth a response for which the affected cell has had a latent endowment, but does not impart upon the cell entirely new properties.[36] (2) The "inductive" effect can, in favorable cases, be shown to be reciprocal; that is, both adjacent tissues are subject to each other's influence, although usually one partner is in a less responsive condition. (3) The effect is transmitted by contact. The original supposition that it is mediated by a single chemical entity has not been confirmed.[34] (4) The "inductive" exposure need only last for a relatively brief period (minimum: a few hours), after which the affected tissue continues the induced course on its own.

To these points, we may add a brief reference to a highly suggestive, though cursory, observation. In some cases, the first sign of an "inductive" influence is a marked re-orientation of the cells of the "induced" layer relative to the "inducing" substratum in such a manner that the cell axes of the former become aligned with those of the latter. Part of this phenomenon could be ascribed to tensile stresses between the two intimately adhering layers. On the other hand, it is quite possible that we are faced with the results of a potent alignment effect on polar molecular groups in the "induced" cells; that is, with a direct index of an orderly molecular regrouping, set in motion through contact with the "inductive" substratum. A closer study of the fine-structural reorganization that accompanies "induction" phenomena is urgently needed. It may produce a direct test of the speculative interpretation tentatively set forth in the following.

We will assume that the capacities for the various courses of differentiation potentially open to a given cell ("differentiation potencies") are based on the presence in that cell of groups of molecular key species which can set up master patterns, each a different one, to which the rest of the cell content will then conform. Ectoderm cells of an early amphibian gastrula, for instance, have actually been found capable of giving rise, under normal or experimental conditions, to epidermal cells, pigment cells, nerve cells, gland cells, muscle cells, notochord cells, pronephric cells, etc.[28] We assign to them, accordingly, a corresponding variety of master compounds of specific configuration. These key molecules need not be present as such from the beginning, but may themselves have their ontogenic history (see above in the section of ontogeny of affinities). None of these key species can gain dominating influence on the fate of the cell so long as they lie all intermingled in the interior. But as soon as one of them succeeds in occupying the surface to the exclusion of the others, it gains a dominant position from which to influence the further course of events in the cell in accordance with our earlier statements.

Thus, we submit that when an ectodermal cell turns into a neural cell, the decisive initiating step is the selective condensation along its surface of the key molecules for neural development which up to then had been mixed with the other species in the interior. Conversely, if the surface of the same cell were to become settled by the key molecules for lens development, this would set the cell on its course toward becoming a lens cell. Embryonic determination of cell fate would thus consist essentially of the accumulation in the cell surface of selected species of master compounds. Any factor that makes such surface segregation possible, thereby becomes a "determining" factor, and in line with the general theme of this article, we might again turn to the principle of steric molecular interlocking across cell boundaries as a possible mechanism.

Let us suppose that the surface of the "inducing" substratum is saturated with a certain species A of polarized molecules of such configuration as to match precisely one single component α of the molecular populations α, β, γ, δ, etc., of the overlying ectoderm cells. Due to their complementary shapes, these two types, A and α, would form strong unions (see Fig. 1A). Thus, given a certain degree of mobility of the cell content, all the α units will gradually be trapped along the surface exposed to A, just as a film of antigen traps antibody molecules. Faced with a different substratum, containing key molecules B complementary to β, the same ectoderm cell would have become covered with a β layer, furnished again from its own stock, and thus become turned into a wholly different course of differentiation.

Progressive determination would occur through a succession of such steps. A given contact situation would bring a certain key species to the surface. Its residence there would affect the chemical processes in the interior, entailing presumably further regrouping along internal interfaces, setting off a chain of effects which will reach the nucleus and chromosomes, whose reactions, in turn, will rebound on the chemical composition of the cytoplasmic population. As a result, new compounds will arise, and when the cell is later faced with a new contact substratum, this may attract some of these new species, initiating the next phase of differentiation, and so forth. At any one stage, the cell will thus have only a limited assortment of specific key species, and its reaction to "inductive" surface contact will therefore vary with time. This is the molecular version of what is usually referred to as the development of responsiveness, or "competence,"[33] in embryonic cells.

A further possibility to be kept in mind is that some "induction" effects may involve actual changes in the morphology of the molecules exposed to the "inducing" surface. It is conceivable that specifically shaped molecules of the "inducing" substratum would impose conforming shape on the adjoining molecular layer of the "induced" cell. That such an impression of specific properties from one molecule to another is feasible is demonstrated by the mechanism of antibody formation, and at least suggested by the fact reported above, that muscles and sense organs impart highly specific characters upon the nerve cells with which they are connected. But we have no way of predicting how common this type of "infective induction" may be in development in general.

Any discussion of "inductions" must take into consideration the many instances in which effects normally exerted by adjacent cells can be experimentally duplicated by rather trivial inanimate agents. Such agents are frequently referred to as "dead organizers." The name is highly objectionable, but the facts as such are on firm ground.[34] There are several ways of reconciling them with our concept. Either the agent in question acts in "skeleton key" fashion, i. e., has key properties in common with the natural conditions; or it provides merely physical conditions for the attraction into the cell surface of a single segment of the molecular population, not necessarily a selected one, thus setting the cell on some accidental course of differentiation; or it may cause the responding cells to release certain specific compounds which, after adsorption to the outside, would then operate somewhat in the fashion of the intermediate layer represented in Fig. 1C. There may be other possibilities, but in view of the lack of pertinent information it would seem futile to enlarge upon this matter at the present time.

The hypothesis of induction here advanced explains satisfactorily (1) the specificity of inductions, depending on the contact substratum; (2) the "evocative" character of "induction" in the sense that it can only operate through the instrumentalities preformed in the responding cell; (3) the "exclusiveness"[86] of cellular differentiation (i. e., the fact that once a cell has entered a given course of specialization, other courses are automatically suppressed), as saturation of the surface with one selected molecular species automatically precludes others from assuming the same vantage position; (4) the potentially reciprocal character of "induction," inasmuch as a surface with a settled α-population could attract A units from within a less consolidated cell containing A, B, C, D, etc., just as the A-surface in the above example attracts α units; (5) the transmission of "induction" effects by contact only; (6) the initial reversibility ("modulation" phase; see above) of the "inductive" effect; (7) the irreversibility of the effect after the critical surface condition has prevailed for a critical length of time; this is the period required for a given surface population to establish permanent chemical changes in the cell.

The hypothesis presupposes a high degree of macromolecular mobility in the intracellular matrix. We know that during mitosis there is free streaming of cytoplasm, so that at this stage at least the cell content would parade, as it were, past the surface and enable the latter to recruit selected key species. However, a great many potent "induction" effects (e. g., neural plate, lens) seem to be quite independent of cell division and to proceed during stages in which most of the responding cells are mitotically inactive. It is for these cases that one will have to ascertain whether or not intracellular liquidity is high enough to permit the comprehensive macromolecular reshuffling postulated by our theory.

Growth

The foregoing discussion has been focussed on cellular *differentiation*, that is, changes in the complexion of the molecular populations, but has ignored *growth*, that is, the increase in their size. Yet, growth confronts us with the same fundamental problem of specificity: Why and how do the various molecular populations that eventually distinguish one cell type from another continue to reproduce more and more of their own kind?

Simple as it is to deal with growth in formal terms, in which the cell is treated as a unit, the problem assumes forbidding complexity when viewed on the molecular level. Of the molecular species entering the cell and available for its growth, some retain their identity (water, electrolytes, etc.) while others are combined into new compounds. Of the synthetic products, some are rather ubiquitous and trivial, others highly specific for a given species or organ or cell type. It is the latter which present the crux of the problem of growth, as they can evidently not be synthesized from simpler ingredients except with some of their own kind present as models. Current studies and speculations on protein synthesis, virus reproduction, gene multiplication, cell morphology, and enzymology are aiming at some tangible scheme that could explain "self-reproduction" of these highly specific key systems of the growing cell. It is noteworthy that in these speculations, the concept of "templates" or molecular master patterns assumes increasing prominence. The cue is taken from immune reactions, in which cells turn out large amounts of antibody in the presence of an antigen template. Guided by this analogy, I started in 1938 a series of experiments to test whether specific antigenic systems of different organs exercise specific catalytic functions in the growth of the respective organs. The work* was done with the collaboration of Dr. Dan H. Campbell, who carried out the immunological part, and Dr. Sewall Wright, who did the statistical calculations. Other commitments and war research prevented the completion of these studies, and only a preliminary abstract has been published thus far.[44] Being pertinent to our present discussion, the results may be briefly reported here.

---

* Aided by the Dr. Wallace C. and Clara A. Abbott Memorial Fund of the University of Chicago.

It is common knowledge that organ-specific antibodies exert deleterious effects on homologous organs of the *mature* organism. Our experiments were designed to explore possible specific actions of such antibodies on the growth of *embryonic* organs. The procedure was as follows.

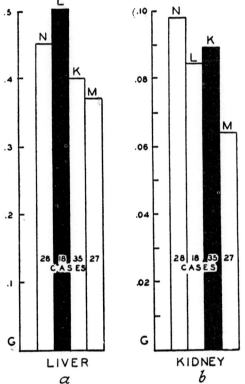

Autolyzed suspensions of three organs of adult chickens, namely, liver, kidney, and pectoral muscle, were injected into three groups of guinea-pigs over a period of 47 days. Ten days after the last injection, blood serum was recovered from the guinea-pigs, supposedly containing, among others, specific antibodies against the injected organ substances. We shall call these antisera "L," "K," and "M," indicating the liver-, kidney-, and muscle-injected series respectively. These antisera were then injected into chick eggs of ages ranging from sixty hours to eight days of incubation. Each egg received only a single injection of 0.4 cc. of one of the antisera, deposited near the embryo. Treated

FIG. 2. Weights of livers (a) and kidneys (b) of chick embryos on the 20th day of incubation, which had been injected during early development with liver- (L), kidney- (K), and muscle- (M) antisera, and of normal controls (N).

embryos and normal controls were fixed at various ages, but only those of the oldest group, sacrificed on the twentieth day of incubation, have thus far been studied. This group is composed of 28 normal controls, 35 K-embryos, 18 L-embryos, and 27 M-embryos. Their livers and kidneys were weighed by a standard procedure. The M-series serves as test of the general effects of antisera injection, while the K- and L-series were intended to reveal any specific effects on homologous organs. Owing to the great variability of organ weights in both normal and treated embryos, the data had to be evaluated statistically.

Figure 2 shows the average weights of livers and kidneys in controls (N) and treated embryos. Total weights of injected embryos averaged considerably below those of the controls, a fact which is evidenced in the smaller livers in the K- and M-series and the reduced kidneys in the L- and M-series. In contrast to this general growth depression, the one organ type in each series for which the injection had been specific shows evidence of positive growth stimulation. In the case of liver, this has led to an absolute increase of ten per cent above normal (Fig. 2a), while in the kidney (Fig. 2b) the specific stimulative effect has not been large enough to offset the unspecific depressive effect of the treatment. The organ-specific antisera thus seem to have specifically promoted the growth of the homologous embryonic organs. This is clearly

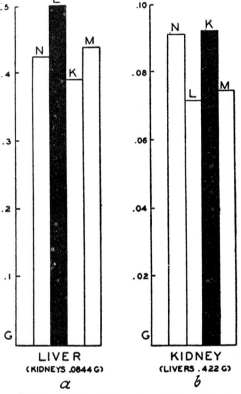

FIG. 3. Weights of liver (a) and kidney (b) of the same group of embryos as in Fig. 2, adjusted by regression coefficients for normal variability.

illustrated by the fact that the livers of L-embryos are larger than those of K-embryos, whereas the kidneys are larger in K-embryos than in L-embryos. By statistical calculations, one can determine the average liver weight for a given mean weight of kidney, and the average kidney weight for a given mean weight of liver (regression coefficients). These values, graphed in Fig. 3, reveal the following facts. Kidneys of all embryos injected with non-homologous (L, M) sera (Fig. 3b) are much more markedly reduced in size as compared to normals (N) than are the livers of embryos which had received non-homologous (K, M) injections (Fig. 3a). On the other hand, the livers of the L-embryos (Fig. 3a) and the kidneys of the K-embryos (Fig. 3b) are substantially increased relative to the reciprocal combinations. The adjusted liver weights are 29 per cent larger in the L-embryos than in the K-embryos, while almost the reverse holds for the kidney weights, those in the K-series being 28 per cent higher than the ones in the L-series. The statistical significance of the observed differences between the L- and K-series is .00001 for liver and .001 for kidney; that is to say, the probabilities of obtaining the results by mere chance are 1 in 100,000 in the former, and 1 in 1,000 in the latter, which makes the results appear as of high statistical significance. Two in 100 is conventionally considered the safety limit.

One may conclude, therefore, that some distinctive biochemical principles of chicken liver and kidney extracts call forth corresponding organ-specific products in the guinea-pig, which, in turn, when transmitted through serum, affect the growth of the homologous structures in the chick embryo. The fact that the specific effect consisted of stimulation rather than depression is at first surprising. Apparently, the manner in which antibodies affect the physiology of mature cells differs from their mode of action in growth. Only the general growth depression observed in the experiments corresponds to the conventional type of antibody action. The specific growth promotion of the homologous organs, on the other hand, is in an altogether different category. If corroborated, it would prove that antibodies to a given organ protein can act as catalysts in the synthesis of more of that particular protein. This would lead us directly to a "template" concept of growth, with molecules of complementary configuration acting reciprocally as moulds for each other's synthesis. In that case, the experimental transfer with double reversal, from chicken to guinea-pig to chick, would actually have been but a complicated version of what happens within the organism itself without transfer. Each differentiating cell would contain complementary key compounds, each of which would act as mould for the other. However, their respective syntheses would have to be assumed to occur at different rates, so that one would always prevail numerically.

Growth rates, according to this concept, would be governed by the concentrations in which the two complementary systems would be present and by the extent to which they would become conjugated and thereby inactivated as specific catalysts. Evidently, if some of the specifically configurated portions of these compounds are liberated from their source cells, their concentration in the medium could exert a growth-controlling influence on distant homologous cells. It is not inconceivable that many examples of "compensatory hypertrophy," which cannot be explained by the "functional overloading" of the residual tissue, go to the credit of such systemic balances.*

In order to test this principle more directly, I had a series of experiments carried out with the technical assistance of one of my students, Hsi Wang.† Essentially, it was a repetition of the preceding series, but leaving out the guinea-pig as an intermediary. Fragments about 0.5 mm.[3] in size were taken from livers of 6-day chick embryos and implanted in the area vasculosa of 4-day hosts. Substances from the grafted material could thus be carried by the blood stream into the host embryos. The latter were allowed to develop further for periods ranging from two to nine days after the operation (six to thirteen days of total age), at which time the weights of the whole embryos as well as of their livers were determined. The results, based on 137 cases with liver implants and 107 controls, were very striking.

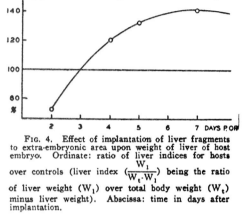

FIG. 4. Effect of implantation of liver fragments to extra-embryonic area upon weight of liver of host embryo. Ordinate: ratio of liver indices for hosts over controls (liver index $(\frac{W_1}{W_t \cdot W_1})$ being the ratio of liver weight (W_1) over total body weight (W_t) minus liver weight). Abscissa: time in days after implantation.

Average body weight (minus liver) of the experimental lot was 10 per cent below that of the controls, indicating a slight general depression of growth in consequence of the treatment. The livers of the host embryos, however, were greatly enlarged. In extreme cases, they were twice to three times as large as the largest of the controls. The time course of this effect is illustrated in Fig. 4. Liver weights were expressed in percentage of body weight, and the averaged ratios of these values for the hosts over the controls were plotted against time, counted from the implantation of the grafts.

---

* It will be of interest to explore the relation between this concept and the theory of "antihormones."[5]

† Only a brief preliminary abstract of these experiments has been published.[4]

The graph reveals that after an initial depression during the first two days, the growth of the host livers greatly overshoots that of the controls, the excess mounting steadily up to the seventh day, when it reaches an average of over 40 per cent. After that, the magnitude of the effect seemed to decline.

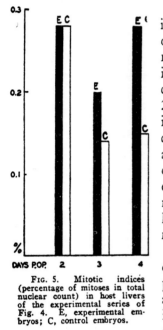

FIG. 5. Mitotic indices (percentage of mitoses in total nuclear count) in host livers of the experimental series of Fig. 4. E, experimental embryos; C, control embryos.

The observed enlargement of the liver is reflected in both the size and mitotic rate of the constituent cells. Cell size, determined for 30,000 sample cells, was larger in the host livers by 9 per cent after two days, 10 per cent after three days, and 20 per cent after four days. The mitotic index, determined for a total of 815,000 cells was the same for experimental host and control livers on the second day, but on the third and fourth days the former exceeded the latter by 43 and 87 per cent respectively (Fig. 5). Evidently, cellular hypertrophy antedates the increase in mitotic activity.

The mechanism of this effect is a matter of conjecture. Either specifically shaped parts of liver cell proteins can act directly as nuclei for further synthesis when entering the appropriate environment, which is another liver cell, or they act merely as models for moulds which then would turn out more of the original product. The latter alternative would bring the experiments with direct liver implantation and those using liver antisera to a common denominator. However, our information is far too sketchy for detailed speculations.

Moreover, the specificity of the effect is not absolutely sharp. Not only is the growth of other organs aside from liver affected by liver implants, but liver enlargement can likewise be provoked by implants other than liver, although the homologous effect is always strongest. In most embryos with enlarged livers, for instance, the kidneys were also somewhat enlarged; this could perhaps be regarded as a secondary functional effect. On the other hand, increased liver growth was observed not only in embryos with grafts of liver, but also, in descending order of intensity, with grafts of blood clots, skin, mesonephros, and perhaps muscle, none of which, however, approached liver tissue in effectiveness. Sham grafts of paraffin were wholly ineffective.

In conclusion, these experiments contain strong indications that substances released from the grafts and carried by the blood stream into the embryo, exert a catalyzing effect on the growth of the homologous tissue. But the admixture of the less localized effects just mentioned leaves the interpretation somewhat in doubt. The main purpose of presenting the results here has been to call attention to a promising line of work which is badly in need of systematic and intensified pursuit.

Pending verification by such future work, the whole concept of growth advanced in these pages remains hypothetical. It has points in common with the scheme of gene reproduction suggested by Sterling Emerson,[9] which likewise resorts to immuno-chemical analogies. In its emphasis on the existence of molecules of complementary configuration, our concept has points of contact with Tyler's thesis of "auto-antibodies,"[32] which has not, however, been explicitly applied to growth. Yet, for the time being, there seems to be nothing more to these convergences than a common conviction that the phenomena of biological specificity have a common stereochemical foundation: reproduction to be based on the ability of a compound to serve as a model for the synthesis of more of its kind; adaptation, on the ability of a compound to impose a conforming configuration upon other compounds; and selectivity, on the interlocking of matching compounds.

This past discussion leans heavily on current concepts of immunochemistry, particularly those developed by Pauling.[24] It may be premature to tie the phenomena with which we have been dealing too closely to the antigen-antibody model. Rather than trying to force all biological specificity into the immunological compartment, we might have to consider the latter as merely a special case of a more universal biological principle, namely, *molecular key-lock configuration as a mechanism of selectivity*, whether involving enzymes, genes, growth, differentiation, drug action, immunity, sensory response, or nervous co-ordination.

Conclusion

The purpose of this article has been to point out how some problems of specificity in development can be resolved into terms of molecular theory. Some of the premises and conclusions are fairly well substantiated, others are in need of experimental validation. From a pragmatic standpoint it is immaterial how many of the more detailed suggestions that were tentatively advanced will actually be borne out by future work, so long as the main line of thinking followed in our discussion proves of value and stimulates such future work.

There can be little doubt that many of our statements will have to be revised, as more facts become known. What we have called "contact" relations, for instance, might very well turn out to be "proximity" relations, still operating at close range, but exceeding the effective limits of the more common intermolecular forces. Or in stressing selective surface adsorption as an ordering principle in the rallying and sorting of molecular species, we may have unduly neglected some faculty of self-sorting of mixed molecular populations (as in the formation of tactoids[1]). The possibility of specific interactions through effects of radiations has not even been mentioned. Also, some phenomena bearing signs suggestive of specificity may yet find a more simple mechanical, electrical, or colloid-physical explanation.

Yet, these and similar reservations notwithstanding, our discussion will have served to illustrate at least the feasibility of breaking the rather abstract notions of specificity in development down into concrete and verifiable issues. We thus prepare what has been a domain of purely formal description for precise analytical investigation at the molecular level. In preparing the transition to this level, the biologist is prone to overstep his competence. Thus, in stealing a leaf from the chemist, I have risked the charge of trespassing on foreign and unfamiliar ground. I have done this in the conviction that, in science, a step forward, even in the wrong direction, is better than stagnation.

REFERENCES

1 Bernal, J. D   Structural units in cellular physiology in "The Cell and Protoplasm." Publ. Am. Asso. Adv. Sci. No. 14, 1940, 199.
2 Bloom, W., M. Bloom, and F. C. McLean: Calcification and ossification. Medullary bone changes in the reproductive cycle of female pigeons. Anat. Rec., 1941, *81*, 443.
3 Bronsted, H. V.: Entwicklungsphysiologische Studien über *Spongilla lacustris* (L.). Acta Zool., 1936, *17*, 75.
4 Clark, E. R., and E. L. Clark: Microscopic observations on the growth of blood capillaries in the living mammal. Am. J. Anat., 1939, *64*, 251.

5   Collip, J. B., H. Selye, and D. L. Thomson: Anti-hormones. Biol. Rev., 1940, *15*, 1.

6   Cooper, R. S.: Adult antigens (or specific combining groups) in the egg, embryo and larva of the frog. J. Exper. Zool., 1946, *101*, 143.

7   Dawson, A. B.: Changes in the lateral-line organs during the life of the newt, *Triturus viridescens*. A consideration of the endocrine factors involved in the maintenance of differentiation. J. Exper. Zool., 1936, *74*, 221.

8   DuShane, G. P.: The embryology of vertebrate pigment cells. Part I. Amphibia. Quart. Rev. Biol., 1943, *18*, 109.

9   Emerson, S.: Genetics as a tool for studying gene structure. Ann. Missouri Bot. Gar., 1945, *32*, 243.

10 Gutmann, E.: Reinnervation of muscle by sensory nerve fibers. J. Anat., 1945, *79*, 1.

11 Harrison, R. G.: Experimentelle Untersuchungen über die Entwicklung der Sinnesorgane der Seitenlinie bei den Amphibien. Arch. f. mikr. Anat. u. Ent., 1904, *63*, 35.

12 Henle, W., G. Henle, and L. A. Chambers: Studies on the antigenic structure of some mammalian spermatozoa. J. Exper. Med., 1938, *68*, 335.

13 Holtfreter, J.: Gewebeaffinität, ein Mittel der embryonalen Formbildung. Arch. exper. Zellforsch., 1939, *23*, 169.

14 Holtfreter, J.: Experimental studies on the development of the pronephros. Rev. Can. de Biol., 1944, *3*, 220.

15 Holtfreter, J.: Structure, motility and locomotion in isolated embryonic amphibian cells. J. Morphol., 1946, *79*, 27.

16 Landsteiner, K.: *The Specificity of Serological Reactions*. 2nd ed., 310 pp., Harvard Univ. Press, 1946.

17 Langmuir, I.: Molecular layers. Pilgrim Trust Lecture. Proc. Roy. Soc. London, Ser. A, 1939, *170*, 1.

18 Lewis, W. H.: The rôle of a superficial plasmagel layer in changes of form, locomotion and division of cells in tissue cultures. Arch. exper. Zellforsch., 1939, *23*, 1.

19 Lillie, F. R.: *Problems of Fertilization*. 278 pp., Univ. of Chicago Press, 1919.

20 Lillie, F. R., and H. Wang: Physiology of development of the feather. VII. An experimental study of induction. Physiol. Zool., 1944, *17*, 1.

21 Loeb, L.: *The Biological Basis of Individuality*. 711 pp., C. C Thomas, 1945.

22 Moore, C. R., and D. Price: Gonad hormone functions, and reciprocal influence between gonads and hypophysis with its bearing on the problem of sex hormone antagonism. Am. J. Anat., 1932, *50*, 13.

23 Oppenheimer, J. M.: The anatomical relationships of abnormally located Mauthner's cells in Fundulus embryos. J. Comp. Neurol., 1941, *74*, 131.

24 Pauling, L.: Molecular structure and intermolecular forces. [In Landsteiner's *Specificity of Serological Reactions*, pp. 275-93, Harvard Univ. Press, 1946.]

III. ON GROWTH AND AGING   307

25   Schmitt, F. O.: Some protein patterns in cells.   Third Growth Symposium. Growth, 1941, 5, (Suppl.), 1.

26   Schoenheimer, R.: *The Dynamic State of Body Constituents.* x + 78 pp., Harvard Univ. Press, 1942.

27   Schulman, J. H.: Monolayer technique.   In Bourne's *Cytology and Cell Physiology*, Oxford, 1942.

28   Spemann, H.: *Embryonic Development and Induction.*   401 pp., New Haven, Yale Univ. Press, 1938.

29   Sperry, R. W.: Visuomotor coordination in the newt (*Triturus viridescens*) after regeneration of the optic nerve. J. Comp. Neurol., 1943, 79, 33.

30   Stone, L. S.: Primitive lines in Amblystoma and their relation to the migratory lateral-line primordia. J. Comp. Neurol., 1928, 45, 169.

31   Taylor, A. C.: Selectivity of nerve fibers from the dorsal and ventral roots in the development of the frog limb. J. Exper. Zool., 1944, 96, 159.

32   Tyler, A.: Specific interacting substances of eggs and sperm. Western J. Surg., Obst., & Gynec., 1942, 50, 126.

33   Waddington, C. H.: *Organisers and Genes.* 160 pp., Cambridge, England, The University Press, 1940.

34   Weiss, P.:   The so-called organizer and the problem of organization in amphibian development.   Physiol. Rev., 1935, 15, 639.

35   Weiss, P.: Selectivity controlling the central-peripheral relations in the nervous system. Biol. Rev., 1936, 11, 494.

36   Weiss, P.: *Principles of Development.*   601 pp., Henry Holt & Co., New York, 1939.

37   Weiss, P.: Self-differentiation of the basic patterns of coordination. Comp. Psychol. Monogr., 1941, 17, 1.

38   Weiss, P.: Nerve patterns: The mechanics of nerve growth.   Third Growth Symposium.   Growth, 1941, 5 (Suppl.), 163.

39   Weiss, P.: Lid-closure reflex from eyes transplanted to atypical locations in *Triturus torosus*: Evidence of a peripheral origin of sensory specificity. J. Comp. Neurol., 1942, 77, 131.

40   Weiss, P.: The technology of nerve regeneration: A review. Sutureless tubulation and related methods of nerve repair. J. Neurosurg., 1944, 1, 400.

41   Weiss, P.: Experiments on cell and axon orientation in vitro: the role of colloidal exudates in tissue organization. J. Exper. Zool., 1945, 100, 353.

42   Weiss, P., and M. V. Edds Jr.: Sensory-motor nerve crosses in the rat. J. Neurophysiol., 1945, 8, 173.

43   Weiss, P., and A. C. Taylor: Histomechanical analysis of nerve reunion in the rat after tubular splicing. Arch. Surg., 1943, 47, 419.

44   Weiss, P., and H. Wang: Growth response of the liver of embryonic chick hosts to the incorporation in the area vasculosa of liver and other organ fragments. Anat. Rec., 1941, 79 (Suppl.), 62.

45   Willier, B. H.: Experimentally produced sterile gonads and the problem of the origin of germ cells in the chick embryo. Anat. Rec., 1937, 70, 89.

46   Wrinch, D.: The native protein theory of the structure of cytoplasm. Cold Spring Harbor Symp., Quant. Biol., 1941, 9, 218.

Reprinted from BIOLOGICAL SPECIFICITY AND GROWTH. Editor: E. G. Butler, Princeton University Press, Princeton, New Jersey, 195-206, 1955.

CHAPTER 12

SPECIFICITY IN GROWTH CONTROL

BY PAUL WEISS[1]

I N ESSENCE, this chapter is a sequel to the general discussion of "Specificity in Development and Growth" given before the Society for the Study of Development and Growth in 1945 (P. Weiss, '47). Of the problems raised then, that of growth control will be singled out here for special consideration. The account will be confined to illustrative examples, with no attempt at reviewing the subject comprehensively.

"Specificity" is understood here as that property of two interacting systems, A and B, which permits A to react to B with some degree of selectivity. Biological specificity, as previously outlined, is a basic property of living systems and is most prominently displayed in drug responses, hormone actions, gene effects, immunological reactions, host-parasite relations, and developmental mechanisms. A common denominator in biochemical, presumably stereochemical, terms is indicated. Translated to these terms, specificity in growth confronts us with four separate issues: (1) the fact that the various cell strains of the body are different biochemically; (2) the possibility that biochemical distinctiveness may be not only a by-product of differentiation but also an instrumental factor in growth; (3) the possibility that this may constitute a mechanism for the humoral coordination and regulation of growth processes throughout the organism; and (4) the hypothesis that this regulatory function presupposes the generation in each cell strain of paired compounds of complementary configuration, after the antigen-antibody scheme.

I. THE BIOCHEMICAL MOSAIC

It has been stressed on previous occasions (Weiss, '49) that cells are "speciated" into biochemically diverse strains, indeed strains of much greater diversity and subtlety than are morphologically discernible. My attention was first drawn to this fact by the discovery of the

[1] Department of Zoology, University of Chicago. Present address of the author is: Rockefeller Institute for Medical Research, New York, N.Y. This chapter is dedicated by the author to Professor F. Baltzer of the University of Bern, Switzerland, on the occasion of his seventieth birthday. It represents a condensed version of the paper of the same title presented at the 1953 Growth Symposium.

Original work referred to in this chapter has been aided by grants from the Wallace C. and Clara A. Abbott Memorial Fund, University of Chicago; the American Cancer Society upon recommendation of the Committee on Growth of the National Research Council; and the National Institutes of Health, Public Health Service.

strict constitutional specificity of individual muscles and skin territories (Weiss, '52), but all the localized effects of hormones and drugs on different predisposed cell groups point to the same conclusion. Different organs are thus composed of cell strains of specifically different biochemical constitutions. On the other hand, organs occurring in pairs or other multiples may be assumed to be composed of biochemically similar cells. Such similarity of composition might provide a means of selective chemical communication between homologous cell groups regardless of spatial separation.

The existence of some such active interrelation between paired structures is indicated, for instance, by the observation that injury to a given nerve is often followed by an involvement of the symmetrical nerve (Greenman, '13, Nittono, '23, Koester, '03, Tamaki, '36). It is more definitely evinced by the compensatory growth reactions of one of a pair of organs after removal of the other.

II. HOMOLOGOUS COMPENSATION

A "spontaneous" spurt of growth of the residual parts of a partially removed organ system has been described for many objects: contralateral appendages in annelids (Zeleny, '02, '05); claws in Crustacea (Przibram, '07); kidneys in mammals (Golgi, 1882, Ribbert, '04, Arataki, '26, Rollason, '49); testes in mammals (Ribbert, 1895) and fishes (Robertson, '54); lungs (Haasler, 1891); and orbital glands (Teir, '51). Liver regeneration following partial ablation (e.g. Brues, '36) and blood cell regeneration after hemorrhage are in the same category. In all these cases the response is essentially confined to the homologous tissue and consists primarily of intensified reproduction of homologous protoplasm, which may take the form of cellular hypertrophy, hyperplasia, or regeneration. Necrosis may produce similar effects as extirpation.

The explanation of such compensatory growth has often been sought in the excessive functional load placed upon the remaining fraction of the organ system. However, neither the invertebrate cases nor the testes and orbital glands of the above list fit such an explanation. I, therefore, considered it more likely that these phenomena are manifestations of a much more general principle, namely the active maintenance of the total mass of each organ system in an equilibrium state and the return to that state after disturbance by virtue of a chemical communication system in which specific releases from each cell type, circulating in the body, would inform the homologous cell types of the state of their total mass. A test of this theory requires the demonstration that, (1) compounds produced by

[196]

a given cell type have some selective effect on the same cell type, and (2) that this homologous effect is instrumental in the regulation of growth.

III. HOMOLOGOUS ORGAN-SPECIFIC EFFECTS

Evidence for the first point is contained in the observations of Danchakoff ('16) and of Willier ('24), who found enlarged spleens in chick embryos whose chorio-allantoic membranes had received spleen grafts. Weiss and Wang ('41), unaware of these results but corroborating them fully (see Weiss, '47), found that minute fragments of liver incorporated in the extraembryonic area of a chick embryo caused the host liver to grow to excessive dimensions. Implants of other tissues either had no effects or their effects were much less marked. The results with spleen were later confirmed and expanded by Ebert ('51). Along the same line, balancers implanted into the body cavity of urodele larvae affect the resorption of the host balancers specifically (Kollros, '40), as shown by the absence of a similar interference from implanted gills. Thus, there are definitely specific chemical effects transmitted humorally from a given type of organ to other cell groups of the same type, and these effects entail alterations of growth and size. Contrary to the inference from compensatory growth after partial removal, however, the experimental addition of tissue in these experiments produced no decrease, but rather a further increase of the homologous host tissue. Therefore, while the specificity of the growth reaction was clearly demonstrated, the sign of the reaction was paradoxical. This in itself was a significant revelation and we shall return to it below.

IV. IMMUNOLOGICAL REVERSAL OF ORGAN-SPECIFIC COMPOUNDS

My original concept had been that: (1) cell populations of a given type keep their total mass in check by producing, as they grow, compounds which would repress the growth process in proportion to their concentration ("feed-back" fashion); (2) this self-inhibition would be due to steric complementariness between these inhibitor compounds and the specific catalysts of growth in each cell type; and (3) the assumed complementariness might be of the antigen-antibody type. On this assumption a series of experiments was started in 1938 to test the effects that antibodies to organ-specific compounds, rather than the compounds themselves, might exert on the embryonic growth of the homologous organs (Weiss, '39; see Weiss, '47).

The fact that antibodies to organ extracts may affect corresponding

[197]

organs selectively, mostly by damaging homologous cells, has been clearly demonstrated for the lens (Guyer and Smith, '18), the kidney (Smadel, '36, Pressman, '49), and others. Our injections of anti-liver and anti-kidney sera into chick embryos, however, entailed again larger sizes, rather than damage, of the homologous host organs. This could of course be ascribed to an initial damaging effect followed by repair with overcompensation. At any rate the specificity of the effect, if not its sign, seemed to have been ascertained, and the demonstration that organ-specific compounds did not lose their specific effectiveness by immunological (steric?) reversal seemed to give validity to the concept that these experiments were intended to test.

Although war work had interrupted this investigative program, its main premises and theoretical foundations were summarized in my discussion of specificity at the Growth Symposium in 1945. Purely as a guide in planning further research and in trying to reconcile the paradoxical results of the past, I have adhered to the particular concept of growth control that I had outlined previously (Weiss, '47, p. 272-273; '49, p. 180-181). Its pragmatic value has proven itself by bringing disparate results to a common denominator and by suggesting the design of new experiments reported further below.

V. A CONCEPT OF SPECIFIC GROWTH CONTROL

This concept is based on the following suppositions.

(1) Each specific cell type reproduces its protoplasm, i.e. "grows," by a mechanism in which key compounds that are characteristic of the individual cell type act as catalysts. The postulated cell-specific diversity of compounds is the chemical correlate of the "differentiation" of cell strains. Growth rate is proportional to the concentration of these intracellular specific catalysts (or "templates") in the free or active state. Under normal conditions these compounds remain confined within the cell.

(2) Each cell also produces compounds ("antitemplates") which can inhibit the former species by combining with them into inactive complexes. These may be turned out as direct by-products in the process of protoplasmic reproduction or be secondary differentiation products. They may be steric complements to the former or matched to them in some other fashion. The only prerequisites are: (a) that, contrary to the specific templates, they are released from the cell and get into the extracellular space and into circulation; (b) that they carry the specific tag of their producer cell type which endows them with selective affinity

[198]

for any cell of the same type; and (c) that they are in constant production so as to make up for their extracellular decomposition and final excretion.

(3) As the concentration of "antitemplates" in the extracellular medium increases, their intracellular density, hence inactivation of corresponding "templates," will likewise increase; in short, growth rate will decline in all cells belonging to that particular strain bathed by the common humoral pool. When stationary equilibrium between intracellular and extracellular concentration is reached, growth will cease. This mechanism results in a sigmoid growth curve for the total mass of each organ system (see Morales and Kreutzer, '45), and the familiar sigmoid curve for the whole organism would essentially be an aggregate of similar curves for the individual constituent organ systems.

This general concept offers a rational explanation for both the self-limiting character of growth in a confined medium (organism or culture) and the homologous organ-specific growth reactions after experimental interference. As can readily be seen, each interference will have to be examined in a dual light as to its effects on the concentration of both "templates" and "antitemplates," since it is the ratio of both that determines growth rate. The following conclusions can immediately be deduced from this scheme.

(a) Removal of part of an organ system removes part of the sources of the corresponding types of "templates" and "antitemplates." Since the former, according to our premise (1), have been in intracellular confinement, neither their former presence nor their recent loss are perceptible to other cells of the system. This is not so for the "antitemplates," which are in circulation and a reduction of whose production source would promptly be recognized by their lowered concentration in the extracellular pool. According to points (2) and (3) this would shift the intracellular ratio of "templates" to "antitemplates" temporarily in favor of the former, causing automatic resumption of growth till a steady state is restored—to all intents a "compensatory" growth reaction.

(b) Addition of a part should have opposite effects depending on whether or not its cells survive, or rather on the ratio of surviving to disintegrating cells. If all cells survive, the net effect would be an increased concentration in the circulation of the particular "antitemplates," hence a reduction in growth rate of the corresponding host system, provided it is still in a phase of growth (actual regression after growth has ceased need not be expected). On the other hand, cells that disinte-

[199]

grate release into the extracellular space a complement of specific "templates" that would otherwise never have escaped. Assuming that these, according to point (2), combine with or otherwise trap homologous "antitemplates," their presence in the pool will entail a temporary lowering of "antitemplate" concentration—hence again a spurt of growth in the homologous cell strains of the host. The simultaneous release of "antitemplates" from the disintegrating cells would have to be assumed to be insufficient to cancel this effect because of their faster metabolic degradation (see point 2c). An alternative possibility is that "templates" freed from cracked cells are directly adopted by homologous cells, where they would temporarily increase the intracellular concentration of growth catalysts—hence growth rate. In either scheme the release of cell content would accelerate homologous growth by increasing the intracellular ratio of "templates" to "antitemplates"—in the former case by reducing the denominator, in the latter case by increasing the numerator. It can be seen that in terms of this interpretation partial necrosis of an organ will have the same effect as partial removal, and that implantation of a fragment followed by some degeneration, as well as the injection of cell debris, are merely further variants of the same procedure.

To test the validity of this concept the following series of experiments were undertaken. In contrast to our earlier attempts, the assay of growth responses by measurements of size attained after a given period was abandoned as too unreliable; not only do such measurements fail to distinguish between the specific components of an organ and its content of connective tissue and blood, but also initial growth reactions can easily be missed due to secondary regulations or even overcompensations. Instead the mitotic index was introduced as a more sensitive and reliable criterion. A large series of experiments on the stimulation of mitotic activity in amphibian skin (Weiss and Overton; mostly still unpublished) has conclusively shown that cell division is secondary to cell growth; hence, if present it can be used as an index of protoplasmic increase. The same conclusion can be drawn from the precession of the increase of mass over that of cell number in liver regeneration (Brues, Drury, and Brues, '36).

VI. COMPENSATORY HYPERPLASIA WITHOUT FUNCTIONAL OVERLOAD

One of the inferences from our theory is that the "compensatory" growth of one member of a pair of organs after the removal or destruction of the other would be attributable to the disturbance of the described

[200]

chemical equilibrium rather than to the burden of augmented functional activity. In view of the well-established compensatory hypertrophy of the remaining kidney after unilateral kidney removal, this experiment was repeated with the embryonic metanephros of the chick at a stage prior to the onset of its excretory function, the latter function being still fully exercised by the mesonephros. Wayne Ferris in our laboratory cauterized one metanephros in 12- to 13-day embryos and counted the mitotic response in the undamaged residual kidney fixed within 2 days after the operation. On the basis of a total count of 12,000 mitoses in 12 controls and 15 experimental cases, 4 of them sham operations, an average increase of 70 per cent was noted on the unharmed side that could be definitely identified as a response to the destruction of kidney tissue rather than to injury as such. In fact the effect was confined to the specific epithelium while the connective tissue stroma remained unaffected.

This demonstration of direct compensatory growth reactions resulting from disturbance of the intracellular-extracellular balance of complementary organ-specific compounds in no way rules out the occurrence of true "functional" hypertrophy as a result of overload; the relative roles played by the two processes will presumably vary from object to object. Moreover, in the endocrine system additional compensatory regulations arise from the reciprocity of hormone relations between different glands.

The direct chemical balance reaction illustrated above may turn out to be a ubiquitous and general principle to which functional and hormonal effects would be merely superimposed. Several considerations indicate its general validity. For instance the observation of an increase in the undamaged liver of one partner of a pair of parabiosed rats following partial removal of the liver of the other partner (Bucher, Scott, and Aub, '51; Wenneker and Sussman, '51) lends itself to the same interpretation.[2] In fact the correlation of liver regeneration rate with blood flow (Flores, '52) and the increase of this rate when the blood is diluted (Glinos and Gey, '52), thus reducing the concentration of our hypothetical liver-"antitemplates" in the circulation, add further support to our interpretation.

---

[2] Further confirmation has recently been produced by Friedrich-Freksa and Zaki (*Ztschr. Naturf. 9b,* 394-397) who found mitotic spurts in the livers of normal rats injected intraperitoneally with serum from partially hepatectomized animals.

[201]

III. On Growth and Aging   315

VII. INJECTION OF CELL DEBRIS

Homologous growth effects were also obtained by the direct injection into the embryonic blood stream of triturated cell masses from kidney or liver, either fresh or after freezing and thawing. These experiments, carried out by Gert Andres in our laboratory (in press in *Journal of Experimental Zoology*) and involving a count of 86,000 mitoses, demonstrated that the mitotic ratio of host kidney to host liver is within a day significantly ($P < 0.001$) raised by the injection of kidney material, and lowered by liver material, each organ debris exerting an homologous effect.

Comparable results on a smaller scale have been reported for the orbital glands of rats following intraperitoneal injection of homologous extract (Tcir, '52).

VIII. HOMOLOGOUS ORGAN EXTRACT IN TISSUE CULTURE

In the light of our theory, the common observation of a "growth-stimulating" effect of embryo extract on the proliferation of tissue cultures would be accounted for by the fact that, since embryo extract contains cell debris of all organs, the growth of any tissue explanted in it would be favored. In order to put this contention to a test, large series of tissue cultures of kidney and heart were set up in paired media, one containing extract of the complete embryo, the other, extract from which the homologous organ was omitted. The experiments, carried out with Ilse Fischer, showed remarkable differences between the two sets of conditions.

In the kidney experiments (6,300 cultures of 12-day mesonephros and 2,335 cultures of 9- or 17- to 20-day metanephros), the frequency of tubule differentiation was used as a criterion. In the presence of kidney extract this frequency was greatly reduced. Since differentiation in tissue culture is generally conceded to be in some inverse relation to the intensity of proliferation, this result could be interpreted as an homologous growth stimulation by the kidney debris in the medium.

In the heart cultures the differentiation of new myofibrils, evidenced by continued pulsation in successive transfers with subdivision, was used as a sign of depressed growth. In a first series of 978 paired cultures, with and without extract of 5- to 6-day hearts, a large preponderance of pulsation was found in the absence of heart extract, signifying presumably reduced growth. In a later reinvestigation (with Margaret W. Cavanaugh) it was noted, however, that the effect could not be

[202]

definitely established unless the embryo extract was from embryos older than about 9 days. After this time it made a great deal of difference for heart cultures whether or not heart extract was present in the medium. Such an age effect was previously reported by Gaillard for tissue cultures in general and by Ebert ('51) for the growth stimulation of spleen by chorio-allantoic spleen grafts. Possibly the "growth-promoting" potency ascribed by Hoffman and Doljansky ('39) to heart extract can be related to the fact that their standard assay objects were heart fibroblasts.

An extension of our experiments to tissue cultures of skin and thyroid has thus far given no comparable results, presumably because of the lack of sharp criteria. Even so, the use of tissue culture for the further analysis of these homologous growth interactions between specific cell types and corresponding cell constituents in the culture medium seems to hold much promise. Whether cell compounds in the extract promote homologous growth by being directly incorporated into the corresponding cells, or by neutralizing homologous growth inhibitors in the medium, is still unresolved.

IX. IMMUNOLOGICAL EXPERIMENTS

After these varied reconfirmations of the effect of cell extracts on homologous growth, it became particularly intriguing to resume our original attempts to secure similar effects with antibodies against specific organ extracts. As mentioned above, we had found livers and kidneys of chick embryos, treated with liver- or kidney-antisera, respectively, to be significantly larger than after treatment with non-homologous antisera. A repetition of the liver experiments confirmed the size increase but proved it to be due to hemorrhages from specifically damaged liver vessels. The homology of the action is still evident but its relation to growth is unsubstantiated. Injection of anti-lens serum likewise failed to produce any appreciable effect on the growth of the host lenses (work with Audrey Peterson).

After these failures to detect specific growth alterations from anti-organ sera, it seemed necessary to turn back one step and ascertain whether or not at least the preferential incorporation of anti-organ sera in homologous embryonic organs could be proved. Immunological reactions of specific embryonic tissues to homologous antibodies have been demonstrated (Burke et al., '44; Ebert, '50; Grunwalt, '49), and it was rather obvious to try to demonstrate the selective absorption directly by using isotope-tagged antibodies. Unfortunately five years of continued efforts in that direction have failed to produce the expected re-

[203]

sults (immunological work with the aid of D. H. Campbell, California Institute of Technology, and Robert Petzold; experiments and isotope assays aided by Gert Andres, Howard Holtzer, Margaret W. Cavanaugh, Evelyn R. Mills, and James Lash).

Antibodies tagged with C_{14}-glycine injected into the yolk sac were concentrated in the embryo, but proved to be of too low titer and radioactivity for a test of selective distribution. We then turned to purified antibody preparations (gamma globulin fraction) tagged with I_{131} according to Pressmann and injected directly into embryonic veins by the technique of Weiss and Andres ('52). Radiation assays proved that the injected material was differentially distributed throughout the embryonic tissues, with blood, heart, and kidneys showing the highest concentrations and other organs following in a certain order. However, there was no evidence of greater absorption of a given organ-antibody by the homologous embryonic organ. It remains undecided whether these failures are due to technical imperfections or actually prove that organ anti-sera are not absorbed in demonstrably larger quantities in the homologous cells of the embryo. It is quite possible, of course, that if growth promotion by these antibodies were of a catalytic nature, even amounts too small to be detectable could exert potent effects.

X. CONCLUSIONS

From this brief survey it seems that the existence of a cell-type specific chemical mechanism of correlating growth processes among homologous cell types must not only be postulated but also also may be regarded as conclusively demonstrated. As for the nature of this mechanism, my own concept or theory, advanced on previous occasions and reiterated here, has proved its value as a guide but must not be taken to have been either proved or disproved. In its current form it is probably too simple to be wholly correct; further work will amend or even replace it. But the general idea of selective chemical communication among cells of identical types by direct exchange of protoplasmic type-specific compounds can hardly be questioned in view of the large evidence in its favor. My earlier detailed suggestion that the complementary systems of growth-catalyzing and growth-repressing compounds are of the antigen-antibody class has found no direct support in our further immunological studies; but it has not been definitely ruled out. Immunoembryology has made vigorous strides of late, but most of that work has been devoted to the detection and tracing by immunological techniques of the appearance of certain antigens during differentiation, e.g. Cooper

[204]

('50), Woerdemann ('53), ten Cate ('50), Schechtmann ('52), and Clayton ('53), rather than to the possible instrumental role of these systems as determining and regulatory factors, as I had originally proposed. This latter view, also adopted by Tyler ('47) in an extension of his earlier ideas on "autoantibodies" to problems of growth, therefore remains open to question, but it still deserves to be kept in mind, if only as a model.

BIBLIOGRAPHY

Arataki, M. 1926. *Am. J. Anat. 36,* 437-450.
Brues, A. M., D. R. Drury, and M. C. Brues. 1936. *Arch. Path. 22:* 658-673.
Bucher, N. L. R., J. F. Scott, and J. C. Aub. 1951. *Cancer Res. 11,* 457-465.
Burke, V., N. P. Sullivan, H. Peterson, and R. Weed. 1944. *J. Infect. Dis. 74,* 225-233.
ten Cate, G., and W. J. Van Doorenmaalen. 1950. *Proc. Kon. Ned. Ak. Wetensch. 53,* 894.
Clayton, R. M. 1953. *J. Embryol. Exp. Morph. 1,* 25-42.
Cooper, R. S. 1950. *J. Exp. Zool. 114,* 403-420.
Danchakoff, V. 1916. *Am. J. Anat. 20,* 255-327.
Ebert, J. D. 1950. *J. Exp. Zool. 115,* 351-378.
Ebert, J. D. 1951. *Physiol. Zool. 24,* 20-41.
Flores, N. 1952. *C. R. Soc. Biol. 146,* 589-591.
Glinos, A. D., and G. O. Gey. 1952. *Proc. Soc. Exp. Biol. Med. 80,* 421-425.
Golgi, C. 1882. *Arch. ital. biol. 2.*
Greenman, M. J. 1913. *J. Comp. Neurol. 23,* 479-513.
Grunwaldt, E. 1949. *Texas Reports on Biol. Med. 7,* 270-317.
Guyer, M. F., and E. A. Smith. 1918. *J. Exp. Zool. 26,* 65-82.
Haasler. 1891. *Centralbl. allg. Path. path. Anat. 2,* 809.
Hoffmann, R. S., and L. Doljanski. 1939. *Growth 3,* 61-71.
Kollros, J. J. 1940. *J. Exp. Zool. 85,* 33-52.
Köster, G. 1903. *Neurol. Centralbl. S. 1093.*
Morales, M. F., and F. L. Kreutzer. 1945. *Bull. Math. Biophysics 7,* 15-24.
Nittono, K. 1923. *J. Comp. Neurol. 35,* 133-161.
Pressman, D. 1949. *Cancer 2,* 697-700.
Przibram, H. 1907. *Roux' Arch. 25,* 266-343.
Ribbert, H. 1895. *Roux' Arch. 1,* 69-90.
Ribbert, H. 1904. *Arch. f. Entwicklgmech. 18,* 267-288.
Robertson, O. H. 1954. In publication; personal communication.
Rollason, H. D. 1949. *Anat. Rec. 104,* 263-285.
Schechtman, A. M., and H. Hoffman. 1952. *J. Exp. Zool. 120,* 375-390.
Smadel, J. E. 1936. *J. Exp. Med. 64,* 921-942.
Tamaki, K. 1936. *J. Comp. Neurol. 64,* 437-448.
Teir, H. 1951. *Commentationes Biol. 13,* 1-32.
Teir, H. 1952. *Acta pathol. et microbiol. Scand. 30,* 158-183.
Tyler, A. 1947. *Growth 10* (suppl.), 7-19.

[205]

Weiss, P. 1939. *Anat. Rec. 75* (suppl.), 67.

Weiss, P. 1947. *Yale J. Biol. Med. 19*, 235-278.

Weiss, P. 1949. *Chemistry and Physiology of Growth,* pp. 135-186, Princeton Univ. Press.

Weiss, P. 1952. *Publ. Ass. Res. Nerv. Ment. Dis. 30,* 3-23.

Weiss, P., and G. Andres. 1952. *J. Exp. Zool. 121,* 449-488.

Weiss, P., and H. Wang. 1941. *Anat. Rec. 79,* 62.

Wenneker, A. S., and N. Sussman. 1951. *Proc. Soc. Exp. Biol. Med. 76,* 683-686.

Willier, B. H. 1924. *Am. J. Anat. 33,* 67-103.

Woerdeman, M. W. 1953. *Arch. Neerland. Zool. 10* (suppl.), 144-162.

Zeleny, Ch. 1902. *Roux' Arch. 13,* 597-609.

Zeleny, Ch. 1905. *J. Exp. Zool. 2,* 1-102.

Reprinted from
THE ART OF PREDICTIVE MEDICINE
By WEBSTER L. MARXER, M.D.
and GEORGE R. COWGILL, PH.D., SC.D. (HON.)
CHARLES C THOMAS, PUBLISHER
Springfield, Illinios, U.S.A.

CHAPTER 13

DETERIORATION IN CELLS[*]

PAUL WEISS[†]

WE REALLY DON'T KNOW enough about cells to talk conclusively about deteriorative trends in them. But we *do* know enough about cells to know that some of the things said about them with regard to their true nature and mechanisms are quite unrealistic. Some of the naive notions current in present thinking, and partly also in research, could readily be, if not put back on the right track, at least blocked on the wrong track. In that regard I could give you a long list of misconceptions that can already be rectified in the light of what we do know about cells and an infinitely longer list of what we don't know about cell life.

In an essay which I sent in prior to this meeting, I put down a brief summary of generalizations to prove that the "generalized" cell of which we carry a rigid mental picture does not exist. Cells, just like individuals, fall into different classes possessing crucially different properties, which one must learn empirically and can't deduce. We can't deal with cells across the board as if they were

---

[*] Work supported in part by Grant No. CA-06375 from the National Cancer Institute (National Institutes of Health of the United States Public Health Service).

[†] At the time of the Symposium, Dr. Weiss was a member and professor of the Rockefeller Institute, New York, New York. On October 1, 1964, he became Dean and University Professor of the Graduate School of Biomedical Sciences of the University of Texas, Houston, Texas.

all identical, like tin soldiers. Cells differ constitutionally according to the types to which they belong. Moreover, one and the same cell changes from moment to moment; and furthermore, every cell individual of a given type is somewhat different from every other individual cell, even of the same type. Therefore, there is a three-dimentional scale of variability implied in the abstract *cell*.

Among the changes of cells from moment to moment, some are repetitive and cyclic. Such fluctuations are only ripples on a basic carrier wave which is not cyclic and recurrent but which leads steadily, definitely and inexorably to the eventual termination of the life of that individual cell, either by death or by dividing.

Enlarging on these premises, we arrive at the following conclusions.

1. Since constitutional properties are totally different among different cell types, no single cell type can automatically serve as a model for the rest.

2. It is particularly important to distinguish between the cells that are permanent individual residents of the mature body (e.g., nerve cells, muscle cells, bone cells) and those that are being continuously or periodically reproduced (e.g., bone marrow, skin, intestinal epithelium). Progressive changes in the former class afflict cell *individuals,* in the latter class, cell *strains.*

3. All cell classes may undergo changes which impair the optimal functioning of the body. Whether such changes originate in single cells, cell groups or systemically throughout the body, whether they are due to extraneous influences or to the inevitable internal wear-and-tear, and whether they remain locally confined or involve ever-wider areas, in no instance will their effects on the organism become detrimental unless and until they have added up to a critical momentum at which the *harmony* and *coordination* within the cell *population* of the body is seriously disturbed. In short, except for irreparable genetic deficiencies, the organism suffers telling damage only from critical *disharmonies* in intercellular *relations.*

4. Short of that critical level, there is so much plasticity and power of compensation, adaptation and repair latent in cells and

tissue systems that the equilibrium of the organism is essentially safeguarded and preserved despite the innumerable deviations caused by internal and external contingencies to which the members of the cell population are continually subjected.

5. In view of these facts, positive anti-deteriorative measures must center on *optimizing the multiple natural conservative and restorative devices of the body.* This implies: (a) the development and widening of the adaptive range of each body constituent by constant practice within stress limits; (b) the detection and identification of potentially hazardous tissue disharmonies early enough to be averted or counteracted; (c) after irreversible impairment of one part or function of the system has occurred, the maximizing of the faculties of substitution, vicariousness and compensation of the remaining parts and functions of the system.

Aside from these generalities, the type of measure will vary not only from tissue to tissue, but also with the degree of our knowledge of the properties of the various cells, tissues and organs and of their interrelations. This is at present the most serious bottleneck. Pathology deals chiefly with those later stages of tissue alterations which are detectable by the relatively crude means at our disposal, that is, with those late stages which countless adverse modifications arising from subtle fluctuations of "normal" cells have piled up to the point where they yield a grossly disharmonious situation. The statistical emergence of the "abnormal" from "normal" variability is largely still a twilight zone in which more research is urgently needed. For it is in this area that we can learn how to exploit "normal" variability selectively for the opportunities it offers for improving the body's resistance to deterioration, in contrast to letting them build cumulatively toward pathological results. The full range of our power to influence the fate of the various cell populations for better or worse is barely realized, let alone taken advantage of.

To quote examples from three different cell types, let us first look at the nonreproductive cells of the central nervous system—the neurons. Even though in higher mammals, including man, multiplication of nerve cells ends shortly after birth, each nerve cell renews its own substance constantly at a very fast rate—the cell body synthesizing a mass of nerve substance equal to its own volume about once every day. The cell individual is continuously regrowing itself around the nucleus, moving the newly synthesized products into its nerve fiber, in part to replenish metabolic loss, in part for export into muscles and sense organs. We know and we understand now that the rate of synthesis in these cells varies quantitatively with the load of functional requirements placed upon it in both the ascending and descending directions; it can, perhaps, even adjust itself qualitatively to specific influences, as in the phenomena of learning and of neural sensitization and idiosyncrasies. The reader should realize that the nerve cell is not a static body built by the embryo once and for all, to function henceforth statically like Atlas carrying the globe on his shoulders, but is in a state of constant dynamic renovation. Also many nerve cells wear themselves out and die during the life of the individual. We are losing about 10^3 nerve cells every day, irreparably, for post-embryonic nerve cells can no longer reproduce; they do not undergo mitotic division. But, since we have 10^{10} or 10^{11} cells in our brain, the total loss in a lifetime remains well below 1 per cent. Since the rate of internal self-renewal, and hence the vigor of the nerve cell, has been shown to vary with the functional load to which the neuron is exposed, the degree and even kind of activity with which we utilize our nervous system assumes major importance as a deterrent to the degenerative decimation of its cell population.

There are other cell types, however, in which it makes no sense to talk about the aging of the cell as an individual. The intestinal cell has an average lifetime of about a day or two, and then it is shed. With such rapid population turnover, it doesn't make any difference how long it would take for the individual cell to die. The skin epidermis has a somewhat longer life expectancy, but still one that measures only in terms of weeks. Again, blood cells of hemopoietic organs in general have different life spans, but still of limited duration. In those cases, therefore, deteriorative trends pertain not to single cells but to the whole lineage, the whole strain.

Now, in a proliferating strain of this kind, there occur changes that remain confined to a given cell individual; they are not necessarily harmful, because that individual won't live long, and they may be wiped out as soon as it divides into daughter cells. But there are other changes that may leave behind residual alterations. If there is such a residual change, then it may in turn be cancelled out in the next cell division by a change of opposite sign. However, if there is a buildup of synergic changes in successive cell generations, of which there are hundreds and thousands in our lifetime, as a result of which minute, almost infinitesimally small deviations become cumulative and pile up, then there may come a time when trouble arises. Such trouble may arise from a cell becoming nonfunctional, hypo-functional or hyper-functional, so that it no longer matches other cells that depend on it reciprocally. So it is a breakdown, not necessarily of the vitality of a given cell strain, but of the *harmonious relations among the various cell strains* of the body, which causes the detectable trouble.

Another illustration is the cellular basis of the immunological defense mechanisms of the body. Even though the individual antibody-forming cells have limited life spans, the phenomena of both acquired immunity and acquired tolerance have demonstrated the perseverance of chemical "memory" traces in the respective proliferating cell strains, which again emphasizes the importance of appropriate cell training (e.g., the epidemiologically demonstrated value of subclinical infections).

The third example pertains to intercellular products, which form the bulk of the connective and skeletal tissues. They consist of a fabric of collagen fibers embedded in mucopolysaccharide ground substances, with or without calcareous incrustations. Subserving mainly the mechanical functions of transmitting and withstanding pressures and tensions, their architecture is qualitatively and quantitatively adapted to the patterns of strains and stresses operating upon them. The molecular and cellular mechanisms involved in their adaptive reactions are among the best understood "functional adaptations." Excessive stress may entail rupture. But failure to exercise them contributes demonstrably to their atrophy and disorganization. Stress, mechanical stress, is absolutely essential, not only for the production of the tissue matrix, but particularly for its organization, for its orderly texture and architecture. The same amount of tissue mass can appear in a functionally useful pattern or in a wholly inadequate form. A rope that has flaws will break, and this is precisely what can happen in the connective tissues if a tendon, for instance, isn't constantly kept in a certain moderate state of activity. Here, likewise, we know that both over- and under-action will be equally harmful. The normal structural properties of the tissue are designed for an intermediate range, beyond which lie the hazards of harm by too little or too much.

At this juncture, let me again call your attention to the mutuality of relations among cells. Earlier in this Symposium, the trophic effects of the peripheral nervous system on tissues were mentioned. There is also an unexpected reciprocal effect of connective tissue on nerves but with an opposite sign. That is, hypertrophy of the connective tissue can encroach on the health of nerve fibers. It can clamp down on them and interfere with their functioning and survival, so the hypertrophy of one type of tissue can lead to the atrophy of another type. It is this kind of reciprocal relations which, given time, I could illustrate by numerous examples. Fortunately, our body has also built into it innumerable mechanisms that stabilize the relations among the tissues. We are beginning to learn about them, but as we still don't know much about them, an enormously wide field is open for future research.

All of these examples, which could be amplified, carry one basic lesson. This is: Regardless of the specific manner in which cells react to their environment and to each other according to their types, they are all endowed with a certain adaptive latitude which represents the margin between superior and inferior performance, vitality and endurance. To learn to keep the net balance of the total complex network of cellular activities as potent as possible within that range is the most natural method of minimizing the deteriorative trend in tissues; the trend itself, however, is a natural and inescapable phenomenon which can only be retarded and mitigated but never abolished. Intensive study of cell biology will have to reveal how individual cells can be stimulated or trained to display their maximum desirable potentialities and how cell strains in continuous proliferation can be influenced so that the more desirable combinations of their offspring will be given a selective advantage over the less desirable ones.

Some practical studies involving cell biology occur to me as worthy of mention here. If we knew a little more about the subtle long-range changes in cell populations with age—and this may come soon—then we could do more in the way of using sample cells from the body of a patient for actual diagnostic assays. To some extent we do this now, but we could do much more. One could, for instance, easily scrape cells from the mucous membranes of the oral cavity, as well as from various other parts of the body, without harm. Studies of such cells, in addition, of course, to determinations of the composition of the blood plasma etc., might furnish data valuable as indices of general deteriorative status. In the longitudinal study-project which we have been asked to consider, storage of such information from presumably healthy people in a computer-setup would permit it to be used later for comparison along with data pertaining to many other parameters. This information could prove to be very significant in the detection of important "predictive parameter patterns." This is a completely undeveloped field.

Similar external tests are provided by the characteristic ridges of fingernails, which are beautiful recorders of rhythmic or arhythmic fluctuations in our metabolic and hormone activity. This field is likewise waiting for systematic investigation. It seems to offer hints of things to come. A large amount of information might presumably be read from these self-registering records. Perhaps it too could prove useful for the "parameter patterns" we hope to discover.

In mentioning these possibilities, I am simply calling your attention to the fact that here lie areas which are not the ambitious glamorized fields in the life sciences, but which are perhaps infinitely more useful in the practical sense and which should be entered by investigators with imagination. I wish we could persuade more of the young people to go into these fields; they would reap an immensely rich yield of useful knowledge.

In conclusion, I would like to point out that an organism does

III. On Growth and Aging 327

not live in a vacuum but in a changeable environment, of which the balance sheet of optimal conditions cannot be stated except in relation to man's actual environment, both physical and social. Therefore, to combat deteriorative trends in the body rationally and methodically, the conclusions and lessons of cell biology, organismic biology and environmental biology will have to be drawn upon conjointly. This calls for immensely intensified work in all these branches and for continued integrated evaluation of their results. This endeavor is still in its infancy.

Reprinted from, PERSPECTIVES IN EXPERIMENTAL GERONTOLOGY, SHOCK
CHARLES C THOMAS · PUBLISHER

CHAPTER 14

AGING: A COROLLARY OF DEVELOPMENT

PAUL WEISS

PREAMBLE

THIS ESSAY is written in full realization of its inadequacy. Yet a request to contribute to a volume commemorating the eightieth birthday of a friend committed to the study of gerontology, whose very youthfulness belies the stigma of the aging process, could not be declined simply for lack of qualification. A commentary by one who has never studied specifically the later part of the life span may seem presumptuous. Yet, it may not be wholly unwarranted to try to raise the sights from specific factual data to the broad overall perspective of biological processes in general from which they must be viewed and rated. And viewed from that perspective, the *aging* process, so called, appears but as a conveniently, but arbitrarily, delineated aspect of the process of *development*. The purpose of the following brief essay, therefore, is to place it back into that context.

Development consists of a continuous succession of processes of change and transformation, going on incessantly in uninterrupted sequences throughout the life span of an individual from egg until death. True, the rate of change varies markedly for different periods, and from tissue to tissue, and the *average* of these kinetics taken over the body as a whole can be said, in general, to decline with time. However, those changes of kinetics which strike the superficial observer or the self-observing subject as sudden and abrupt, reveal themselves on closer inspection as continuous and gradual, without the sharpness of demarcation which our conventional terminology seems to denote.

In this light the aging process appears as an integral facet of the continuous progress of development of an organism. It cannot be dealt with as if it were a separate and separable encumbrance, superimposed upon the ordinary processes of life—a sort of extraneous contamination, like a bacterial infection, that could be stripped off film-like once it has formed, or hopefully, be totally averted. Aging is not only with Minot (1908) the price we pay for our developmental differentiation: it *is* differentiation. On these grounds, I have given it at least "honorable mention" in my text on *Principles of Development* (1939, pp. 27-35), and nothing discovered in the quarter of a century and more since has weakened the argument.

A BRIEF SYNOPSIS OF DEVELOPMENT

The more we learn about development, the harder it becomes to compress its essence into a brief essay. The following synopsis, therefore, is crude and sketchy. Its main objective is to give concrete meaning to the loose term "differentiation," the substance of which is the source of "aging."

The egg, as the link between successive generations, reflects this role in its dual constitution: as both a highly specialized cell of the somatically differentiated maternal gonad and the germinal primordium of the new offspring. It carries over, accordingly, two sets of properties. One is the set of chromosomes containing the genome, now identified with specifically arrayed sequences of nucleotide pairs in strands of DNA, which at fertilization are combined with the paternal genome. The other set is the map of cytoplasmic organization which the egg derives directly from its residence in the ovary. Although this cytoplasmic pattern has often been eclipsed by monopolistic attention to the genome, it is important to restore it to its determinant role as a primordial framework of organization with which the genome is bound to interact. Its firm physical bearings lie presumably in a surface mosaic in the egg cortex, in which fields of different chemical composition and physical constitution are mapped out in typical configuration. During the subsequent cleavage of the egg by cell division, the various blastomeres thus receive disparate portions of this surface mosaic while all of them receive nuclei possessing the same full genic complement. As a result, nuclei that are erstwhile equivalent and interchangeable come to lie, from the very first, in different cytoplasmic environments. This sets the stage for *differential* nucleo-cytoplasmic interactions in different regions of the germ. Each cell genome thus is exposed from the very start to a system of different conditions arrayed in a specific pattern which is as much a part of the basic blueprint of the new organism as is the orderly array of genes. In further consequence, the interactions between the two systems yield different products and diverse effects in the different regions of the germ.

The evidence that the genome as such remains throughout life essentially identical in all cells of an organism (except for sporadic somatic mutations) is compelling. Yet, despite this essential constancy of their genic content, cell lines diverge progressively in the development of higher organisms. It is this process of diversification—"differentiation" in the strict sense—which is basic to our understanding of aging, as follows.

The brilliant progress of molecular cytogenetics, resting heavily on microbial evidence, has led to the concept of a direct transcription and subsequent translation of the DNA code, through various RNA intermediates, to the orderly sequence of amino acids in the synthesis of cell proteins. Since in bacteria the generative and somatic cell is one and the same, bacteria obviously cannot furnish a model for the puzzling problem of differentiation in higher forms. For in the latter, a great diversity of qualitatively different cell types emerge during ontogeny. Just how many, is indefinite. The only sure fact is that estimates based on morphological criteria vastly underrate the real magnitude. For a realistic perspective, the reader may be referred to an earlier summary account (Weiss, 1953). A sharp distinction must be made between the differentiation of an individual cell ("cytodifferentiation") and the differentiation among cell *strains*. Cytodifferentiation refers to the elaboration by a given cell, mostly the terminal descendant of a cell line, of specialized characters or products by which we have come to identify that cell as belonging to a given type; such critical products are, for instance, myofibrils, secretions, blood pigments, antibodies, pigment granules, and so forth. *Strain* differentiation, by contrast, denotes the splitting of the progeny of cells of common origin into branch lines of different "potencies," that is, restricted repertories of performance and production faculties differing qualitatively from branch to branch. The fact that the branch lines are capable of passing on their differentials, once established, to their descendants indefinitely even in a common indifferent environment (e.g., in tissue culture) proves that strain differentiation connotes some indelible change which can be propagated through many cell generations without attenuation.

Admittedly, we still miss the key to the explanation of either cytodifferentiation or strain differentiation, although the former is beginning to yield. In each cell, only a selected fraction of the full complement of genes which it contains is supposed to be active in a given instant, the rest being effectively blocked (the blocking effect recently ascribed to histones). This evidently presupposes a rather well-structured *extragenic* machinery to do the proper "activating" and "blocking." Such a view is perhaps conceivable as long as one focuses on a single *individual* cell in cytodifferentiation. But its extension to *strain* differentiation presents difficulties. Consider that a thyroid cell can either grow and divide, giving rise to another thyroid cell, or alternatively switch to manufacturing thyroxin and colloid. The same extragenic models (or "templates," as I have termed them) would therefore at one time have to induce the genomes to fashion more of their own kind needed for further self-perpetuation of that special cell line while at another time, they would have to trigger the machinery for the synthesis of a special product. It seems that cytoplasmic RNA units are somehow involved in the extragenic template systems, but if so, one must postulate that they come in as many distinct molecular forms as there are self-perpetuating types of differentiation. The immense diversity of differentiations seems to militate against explanations of differentiation in terms of somatic mutation or of induced enzyme synthesis.

To put it candidly, the detailed mechanisms of differentiation are still wholly conjectural. Nevertheless, some general rules which bear on the aging problem have emerged. They can be summarized roughly as follows.

1) *The Egg Mosaic*

For brevity, we lump all intracellular, but extragenic, matter into a single compartment. The primary topographic pattern of cytoplasmic egg districts, outlined above, places the equal genomes of cleavage nuclei into preformed unequal intracellular environments. The ensuing "epigenetic" interactions between each genome and its local environment elaborate these primordial differences. Thus, from the very start, the genome is a captive of its intracellular environment. Both are in "cross talk": the various environments "demand" and each genome "responds" with a matching answer provided for in its "code," thus altering its own intracellular environment. Where this process is slow relative to the rate of cell division, a multitude of cells of comparable constitution—identical genomes in identical environments—arise. The subsequent, strictly "epigenetic," differentiation among their progeny, however, poses a new problem.

2) Epigenetic Differentiation

The answer lies in the group dynamics of the cell population, which creates secondary inequalities among the erstwhile equal members of a group. As the extragenic *intra*cellular environment determines the genic response, so the *extra*cellular environment, in turn, affects the intracellular system in various degrees of specificity. For any given cell, all other cells are part of the "extracellular environment." The interactions between contiguous cells (by contact) or distant cells (through diffusing products) thus introduce a new major diversifying principle. When interactions between two cell types, though in theory mutual, are conspicuously unilateral, they are often referred to as "inductions."

3) Nature of Cell Interactions

The kinds and modes of operation of cell interactions are too varied to submit to any common formula (Weiss, 1963). They may involve transfer of substances, electric polarization, mechanical deformations, osmotic or pH changes, activation or stoppage of movements or of energy delivery, redistribution of cell content, and above all, such specific changes in the *intracellular* environment as may evoke an altered response from the genome.

4) Progressive Dichotomous Differentiation

Schematically, therefore, if we designate the genome as **G**, its intracellular environment as I, and the extracellular environment (e.g., a cell of another type) as E, the following chain of events takes place:

$$I_1 \to G \to I_1; \quad E \to I_1 \to I_2; \quad I_2 \to G \to I_3; \quad E \to I_3 \to I_4; \quad I_4 \to G \to I_5; \text{ etc.}$$

Note that E and G can communicate with each other only indirectly through the mediation of I, in which G lies entrapped. If, in addition, we let E go through a series of modifications E', E'', E''', $\cdots$, we gain a realistic, if diagrammatic picture of differentiation as a stepwise process of progressive diversification inasmuch as the internal changes in one cell create a changed external environment for its neighbors. The stabilization of self-perpetuating cell types $I_n \rightarrow G \rightarrow I_n \rightarrow G \rightarrow I_n \cdots$, mentioned earlier, further complicates the picture.

This mode of development contains several potential sources of aberrations. 1) G in a particular cell line may suffer a somatic mutation to G'; the whole descendant cell generations will then show altered behavior. 2) The process $I_1 \rightarrow I_2$ might deviate to $I_1 \rightarrow I_2'$, leaving two possibilities: a) $I_2' \rightarrow G \rightarrow I_3$ (the genome failing to discriminate between I_2 and I_2' as if $I_2 \sim I_2'$); the error would then remain confined to the affected single individual cell without entailing lasting effects on the descendants. b) $I_2' \rightarrow G \rightarrow I_3'$; consequently, $I_3' \rightarrow G \rightarrow I_3' \rightarrow G \rightarrow I_3'$, etc.; the error would thus be perpetuated throughout subsequent cell generations. c) $E \rightarrow I_3' \rightarrow I_4'$; this would in general aggravate the deviation. d) $E \rightarrow I_3' \rightarrow I_4''$, where $I_4'' \sim I_4$; this would correct the error and bring the cell line back on the track. Net error relevant to strain differentiation thus is the sum of self-perpetuating deviations of positive and negative signs. Since the probability of each positive deviation being offset by one of opposite sign is very low, errors of types (b) and (c) are apt to become compounded as cells continue to proliferate.

5) Developmental Variance

The outlined sequential origin of terminal "characters" of the body through numerous seriated interactions, in each of which the genome functions as reactor, but never as an autocratic actor, rules out any concept of the developmental course as following rigidly and microprecisely laid out linear tracks. In contrast to the high degree of precision in the specific array and reproductive duplication of the genes, the extragenic system, with which they must continuously interact, is far more variable due to the random fluctuations of both the outer and inner environments. Consequently, the respective interactions and their results are equally variable. **Development** thus becomes a probabilistic, rather than micro-deterministic, phenomenon. The range of variability—the margin for error—is held within tolerable bounds a) by the evolutionary eradication of non-viable excesses; b) by stabilizing dynamic devices of cells and organ systems, such as the regulatory functions of the nervous system, of hormones, of "homeostatic" mechanisms, of growth correlations, and so forth; and c), above all, by the composite network character of living systems, as follows.

6) Network Dynamics

Development does not proceed as a bundle of separate linear single-tracked courses. Rather, the program of development calls for many secondary linkages and cooperative interactions between cells (and products) of diverse origins and different prior histories. As developmental diversification increases, so does the number and intricacy of interdependencies. But the outlined variations of erstwhile independent component courses and kinetics also become compounded as time goes on. Thus, while mutual dependence grows, the probability of the actual occurrence of the scheduled linkages declines. This would seriously jeopardize development, were it not for the provision in the developmental program of *multiple pathways* converging toward common unified results. Instead of linear deterministic chain reactions, we discover a multidimensional network system of dynamic interactions, intricately

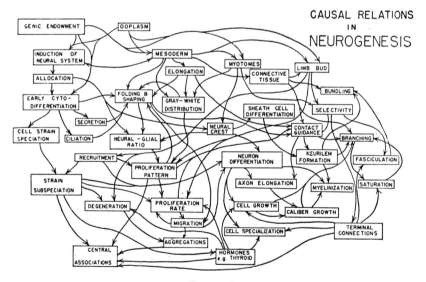

FIGURE 1

branched and anastomosed. As illustration, I reproduce in Figure 1 from an earlier publication (Weiss, 1955) the network of dynamic interactions and dependencies involved in the development of the nervous system. Every arrow represents an activity for which a physical or chemical effect upon another branch has been experimentally demonstrated. For intracellular events likewise, the linear chain reaction concept must be replaced by network theory, as is clearly reflected, for instance, in the diagram by Nicholson (1963) showing the net of alternative pathways in an intracellular metabolic system (Fig. 2). The point is that such networks endow the respective systems with high degrees of elasticity; if any one branch line lags or fails, the remaining bypasses can still produce a viable. if slightly modified, result.

7) Commentary

All the preceding principles from (2) to (6) continue to operate throughout life, even though with increasing restrictions. The programming of the different courses of differentiation leads to the terminal variety of tissues with radically different properties, each continuing to grow, to react and to produce according to its own type and rate. Many cells remain in pre-terminal "bipotential" conditions, in which they can either reproduce more cells of their own kind or transform by cytodifferentiation into sterile workers of shorter or longer life expectancies. Times from reproduction to death vary from a few days (e.g., intestinal mucosa) to weeks (e.g., epidermis; blood elements) to years (e.g., bone cells). Other cell types grow continuously without dividing (e.g., neurons, which burn and discharge what they grow). None of them truly rest. And as they keep being exposed to the fluctuations of the internal and external milieu, the harmony of their dynamic and material interdependency relations continues to be subject to the danger of disruption by excessive deviations from the range of tolerable divergence.

AGING

There is nothing in this picture of development to suggest any abrupt discontinuity, whether initiation or cessation, from which one could date the onset of "aging." *Growth* goes on steadily. Its

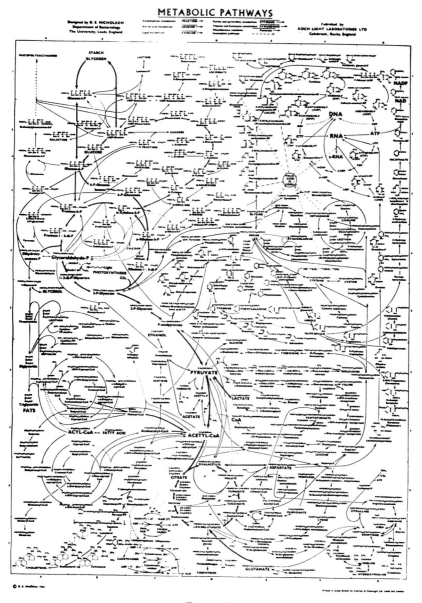

FIGURE 2

rate declines gradually but not necessarily because the reproductive units become less efficient, but chiefly because differentiation progressively reduces the generative compartments of cells and tissues by conversion into products (e.g., fibers) (Weiss and Kavanau, 1957). Conversely, *cell degeneration* and *cell death* are by no means peculiar of older ages; they occur extensively in embryonic stages (Glücksmann, 1951). Progressive changes in *composition and physical properties* of tissues, particularly of the more inert extracellular components, (e.g., the fiber-to-ground substance ratio in tissue matrices) are likewise traceable to early life. So, really all the *elemental* attributes of what in later life one is accustomed to consider as the aging syndrome, are common attributes of the entire developmental process. It is correct, therefore, though commonplace, to describe aging in terms of common language, as simply a function of age. But then, what is aging in terms of developmental biology?

To summarize the answer, let us turn back to the seven criteria of differentiation enumerated in the preceding section. In (4) I have pointed to the sources and the compounding of errors, as life —i.e., development—proceeds. Each body component thus accumulates its own unique history of deviations. These of themselves need not be detrimental to the organism as a whole. What really troubles the organism is rather the increasing disharmony of *mutual relations* among the various component activities, a loss of integration. As pointed out in (6), numerous component processes must constantly cooperate for conjoint effects, and like a team of horses, some slow, some fast, hitched to a carriage, they are held in check by control reins as hinted at in (5,b). Now, if the same horses were to be released, they would disperse according to their speeds, as in a race and cease to act as a team. The same happens in the body if either the component processes get too far out of step or the checking devices lose their hold on them. The independent variance of the components discussed under (5) tends to have just such an effect. It cumulatively magnifies what were initially innocuous discordances (e.g., of timing or chemical specificity) to that critical degree of discrepancy where the erstwhile conjugated components will cease to mesh. The coordinating systems, as body parts, must share this fate. If, following our earlier analogy, we compare the system of interdependent component processes to an elastic net-work, more and more meshes will thus be strained beyond the stress limit and snap. More and more of the alternate bypasses formerly available to vicariate and compensate for lost interaction pathways will thereby be put out of commission. This then is the crux of the phenomenon which manifests itself to introspection, observation and measurement alike as *decline of plasticity, adaptability and efficiency* with age.

Aging, in biological perspective, therefore, must be regarded as basically a matter of disturbed normal *relations;* and not as a product of *"agents,"* such as "aging factors," "aging substances," "metabolic slags," etc., as it has at times been pictured. The latter misconception would not only, with Ponce de Leon, cruelly mislead public thinking, but misguide research. In my opinion, research on the biology of aging will always be auxiliary to research in *developmental biology* in general. Since time magnifies the effects of critical disturbances of interactive relations, the study of the aged will favor the discovery of such relations; this then will stimulate and guide the backtracking of those relations through their prior life history, when their ever present fluctuations and distortions had not yet become grossly disruptive, hence were accepted as "normal." Our dismally deficient knowledge of "normal" development, aggravated by our ready acceptance of words as substitutes for knowledge, might well receive as much enrichment through the tracing back of change from old to young as through the customary tracing forward from the egg.

Regardless of which end he chooses to proceed from, the searcher will be disappointed if he looks naively for "single causes" and strictly "deterministic" models. As development is a *probabilistic network of multifactoral dynamics,* so is aging. Szilard (1959) has correctly recognized the probabilistic nature of the aging process, although his specific model, derived from microbial concepts, is far too monotonic—indeed, too fatalistic—to fit the highly diversified developmental dynamics of metazoan organisms, including man. For unpreventable though the statistical deterioration of individual cell lines be, the network character of their interrelations enables the higher organism to stem impending disintegration within limits by appropriate compensatory and substitutive counter measures. To keep the latter faculties in training by practice throughout life is therefore the biologist's rational design for living: if aging is inevitable, we can at least retard its pace.

REFERENCES

GLÜCKSMANN, A.: Cell deaths in normal vertebrate ontogeny. *Biol. Rev.,* 26:59, 1951.

MINOT, C. S.: *The Problem of Age, Growth, and Death.* New York and London, Putnam, 1908.

NICHOLSON, D. E.: Metabolic pathways. Colnbrook, England, 1963.

SZILARD, L.: On the nature of the aging process. *Proc. Nat. Acad. Sci., USA,* 45:30, 1959.

WEISS, P.: *Principles of Development.* New York, Holt, 1939.

WEISS, P.: Nervous system (neurogenesis), Eds. B. H. Willier, Paul Weiss, and Viktor Hamburger. In: *Analysis of Development.* Philadelphia, Saunders, 1955, pp. 346-401.

WEISS, P.: Cell interactions, In: *Proc. Fifth Canad. Cancer Conf.,* New York, Academic Press, 1963.

WEISS, P., AND KAVANAU, L.: A model of growth and growth control in mathematical terms. *J. Gen. Physiol.,* 41:1, 1957.

IV. ON FORM AND FORMATIVE PROCESSES

[Reprinted from INTERNATIONAL REVIEW OF CYTOLOGY, 1958, Vol. 7, pp. 391–423]

CHAPTER 15

CELL CONTACT*

By PAUL WEISS

(From The Rockefeller Institute)

I. PREFACE

This article presents a slightly modified version of a paper delivered by invitation at a Plenary Session of the Ninth International Congress of Cell Biology at St. Andrews on August 30, 1957. Although the original assignment may have implied a mandate to give a general survey of all cellular interactions mediated by contact, this subject is far too vast to be reviewed even cursorily, let alone critically, within the allotted time. Contact, or "contiguity," as it is often called in reference to cells, is involved in so many and such diverse biological phenomena, from virus infection to the synaptic transmission of nerve impulses, that the best one can do within the given limitations is to concentrate on only a few key examples of issues in need of clarification. I have chosen the following three: (1) the contact relations between cells and their physical substrata; (2) the mutual reactions of cells on contact with one another; and (3) the transmission of specific agents and influences from one cell to another by direct contact. Yet, in respect to all of them, solid information is so scanty and provisional that a sheer compilation would have produced more bewilderment than enlightenment. This naturally raises the question of just why there has been such a scarcity of interest and methodical research in this important area. The answer apparently lies, at least partly, in the fact that the problems have not been made sufficiently explicit and

* This investigation was supported by research grants from the National Cancer Institute (National Institutes of Health, Public Health Service) and the American Cancer Society.

IV. ON FORM AND FORMATIVE PROCESSES

stated in sufficiently concrete terms to invite analysis. It therefore seemed more profitable, instead of reviewing the fragmentary developments of the past, to help set the stage for a more concerted advance in the future by out-lining more sharply and tangibly the problems to which further thought and research must be directed.

That, then, has become the main object of this paper. Much of it will be simply an attempt to give greater precision to pertinent questions, with no pretense of having any conclusive answers. If, in crystallizing these questions, I draw disproportionately on illustrations from my own laboratory, this is not because of any illusion that they are more instructive than others, but solely because of the immediacy of my acquaintance with the detailed facts in each case, which seems quite essential in virgin territory, where relative remote-ness from first-hand factual experience carries the danger of getting lost in fiction. In giving in to all these considerations, the article has turned out to be an essay, fraught with personal views, rather than a truly "objective" survey that pits view against view without arbitration. It is hoped that the resulting gain in penetration will in some measure make up for the loss of surface coverage.

II. INTRODUCTION

Before turning to the facts proper, the meaning of "contiguity" itself needs some clarification. After all, our concepts of cell surface and of cellular inter-actions by surface contact date from an era dominated by the microscope with its limited resolving power. Contiguity refers to the extreme of spatial proxi-mity between two bodies: It was thought to exist wherever the surfaces of the two bodies gave the appearance of being merged into one, when viewed under the microscope. With the extension of our techniques and our think-ing to submicroscopic and molecular dimensions, however, the concept of sur-face contact has become in need of revision, as microscopic failure to discern interstices can no longer be taken as proof of their absence. Then, what is cell contact in this new dimension?

The following operational definition will serve our present purpose. We shall consider a cell in contact with another body not only if the two surfaces are in direct apposition, but also if they are separated by a narrow space oc-cupied by a molecular population whose free mobility is restrained. Not being subject to random dispersion, such molecules can form temporary links between the two surfaces and thus transmit effects that would be ob-literated by randomizing free diffusion. Cell surfaces can be regarded as two-dimensional solid bodies in which surface tension, adsorption, chemical bonds, etc., hold the resident molecular populations in relatively stable configura-tions, thus permitting some degree of structural order. Any effect that is predicated on specific correspondences between localized surface spots, therefore, presupposes that, if the two surfaces are not in intimate apposition, the intervening space has sufficient physical stability as not to blur the specificity of those correspondences.

Diagrammatically, the situation is as follows (Fig. 1): Portions of two cells (heavily stippled) face each other with surfaces occupied by oriented macromolecules (pin-shaped). Beyond each surface, in the ambient medium, there is a molecular cloud, the free mobility of which is restrained (by adsorption, chemical and osmotic interactions, etc.) within close range. With surfaces far apart, as in *a* (upper left), most of the interstice is beyond that

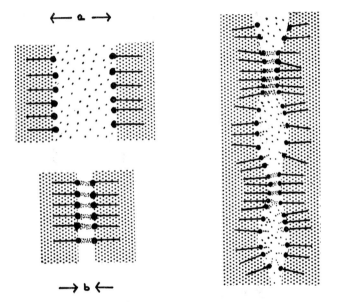

Fig. 1. Diagram of molecular constellations between surfaces.

critical range, and hence permits free circulation. With decreasing distance, however (*b* in lower left), the two spheres of influence eventually merge, and the molecular population of the interstice, removed now from random turbulence, can become stabilized, either just as cement or even as orderly transmitter of specific patterned cell-to-cell influences. Since surfaces cannot be microprecisely uniform and identical all over, the right-hand diagram would come nearer to reality in that ordered patches would alternate with disordered ones. Total interaction would vary with the statistical probability of coincidences between ordered spots at confronting sites. Such probabilistic considerations can account for the known intensity gradations of surface interactions, as in adhesiveness, agglutination, and inductive transmissions.

IV. ON FORM AND FORMATIVE PROCESSES   345

Instead of stronger or weaker interacting forces, we may have to think of more
or fewer interacting sites.

Now, why introduce a "contact interspace" at all? Because dynamically
a cell does not end at its geometric outline. I suspect that the cell surface
is rarely naked, but rather is covered with a coat of cellular exudate and
adsorbed matter—mostly submicroscopic, but sometimes attaining microscopic dimensions, as in slime cells or the surface coat of amphibian eggs

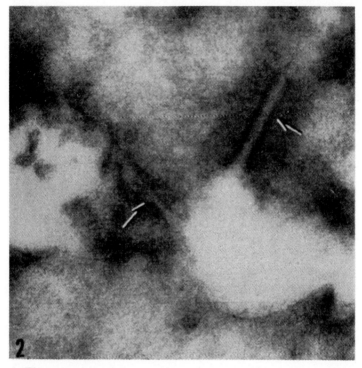

FIG. 2. Electron micrograph of contact points (at arrows) between neighboring
cells of larval amphibian epidermis (nodes of Bizzozero). Osmium tetroxide fixation.

(Holtfreter, 1943). If this be true, all contacts would be effected through
enveloping films. Electron microscopy tends to support this notion. For
instance, the outlines of neighboring epidermal cells at contact points are
separated by a structureless space of a few millimicrons (e.g., Selby, 1956)
(see Fig. 2).

Another example of such intermediary layers is furnished by the coat which
lines the underside of the epidermis of amphibian larvae (Weiss and Ferris,
1954) (Fig. 3). It connects the cell (upper part of picture) to the basement
membrane underneath. Of the three parallel lines marking the border, (1)

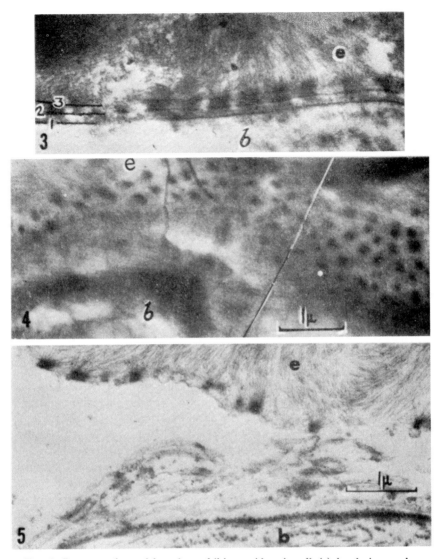

FIG. 3. Lower surface of larval amphibian epidermis cell (e) bordering on basement lamella (b) (explanation in text).

FIG. 4. Oblique section through border between epithelial cell (e) and basement lamella (b), showing distribution of bobbins.

FIG. 5. Epidermal cell (e) detached from basement lamella (b) in vicinity of skin wound made 1 hour previously. Note freely exposed bobbins and cement substance in space between cell surface and (granular) border of basement lamella.

IV. ON FORM AND FORMATIVE PROCESSES

represents the top of the basement membrane, and (2) the lower surface of the epidermal cell, whose exoplasmic cortex extends from (2) to (3). What concerns us here is the extracellular layer between (1) and (2), which is roughly 500 A. thick and corresponds to the barrier at which, according to Ottoson *et al.* (1953), the bioelectric skin potential originates. It consists of a seemingly structureless substance in which a single layer of osmiophilic granules lies embedded; its ground substance dissolves readily in salivary amylase. This layer not only cements the epidermal cell to the basement membrane but also mediates specific inductive influences reciprocally between these two components: The basement membrane elicits the formation of "bobbins" (see below) in the epidermal cell along the common "contact" surface, and the epidermis, in turn, imposes lamination on the erstwhile matted fabric of the basement membrane (Weiss and Ferris, 1956). It is evident, therefore, that a lining of the given constitution and size constitutes no barrier to specific "contact" interactions of the given kind. Similar submicroscopic linings have been found interposed between microscopically contiguous structures known to be in specific interaction—for instance, at the synaptic junctions of neurons (Palay, 1956). It is not impossible, therefore, that this may be a common feature of "cell contact." Whether it is universally present, only the future can tell.

Figure 3 adds another detail to the submicroscopic revision of microscopic notions of cell surface. Customarily, the cell surface is viewed as an entity, and its properties, determined as averages for whole cells, are prorated evenly over its fractions as if each and any fraction were a fair sample of the whole surface. Heterogeneity, such as the protein-lipid mosaic (e.g., Danielli, 1954), is envisaged as of molecular grain size. The epidermal cell surface in Fig. 3, by contrast, presents the aspect of a much grosser mosaic, composed of patches of distinctly different properties. Spanning the exoplasm between (2) and (3), one notes bobbin-shaped bodies (less than 200 millimicrons high), which after osmic fixation show two electron-dense plates, separated by a lighter belt (Weiss and Ferris, 1954). Tangential or oblique sections (Fig. 4) show that they are spaced in a rather regular pattern. They are strictly confined to that part of the cell surface which faces the basement membrane and must not be confounded with the more diffuse single-plated condensations (nodes of Bizzozero) at the junctions between two epidermal cells (Selby, 1956; Montagna, 1956), with which they have in common that they serve as anchor points for a peculiar endoplasmic fiber system (see Fig. 2). Their outer plates are integral parts of the cell surface, for they lie freely exposed whenever the cell detaches itself from its substratum, as after a skin injury (Fig. 5). By immersing fresh skin in various enzyme and other solutions prior to fixation, these bobbins could be chemically dissected. Pancreas lipase removes their strong osmiophilia, which indicates high lipid content of the dark plates, whereas the lighter belt swells greatly in distilled water, thus proving its strong hydrophilia. The whole bobbin structure is rather firm, as it persists unaltered in salivary amylase (Fig. 6), which destroys all other parts of the cell surface and erodes the interior. As mentioned before, bobbins are induced along the underside of the epidermal cell by its contact with substratum; Fig. 7 gives an example of their appearance *de novo* in a cell which has recently

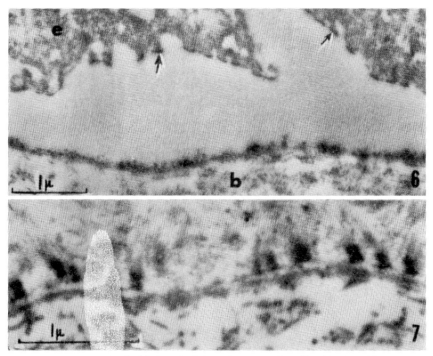

FIG. 6. Epidermal cell (e) detached from basement lamella (b) after 1-hour treatment in salivary amylase, followed by osmium tetroxide fixation. Note persistence of bobbins (examples at arrows!) in heavily disintegrated cell surface.

FIG. 7. Reappearance of bobbins in epidermal surface over basement lamella reconstituted in a 14-day wound.

settled on a raw wound. Skipping all further detail, the point seems clear that surfaces can have a much higher degree of structural and chemical diversity than one commonly envisages and that the respective surface mosaic can be of decidedly supramolecular, and indeed particulate, grain size.

This point bears directly on our topic, for we have strong indications that the osmiophilic plates are organs of attachment firmer than simple adhesion. If we cause these epidermal cells to shrink, for instance, by fixation in formalin or by hypertonic NaCl, the only spots at which their surface remains

IV. ON FORM AND FORMATIVE PROCESSES   349

firmly pinned to the substratum are the bobbins; all the rest of the surface sags away (Fig. 8). Such facts caution against allocating adhesiveness—or any other property, for that matter—evenly over the entire surface of a cell. We should adopt instead a concept of "active spots" as carriers of such properties. Mostly, they will not be so neatly delineated as in our present example, but will be more variable in frequency and distribution, depending perhaps on composition and state of the triple system cell-substratum-medium.

So much by way of introduction. Clearly, none of these observations could have been predicted from general considerations. Many, many more will have to be gathered by detailed labor before we can venture ambitious gener-

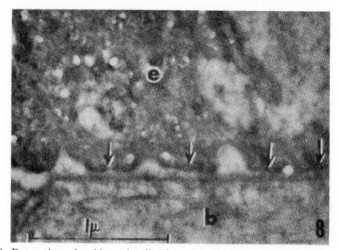

FIG. 8. Retraction of epidermal cell (e) surface from basement lamella (b) after NaCl treatment, with bobbins (at arrows) mostly remaining attached.

alizations. This profession of comparative ignorance will remain the keynote of the account to follow.

III. CELL-TO-SUBSTRATE RELATIONS: CONTACT GUIDANCE

Let us first consider contact relations between cells and noncellular bodies in their environment. If small in proportion to the cell, the body may be engulfed, as in phagocytosis. If large, the cell may spread out along its surface. If one views this spreading as a vain attempt of the cell dwarf to engulf the substrate giant, one recognizes that engulfment and spreading are fundamentally alike. We shall deal here only with the latter phenomenon.

A systematic study of the effects of the physical configuration of the contact substratum on the morphology and behavior of cells and nerve fibers in tissue culture (Weiss, 1929, 1934, 1945; Weiss and Garber, 1952) has revealed a rather general principle which I have called "contact guidance." Here is a very sketchy summary of its essential features. We speak of contact guidance when a cell on a given oriented substratum assumes a corresponding orientation and moves along that line. The phenomenon is complex and can be dissected into at least five different and separable processes: (1) the imposition of orientation upon the substratum; (2) the adhesion of the cell to the substratum; (3) the assumption by the cell of a conforming orientation; (4) the motility of the cell; and (5) the actual locomotor advance in one, rather than the reverse, of the two possible directions along the line of orientation.

1. Adhesiveness

The first point, of just how a substratum acquires nonrandom structure, need not concern us in the present context. As for the second point, the application or adhesion of the cell to a substratum, it should be borne in mind that it implies more than simple contact, differing from the latter by the tendency of the area of contact to attain maximum extent, as a result of which the cell spreads. For such transitory adhesion to become permanent attachment seems to require additional cements. Adhesiveness is not an independent property of the cell but is a function of both the substratum and the medium. A model of this triple dependence is provided by the spreading of oil on the interface between air and water, which likewise is a function of all three phases. Actually, such surface-tension models have been suggested as explanations of cell adhesiveness (Rhumbler, 1914), but, since they can hardly account for the high degree of selectivity noted in cell-to-substratum and cell-to-cell adhesion (see below), the resemblance is presumably a purely formal one.

At the same time, it must be conceded that our factual information on adhesiveness is still far too meager and inconsistent to serve as basis for any valid theory. Wetting power is definitely involved, but not in a simple relation. For instance, when fibroblasts are evenly disseminated over a glass slide speckled with islands of cholesterol, they spread readily over the glass but fail to settle on the cholesterol patches (Fig. 9); yet, if the experiment is repeated with other hydrophobic surfaces, e.g., paraffin, they do adhere to the latter. Lacking systematic quantitative studies of the matter, it would be difficult to make a conclusive case for a direct and simple connection between adhesiveness and wetting power. Even on readily wettable surfaces, adhesiveness is subject to gradations: For example, cells in tissue culture adhere easily to either glass or fibrin. But if they are exposed to both glass fibers and fibrin stands in competition with each other, they cling much more firmly to the former (Weiss, 1945). According to preliminary observations (Weiss, unpublished observations), they also apply themselves more strongly to certain metals (e.g., tantalum and columbium) than they do to others. Holtfreter (1947) has compiled further significant data on adhesiveness (see also the review by Devillers,

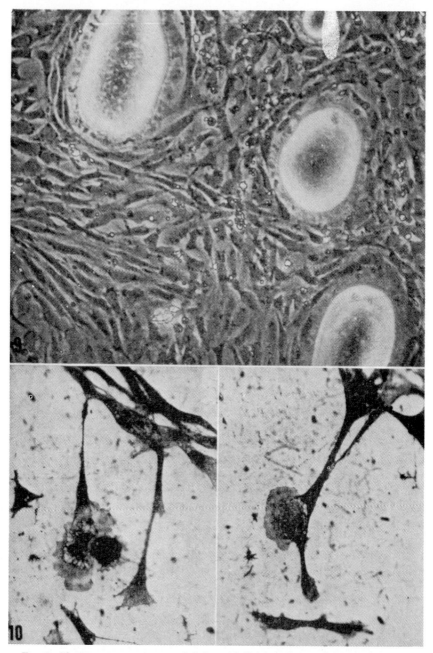

FIG. 9. Phase-contrast picture of 1-day-old living culture of suspension of skin cells (from 9-day chick embryo) on glass slide with specks of cholesterol (four such islands are visible in picture).

FIG. 10. Schwann cell emigrating from chick spinal ganglion transforming into "macrophage" on glass surface. (Weiss, 1944.)

1955). Despite all these efforts, there is as yet no unifying theory. The role of medium is even less known than that of substratum. We have often noted that horse serum tends to favor cell spreading in culture much more than do other sera. But why it should act in this manner is wholly obscure.

In spreading, the cell surface becomes greatly enlarged. That this is not simply a matter of passive interfacial pull, comparable to capillary action, but requires the active participation of the cell, has been shown for the nerve fiber, whose free advance occurs by active cellular thrust (Weiss and Hiscoe, 1948; Young, 1945), the interface furnishing direction, but not force. The same holds true for other cells (Holtfreter, 1946), as we had occasion to observe in time-lapse films of freely suspended spherical cells: They were seen to flatten out on the surface of the glass slide not in a continuous steady creep, as would be expected if they were being drawn out passively, but stepwise in rhythmic pulses originating within the cell body.

Taken together, these facts confirm the active role of the cell in the expansion of its surface that accompanies adhesion, and force us to discount explanations in terms of simple surface tensions, as in the air-oil-water system. They do stress, however, in common with the latter model, the fact that if a spreading substance (whether cell or oil) replaces another substance (culture medium or air) from the surface of a third (glass or water), the properties of all three phases of the respective systems contribute to the effect.

2. Cell Orientation by Contact

The third component of contact guidance, cell orientation, is partly a direct consequence of differential adhesion. When a cell applies itself to a plane surface, its free edge spreads centrifugally until adhesive and protrusive forces are checked by cohesive and contractile ones: The cell assumes the shape of a disk. A cell applying itself to a thread or fiber, on the other hand, will envelop it, thus becoming linearly extended into a spindle. Passing from a linear fiber to a planar surface, a spindle should therefore be expected to transform into a disk. Such morphological accommodations of cells to the varying configuration of the contact substratum, which fall under the general heading of "modulations" (Weiss, 1939), have, in fact, been repeatedly demonstrated—for instance, in the transformation of spindle-shaped connective tissue cells (Fischer, 1927), sheath cells of nerve (Weiss, 1944; Weiss and Wang, 1945), or muscle cells (Chèvremont, 1943) into disk-shaped, phagocytic cells of macrophage character. Figure 10 shows examples of the process under way. Incidentally, just as it is incorrect to refer to all spindle cells as "fibroblasts," i.e., truly collagen-forming cells, so it would be unwarranted to regard them after their transformation to the flat phagocytic form as having been reduced to a single common type of "macrophage." The fact that cells of different types can adopt similar morphological and functional features in response to a common environment definitely does not signify that in doing so they lose the original type specificity of their respective strains. Similitude in structural regards is no test of similarity of constitution. The sooner the unreliability of purely morphological criteria for the characterization of cell type is generally recognized, the faster will some of the terminological and conceptual controversies that have confounded the biological and pathological literature disappear.

It will be noted in Fig. 10 that the spreading portion of the cell border is neither circular nor smooth, but irregular and serrated. This is because neither the cell surface nor the substratum is ideally uniform. A plane surface is crisscrossed with linear singularities—microscratches—to which the cell edge adheres preferentially and along which it is further protruded by centrifugal thrust. The fewer such singularities favoring adhesion the surface contains, the fewer dominant radii can the cell project. We are thus faced with a graded series of cell forms, from the disk shape at one extreme, through multipolar forms, down to the strictly bipolar form, the spindle cell, reflecting the presence of only a single guide line. It has actually been possible to describe and explain the multiform morphology of cultured cells on such a quantitative scale (Weiss and Garber, 1952; Garber, 1953).

A bipolar cell thus is one in contact with a single adhesive line (thread or groove); a disk-shaped cell, one in contact with a net of countless intersecting ones. The former spreads at the same time in two opposite directions, becoming axially stretched; the latter extends its free edge in all directions, thus flattening its body out. Cell orientation is therefore essentially a process of deformation. In my original experiments demonstrating this effect, I had been using tissue fragments growing in clots of blood plasma whose constituent fibrin strands had been forced into a common orientation (Weiss, 1929). For a deeper analysis of the underlying mechanism, however, this method suffered from several shortcomings: The test cells had already some polar shape even before starting to emigrate; their axial orientation could not be divorced from their migratory movements; and there was poor control of the degree of precision with which guide structures could be imposed on the medium. In more recent experiments (Weiss and Taylor, 1956) these defects could be largely removed. For test cells we have turned to the rounded-up single cells obtained by trypsin dissociation of embryonic tissues (Moscona and Moscona, 1952); for linear guide tracks, we make use of a grid of uniform and strictly parallel collagen fibers conveniently furnished ready-made by the interior layers of fish scales. When a suspension of round cells is seeded out over such an oriented matrix, each cell promptly stretches out parallel to the striations of the contact surface (Fig. 11). Currently, we are using fine gratings engraved on glass slides (Fig. 12), with much the same results.

Now, watching the elongation in progress (Fig. 13 pictures four successive stages of this process in the same cell), one observes the following sequence:

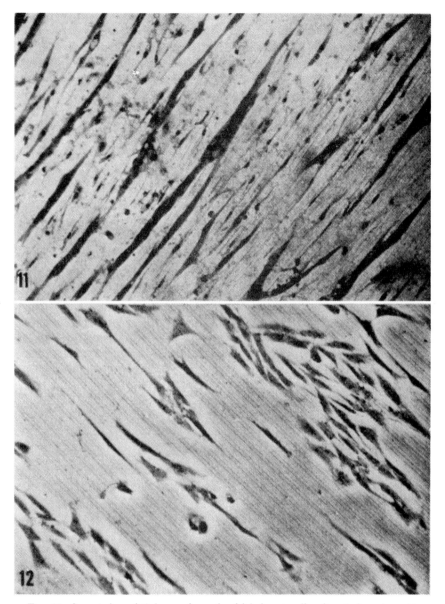

FIG. 11. Suspension of 6-day embryonic chick heart cells after 4 days in culture on fish scale. Hematoxylin stain; cell nuclei dark; striations in background are collagen fibers in scale.

FIG. 12. Phase-contrast picture of living culture of suspension of 8-day chick embryo skin cells on finely grooved glass.

IV. ON FORM AND FORMATIVE PROCESSES   355

First, the settling cell develops a mobile exoplasmic ruffle almost all around its circumference and by means of it can expand more or less uniformly over both lined and unlined portions of the substratum; then a process of surface consolidation sets in which gelates and immobilizes all the cell margin with the exception of those two narrow sectors where the ruffle crosses the guide line. As these two sectors remain fully motile and keep advancing in opposite directions, the cell is pulled out between them, thus becoming more and more elongate. According to these observations, orientation can be as much

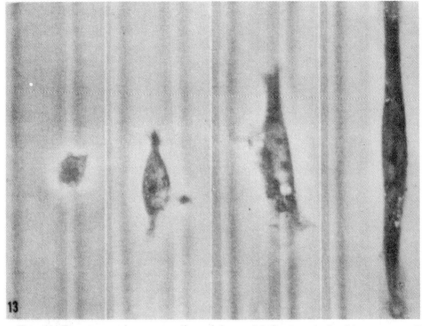

FIG. 13. Four successive stages (from left to right) of a single skin stroma cell of chick embryo, elongating on grooved glass slide (phase contrast).

a result of mobilization in the direction of the guide line as of immobilization in all other directions.

In conclusion, one can visualize the process of orientation as follows: Let us start from a spherical cell. To enlarge the surface of a sphere requires energy. This the cell can furnish from its own metabolic resources, which according to some authors involves the intervention of ATP (Lettré, 1954; Goldacre, 1952). Since, at a fixed volume, expansion in any one dimension can occur only at the expense of other dimensions, it remains to define the direction in which the deforming forces will find an outlet. An ideal sphere with an ideally uniform surface in an ideally homogeneous environment would remain spherical despite increased internal pressure. Actual deformation therefore presupposes the appearance of local differences in the pressure distribution along the surface. Once such inhomogeneities have arisen, whether as expressions of internal instability of the cell or of local differences in the environment, they will entail protrusions of cell processes at the weaker points, with concomitant retractions of the more resistant portions. The cell surface springs temporary leaks, as it were, into which the cell content herniates. Random occurrence of these irregularities gives rise to the familiar "bubbling" of the cell surface during the unstable phase of daughter cells just after mitosis. Freely suspended single cells show the same phenomenon. But after contact with a substratum has been made along a broad front, surface activity becomes chiefly confined to the edge of the contact area. Evidently, this line marks a zone of greatest surface lability and structural weakness, visibly expressed in the continuous agitation of the mobile ruffle. Now, when linear structures such as fibers, grooves, or scratches cross this margin, singularly weak spots, a sort of microperforations, seem to arise at the intersections, thus favoring protrusion of cell content at these points to the competitive exclusion of other parts of the border; the most unstable portions of the cell surface drain motility off the stabler ones.

According to this picture, contact guidance in its simplest form would operate essentially by localizing the cellular thrusts, for instance, by placing the parts of the cell surface over the unlined substratum at an adhesive disadvantage relative to parts intersected by grooves or fibers, while the full motive force—point 4 of our list—resides within the cell itself. But although such trivial microstructural differences as that between a scratch and a smooth surface furnish a true, if elementary, model of how differential adhesiveness can act as guide, the living organism, in which contact guidance is often highly selective, seems to have additional and much subtler devices at its disposal, which we shall deal with below.

3. Cell Locomotion as Contact Reaction

Now, in passing to the next point, it helps to describe contact guidance in railroad terms: the substratum furnishes the rails, and the cell represents the train, providing its own steam, with the pseudopodial tip as locomotive. But a bipolar cell on a linear track has two engines, one at each end, and pulling in opposite directions they stall the train. This is indeed the picture one obtains if one seeds out a sparse suspension of single cells evenly over a substratum in tissue culture: each cell becomes elongate, but remains stationary—that is, motile at both ends, with no resultant displacement. Orientation as such does not imply oriented movement. In order for a cell to get actually displaced, one of its ends must out-pull the other. The question raised by point 5 of our list is then: What causes such inequalities of strength between the opposing engines as to turn bipolarity into anterioposteriority? Contrary to the behavior of uniformly dispersed cells, which remain stationary, the cells emigrating from an explanted tissue fragment, as is well known, do move away from the center. Why? Formal answers have been given in terms of attractions, repulsions, taxes along chemical or electrical gradients, etc. The empirical answer which we have recently found through time-lapse microcinematography (with A. C. Taylor), and which I shall now present, conforms to none of these; once again, the answer lies in a true contact effect.

In essence, it is as follows: Any chance contact between the expanding pseudopods of two free mesenchyme cells evokes a shocklike stoppage and contractile response in the colliding ends of both, which deprives these ends suddenly of their motility and of their firm grip on the substratum, and leaves each cell then to be tugged only by its other, unaffected end. As a result, the two cell bodies are forced apart till eventually their tenuous connection snaps. This process resembles closely the separation of daughter cells after mitosis, and the mechanism underlying both may very well be basically the same. The retracted end of either cell then remains immobilized in the shape of a tapering tail for about 15 to 20 minutes. During this period, the other pseudopod keeps moving, and, being now unopposed by counterpull, tows the cell body away from the collision point, the two cells of such a pair thus traveling away from each other in opposite directions. Gradually, however, after its temporary paralysis, the retracted end recovers its motility and grip and builds up progressively increasing counterpull to the opposite, freely advancing end. The result is that after a period the cell becomes stalled again at a new position, which it will hold till the next encounter produces another one-ended paralysis, and so on.

Now, in a cell population whose density grades off from a center, the probability that any one cell will collide with another cell, and hence the average frequency of mutual contact and reciprocal withdrawal, is evidently higher in the central than it is in the peripheral direction. And the gradual centrifugal shift of the population, starting from the free edge of a tissue, is simply the statistical result of this radial gradient of collision probabilities. Figure 14 shows a sample sequence of random encounters and snapping responses of a given cell at the front line of a culture. Recorded individually, cells have always been seen to move as readily and as rapidly toward an explanted fragment as away from it. Only the paralyzing encounters are more frequent at their near, than at their far, sides. The situation is quite analogous to molecular diffusion in systems with concentration gradients. In a formal sense, Twitty and Niu (1954) were close to this solution of the problem in assuming "repulsion" between pigment cells, although they missed out on the mechanism, which, as is now realized, is one of reciprocal localized inhibition rather than of active repulsion.

As the contact retraction resembles the reflex withdrawal of a tentacle on tactile stimulation, one thinks immediately of an ATP-energized contractile machine, similar to Goldacre's (1952) suggestion for contractility in amoebae. Actually, by adding ATP to cultures we could crudely simulate a reversible snapping response. But this takes us beyond our topic.

Significantly for our topic, the snapping response occurs only between the solitary cells of mesenchyme type. Epithelial cells react in the contrary sense; instead of withdrawing on contact, they draw together more closely, enlarging, rather than reducing, the area of common contact. Thus, after any chance contact, they tend to remain stuck to each other and by repeated such contacts trap one another, adding up to large islands and clusters (Fig. 15). This differ-

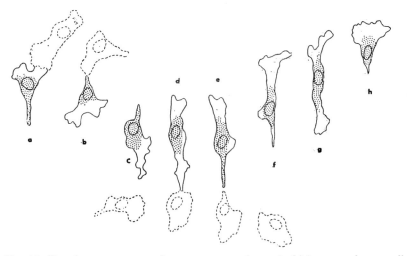

FIG. 14. Four-hour sequence of contact retractions of chick mesenchyme cells (traced from frames of motion-picture film strip at irregular intervals).

ence in response between mesenchyme and epithelial cells is an impressive demonstration of how tissue structure is founded on cell properties, and how differential contact reactions function as constructive agents in morphogenesis.

IV. CELL-TO-CELL REACTIONS: THE PROBLEM OF SELECTIVITY

To react differentially, cells must be able to tell each other apart according to kinds—which brings us at last face to face with the problem of specificity in cell contacts. One of the basic facts of development and tissue repair is the selectivity with which cells of a given kind associate, combine with others, and settle at preferential sites. Such selective associations can arise in two ways. Either the cells are guided to predestined locations by selective tracks, or they arrive at random, but sort themselves out, like to like, at the terminal assembly points. Experiments in which either the tracks were disarranged or the assem-

kind, while bypassing any alien one. In the latter instances, it became quite clear that an epithelial sheet keeps on the move, with concomitant cell multiplication, until integral continuity with a matching kind of epithelium has been restored. "Matching" refers either to cells of the same type or to complementary cell types that normally live in close association, as, for instance, nerve fibers and sheath cells (Abercrombie *et al.*, 1949). Mechanical continuity alone does not satisfy the conditions under which the system will come to equilibrium. It seems to require the full adaptive harmony of relations between each cell and (1) its neighboring cells, (2) its substratum, and (3) its outer medium, to arrest that cell in its roaming and proliferation. I have termed this state of complete fitting of a cell into its ecological niche within the organism "coaptation" (Weiss, 1950). Like other such terms, "coaptation" is meant to be no more than a convenient label for a phenomenon the reality of which has been established but which remains in need of further analytical resolution. As I have pointed out before (Weiss, 1949a), the emigrative, infiltrating, and metastatic power of cancer cells is presumably attributable directly to loss of surface coaptation. This makes it all the more vital to devote more intensified study to the specificity of surface contact relations.

Such, then, are the meager facts which stake out the problem of selectivity in cell-to-pathway, cell-to-matrix, and cell-to-cell association. Viewed soberly, they add up to little more than a descriptive record. They tell us that cells recognize matching environments and cellular kinships and respond discriminately, but just how is still obscure. All present evidence indicates that the discrimination is based on a contact sense, and hence must reside in surface criteria. On this premise, I suggested (Weiss, 1941, 1947), that selective adhesion might be due to the steric interlocking of complementary molecular fringes in the two apposed surfaces—with or without the mediation of an interlining acting as a zipper in the sense of Schmitt (1941). Conforming molecular populations would link up, but nonconforming ones would remain disjointed. Devillers (1955) has called this suggestion "terribly speculative," and so it is. One thing that can be said in its favor is that similar mechanisms of steric conformance and hindrance have been familiar from enzyme-substrate relations and antigen-antibody combinations, which may be the reason why my suggestion of tissue formation by molecular conformances has gained some currency, and was, for instance, adopted by Tyler (1947), who had already, following Lillie, come to view the union of egg and sperm in immunological terms. Although I still suspect, however, that cell affinities and immunological reactions have some common denominator, I doubt whether a static mechanism of molecular linkages of the suggested kind can hold the solution to our problem.

This reserve is due partly to 'the lack of conclusive evidence. The success Spiegel (1954) and Gregg (1956) have reported in their attempts to affect cell aggregation in sponges and slime molds, respectively, by species-specific antibodies, can only prove, at best, that surface molecules which are involved in cell adhesion also have antigenic properties, but not that these very same properties are instrumental in the normal coupling of cells. Recent attempts of our own to interfere with the selective association of vertebrate cells by organ-specific antisera have failed.

But the major source of skepticism lies in recent direct observations of cell encounters. After Moscona's demonstration that in a mixture of cells of different types those of identical type tend to clump together, we (Weiss and Taylor, unpublished data) have now tried for two years to detect direct signs of this selectivity of association by phase-contrast time-lapse microcinematography of homologous and heterologous pairs of epithelial cells in liver-to-liver, lung-to-lung, lung-to-liver, kidney-to-kidney, kidney-to-liver, etc., encounters. These observations seem to dispose rather convincingly of any notion that homologous cell aggregations come about by direct agglutination. Cells in homologous, as well as heterologous, groups keep shifting about one another in a manner that rules out stable mutual attachments as a primary event in aggregation. The only difference between the homologous and heterologous groups seems to be that homologous cells tend to retain mutual contact along a broad front (Fig. 15), whereas heterologous combinations tend to break up soon after contact (Fig. 18). As indicated before, the manner in which cells on encounter translate their discriminative reaction into either a closer merger or an active secession is unknown. Therefore, in order not to miss any possible clues, it is worth contemplating that the effect may be mediated by a specific cellular exudate rather than by the cellular surfaces themselves. That is to say, homologous exudates around homologous cells would become confluent, thereby confining the cells in a common envelope, whereas nonhomologous exudates would fail to merge, keeping the cell territories apart. In general, it seems that type-specific segregation is not a matter of primary nonfusion but one of secondary disjunction. We have seen no evidence thus far that type-specific aggregation implies anything more positive than simply the absence of segregation, which, if confirmed, would place this case in a different class from that of the slime mold, where aggregation is achieved by an active principle (see Shaffer, 1957).

One additional cautionary note is called for, however. As culture medium in these experiments, we had to use horse serum, which, as mentioned above, promotes cell spreading. Conceivably, this may have obscured some relevant features of contact relations. Moreover, in all our observations on contact relations, whether of mesenchyme or epithelial cells, we have gained the impression that the discriminative reactions of the cells are most pronounced in the mobile sectors of the cell border (pseudopodial tips; ruffles), whereas the more consolidated parts of the surface seem to be less sensitive. This should caution us, once more, against assigning surface properties unreservedly to the entire surface, but whether or not these observations made under the peculiar condi-

IV. On Form and Formative Processes   363

tions of tissue culture are otherwise directly applicable to the situation in the intact body remains an open question. The nature of the incompatibility that disrupts heterologous cell unions remains likewise undefined, and pending further observations and experiments conjectures about the mechanism of cell contact selectivity will remain ill-founded.

V. TRANSCELLULAR TRANSMISSIONS BY CONTACT

The great uncertainty in these matters also confounds the problem of the transmission of those specific effects from cell to cell that are commonly lumped under the term "inductions." As the term covers a motley of diverse phenomena, it would be poor logic to postulate that they all have a single common mechanism. Accordingly, to ask whether "inductions" do or do not depend on contact between the interacting cells, in terms of an alternative, makes no sense. Each particular instance will have to be decided empirically. Of concern to us here is merely the fact that some inductive effects are transmitted by contact, or better still, in view of the "assembly-line" character of all complex morphogenetic products, that some components of some effects are.

Lens induction by the eye cup, for instance, has been pointed out long ago by Lewis (1907) to require mutual contact. After experimental restriction of the contact area between the retinal layer and the overlying epithelium (McKeehan, 1951), the lens-forming region of the latter is commensurately reduced, and the appearance of one of the characteristic lens proteins is likewise coextensive with the actual contact area (see Ten Cate, 1956). How intimately the interacting layers are fused during induction is shown by the fact that during the critical period they cannot be pried apart (McKeehan, 1951). Preliminary electron-microscopic observations of the junction between the axial mesoderm and the neurulating ectoderm in the chick embryo (Weiss and Nieuwkoop, unpublished observations) have also revealed signs of rather tight fusion of the two layers during the inductive period. Balinsky (1956) has recently submitted that basement membrane insulates the epidermis against such inductive influences. If contact is understood in the broadened sense of the introductory remarks above, even the effects mediated between cell layers across porous sheets (Grobstein, 1954) might qualify more nearly as contact reactions than as diffusible effects.

Many instances of postembryonic inductions also bear the marks of contact interactions. The problem of artificial "induction of bone" (e.g., Lacroix, 1951), which for some time has been rather controversial, seems now to have been resolved in favor of nondiffusible contact agents, at least as far as the induction of true organized bone tissue is concerned (McLean and Urist, 1955). Our own experiments, in which the inductive power of devitalized (frozen-dried and rehydrated) tissues in the developed organism was tested, bear out the contact concept, as new cartilage (see Plate II, in Weiss, 1949b), bone (Fig. 19), and corneal stroma (Weiss and Taylor, 1944) have been found to be formed by the

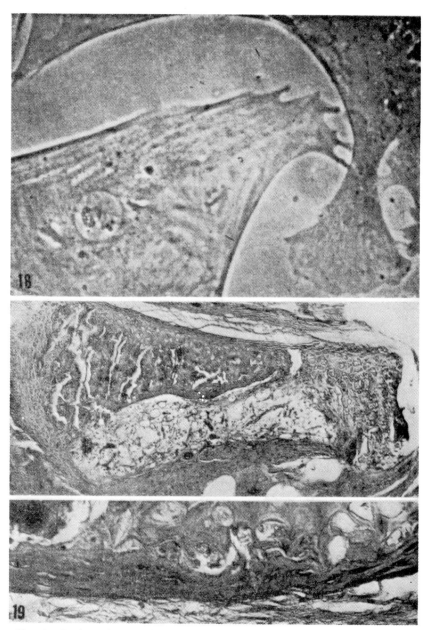

Fig. 18. Reciprocal retraction of margin of mouse lung cell (upper right) and mouse liver cell (lower left) after contact *in vitro*. (From phase-contrast motion picture film.)

Fig. 19. Induction of live bone and bone marrow in the interior of a devitalized (frozen in isopenthane at −159°C., dehydrated *in vacuo* at −40°C., then rehydrated) metatarsal rat bone transplanted into the foot of a rat. The new bone can be seen inside the shell of the partly sequestered graft. Lower picture: part of the new bone at higher magnification. (Weiss and Taylor; picture not previously published.)

host body only in intimate apposition to the corresponding devitalized fragments.

On the other hand, it seems equally certain that other conditioning effects in the chain of events labeled "induction" do not require contact, and hence must be ascribed to diffusible agents. This is obvious for "hormonal" action, but may hold true more widely. In embryos, cartilage induction by spinal cord has likewise been reported to be attainable from a distance (Holtzer, 1953), without microscopically discernible structural connections, and the extensive studies on induction of embryonic systems *in vitro* have also been adduced as evidence for the diffusible nature of "inductive agents" (see Holtfreter and Hamburger, 1955). Yet such facts merely place renewed emphasis on the composite nature of inductive agents, without disputing the indispensability of contact in certain of the component steps.

Nor can we expect that contact interactions will be all of one kind. Contact may signify transport of substance as well as transmission of impulses and ordering influences. Brachet (1950) has proposed substance transfer in neural induction, whereas I (Weiss, 1950) have interpreted the elongation of lens cells in the axial direction of the inducing retinal cells as a sign of the transmission of molecular orientation. I see no reason why these two views need be mutually exclusive. Indeed, their synthesis leads to a concept of rather general applicability, which is here presented by way of concluding note.

Let us turn back to the spindle cell which on striking a flat surface transforms into a disk-shaped "macrophage." There are two features in such a cell which are significantly related: One is that particles (e.g., carmine granules) are phagocytized solely by that portion of the cell which has already assumed disk shape, but not by the spindly portion (Weiss, 1944); the other that, according to silver impregnations, major endocellular fiber systems are oriented radially in the disk-shaped portion, but tangentially along the spindly part—that is to say, perpendicular to the cell border in the disk, parallel to the border in the spindle. Or, speaking technologically, in the former they are in "open-gate" position, letting the particles pass; in the latter, in "barrier" position, keeping the particles out (see also a similar suggestion for osteoclasts by Hancox, 1949). If we further let transfer, not only of particles, as in phagocytosis, but of macromolecules as well, be conditional on the opening of microchinks in an otherwise tight fibrous armor along the cell surface, then the crucial role for all surface interactions of a local reorientation of the molecular fringe from a tangential to a radial—from barrier to open-gate—position would be self-evident. Molecular reorientation along contact surfaces, such as I postulated previously, would then be but a trail blazer for the passage of substances, the total process of transfer consisting of two steps: a conditioning and an implementing one.

More broadly, I would then suggest the following concept for further exploration (Fig. 20): the cell surface to be protected against extrinsic disturbances by tangential positions of chain molecules; these border molecules then to be forced, either by attraction from matching end groups of molecules on the other side of the fence or by electric fields, into radial, reactive, positions; this reorientation creating wide intermolecular pores, or microleaks, which then can serve as channels for substance intake and extrusion—the wedging into the cell of macromolecules and particles, as well as the escape from the cell of large cell products. Membrane pores, such as have been suggested by Danielli (1954), would then be not structural fixtures, but casual and transitory formations. Molecular specificities acting across cell borders need only be involved in the first phase, serving as the key, as it were, to open the door for subsequent

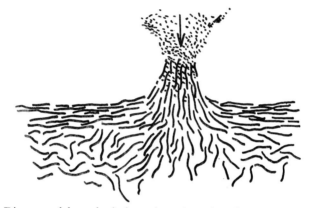

Fig. 20. Diagram of hypothetical reorientation of surface molecules opposite a source of "stimulation."

unspecific deliveries. The specific component of the transfer process need exercise no major force, and the energetic component need have no specificity.

Such a dualistic concept would solve many other biological puzzles. It could explain, for instance, how a relatively few agglutinin molecules can manage to tie a large bunch of wriggling spermatozoa firmly together with their heads, or lug huge red cells along: if all that the specific molecules had to accomplish were merely to trigger a much larger discharge of unspecific binding cements, the paradox would disappear. As another example, the selectivity in the transmission of impulses in the neuronal network of the central nervous system could similarly be conceived of as the discriminative channeling or barring of passage for the unspecific electric signal across the synapse, depending on whether or not the macromolecules on either side are of a matching kind and consequently either do or do not establish a common orientation in the conducting substratum, this of course on the premise that orderly molecular alignments ("microstructures") are prerequisite for impulse propagation.

IV. On Form and Formative Processes

As was indicated in the introduction, the possible presence of an interstice of a few hundred Angstroms between interacting surfaces need not interfere with the specificity of the interaction. All one would have to assume is that the space is occupied by appropriate types of macromolecules of that particular length, which in spanning the gap would act much like pontoon bridges mediating traffic from shore to shore. If time would allow, I could further illustrate how, once initiated by specific correspondences, the open-door array of surface molecules would be reinforced and maintained by the less-specific cross-flow of substance or electricity until it is actively interrupted, and how, on this basis, a common denominator could be found between such apparently diverse contact interactions as transcellular induction and synaptic transmission of nerve impulses.

VI. CONCLUSIONS

Our course in this sketchy commentary on contact reactions, which started out from considerations of molecular constellations in surfaces, has now brought us full cycle back to the same point. Perhaps this fact in itself summarizes the major lesson from our discourse, which also contains a device for the future—that real progress in this field hinges on our ability of shuttling in our research and of shifting our viewpoints back and forth between the microscopic and submicroscopic levels. The time has passed for relying on solitary preoccupation with either one or the other. But although it would seem platitudinous in these days to stress the contributions of submicroscopic advances to the rectification and supplementation of views gained solely from the microscopic dimension, it is not always pointed out with equal strength that only the disciplined study of the actual key objects, the living cells—and the plural is being used here advisedly—can give realistic orientation and precision to the questions to which the molecular biologist is to direct his analytical search. In this sense, use of any one cell type—whether erythrocyte, fibroblast, or sea urchin egg—as source of partial information is profitable. But to elevate any one of them to the status of generalized standard cell would be utterly misleading in guiding the construction of molecular models of contact relations, where, as we have seen, the surface differences among cells are often more crucial for a response than are their common properties. It is for this reason that I have skirted here the more general colloid-chemical accounts of cell surface constitution, interactions and adhesiveness, the role of ions, etc. (see Heilbrunn, 1952); for none of them seems to contain the clue to the understanding of those subtle specific differences of surface complexion which cells can tell so sharply and which we can tell only indirectly from the cells' behavior. To throw more light into this darkness is beyond any one specialized discipline and calls for the combined efforts of physiologists, cytologists, embryologists, pathologists, immunologists, pharmacologists, and molecular biologists.

Much of the content of this article has a bearing on the cancer problem in that it tries to translate into factual terms such symbolic notions as "harmony" or "disharmony" in intercellular relations. In an earlier publication (Weiss, 1949), I had referred to the crucial role played in the chain of events leading to malignancy by the development of incompatibility (loss of coaptation) between the aberrant cell and the rest of the cell population of its mother tissue. The origin of such incompatibilities is part of the larger problem of carcinogenesis and need not concern us here in detail; basically it is a matter of a modified course of differentiation, of which the subsequent unrestrained growth is but a secondary consequence. The primary thing is that the particular cell strain has become constitutionally different from its sister strains beyond the range of adaptive tolerances provided for in normal development. The number of signs by which an observer can tell this newly acquired difference of character will depend on the diversity and discriminative power of his techniques, but at any rate, on common biological considerations, one must expect *multiple* expressions of the altered constitution, revealing themselves in modifications of composition, metabolism, nutrient requirements, reactive behavior, sensitivity, etc. All these signs can serve as convenient indicators; but only some of them signify properties instrumental in rendering the progeny of that strain not only different, but malignantly so. And among these latter signs, changes in surface constitution surely deserve prime attention.

We have reported above how conformance or nonconformance of surface constitutions between corresponding or noncorresponding cell types determines whether or not two normal cells join or separate. We have pointed to the surface selectivity which permits a cell to follow a prematched pathway or occupy a prematched site. The acquisition of such selectively discriminative surface properties is an outcome of the normal course of differentiation (Weiss, 1953). Therefore, by extrapolation, the biologist would range the cancer cell right alongside its normal sister cell strains in ascribing to it just one additional differentiative change, yet one occurring in a direction not provided for in the standard pattern of coaptive interdependencies among the different parts of the body. Having lost their coaptive bonds that held them tied to home base, the altered cells will thus become mobilized, roam, and metastasize, that is, facultatively assume "autonomous" colonial existence at any new site which is compatible with their altered constitution, even though it would not have accepted a cell of the normal counterpart strain. Surface changes, as a corollary of far more profound transformations, thus are presumably the ones that permit the cancer cell to engage in the most pernicious act of its career—metastasis. Coman (1954) has stressed this point compellingly, and furthermore has tried to identify the crucial change with calcium deficiency. Without going into the conclusiveness of this explanation, I can only repeat that experiences such as have been related in the present paper point to much subtler and more dis-

IV. On Form and Formative Processes   369

criminative determinants of intercell contact relations than a grade scale of the concentrations of a single ion could offer. Of course, mindful of the known cementing function of calcium, one might conceive again, as above in inductive transfer, of a dual mechanism, with steric properties of the surface molecules determining the specificity, and calcium contributing to the holding force; but at present this is mere speculation.

Although the dissociation of the cancer cell from its mother colony thus could be viewed as of the same order as the nonassociation among normal cells of different types, there is still one striking distinction: Mobilized normal cells, in those cases that have been examined, after recovering coaptive relations (e.g., epithelia meeting their own kind), cease to proliferate, whereas cancer cells under comparable conditions evidently do not. This could be interpreted to mean that the latter do not restore as intimate mutual surface relations as do normal cells, but factual information is too scanty for even a tentative decision between this and several alternative possibilities, many of them surely still unsuspected.

It would be presumptuous for one not expert in the cancer field to go beyond these general comments. Yet acquaintance with the field of normal differentiation and growth forcefully draws attention to the close relation between loss of specific surface contact on the one hand, and mobilization and proliferation on the other, as a problem in need of more penetrating investigation in connection with the cancer problem. What has been presented in the foregoing ought to have demonstrated in a modest way that penetration beyond mere verbal statements of "harmony" or "disharmony" is feasible and profitable.

VII. REFERENCES

Abercrombie, M., Johnson, M. L., and Thomas, G. A. (1949) *Proc. Roy. Soc.* **B136,** 448.

Andres, G. (1953) *J. Exptl. Zool.* **122,** 507.

Balinsky, B. I. (1956) *Proc. Natl. Acad. Sci. (U.S.)* **42,** 781.

Brachet, J. (1950) "Chemical Embryology." Interscience, New York.

Burns, R. K. (1955) *in* "Analysis of Development" (Willier, Weiss, and Hamburger, eds.), p. 462. Saunders, Philadelphia.

Chèvremont, M. (1943) *Arch. biol. (Liège)* **54,** 377.

Chiakulas, J. J. (1952) *J. Exptl. Zool.* **121,** 383.

Coman, D. R. (1954) *Cancer Research* **14,** 519; **15,** 541.

Danielli, J. F. (1954) *in* "Recent Developments in Cell Physiology" (Kitching, ed.), p. 1. Academic Press, New York.

Devillers, Ch. (1955) *in* "Problèmes de structures, d'ultrastructures et de fonctions cellulaires" (Thomas, ed.), p. 139. Masson, Paris.

Fischer, A. (1927) *Arch. exptl. Zellforsch. Gewebezücht.* **3,** 345.

Ford, C. E., Hamerton, J. L., Barnes, D. W. H., and Loutit, J. F. (1956) *Nature* **177,** 452.

Garber, B. (1953) *Exptl. Cell Research* **5,** 132.

Goldacre, R. J. (1952) *Symposia Soc. Exptl. Biol.* **6,** 128.

Gregg, J. H. (1956) *J. Gen. Physiol.* **39,** 813.

Grobstein, C. (1954) *in* "Aspects of Synthesis and Order in Growth" (Rudnick, ed.), p. 233. Princeton Univ. Press, Princeton, New Jersey.

Hamburger, V. (1929) *Wilhelm Roux' Arch. Entwicklungsmech. Organ.* **119,** 47.

Hancox, N. M. (1949) *Biol. Revs. Cambridge Phil. Soc.* **24,** 448.

Harrison, R. G. (1904) *Arch. mikroskop. Anat. u. Entwicklungsgesch.* **63,** 35.

Heilbrunn, L. V. (1952) "An Outline of General Physiology." Saunders, Philadelphia.

Holtfreter, J. (1939) *Arch. exptl. Zellforsch. Gewebezücht.* **23,** 169.

Holtfreter, J. (1943) *J. Exptl. Zool.* **93,** 251.

Holtfreter, J. (1944) *Rev. can. biol.* **3,** 220.

Holtfreter, J. (1946) *J. Morphol.* **79,** 27.

Holtfreter, J. (1947) *J. Morphol.* **80,** 25.

Holtfreter, J., and Hamburger, V. (1955) *in* "Analysis of Development" (Willier, Weiss, and Hamburger, eds.), p. 230. Saunders, Philadelphia.

Holtzer, H. (1953) *J. Exptl. Zool.* **123,** 335.

Hooker, D. (1930) *J. Exptl. Zool.* **55,** 23.

Korschelt, E. (1931) "Regeneration and Transplantation," Vol. 2. Berlin.

Lacroix, P. (1951) "The Organization of Bones." Blakiston, Philadelphia.

Lettré, H. (1954) *Naturwissenschaften* **13,** 306.

Lewis, W. H. (1907) *Am. J. Anat.* **6,** 473.

McKeehan, M. S. (1951) *J. Exptl. Zool.* **117,** 31.

McLean, F. C., and Urist, M. R. (1955) "Bone." University of Chicago Press, Chicago.

Montagna, W. (1956) "The Structure and Function of Skin." Academic Press, New York.

Moscona, A. (1957) *Proc. Natl. Acad. Sci. (U.S.)* **43,** 184.

Moscona, A., and Moscona, H. (1952) *J. Anat.* **86,** 287.

Ottoson, D., Sjöstrand, F. S., Stenström, S., and Svaetichin, G. (1953) *Acta Physiol. Scand.* **29,** Suppl. **106,** 611.

Palay, S. L. (1956) *J. Biophys. and Biochem. Cytol.* **2,** Suppl. 193.

Rhumbler, L. (1914) *Ergeb. Physiol.* **14.**

Schmitt, F. O. (1941) *Growth* **5,** Suppl. 1.

Selby, C. C. (1956) *J. Soc. Cosmetic Chemists* **7,** 584.

Shaffer, B. M. (1957) *Am. Naturalist* **91,** 9.

Spiegel, M. (1954) *Biol. Bull.* **107,** 130.

Taylor, A. C. (1944) *J. Exptl. Zool.* **96,** 159.

Ten Cate, G. (1956) *Verhandel. Konink. Ned. Akad. Wetenschap. Afdel. Natuurk. Sect. II* **51,** No. 2.

Twitty, V. C., and Niu, M. C. (1954) *J. Exptl. Zool.* **125,** 541.

Tyler, A. (1947) *Growth* **10,** Suppl. 7.

Urso, P., and Congdon, C. C. (1957) *J. Hematol.* **12,** 251.

Weiss, P. (1929) *Wilhelm Roux' Arch. Entwicklungsmech. Organ.* **116,** 438.

Weiss, P. (1934) *J. Exptl. Zool.* **68,** 393.

Weiss, P. (1939) "Principles of Development." Holt, New York.

Weiss, P. (1941) *Growth* **5,** Suppl. 163.

IV. On Form and Formative Processes

Weiss, P. (1945) *J. Exptl. Zool.* **100,** 353.

Weiss, P. (1947) *Yale J. Biol. and Med.* **19,** 235.

Weiss, P. (1949a) *Proc. Natl. Cancer Conf. 1st Conf.* 50.

Weiss, P. (1949b) *in* "Chemistry and Physiology of Growth" (Parpart, ed.), p. 135. Princeton Univ. Press, Princeton, New Jersey.

Weiss, P. (1950) *Quart. Rev. Biol.* **25,** 177.

Weiss, P. (1952) *Science* **115,** 293.

Weiss, P. (1953) *J. Embryol. Exptl. Morphol.* **1,** 181.

Weiss, P. (1955) *in* "Biological Specificity and Growth" (Butler, ed.), p. 195. Princeton Univ. Press, Princeton, New Jersey.

Weiss, P., and Andres, G. (1952) *J. Exptl. Zool.* **121,** 449.

Weiss, P., and Ferris, W. (1954) *Exptl. Cell Research* **6,** 546.

Weiss, P., and Ferris, W. (1956) *J. Biophys. and Biochem. Cytol.* **2,** *Suppl.* 275.

Weiss, P., and Garber, B. (1952) *Proc. Natl. Acad. Sci. (U.S.)* **38,** 264.

Weiss, P., and Hiscoe, H. B. (1948) *J. Exptl. Zool.* **107,** 315.

Weiss, P., and Taylor, A. C. (1944) *Anat. Record* **88,** *Suppl.* 49.

Weiss, P., and Taylor, A. C. (1956) *Anat. Record* **124,** 381.

Weiss, P., and Wang, H. (1945) *Proc. Soc. Exptl. Biol. Med.* **58,** 273.

Wintermann-Kilian, G., and Ankel, W. E. (1954) *Wilhelm Roux' Arch. Entwicklungsmech. Organ.* **147,** 171.

Young, J. Z. (1945) *in* "Essays on Growth and Form" (Le Gros Clark and Medawar, eds.), p. 41. Clarendon Press, Oxford.

Reprinted from REVIEWS OF MODERN PHYSICS, Vol. 31, No. 1, 11–20, January, 1959
Printed in U. S. A.

CHAPTER 16
Cellular Dynamics

PAUL WEISS

The Rockefeller Institute, New York 21, New York

BIOLOGY is now in an exciting phase marked by the confluence of different disciplines of research in the attack on focal problems. But the enthusiasm raised by the spectacular results of combined physical and chemical approaches to biology sometimes has outraced people's ability to keep pace conceptually with the technical developments. As a result, we frequently try to fit our questions to the very limited answers which our fragmentary knowledge has been able to provide, instead of boldly facing the much broader questions posed by living systems and phrasing them in such a way that still more penetrating answers may be obtained in the future. In order to do this, one needs to focus on the real living objects, rather than on the somewhat fictitious and oversimplified models that one is prone to formulate, as intended targets for physical and chemical attack. Models are necessary, but they must bear more than a coincidental resemblance to the real object if they are to serve as meaningful aids to analysis.

As the result of this extensive use of overly simple models, notions about the cell have become at times slightly vague and unrealistic. It would be presumptuous in one single chapter to try to do more than to just give a few illustrative examples of what the real cell is like. The best that can be hoped is to show the change that has occurred in our thinking about the cell from the static to the dynamic—that is, from static organization to organized behavior. Much of the knowledge of what the cell is has come from ruling out erroneous conceptions of what it is not. Progress has come from narrowing the margin of error. By being exposed to a few examples of the living cell in action, the reader can judge for himself whether or not his mental picture of the cell corresponds to the real thing.

The first example deals with one of the most prominent characteristics of the cell—its shape. Figure 1(a) shows a textbook picture of a particular cell found in the cerebellum. One sees that the cell body has elaborate ramifications. This is the way one usually learns about a cell—through pictures in a book; and since the picture looks the same in all of the thousands of copies of a textbook of microscopic anatomy, one forms the notion that all such cells are like tin soldiers stamped out according to a standard pattern. Thus, the mental habit of cell form as something static and rigid becomes ingrained. The truth is, however, that no two cells are ever strictly alike, nor is any one cell quite the same at different times of its life history. It is this history which the static textbook picture fails to reveal. To stress this fact, Fig. 1(b) shows by comparison a Chinese brush

drawing of some shrub. Here common experience tells us that the bush has not been stamped out in the shape in which one finds it. It has grown into that shape from seed. So, what one sees as pattern in the shrub is merely the residual record of prior activities of that particular protoplasmic system. In other words, shape is simply an index of antecedent processes by which that shape has come about.

Something else is lacking, however, in both of the pictures besides the account of prior events. The objects are portrayed against a blank background as if they were in a vacuum. Again, in the case of the plant, one knows from daily life how vital the invisible air is for its existence as a provider of chemical necessities.

Now, in the case of the cell, the medium is involved even more intricately: it provides not only chemical components for nutriment, but also a physical framework that integrates the separate cells into a structural continuum. The existence of this continuum usually remains unrecognized because staining techniques deliberately leave the substratum out of sight. Yet, to the classical morphologist only seeing was believing and, what is worse, not seeing amounted to not believing. This attitude is undergoing radical change.

Processes as such are not visible. What is visible is a constellation of elements at different stages of the process. Visible form, a pattern at any one stage, must be viewed as the product of antecedent formative processes. The cell thus appears as a system of highly complex, but ordered, molecular populations grouped in a hierarchy of supramolecular complexes, in constant interactions among themselves and with their environment (that is, the space beyond the cell border), leading to features, some permanent, others transitory, the visible expression of which is recorded as shape. If the environmental conditions are reasonably constant for a group of cells of the same type, the behavioral history of the latter will be reasonably similar so as to end up

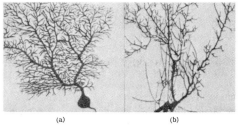

(a)   (b)

FIG. 1. Comparison between a nerve cell (a) and a plant (b).

IV. ON FORM AND FORMATIVE PROCESSES   373

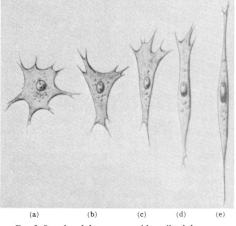

(a)   (b)   (c)   (d)   (e)

FIG. 2. Samples of shapes assumed by cells of the same
connective-tissue cell strain in tissue culture.

with reasonably similar and classifiable shapes of the
sort that have made it possible for sciences of microscopic
anatomy and microscopic pathology to develop. As soon
as there is a change in the conditions, the behavioral
response of the cell likewise changes and the familiar
shape derived from normal standard conditions ceases
to be a diagnostic sign.

A most dramatic illustration of this situation is seen
when cells are taken out of their normal site in an organism and transferred into an extraneous medium in tissue
culture. As an example, a time-lapse phase-contrast
microcinematograph* of human-liver cells spread on
glass in horse serum (film made in my laboratory by
A. Cecil Taylor and Albert Bock) shows up impressively the lack of fixity and the incessant reshuffling
of cell content and contour. No static description can
do justice to this vivid record of ever-changing activity.
These liver cells in culture look quite different from
those one would be used to seeing in stained sections
through an intact liver. Except for shape, however,
they still possess most of the essential properties of liver
cells. In a third type of setting, for instance, in suspension, they would assume still other shapes.

Thus, one realizes that there is no way of getting a
fully valid description of a cell except by studying its
behavior under as wide a spectrum of conditions as is
feasible. Cells of different kinds behave differently.
While the transfer to tissue culture alters their morphological expressions markedly, they do retain their
constitutional distinctions of behavior.

In conclusion, one is led to the thesis that cell shape
is the result of a distinctive behavioral reaction of a
living cell to its environment.

To make this concrete, consider a specific example.

_____

* The motion pictures referred to in the text were shown at the
Study Program in Biophysical Science in conjunction with the
lecture on which this article is based.

A colony of fibroblast cells cultured in dilute blood
plasma yields a wide spectrum of shapes. The same cell
can appear in any of the series of forms pictured in
Fig. 2, ranging from the bipolar spindle at one extreme
[Fig. 2(e)] to the multipolar star at the other [Fig. 2(a)].
The shape thus depends upon the number of directions
in which the cell border shows radial extensions. The
tips of these extensions are the active mobile organs of
the cell. They push outward and thus distort the originally rounded surface of the cell. Evidently, if there are
only two processes tugging in opposite directions, the
cell body is drawn out between them into the shape of
a spindle [Fig. 2(e)]. If there are three major protrusions, the cell assumes a tricornered shape [Fig. 2(b)],
and with even more processes along its circumference
it approaches more and more a star shape [Fig. 2(a)].

Must one accept this spectrum merely as a given
descriptive fact, or can it be explained causally in the
way physical systems are treated? The answer is that,
to a certain extent, the whole series can be expressed in
terms of a single function derived from a study of cellular behavior. Cells do not live in a structural vacuum
as is the illusion created by standard histological preparations, which, by stressing only those features which
happen to be stained, obliterate the structural continuum within which the cells reside. In tissue culture
in a blood-plasma clot, for instance, this continuum is
provided by a network of fibrin fibers—aggregates of
molecular chains of varying diameter from submicroscopic to microscopic dimensions, the meshes of the
network being filled with serum (Fig. 3). It is in this
fibrous jungle that these cells live and move, applying
themselves to the interfaces between the fibers and the
liquid medium as to a trellis. As was mentioned before,
the shape of these cells is determined by the number
of protrusions from their surface. One can go one step
further and prove that the number of such processes,
in turn, is a function of the fibrous constitution of the
medium.

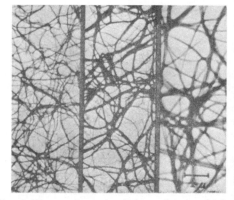

FIG. 3. Electron micrograms of plasma clots coagulated at
different pH values (from left to right: alkaline, neutral acid).

The relevant interaction is between the cell surface and the fibers in its microenvironment. To understand such surface reactions, one must give up in the first place the outdated notion that the cell surface is a sort of static cellophane-like bag. This may be true of some specialized cell types, for instance, the cellulose membrane of a plant cell, the capsule of a bacterial cell, or the envelope of a red blood corpuscle. But in most types of cells, the surface is far from stable and is by no means of identical composition and state all over the cell. In tissue culture, this state of disequilibrium manifests itself in the continual thrusting forth and withdrawal of surface processes at the expense of cellular energy, showing great variations of the contractile force along the surface. Temporarily weaker points along the surface thus become outlets for thrusts. Such microleaks or "herniations" may occur at random or they may be determined systematically by outside factors, of which one of the most important is the encounter of a fibrous-liquid interface with the cell surface. The fiber contact, in a sense, pricks the cell surface locally. The strength of the resulting strain can be shown to vary with the size of the fibers. Hydrodynamic, viscous, and elastic competition for outflow favors fibers which have a larger diameter (Fig. 4). Consequently, the prevalence of a few major protrusions over minor ones may be expected to be the greater, the larger the average fiber size is in the medium.

These predictions have been tested (jointly with B. Garber) by culturing cells in plasma clots containing fibrin fibers of different average dimensions. The average diameter of such fibers is a function both of pH (Fig. 3) and of plasma concentration, larger fibers being formed at either lower pH or higher plasma concentrations. It actually was found, in line with expectations, that the ratio of bipolar cells (few processes) over multipolar stellate forms (many processes) increased as a steady linear function as the plasma concentration was raised

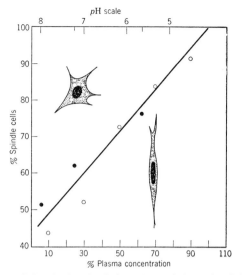

FIG. 5. Distribution of cell shapes in populations cultured in plasma clots of different average fiber sizes.

or as the pH during clotting was lowered (Fig. 5). Other criteria of cell shape, such as the ratios of length over width of the cell bodies and of the cell nuclei, showed correspondingly systematic changes. In other words, the whole gamut of shapes displayed by this particular cell strain could be written in a single formula derived from insight into the mechanisms by which deformations into one or another shape come about. The point to stress is that one gets further by studying realistically the formative process rather than by dwelling upon pictorial samples of forms already achieved.

At the same time, it must be stressed that the formula is a probabilistic one, for to predict just what any individual cell will look like is impossible because of the accidental nature of the details of its surroundings. The microclimate and the microenvironment of the individual cell are unique, unknown in each particular instance, and this establishes a certain degree of variance for the actual expressions within each cellular system which is built into its nature.

In a different medium, the same cell strain would give different responses. For instance, these cells, when suspended in a liquid medium without interlaced fibers, would manifest the inequalities along their surfaces by blunt herniations rather than by the pointed protrusions noted along filaments. As a result, such cells appear to be blistering and boiling along their surfaces as can be seen in cinematographs of unattached single cells, and is well known from the loose cells in the late stages of cell division. Conversely, cells of different tissue types would show morphological responses different from the strain just exemplified. The more a cell is given to producing internal structures that serve as a cytoskele-

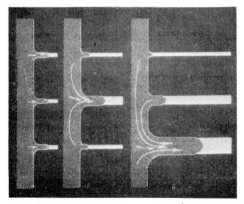

FIG. 4. Microherniations of cell content at intersections of fibrin fibers of various sizes with cell surface. Arrows indicate protoplasmic outflow.

IV. ON FORM AND FORMATIVE PROCESSES

FIG. 6. Effect of regionally varying degrees of tensions on the organization of a fibrin network and, through it, on the morphology and orientation of enclosed cells.

ton—such as in sperm cells, higher forms of protozoans, or muscle fibers—the less will its shape be codetermined by physical constellations in its environment, although even in the most extreme case the polarization of the axiated system of the cell is presumably still a response to external gradients or to other inhomogeneities of the environment. There is an enormous task laid out here for future detailed investigations on the physical factors involved in cytogenesis and morphogenesis.

As a further example of the complexities involved, one can again cite experiments with connective-tissue cells in tissue culture. A long time ago, I became preoccupied with the role of external factors in guiding the oriented movements of cells. In 1927, I succeeded in orienting cells in tissue culture by applying stretch to the blood-plasma clot; the cells assumed a common orientation along the lines of tension. The analysis of this phenomenon over the years has led to the following summary conclusion: The primary effect is on the orientation of molecular chains which become aligned in the direction of the stretch. When cells are contained in such a medium, their shape—that is, behavioral deviation from a sphere—simply reflects the degree of structural organization of the underlying submicroscopic fibrous network. Where all fibers run parallel and, therefore, where one direction only is open to the cells, they naturally become bipolar, all pointing in the same direction (Fig. 6). With decreasing amounts of stretch and correspondingly less rigorous orientation of the fibrin, cell forms grade over from the strictly determined spindles at one extreme to the only probabilistically describable cells of the multipolar sort mentioned in the earlier example. In the present case,

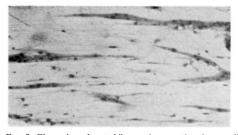

FIG. 7. Elongation of erstwhile-round connective-tissue cells placed on a sheet of parallel collagen fibers from the interior of a fish scale.

however, the average shape does not vary from clot to clot in accordance with the average constitution of the clot, but varies locally within the same clot in accordance with a systematic variation of an extrinsic factor, namely, stretch.

However, the immediately relevant thing for the orientation of the cells is the orientation of the fibrous pathways which they are bound to follow, and it is immaterial that, in those earlier experiments, the agent for producing such fibrous orientation was stretch. Even though the normal organism frequently uses stretch in that capacity, other forces also are at work to produce oriented fiber patterns, which likewise act as guides for cells. The inside of a fish scale, for instance, contains layers of collagen fibers beautifully arrayed in parallel lines, and when loose spherical cells are deposited on such a sheet they immediately assume spindle shapes, with the axes strictly aligned along the fibrous substratum (Fig. 7). Therefore, to carry the analysis

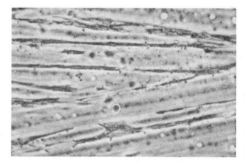

FIG. 8. Phase-contrast photomicrogram of cells which have become elongated along streaks of silicone paste on glass (round white and dark splotches are clusters of cells over nonadhesive silicone where they have been unable to attach themselves).

further, one must study why a round cell on a linear track becomes correspondingly deformed.

A clue to the mechanism operating in this case may be obtained from the following experiment. A glass slide first is streaked with silicone paste or with cholesterol. A loose cell suspension then is placed on top in an appropriate liquid medium. As the cell surface does not adhere to cholesterol but does adhere to glass, the cell becomes drawn out along the striplets of bare glass; hence, it is oriented parallel to the streaks (Fig. 8). Shape is determined here by the differential adhesion of different parts of the cell surface to the substratum. An even more impressive example of this kind is provided by cells which have been seeded out on glass scored with a microlathe (experiments performed with A. Cecil Taylor). These cells become deformed from a spherical to an elongate shape in the direction of the micro grooves (Fig. 9).

What then precisely is the mechanism of this response? To understand it, consider an older experiment in which spindle-shaped cells were made to spread out flat when they were forced to splash against a smooth surface from above. Such a surface offers the cell innumerable directions in which to radiate at the same time. Thus, the periphery of the cell actually expands concentrically, flattening the cell into a disk (Fig. 10). This transformation is accompanied by profound changes in function—the cell becomes phagocytic—and in the distribution of intracellular materials, further emphasizing the intimate dependence of chemical activity on the physical constellation in a cell. Similarly, a freshly isolated cell set down on scored glass spreads out at first in all directions, but only those sectors of the cell border which lie in a linear direction of the substratum retain a foothold and continue to advance. The other sectors are retracted and become consolidated as cell flanks.

Each cell surface thus acquires radically different properties at the ends and along the flank. The two "engines," one at either end, remain active and by extending in opposite directions simply draw the cell out

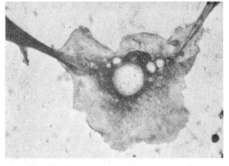

FIG. 10. Spindle cell expanding into a flat disk on impact with glass surface.

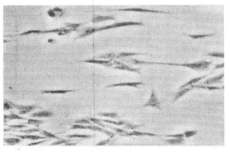

FIG. 9. Cells which have become elongated along the microgrooves of scored glass.

into an elongate form. Such elongation again has further consequences for the cell. For instance, the mitotic spindle preparatory to cell division mostly will become aligned with the long axis of the cell so as to make elongation a major factor in the orientation of cell growth. But because of the opposing tugs, to which this elongate cell is subjected from its two active ends, there is no net forward movement. Such a cell simply shuttles back and forth about a stationary position, comparable to Brownian motion, depending upon which one of the two ends happens to have the upper hand at the moment.

Thus, although the foregoing has brought some deeper understanding of cellular orientation, it tells nothing about the mechanism of cellular locomotion, which remains one of the basic unsolved biophysical problems. Something is known about those cases in which cells have special locomotor organs, such as cilia or flagella, but when it comes to cells moving with their free, unstable surfaces devoid of structural specializations that could be related to motility, ignorance is profound. Cyclosis in plant cells, the gliding of slime molds, the

shift of a sheet of skin to cover a raw wound, the invasion of tissues by metastasizing cancer cells, or the penetration of leucocytes through capillary walls to converge on a focus of infection—in none of these cases is it known just how the cell achieves these movements, except that it is suspected that gelation-solvation cycles or contraction-relaxation alternations may somehow be involved. This is one of the most neglected areas of physical approaches to biology. Not only is the mode of locomotion shrouded in ignorance, but also there is equal uncertainty about the reasons why a free cell, which can extend in many directions, often advances steadily in one direction to the exclusion of others. As was just said, a cell left to its own devices in an isotropic environment strays at random with no net dislocation.

There have been many theories and speculations about the directive movement of free cells. Once again, one can illustrate how progress has come from eliminating among such competing concepts the ones ruled out by factual analysis. Turning again to the sample object of cells in tissue culture, contrary to the scattered population of stationary isolated cells discussed before, the cells of a solid fragment of tissue explanted into culture behave quite differently. They move in droves from the explanted piece into the empty medium, giving rise to the well-known phenomenon of "peripheral outgrowth." Why do they move centrifugally? For a long time, I had considered this question synonymous with that of cell orientation and had invoked the fibrous guide rails as explanations of "oriented movement." But, as was just explained, orientation and displacement are two different things, and for the displacement there has been no crucial explanation. "Tropisms" and gradients of various kinds have been proposed to explain the phenomenon. It has been assumed that cells respond either positively or negatively to differential concentrations of hypothetical "attractive" or "repelling" substances emanating from point sources, even though it has never been possible to demonstrate just how a cell could translate such directional cues into actual convection towards or away from the source. Recent observations on our tissue-culture strain have, however, turned up a wholly different story, and it is this.

The only way to get locomotion of a cell with two motor engines at opposite ends is to stop one of the engines, at least for a while. Then the remaining engine can tow the cell away without opposition. This is precisely what happens in tissue cultures whenever free tips of two cells make accidental contact with each other: the colliding ends become temporarily paralyzed. A wave of retraction runs over the affected processes, which become partly detached from the substratum, and the two cells thus come under the exclusive pull from their remaining motile ends facing away from each other. Thus, they move apart in opposite directions. After some time, the paralyzed ends gradually recover their motility and again take hold on the ground, so that the cells are stalled once more, but at a greater distance from each other.

Extrapolating this process to a cell population with a gradient of densities, such as a tissue explant, it is evident that, of a pair of cells moving apart, the one shifting peripherally has a lower probability of encountering another cell than has the one which moves toward the explant. Statistically, this leads to a prevalence of outward migration even though cells are actually free to move in any direction. Eventually, a situation obtains in which the cells have become so widely spaced that random collisions are no longer likely; at this point further migration ceases. Thus, the only gradient which plays any role is a gradient of population density. The phenomenon is formally comparable to the diffusion of molecules from regions of higher to lower concentration, except that one deals with the random collisions not of molecules but of complex entities which may be treated as units for the purposes of description. Again, a close observation of the behavior of the real living object has brought answers far more concrete than what could be anticipated from such generalities as "attractive" or "repulsive" forces between cells. Parenthetically, it should be stressed that the type of contact separation between cells mentioned above is characteristic of the species of connective-tissue cells here described, as well as of several similar strains, but it does not apply to other cell strains, for instance of the epithelial variety, where reaction of two cells on contact can be just the opposite—namely, their drawing much closer together, provided they are both of the same kind. Further details on this behavior are given by Weiss (p. 449).

As still another instance of the dangers inherent in dealing with the living cell in terms of broad generalities, one may once more cite the behavior of cells of diverse shapes observed in plasma clots of different pH or concentrations. As was reported before, such cells have a variable number of processes, each of which now may be thought of as a train behind an "engine." In such cultures, it is possible to calculate the average rates of advance of the various cell types in a given direction by dividing the total distance spanned in a given period by the time elapsed. Since such cells move neither steadily nor in a straight course, however, any such average value for "rate of advance" is wholly unrepresentative of the true velocity of cell motility. As plasma concentration is increased, the cells tend to become bipolar. This means, inevitably, that the average rate of locomotion increases simply because the course is less tortuous and the cell is stalled less frequently by simultaneous divergent pulls from multiple processes to which multipolar cells are subjected. Consequently, in the lower concentration range, where most cells are multipolar, there is a progressive increase of the average rate as the number of processes declines toward the liminal value of two (Fig. 11). Once the great majority of cells have attained bipolarity, the average rate of advance remains constant, expressing more nearly the true velocity. By comparing Fig. 11 with Fig. 5, it can be seen that, at a plasma concentration above 50%, where the "rate" curve levels off, more than three-fourths of the cell population are actually bipolar. Just as one cannot tell the true speed of a railroad train if the times of departure and arrival at the terminal are the only data available and the frequency and duration of station stops on the way are not taken into account, so the mere establishment of the average rate of locomotion of a cell under various conditions has little practical meaning. Yet, the literature is full of examples in which such average rates have been used to assess the effects of a variety of agents or drugs on cells, without due attention to the effects of these agents on the medium, which may alter the whole setting in which

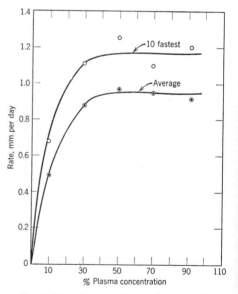

FIG. 11. Rate of progress of connective-tissue cells in plasma clots of different constitution.

the cells move. Without knowledge of these effects, no meaningful comparisons can be made. This illustrates some of the hazards of operating with *average* quantities when one deals with systems such as cells, the composition and behavior of which are inhomogeneous in space and time.

The foregoing discussion has illustrated how a controlled modification of the medium can indirectly modify cell morphology and behavior, including locomotion. It has explained the orienting effect of tensions on the fiber systems of the medium; how such structures evoke conforming organization in the cell population residing in them; and how cell-to-cell interaction and, in the last analysis, population dynamics govern cell locomotion. In this chain of events, an outside experimenter applying tensions to a culture medium appeared as the primary agent. This, of course, immediately raises the question as to what factors serve this organizing function within the living body. The sole agency of the body is its own cells and it was comforting, therefore, to find that cells, by their own activity, can create the type of orderly structural patterns in the intercellular spaces which had been imitated crudely by extraneous tensions. The cells engender in their own environment physical conditions and orderly restraints which then in turn play back on them as guides and regulators of their own behavior. Thus, a further step of complexity is added to the picture. There are innumerable examples, mostly poorly understood, all showing how, through an enormous variety of mechanisms, the same basic principle is served; which is that the cell population, through its products and interactions, sets up conditions modifying the behavior of the enclosed cells, and thus often leading to new settings and interactions which may cause further alterations of the cells, and so forth, in sequences of interactions of ever increasing complexity.

To illustrate this, consider a piece of tissue embedded in a fibrous network and let the boundary of the tissue

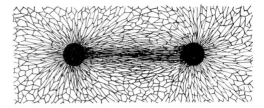

FIG. 13. Effect of two simultaneously contracting centers on a common meshwork.

expand, as happens, for instance, in a vesicle or tube or cyst swelling by the secretion of liquid into its closed lumen. Evidently, as shown in Fig. 12, top, the meshes of the fibrous medium become circumferentially compressed so as to assume a predominantly tangential orientation around the fragment. Cells happening upon such a territory would obviously be forced to circle round and round producing an envelope or tunic. Many of the connective-tissue sheaths and capsules in the meshwork of the body owe their origin to this mechanism. By contrast, if somewhere in the meshwork there arises an area which contracts, the fibrous components will be gathered toward that center in a predominantly radial orientation (Fig. 12, bottom). Cells in such zones then become likewise disposed radially. This type of effect obtains frequently in the vicinity of rapidly proliferating cell groups. It seems that chemical agents, as yet undefined, are being discharged by such cells which cause intensive synaeresis of the surrounding colloids, which means condensation of the fibrous components with loss of bound water, the resulting local shrinkage being much greater than the gain of mass by cell growth. In other words, a purely *scalar* change in a piece of tissue—increase or decrease of volume—can, through the intermediary of a fibrous continuum, translate itself into marked *vectorial* effects, establishing well-defined geometrical and structural patterns. A further degree of ordered complexity is introduced if two or more cell masses are explanted together in a common clot (Fig. 13). Through their joint constriction effects, they deflect the fibrous meshes of the intervening medium into a line connecting the active centers, and this, of course, constitutes a path for direct cell traffic between them. The cells growing out from the two centers simply follow the submicroscopic bridge which has been laid down for them automatically by the tension-engendering chemical activity of their sources (Fig. 14). Here is one primitive example of how chemical action can translate itself into physical organization.

Extrapolating briefly from these model systems of living cells, one may assume that the same sort of intimate interdependence shown here for the microscopic dimension repeats itself both in submicroscopic and in higher supracellular dimensions. One is led to the conclusion that there is a tie between physical structure and chemical activity which is indissoluble and which

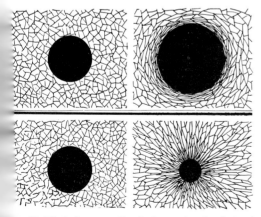

FIG. 12. Effect of an expanding (top) or contracting (bottom) center on the architecture of the surrounding meshwork.

IV. ON FORM AND FORMATIVE PROCESSES 379

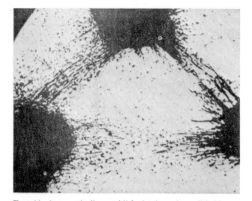

FIG. 14. Automatically established triangular cell-bridge connections between three separate tissue cultures (dark masses) in a common blood-plasma membrane.

it is not enough to assert, but which has to be explored systematically. Where, in order to study a particular metabolic reaction in isolation, the biochemist provides optimum conditions for that process, the cell produces the unique prerequisites for that same reaction only in certain strictly confined localities. It does it through the metabolic products of other chemical reactions going on in another equally confined sample of cell space. Thus, one system feeds another and, in turn, is fed by a third, in a vast system of mutually interdependent, symbiotic, and harmonized partial reactions subsidiary one to another. This interdependence, characteristic of the living state, is what I once termed "molecular ecology." As a field of investigation, it barely has reached infancy. Yet its significance is pointed up by the fact that, in the living cell, the various biochemical partial systems coexist in a common space, without preformed rigid partitions, but rather compartmentalize themselves by structural effects of their own activities of the sort exemplified in crude and elementary fashion by the case just cited. Physical structure and physicochemical conditions limit the types of enzymatic and other chemical reactions that can go on in a given spot —for instance, along interfaces of fibrillar, lamellar, or corpuscular systems—while the resulting reactions, in turn, modify the physical substratum; and by continual interplay of this sort between physical structure and chemical action, the cell system passes first through its progressive developmental transformations and then is stabilized in the steady state of maturity. To some extent, physical structure then is frozen into static arrangements of cytoskeletons, but even then physical structures still are regenerated continuously by cellular activity, leaving at least part of the cellular system in a state of incessant development and self-renewal.

A realistic concept of cell behavior also must take into account the limits set to interactions between distant parts by the formation of compartments within compartments. A diagram of the organism (Fig. 15)

would represent it as a system of concentric shells, with the gene in the center enclosed in the chromosome, which is enclosed in the nucleus, which is enclosed in the cytoplasm, which forms part of a tissue, which forms part of the organism, which is surrounded by the external environment; for simplicity, cytoplasmic particles are omitted. No outer agent can influence any of the inner shells except through the mediation of the shells in between, which may or may not modify that factor during its inward passage. Conversely, products of inner systems may not reach outer shells as such, but may be significantly screened and altered in transit. The arrows in the diagram indicate the complex network of relations that one must bear in mind. Oversimplified mental pictures form when, for instance, the simple statement is heard that a gene "controls" a particular feature of the organism, without allowing that it can do so only through interactions with the outer shells, which in themselves have become progressively modified in their long developmental history by countless chains of interactions with other shells, including, of course, the innermost, the gene. Since it has been my assignment to sketch the living cell as it truly is in all its highly ordered complexity, I feel compelled to caution against the illusion that a simple statement, such as, "a gene controls a character" reflects any similar degree of simplicity in the phenomenon covered by the statement.

A final example projects the principle of interaction between physical structure and chemical activity upward into the realm of supracellular order, a field so baffling in its problems that many an investigator prefers to look the other way when he encounters them. The example is chosen from an almost diagrammatically simple object which, because of this, holds at least some promise of more penetrating analysis by the combined physical and chemical tools now at one's disposal. It refers to the origin of the internal architecture of cartilage. Cartilage is formed by groups of cells producing

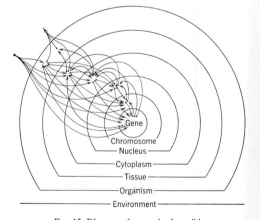

FIG. 15. Diagram of network of possible interactions in an organism.

ground substance between them. That ground substance has been identified partly as a mucopolysaccharide—chondroitin sulphuric acid—and collagen. But cartilage is not just chemical substances; it assumes characteristic shapes and configurations, depending upon its sites, owing to developmental processes under genic control. The nature of the problem becomes clear when one contemplates that the peculiar pattern of convolutions of one's earlobes, which are but covered cartilage, is an individual inherited characteristic. How do such patterns form? Cartilage of the limb is different from cartilage around the eye; the former grows in compact, whorl-shaped masses, the latter develops as a flat plate. It has long been known that, if the cell group that is destined to form limb cartilage is reared in tissue culture, it will grow according to the standard limb pattern, giving rise to recognizable skeletal elements. Similarly, the author observed years ago (jointly with R. Amprino) that the precursor cells of the scleral cartilage around the eye, after explantation as an intact group, go on to form cartilage in the shape of a plate. Evidently, in either case, the explanted tissue complex contained some physical properties that guided the cells contained in it into their respective typical arrangements and growth patterns. The crucial property differentiating between limb and eye cartilage had to be thought of as inherent in the block of tissue as a whole, the further development *in vitro* merely amplifying some distinctive architectural pattern already present in the tissue fragments at the time of their isolation from the embryo.

To test this conjecture, the author (with A. Moscona) recently resorted to a technique that permits one to disintegrate a tissue into its constituent cells and thereby to destroy any intercellular structures and supracellular arrangements that could have acted as guides for subsequent development. Cells can be separated from one another and from their matrix by trypsin, washed in a loose suspension, and then seeded out in a tissue culture where they can aggregate into random clusters. More about the manner of aggregation is reported in another article (Weiss, p. 449). When a piece of prospective limb cartilage and a piece of prospective eye cartilage were dissociated in this manner into their component cells and the cells of each type were permitted then to aggregate in tissue culture and to continue with their actual cartilage-forming activities, it turned out that the cells from the limb site produced cartilage of the typical whorl-shaped massive limb pattern (Fig. 16, top), while the cells that had originated at the eye site produced the plate-shaped laminated structure to which they would have given rise prior to their disaggregation (Fig. 16, bottom). In other words, the "blueprint" of the architecture of the group performance is ingrained in each individual cell and not just carried over into the culture by the block of tissue as a whole. Now, how to convert this figure of speech into concrete terms is problematical. But one may speculate that the architecture lies in the specific

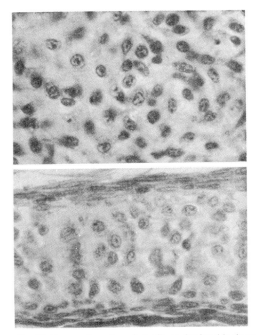

FIG. 16. Cross sections through cartilages developed in tissue culture from dispersed and reaggregated precartilage cells from a prospective limb (top) and from a prospective eye coat (bottom) of the chick embryo, forming whorl-shaped and lamellar structures, respectively.

ground substances secreted by the cells. One would have to suppose that each cell type secretes a complex ground substance of a distinctive pattern of macromolecular stacking or crystallinity of such a kind that it would predispose a planar array of the mass in the form of layers in one case, and a more massive isotropic arrangement in the form of three-dimensional whorls in the other. The cells then would dispose themselves in conformity with these structural patterns of ground substance for which they had furnished the elements in the first place.

This is sheer conjecture. No facts are known that would either support or contradict it. Yet this very uncertainty helps to point up the immensity of ignorance in matters of supracellular organization. If one is convinced that such higher-order organization is to be explained solely in terms of cellular dynamics, then one must raise one's sights to the domain where those phenomena occur, which is no longer the intracellular microcosm.

The study of tissue architecture, used here as an example, may be one of the easier inroads into the maze of perplexing problems presented by the orderliness of the complex living organism, as against the relative disorder and simplicity of its shattered fragments, on which one preferably concentrates, mostly after homogenizing them either physically or conceptually.

IV. ON FORM AND FORMATIVE PROCESSES

The lessons of this story repeat themselves as the viewpoint is shifted from the organization of the body in its structural aspects to the supracellular coordination and ordering systems of its functional activity, whether this be the homeostatic maintenance of blood composition, the integrative action of the nervous system in behavior, or the mobilization of defense and repair mechanisms in response to pathological disturbances. Of course, the operative tools in all of these performances are the individual component cells, which in turn operate through physically ordered subsystems in incessant chemical interaction. But unless one remains cognizant of the fact that the level of the organism is reached from the level of molecular biology not in one single jump over the conceptual gap customarily bridged by the word "organization," and unless one learns to think in terms of a hierarchy of ordered systems, each one with some degree of identity and stability even on supramolecular levels, one obscures rather than enucleates the problems to which thinking and research must be directed. The most impressive feature of a cell is not the constant flux, reshuffling, and variability of its population of molecules and particles (of which our cinematographs presented at the Study Program have given a clear expression), but the fact that, in spite of this ever-present change, each cell remains so remarkably invariant in its total behavior; that indeed, as an entity each behaves so much like millions of other entities equally variable in inner detail, that one comes to recognize them as essentially alike. Such relative invariance of the whole presupposes the harmonious subordination of the behavior of the parts to the conditions of the collective group. It presupposes that the free interactions among the subunits are subject to restraints, the nature and direction of which vary adaptively with the state of the system as a whole. I can think of no more propitious introduction to a biophysics program than by re-emphasizing that the restraints in question are not the sole province of chemistry, especially the stoichiometric branch, but that they include crucial restraints of *physical* nature, with nonrandom molecular arrays a most prominent and analytically most promising feature.

Awareness of the existence of these problems in no way detracts from the pragmatic value of analyzing each elementary component process in its own right and of using whatever models, however simplified, may be found to be constructive as aids to understanding. But it is precisely the phenomenal advances made in the study of the partial and isolated system, in contrast to the dearth of information on the *organized interactions of such systems in the living cell and organism*, that lead me to conclude with a plea for a more balanced effort and more extensive occupation with the latter problems than there are at present. Cellular and tissue dynamics are population dynamics of species of molecules, of cellular subunits, and of cells. Such group dynamics cannot be derived solely from looking at the members of the population in isolation, but only from a study of the ecology and technology of their behavior in the group. It is hoped that this modest effort at presenting the case of the living cell not only has given a fair introduction to the problem but also has shown that practical techniques for its eventual solution are at hand, with physical and chemical approaches indissolubly interwoven.

BIBLIOGRAPHY

P. Weiss in *Chemistry and Physiology of Growth*, A. Parpart, editor (Princeton University Press, Princeton, New Jersey, 1949), p. 135.

P. Weiss and B. Garber, Proc. Natl. Acad Sci. U. S. **38**, 264 (1952).

P. Weiss, J. Embryol. Exptl. Morphol. **1**, 181 (1953).

P. Weiss in *Analysis of Development*, B. H. Willier, P. Weiss, and V. Hamburger, editors (Saunders Company, Philadelphia, 1955), p. 346.

P. Weiss, J. Cellular Comp. Physiol. **49**, Suppl. 1, 105 (1957).

P. Weiss, Intern. Rev. Cytol. **7**, 391 (1958).

Reprinted from REVIEWS OF MODERN PHYSICS, Vol. 31, No. 2, 449–454, April, 1959
Printed in U. S. A.

CHAPTER 17

Interactions between Cells

PAUL WEISS

The Rockefeller Institute, New York 21, New York

INTERACTIONS among cells are the means by which the cell community of the organism establishes and maintains its organizational harmony. They are so numerous and varied that it would take more space than is allotted here just to draw up a reasonably comprehensive list. Yet, knowledge about them is still very scanty. Fascinating progress has been made in the study of some of the biochemical and biophysical components of the living system. But knowledge seems to have grown the faster, the smaller the sample of the living system that was taken under investigation. The major advances were made down at the molecular level. The task of dealing with the larger cellular level, therefore, involves a re-entry into areas of uncertainty, and the course from the introductory article on cell dynamics (Weiss, p. 11) to the present article marks a full circle from relative ignorance through knowledge back to relative ignorance. The study of cells in interaction deflates complacent notions that all the major facets of cell life are truly understood even in principle.

"Interactions between cells" covers practically everything that is going on in organisms, and obviously this account must confine itself to a few crucial examples. There are essentially two ways for cells to interact: either a cell elaborates a diffusible substance which affects another cell at some distance, or a cell transmits an effect to another cell by direct contact. Since mediation by diffusible agents, as in the hormone system, is more widely studied and better understood than are the contact interactions, the emphasis here is on the latter. A more detailed account can be found in the author's recent review article on "Cell Contact."[1]

First, a few words on what is meant by "contact." It has been very easy to define contact between cells in terms of observations with the ordinary light microscope. If microscopically two cell borders came so close as to merge into a single line, there was contact, whereas a microscopic space in between signified lack of contact. Conversely, any interruption of the microscopic outline between two cells was interpreted as "protoplasmic continuity." But evidently, such questions of "contiguity" vs "continuity" between cells are merely questions of the resolving power of the instruments at hand it is not surprising, therefore, that the introduction of electron microscopy has brought a marked change of outlook.

In the first place, electron microscopy has revealed a higher degree of organization at cell surfaces than previously assumed. As an example, one can cite the larval amphibian skin referred to in the introductory article (Weiss, p. 11). The underside of the epidermal cells, which rest on the basement membrane, is dotted at regular intervals with submicroscopic bodies, about 1200 A high, consisting each of two electron-dense round plates connected by a lighter neck.[2] The outer one of the plates forms part of the cell surface. This surface itself is separated from the underlying basement membrane by a gasket-like granulated film of a few hundred Ångströms thickness. The cell surface thus is actually a mosaic of patches of supramolecular order and different chemical and physical properties, which have been partly identified. This was done by applying various enzymes to fragments of live skin before they were fixed in osmium tetroxide and prepared for electron microscopy. In pancreatic lipase, the dark plates lose their osmiophilia, indicating that normally they have a high lipid content. In distilled water, the neck of these bodies swells and breaks, leaving unconnected double plates. Thus, the neck region is hydrophilic. Salivary amylase dissolves the film through which the epidermal cell is attached to the basement lamella, suggesting abundance of carbohydrate. It also erodes the parts of the cell surface between the dense bodies, but leaves the latter intact and protruding, thus identifying them as solid bodies.

Such observations demonstrate that it would be quite mistaken to consider a cell as being equal and uniform over its whole surface. Characteristically, this surface pattern is confined sharply to the portion of the cell that is in immediate apposition to the basement membrane, thus indicating a direct contact interaction between cell and substratum. This supposition is confirmed by experiments in which the contact between the two structures was first broken and then restored, as follows. When part of the skin is injured, the epidermal cells near the wound roll off. The dense bodies, which seem to serve as suckers to attach the cells to their substratum, become detached and are resorbed within a few days. They are re-formed during wound healing as the epidermal cells migrate over the defect, but are precisely confined to that fraction of cell surface now in fresh contact with the substratum. In other words, the cell has a mechanism for producing this specialized apparatus which responds to the interaction between cell surface and the underlying carbohydrate-rich film. As a result, an "induction" on the submicroscopic level occurs. The cell is induced along a geometrically and physically defined surface to display a specific fraction of its synthetic repertory. Other instances of contact induction are given later in this paper.

When two cells are in contact, the problem becomes more complicated. At the contact points between two epidermal cells, single dark plates are present, the well-known nodes of Bizzozero. Electronmicroscopically, they show as a single dark plate in each of the contacting cells with a lighter space in between.[1] Odland[3] has recently scanned this region densitometrically in superior electron micrographs and has found an additional dark line in the center of the light interspace. This would indicate that the molecular array separating the two cell surfaces is not random, but has some degree of spatial order. It may be that the cell surfaces are linked by macromolecules normal to the surfaces of some 100 Å in length, which is the order of width of the light space. Gaps of similar dimensions have been found between practically all cells that make intimate connections with each other, including synapses between nerve cells. It would seem best, therefore, to define two cells as being in contact, operationally speaking, when they are separated by a space whose contents are not subject to random perturbation. Depending on the dimensions, several factors might operate to limit perturbation of the interspace. For example, macromolecular compounds moving out of the cells may form bridges between organized parts and establish special submicroscopic attachments or interaction points between the adjoining cell surfaces. Evidence from tissue culture, obtained over a decade ago,[4] indicates that certain cells, and perhaps all cells, when left to their own devices, surround themselves with characteristic colloidal exudates, which must be taken into account when cell-to-cell relations are analyzed. Such surface coronas, extending for unknown distances into the cellular environment, might be major factors in immune reactions, cellular aggregations into colonies, phagocytosis of specific substances, selective associations among nerve fibers, and the like. It is equally possible that cell surfaces react with each other specifically without such mediators. A few examples are cited in the following.

In experiments on the development of nerve fibers, it has been found that fibers of the same type have a selective affinity to one another—motor fibers applying themselves to motor fibers, and sensory fibers to sensory fibers, and within each class, each subgroup to its corresponding type.[5] Only by virtue of such specific grouping is it possible for a surgeon, for instance, to expose the spinal cord and sever discriminately a bundle of pain fibers or of fibers for deep perception. Such fibers would not run in common bundles unless they had grown out that way, and since they do not develop all at the same time, the older fibers must have served as guides to latecomers of the same character.

The same principle applies, even more subtly, to the relations among the cells within the central nervous system on which the coordination of neural functions depends. Our current concern with the nervous system as a system centers mostly around such matters as geometric properties, electric parameters and time constants—length of interconnections within the network, number and distribution of branches, temporal characteristics of the individual units, synaptic resistances, etc. We generally ignore the fact that there is great biochemical diversity within the system far beyond the gross distinctions of cholinergic and adrenergic portions, which makes it work as something more than a monotonic network of conducting fibers. Developing nerve cells have highly selective discriminatory affinities by which they "recognize" each other and their surroundings. This introduction of the anthropomorphic term of "recognition" is not anything one need apologize for so long as physicists speak of electrons "seeing" each other. Recognition of one cell by another may be based on conforming charge distributions, on conforming molecular groupings, or on as yet wholly unsuspected mechanisms. The problem of explaining such recognition is of course encountered in many other biological phenomena, e.g., in enzyme-substrate relations and in antibody-antigen reactions. With a few notable exceptions,[6] affinity reactions among somatic cells, however, have received little attention and even less methodical study, even though there is ample evidence for the widespread occurrence of such selective behavior.

To cite a specific example, cells can "locate" their proper destinations in the body even if they are deprived of their customary routes for getting there. This has been shown[7] by letting embryonic cells of known destination be distributed at random through the blood stream of a host embryo. In order to be able to identify the injected cells, one chooses cells which carry a marker. We chose the precursor cells of pigment cells which, when introduced into nonpigmented breeds of chicks, would reveal their origin by synthesizing black melanin. It was found that after an injection of such cells into the blood stream, about 8% of the host chicks later carried scattered patches of black pigment cells in their feathers and skin. These pigment cells were always situated in the precise positions where such cells would have resided in the donor animals, and in no case was a pigment cell ever observed to have become lodged where pigment cells do not normally belong. One must conclude, therefore, that donor propigment cells had colonized only specified areas in the host embryos where such cells normally end up, which, in view of the random dissemination of the dissociated cells in the host body, implies the existence of a mechanism for highly selective localization. As few as one or two propigment cells, when arriving by chance at an appropriate site, "recognize" it and settle down to proliferate and differentiate into pigment cells. Cells that miss encountering within the embryo an environment favorable to their type fail to become lodged and to differentiate, and presumably are resorbed.

On the other hand, outside the embryo, on the yolk sac, where specific sites for embryonic cells are lacking, clumps of injected cells get stuck, and their fate led to a further significant extension of these observations. It was discovered[8] that, where random mixtures of embryonic cells from a variety of sources (nerve, muscle, cartilage, glands, etc.) had become lodged on the yolk sac, they gave rise not to indiscriminately mixed structures, as might have been expected, but to quite harmoniously organized organ complexes. Some degree of self-sorting according to types must have taken place among these scrambled cells after they had become trapped.

Further conclusive evidence of "self-sorting" was found in tissue cultures in which random mixtures of suspensions of cells of different embryonic origins —for instance, cartilage and kidney cells—had been combined.[9] These cultures developed compound structures in which islands of pure cartilage were sharply demarcated from blocks of pure kidney tissue, indicating that cells had become re-associated like-to-like. Thus, even *in vitro* cells, which have acquired in their prior embryonic history some biochemical differential related to their subsequent development as either kidney cells or cartilage cells, can assort themselves according to type. Since the cells make contact only with their surfaces, one must assume that some property associated with differentiation into kidney cells or into cartilage cells has imparted type-specific markers to their surfaces for mutual recognition. But how such "recognition" leads to active sorting remains obscure.

How do mixed cell populations achieve this orderly reassortment? Do like cells "attract" each other? Or do they "recognize" each other only after chance encounters? A clue came first from some observations on selective wound healing after the grafting of various kinds of epithelia into skin gaps in the flanks of amphibian embryos.[10] The first coverage of a wound is effected by the migration of epithelial cells of the wound edge over the raw surface. When these advancing cells reach a graft, fusion of the two fronts occurs only if the graft contains an epithelium of a type which normally borders on skin epidermis. In that event, the advancing cells stop in their tracks, so to speak, merge into a uniform sheet with the graft, and no further proliferation occurs. Implants of skin, or of cornea or oral lining—which are normally continuous with skin—are accepted in this manner. But, when a fragment of lung or gall bladder or esophagus is implanted in a skin wound, the migrating skin epidermis is not halted on contact with those foreign-type cells. The epidermal edges glide over or under the graft and continue until they meet with the skin advancing from the opposite side.

Results of this kind have long been known in surgery and lead to the conclusion that, whether we have a plausible explanation for the phenomenon or not, cells of identical or conforming constitution (as epidermis, cornea, and oral lining) will recognize each other as akin and stay put; whereas cells of unlike character develop a reaction which will cause them to separate.

With these facts as background, we proceeded to study the contact reactions between different kinds of cells directly *in vitro*,[1] by taking phase-contrast time-lapse motion pictures of encounters among cultured cells.* If the cells that make contact are all of the same kind—all liver cells, for example—they aggregate and stay together. The free borders of roaming cells are in constant motion; but wherever two like cells touch each other, that portion of their surfaces becomes quiescent. The bond between them, however, is not static; there is no cement which sticks them together as in an immunological precipitin reaction. On the contrary, the cells continually glide over one another, constantly changing their positions relative to each other and, by the time others have joined them, their positions in the group.

On the other hand, if cells of two different types are cultured together (e.g., lung and liver or lung and kidney), the result is quite different. As the loose cells stray about, they collide at random. There is not the slightest indication that they would react to each other's presence unless or until their free borders touch. Neither do like cells "attract" each other, nor do unlike cells avoid each other. Mutual recognition and consequent discriminatory behavior are decidedly contact responses. All cells, whether alike or not, make primary contact indiscriminately. But, whereas like pairs thereafter draw closer together and remain joined, unlike pairs separate secondarily by reciprocal withdrawal of those parts of their borders that had been in fleeting touch. While the mechanism of identification undoubtedly resides in the surface, the interior of the cell seems to be engaged in the withdrawal reaction, giving the phenomenon the general appearance of a "reflex" response with a "sensory" and a "motor" component. The "reflex time" is of the order of 10^3 sec, varying both with the strangeness of the confronted cells and with their respective rates of motility. In conclusion, self-sorting of scrambled cell types results from chance collisions with matching combinations holding on to each other, whereas nonmatching ones do not last.

"Matching" combinations involve either cells of the same type or of complementary types, the latter being types that normally cooperate in the building and functional operation of complex organs. For instance, an active mutual adhesion between nerve processes and enveloping sheath cells has been demonstrated,[11] and our motion pictures reveal a similar marked tendency of macrophages to confine their excursions to within the borders of the large flattened lung cells with which they have been jointly explanted.

The discriminatory response is strictly cell-type specific, but it is not species specific. That is to say, cartilage cells of the chick will reject association with liver or kidney cells of the same chick, but they will combine readily with cartilage cells of a mouse. This fact, derived earlier from reaggregation experiments,[12] has now

---

* These moving pictures, made in collaboration with A. C. Taylor and A. Bock, were shown at the Study Program in Biophysical Science in conjunction with the lecture on which this article is based.

IV. ON FORM AND FORMATIVE PROCESSES   385

been proven visually by the motion pictures. On the other hand, the same cell type may undergo progressive alterations with age, so as to make more mature and less mature cells of the same strain less acceptable to each other than cells of the same age would be. Furthermore, there are signs that the differentials among cell types, which underlie their discriminatory responses to each other, may be subject to gradations in accordance with their ontogenetic relationships.

The details of this principle still remain to be worked out. But what has emerged thus far is enough to force a substantial revision of former excessively static pictures of the organism as a fixed framework of parts shifted into predestined places by a rigidly prescribed system of tracks and schedules and then immured by fibrous cements. Although, grossly, this picture still holds, it gains much greater flexibility in detail from the realization that the eventual pattern of combination, association, and segregation among cell types in typically ordered arrays is not just passively arrived at, but is actively insured and guarded—and restored after disturbance—by a system of subtle mutual conformances and nonconformances with which the various cell types have been endowed as a corollary of their ontogenetic differentiation.

The principle of the self-linking of cells into matched groups by contact affinities and disaffinities has an important bearing on the explanation of the establishment and maintenance of order in the networks of the nervous system. The building up of central and peripheral nerve cables by the selective attachment of nerve fibers of a given type to others of like specificity has already been referred to in the foregoing. The same principle holds for interneuronal relations, as well as for the relations between neurons and their non-neural receptor and effector organs, on an even subtler scale. Since this is one of the most compelling, yet at the same time least recognized, instances of specificity and selectivity in cell-to-cell interactions, it is well to recapitulate briefly the relevant experiments, which I started almost forty years ago (see a recent review[13]) and which have more recently been amplified by my student, Sperry.[14]

During development, countless connections are established among individual neurons and between neurons and peripheral sense organs and muscles. The latitude inherent in the primary developmental mechanisms of neurogenesis[6] is so great that it would rule out any such microprecision in the details of the neuronal circuitry as is usually postulated as basis for coordinated functions. On the other hand, there is equally conclusive evidence to show that the nervous system does emerge from its embryonic phase with a large list of ready-made coordinated performances, which, since learning by trial and error could have played no part in their formation, must be explicable in terms of developmental interactions. The dilemma was solved by the demonstration of secondary processes which adjust the details within the gross primary system of connections

to the unique arrangements of each individual specimen. Although nerves from the same central sources may in different individuals end up in different muscles, the muscles themselves then send a specifying influence back into the centers with the information on just what muscle of what name lies at the end of what line.

The test experiment consisted of transplanting a supernumerary muscle (or a whole set of limb muscles) near a normal limb and hitching it to a random branch of the local nerve supply. This amounted to "tapping" the communication network by inserting an extra receiver. What was observed then was that the transplanted muscle responded in the course of normal activities of the animal always at the precise time and with the same strength as the muscle bearing the same name in the normal limb. Thus, each muscle is called up, as it were, by the central system by its proper name, and if there are several muscles of the same name present in a common innervation district, they all respond simultaneously when the particular code is called. Since the test muscle was inserted arbitrarily, it is clear that it must have established its name-specific correspondences in the nerve centers secondarily. The same correspondence between periphery and centers was also confirmed for the sensory part of the system, first for proprioceptive[15] and tactile[16] sensations and later for the visual field.[14]

It all boils down to the following. Even though it is impossible to distinguish different skeletal muscles by present biochemical methods, the nervous system can tell them apart very well. Whatever the telling difference between muscles, it must be something that can make itself known in a retrograde manner over the motor nerve fiber in such a way that the central apparatus thereby gains exact information about its peripheral terminations. Motor fibers are initially blank and unspecified. They acquire specific identity only after they penetrate identified muscles at the periphery, and the same holds for sensory neurons in regard to their endings. Somehow, the acquired detailed specificity is then transmitted still further back from the primary to secondary neurons, and so on up the lines.

By this device, the neuronal population become progressively specialized into subunits whose member cells can henceforth regulate their own mutual relations by virtue of their acquired distinctive properties. Neurons of corresponding or complementary specificities thus become joined into cooperative systems. The proposition that this implies only linkage of a static morphological order all the way through the nerve centers[14] does not seem to go far enough if one is to explain either the aforementioned experimental results or the general problems of neural coordination.[13] But if one adds the assumptions: that synaptic contact between neurons is merely an enabling, but not a decisive condition for impulse transmission; further, that actual transmission requires specifically matched states of two apposed surfaces; and, finally, that the specificities underlying this conformance are subject to modification by both peripheral and central influences; one comes closer to a satisfactory concept. The nervous system would then emerge as a vast system of resonance circuits, the elements of which would be linked by conforming molecular surface configurations, partly permanently, partly variably in response to changing central and peripheral influences.

From these remarks, it can readily be seen that compatibilities and incompatibilities along the boundaries between neurons are as significant as discriminatory devices as are the signs of surface "recognition" described earlier in this article for other cell types. Indeed, the direct demonstration of cell discrimination in the motion pictures should help to make the corresponding property of neurons acceptable, if not more palatable, to those who had hoped to be able to maintain their oversimplified faith in the essential identity of all neurons so long as they were faced only with the more indirect earlier evidence. In reopening the whole question of specificity in the nervous system, which of late has been lying dormant, our observations on cell encounters in tissue culture thus assume added significance, as they also open the way to a more direct practical study of the nature of the specificities concerned.

The reference to the progressive diversification within nerve-cell populations points logically to the more elementary question of how any primitive cell types, including neurons, acquire their distinguishing characteristics in the first place. To discuss the problem of divergent differentiation would go far beyond the scope of this article. There is one aspect, however, that fits into the context. Differentiation occurs essentially by dichotomies in the courses of transformation of cells with basically identical physical and chemical endowments.[17] Any cell strain is, by virtue of such endowments, initially capable of effecting a limited variety of qualitatively different reactions. Every step of differentiation implies that from that multiple repertory one definite course has become activated in some members of the group, and an alternative course in the other members, the two courses being mutually exclusive. Sooner or later, the different reactions lead to manifest diversity. Development is composed of long series of such steps. Present evidence is that each step has its own mode of triggering mechanism, and that, contrary to earlier illusions, there is no single master agent that could be held generally accountable for the various and successive dichotomous changes. Formally, however, one can distinguish two classes—one in which the differentiating activity plays entirely within the bounds of a group of similar cells, referable perhaps to differential interactions among the members of the group depending on their relative positions; and another in which the reaction of a given cell group is decisively influenced by an extraneous cell group.

This second class is usually termed "induction";

again, different steps of "inductive" influences need have little in common but the name. Some of them operate over distances, others only between adjacent tissues. Evidently, these latter raise the issue of whether true contact interactions are involved. The answer seems to be that in some instances and for certain steps of differentiation, intimate contact between the interacting cells is essential, whereas, in other cases, such direct contact is dispensable, the interaction being mediated by diffusible substances.

Some cases in which direct contact is necessary may be cited as examples. When devitalized (frozen-dried and rehydrated) cartilage is transplanted into the vicinity of a limb bone in amphibian larvae, the implant can induce the formation of new cartilage in contact with its surface.[1] Thus, competent living host cells build on cartilage to the dead cartilage model as bees would build on new honeycomb to an artificial wax honeycomb presented to them. This is a particularly interesting case, because the bodies of the cells in the implant cannot have been the inducing agents directly, as they are enclosed and insulated in the cartilaginous matrix; therefore, the inductive stimulus to surrounding host cells must have originated in the exposed surface of that matrix. Another set of experiments showed the same thing for bone.[1] A frozen-dried and rehydrated metatarsal bone of a rat, inserted into the leg of a host rat, induced there new bone, including new bone marrow, in contact with the old bone. That contact is essential has been confirmed from other sources.[18] Such observations support the conclusion that extracellular materials may play important roles in certain inductive interactions.[19]

The next case is of special interest to students of collagen. The stroma of the cornea consists of layers of collagen in a characteristic regular arrangement. When frozen-dried cornea was transplanted into a corneal defect in rabbits, the grafted stroma as such persisted for many months. In addition, however, it induced along its inner side the formation of several new layers of typical stroma.[20] Thus, the architecture of the dead stroma matrix has in some fashion been transmitted to the induced layers, so that they assumed the characteristic pattern of collagenous corneal lamellae. This result again proves that not only cells but cellular products likewise can influence other cells inductively.

The intimacy of submicroscopic cell contact, in the sense of the introductory remarks, which may be required for "contact inductions" has not yet been clarified. One of the few clues is the fact that, during the period of inductive interaction between eye cup and prospective lens epithelium, the attachment between the interacting cell layers is so firm as to resist mechanical separation.[21] But, it is still undecided whether the specific transmission across the cell borders is simply an orienting or sorting influence on pre-existing molecular populations[22] or whether it involves the actual passage of substance,[23] perhaps prepared by molecular reorientation.[1] There is a vast field here for exact studies on the relations between physical configuration and chemical activities. These relations are bound to remain in obscurity so long as one focuses attention on those cell-to-cell influences that are mediated not by contact, but by remote control through diffusible agents. The latter, although equally real, can tell only part of the story of cellular interaction.

In summing up, to state that all of the specific interactions among cells exemplified in this article are based on specific properties of the interacting cells is a truism. But, it explains why so little is known about the molecular phenomena governing cell-to-cell relations. There just has not been enough preoccupation with the molecular basis of cellular specificity. Unquestionably, cell "recognition" and selective response must eventually be reduced to terms of properties of molecules and molecular populations. But, so far, there has been very little tangible progress in that direction. There have been a few hypothetical suggestions to explain cell-to-cell conformances, none of them quite tenable in the light of recent observations. I had proposed[24,25] that some sort of steric fitting (e.g., by corresponding charge distributions) between complementary molecules might produce bonds or linkages between like cells, whereas unlike cells without conforming molecules in their surfaces to interlock could establish no connections. Yet, this supposition is now clearly contradicted by the fact that like cells, while remaining together, are not fastened but move continuously around each other. This leaves us for the time being without substantial explanation of the phenomena of selective aggregation as shown in the films.

The only thing that seems clear is that these phenomena point to the same general area of specificity that covers immune reactions,[26] enzyme-substrate interactions, pairing of chromosomes, fertilization, parasitic infection, and phagocytosis. The latter clearly involves selective uptake of particles of suitable molecular surface organization. The problem is no different from that met in cell-to-cell encounters, except for the great disproportion of size between the members of the pair in phagocytosis. Presumably, the selective ingestion of macromolecules belongs in here, too, with the discrepancy of size even more pronounced. The opposite extreme is found in a cell which spreads out selectively over a large body of proper substratum, where the cell is very small in comparison to the partner which it tries to engulf. So, from phagocytosis and the uptake of macromolecules, through cell-to-cell contacts with specific "recognitions," up to the problem of selective adhesion of a cell to its substratum, one deals with a continuous spectrum of problems of the same nature. This very fact holds promise of progress, as advances in any one sector of that broad area will shed light on other sectors. On the other hand, cleverly getting around the acknowledgment of specificity in any one sector, whether by theoretical constructs or by unrealistic models, will not strip the other sectors of their aspect of specificity, and, hence, will not relieve the intellectual discomfort engendered by our inability to squeeze the broad subject of specificity into the limited conceptual framework which we have erected from the study of fragmentary vital phenomena lacking that aspect. Specificity, as a real and basic control mechanism of cell behavior, can no longer be relegated to a corner, but must be placed in a central position in cellular and molecular biology. The less that is known about it, the greater is the challenge. If there has been some dodging in the past, it was because of some concealed hope that the whole thing might yet in the end turn out to have been an illusion born of inadequate penetration. Exactly the opposite has happened. The more penetrating the analysis, the more cogent has the evidence of specificity become; witness the motion pictures of cell discrimination. There seems little doubt that once the reality and universality of the problem are generally acknowledged, and the promising techniques at our disposal for its disciplined study are recognized, progress can be rapid. But there is even less doubt that, if the existence of the problem continues to be widely ignored, or even denied, some of the major clues to the understanding of living systems will remain missing.

BIBLIOGRAPHY

[1] P. Weiss, Intern. Rev. Cytol. **7**, 391 (1958).
[2] P. Weiss and W. Ferris, Exptl. Cell Research **6**, 546 (1954).
[3] G. F. Odland, J. Biophys. Biochem. Cytol. **4**, 529 (1958).
[4] P. Weiss, J. Exptl. Zool. **100**, 353 (1945).
[5] P. Weiss in *Analysis of Development*, B. H. Willier, P. Weiss, and V. Hamburger, editors (Saunders Company, Philadelphia, 1955), p. 346.
[6] J. Holtfreter, Arch. exptl. Zellforsch. Gewebezücht. **23**, 169 (1939).
[7] P. Weiss and G. Andres, J. Exptl. Zool. **121**, 449 (1952).
[8] G. Andres, J. Exptl. Zool. **122**, 507 (1953).
[9] A. Moscona, Proc. Soc. Exptl. Biol. Med. **92**, 410 (1956).
[10] J. J. Chiakulas, J. Exptl. Zool. **121**, 383 (1952).
[11] M. Abercrombie, M. L. Johnson, and G. A. Thomas, Proc. Roy. Soc. (London) **B136**, 448 (1949).
[12] A. Moscona, Proc. Natl. Acad. Sci. U. S. **43**, 184 (1957).
[13] P. Weiss, Symp. Soc. Exptl. Biol. **4**, 92 (1950).
[14] R. W. Sperry, Growth Symposia **10**, 63 (1951).
[15] F. Verzar and P. Weiss, Pflüger's Arch. ges. Physiol. **223**, 671 (1930).
[16] P. Weiss, J. Comp. Neurol. **77**, 131 (1942).
[17] P. Weiss, J. Embryol. Exptl. Morphol. **1**, 181 (1953).
[18] F. C. McLean and M. R. Urist, *Bone* (The University of Chicago Press, Chicago, 1955).
[19] C. Grobstein in *Aspects of Synthesis and Order in Growth*, D. Rudnick, editor (Princeton University Press, Princeton, 1954), p. 233.
[20] I. H. Leopold and F. H. Adler, A.M.A. Arch. Ophthalmol. **37**, 268 (1947).
[21] M. S. McKeehan, J. Exptl. Zool. **117**, 31 (1951).
[22] P. Weiss, Quart. Rev. Biol. **25**, 177 (1950).
[23] J. Brachet, *Chemical Embryology* (Interscience Publishers, Inc., New York, 1950).
[24] P. Weiss, Growth **5**, 163 (1941).
[25] P. Weiss, Yale J. Biol. Med. **19**, 235 (1947).
[26] A. Tyler, Growth **10**, Suppl., 7 (1947).

Reprinted from THE AMERICAN NATURALIST, Vol. LXVII,
July–August, 1933.

CHAPTER 18

FUNCTIONAL ADAPTATION AND THE RÔLE OF GROUND SUBSTANCES IN DEVELOPMENT[1]

DR. PAUL WEISS

OSBORN ZOOLOGICAL LABORATORY, YALE UNIVERSITY

WHEN it was first known from the work of Culmann and Meyer that the arrangement of the lamellae in the spongy bones was such as to correspond to the trajectories of tension and pressure acting on the bone, the idea of a direct influence of mechanical function on the organization of the tissues seemed to be substantiated. Even more so, since Wolff (1892) could demonstrate that the internal structure of bones subjected to abnormal strains, as in pathological cases, changes profoundly into a pattern corresponding to the new situation and supposedly adapted to it. This is what Roux (1895) then, called "functional adaptation." Later, similar phenomena have been observed also in the cartilage and in the connective tissue systems (Benninghoff, 1931). Although these phenomena were found in the adult, Roux was ready to proclaim adaptation to the functional requirements of activity as a general principle in morphogenesis. There was, however, one essential point utterly opposed to such an attempt; that is, that the so-called functional constructions, in the embryo, already exist long before there is any corresponding functional activity. So, although there are functional structures, there is certainly nothing of the kind of a functional adaptation, in the original sense, to be found in the embryo. Rhumbler (1914), later, demonstrated that the internal architecture of the embryonic bone could be explained on the basis of assuming that internal tensions arising from the differential surface growth of the bone are the organizing factors. This system of growth tensions incidentally coincides, in its main features, with the tensions

[1] Paper read at the symposium on "Embryonic Determination," before the American Society of Zoologists, Atlantic City, December 30, 1932.

that would be created by an external load, and the so-called functional structures are rather growth structures. This idea has later been corroborated by Triepel (1922), Benninghoff (1925), and others. All these authors still consider tensions to be the agents in the formation of the so-called functional structures, differing from Roux only in that they attribute these tensions to growth instead of to external load.

In a further attempt to approach, experimentally, the problem of functional structures, I started, a few years ago, a series of experiments, the results of which may be outlined as follows: The purpose was to produce functional structures in tissue grown *in vitro* outside the organism; in other words, to imitate in the tissue culture, the conditions within the organism. So, I devised a method by means of which it was possible to grow fibrocytes in a medium subjected to appropriate tension (Weiss, 1928, 1929). A thin membrane of blood plasma was coagulated in a tiny horizontal glass frame and a fibrocyte culture was put in the center. The distribution of tensions in the membrane is a definite function of the geometrical form of the frame. In the case of a triangular frame, for instance, the maximum tensions are directed towards the sides of the triangle. Under these conditions, the direction of the outgrowing cells coincides with the lines of maximum tension. In a triangular frame, we see the cells deviating from the radial direction and converging into three main bunches, directed, as are the principal tensions, toward the sides of the frame. Varied experiments of this kind always yielded essentially the same results. As in the functional structures of the organism, the cells, in their arrangement, duplicate the pattern of tension trajectories. Similar results have since been described by Huzella (1929). I further found that not only the orientation but also the intensity of cell growth is influenced in that the growth[2] proceeded at a higher rate in the directions of stronger tensions.

[2] ''Growth'' of a tissue culture includes both multiplication and outward migration of the cells.

Considering these experiments, one might be led to believe that they furnish decisive proof for the direct action of mechanical factors on the direction and intensity of cell growth.   This direct action could be either an immediate pull or some kind of trophic stimulus in the sense of Roux.   Nothing of this sort, however, holds true.   It can easily be realized that there is no way in which pull could act directly on the cell.   There is, on the one hand,

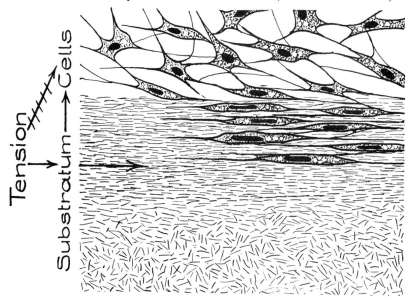

the coagulated medium in equilibrium with the tensional forces to which it has been subjected, and there are, on the other hand, cells freely traveling in or on it.   It is obvious that these traveling cells, themselves, are as little aware of, and affected by, the tension existing in their support as a man lying on the ground is aware of, and extended by, the tension existing in the earth's crust. Unfortunately, this simple fact is sometimes entirely overlooked, so that we find in the literature effects described as tension effects in cases where tension obviously was not in play at all.   Furthermore, some of our experiments bear striking evidence against the view that tension might have acted directly on the cells. If the culture medium, having been coagulated under the influence of tension, is removed from the frame and put on a support so as to release it from the original tensions, we observe, in favorable cases, the cells nevertheless growing out in those directions which mark the former tension trajectories. Tension, therefore, can not be the immediate agent influencing the cells. We found in the experiments the cells following lines of tension which had already ceased to exist. In order to account for this result, we have to assume, first, that tension had had some orienting effect on the medium in such a way as to establish therein a system of material lines coinciding with the tension trajectories, and that these lines, then, formed pathways over which cells might migrate. The second inference is that such pathways really are preferential traffic lines for the cells, no matter whether tension persists or not.

It has been suggested (L. Loeb) and proved (Harrison, Burrows, Carrel) that tissue cells in order to be able to grow and to migrate require some solid structure as support (Harrison, 1914). In demanding a supporting substratum these cells behave like amoebae. We can not, however, enter here into a discussion of the various explanations which have been suggested to account for this peculiar behavior of cells. Terms like "thigmotaxis," although not elucidating the situation, may still be accepted for the sake of description. Probably surface tension relations on the liquid-solid interfaces along which the protoplasm extends play an important rôle in determining the thigmotactic movements. Hence, inasmuch as different kinds of protoplasm are likely to have different surface tensions, we may expect to find given interfacial structures serving as leading structures for one kind of cell while not affecting another kind at all. In any case, it is a fact that the fibrocytes in the tissue culture tend to use liquid-solid interfaces as tracks, the solid phase being represented as fibrin, the liquid phase

as serum. If, now, the particles of the solid phase are
arranged in a definite orientation, the movement of the
cells will necessarily be correspondingly oriented. Thus,
the problem of the trajectorial arrangement of the cells
reduces itself to the problem of a trajectorial arrangement of the fibrin particles. And remembering now that
particles in a colloidal substratum, as has long been
known, can be oriented by mechanical tension (V. Ebner,
1906), we can replace the idea of a direct action of tension on cell growth by such a conception as is represented
in the following schematic picture. The ultramicrones
or "micellae," as we may call them, using the term of
Naegeli, are, in the case of organic colloids, generally of a
definite polarity which in itself may be an expression of
the constitution and arrangement of the high molecular
compounds composing them. We may think of these
micellae as being little rods of sub-microscopic dimensions (Schmidt, 1924). In a fresh colloidal sol, these
little rods are irregularly distributed. Without the intervention of polarizing agents, they will aggregate to
form fibrillar threads, but there will be no definite orientation in the resulting network. If, however, there is
some vector force, for instance tension, acting on the
system, the rods will all be oriented in such a way that
their polar axes nearly coincide with the direction of that
force (Ambronn and Frey, 1926). Usually, from the
further agglomeration of polarized and oriented particles, oriented fibrillar structures result. As Baitsell
(1915) and others were able to observe, fibrin fibers in a
plasma clot subjected to tension attain even microscopic
visibility. It has to be mentioned, however, that in the
case of my experiments, besides a lamellar structure, no
coarse fibrillation of microscopic dimensions could be detected, from which we may conclude that as a guiding
structure for the cells an ultramicroscopic orientation is
sufficient. We will call such an ultramicroscopic structure "ultrastructure." In our frame cultures, for example, the tensional forces acting in three main direc-

IV. On Form and Formative Processes   393

tions cause the formation of a correspondingly oriented metastructure in the clotting medium, and the outgrowing cells simply follow the preestablished pattern. However, as I stated above, not only the orientation but also the growth rate of the cells was affected. As a matter of fact, the particular ultrastructure of the medium can account for this phenomenon, too, since the growth rate depends upon the supply of liquid nutritional substances, and the distribution, circulation and supply of liquid with respect to the growing culture prove to be profoundly influenced by the ultrastructure. The displacement of liquids is facilitated along the lines of micellar orientation. Hence, growth, too, is favored along these lines.

The reason for having reported these experiments at some length is not only that they apparently furnish some clue as to the mechanism involved in the development of functional structures. Their main significance seems to lie in that they unmistakably point toward a more general principle of morphogenesis of which the so-called functional constructions are only special cases. This general principle, which may be called "principle of ultrastructural organization," as applied to the organism, could be formulated as follows: The interior architecture of the body, as expressed by the arrangement of cells and intercellular formations, is to a certain extent determined by the ultrastructural organization of the colloidal continuum which fills the interior of the organism.

Before, however, being entitled to apply to the organism this principle derived from the somewhat artificial conditions of tissue culture, one has to prove that, in the essential points concerned, the conditions in the organism and in the tissue culture are strictly comparable. The best example to prove the comparability is the process of regeneration of tendons. Tendons show the most pronounced functional structure since their cells and fibers are all oriented in the direction of pull.

From surgical experience as well as from experiments of Levy (1904), it can be learned that, after cutting a tendon, the gap between the stumps is soon bridged again by true tendinous tissue with the typical lengthwise arrangement of fibers and cells, provided a longitudinal tensional strain has been allowed to work on the regenerating tissue. In the absence of the oriented pull, the regenerated tissue does not show any oriented structure. The situation is essentially the same as in the tissue culture. There is, at first, a clot of blood formed which connects the stumps. This clot is, then, subjected to tensional stress resulting in a lengthwise orientation of the micellae, and all the immigrating cells, eventually, follow the preestablished oriented pathways. Even more striking perhaps are experiments of Nageotte (1922). He implanted in the connective tissue of adult animals pieces of tendons which had previously been fixed in alcohol, thus offering to the cells a medium with a preformed structure; and, really, cells which happened to penetrate from the surroundings into the dead graft attained, therein, such arrangement as is typical for tendons. Undoubtedly experiments of this sort bear more than a purely superficial resemblance to the tissue culture experiments. The resemblance is even closer if we take into consideration that blood plasma in the organism (Nageotte, 1931) as well as in the tissue culture (Maximow, 1929, Baitsell, 1917) is able to develop true collagenous and argyrophile fibers. Mentioning them, it must be added that they are by no means to be identified with what we consider to be the general guiding structures of cells. Where fibers are differentiated they undoubtedly can be utilized as tracks by the migrating cells. For the rest, ultramicroscopic "ultrastructures" do as well.

It is, of course, obvious that the medium in which the connective tissue normally arises in the embryo is not blood plasma, as in the case of tissue culture or wound healing; however, in all features that concern organization it is similar to blood plasma, similar especially in that its micellae are polarized and hence capable of being oriented along the lines of force. This medium is known in later stages as the ground substance of the various organs. It appears in earlier stages as the ground substance of the mesenchyme and can even be traced back to still earlier, pre-mesenchymal, stages as mesostroma, filling the spaces between the germ layers (Suessarew, 1932). And it may be that there is some substance present even in the unsegmented egg playing an analogous rôle. At least, the way in which Lillie (1909) speaks of the "ground substance" of the egg suggests this possibility. There is still much discussion about the origin and nature of the ground substance. Some believe that there is no ground substance which is not part of living protoplasm, and some others believe that ground substances are only dead secretions of cells, like mucus. Between the two extremes, we find all intermediate shades of opinion. It is not unlikely that some time it will come to be realized that the facts are such as to preclude a uniform solution of the problem. There is one thing, however, to be considered as certain. That is, that the properties on which the formation of ultrastructures is based exist in every protoplasm. We find them at work in the processes of mitosis, in the formation of intracellular tonofibrillae, myofibrillae, neurofibrillae, etc. But it is equally certain that substances which are not protoplasmic, or no longer protoplasmic, as, for instance, some body fluids, possess in that one respect the same properties. For that reason, there is, from the viewpoint of the present problem no need of further discussing the question of whether the ground substances originate as detached ectoplasmic substances or as transformations of protoplasm or as cell secretions of some kind. We may simply record one significant point in which all the different opinions are in accord, namely, that the ground substances form a continuous system all through the embryo, from the earliest stages on. This proves that a material which is suitable to ultrastructural organization and comparable to the medium of tissue culture is present in embryogenesis.

As to the further assumption that cell arrangement depends upon structures in the ground substance, normal embryogenesis also furnishes appropriate examples. The chorda sheath,[3] the cornea and the vitreous body of the eye, for instance, at first, consist only of a cell-free colloidal layer into which cells later immigrate, proceeding along the preestablished pattern of the ground-substance. Therefore, one has to admit that, in the points under consideration, conditions in the tissue culture and in the embryo are strictly comparable.[4]

Now, however, in the embryo, what are the forces which shape the ground substances, which bring about such ultrastructural organization as has been artificially induced by mechanical factors in our *in vitro* experiments? Are the factors in the embryo mechanical as well? It has already been mentioned above that many of the organ structures considered as functional structures are merely growth structures. They are caused by internal tensions resulting from the surface growth of the organ. In these cases, tension is undoubtedly the factor which impresses an oriented structure upon the ground substance. Wherever this is the case, the particular pattern of tension trajectories is determined by the peculiar form of the growing surface. This latter plays a rôle similar to that of the frame in our experiments. If, in the simplest case, the contour of the growing organ is circular, a radial structure is to be expected. That is exactly what we find in the tympanic membrane of the ear. As the

---

[3] V. v. Ebner, *Zeitschr. f. Zool.*, 62, 1896.—The chorda sheath shows a very marked ''functional structure.'' *Cf.* Tretjakoff, *Zeitschr. f. Zellforschg. u. mikr. Anat.*, 4: 266, 1927.

[4] Harrison was probably the first to emphasize this similarity; in 1914 (*Jour. Exp. Zool.*, 17, 521) he states: ''. . . since it has been shown that most embryonic cells are stereotropic, and that such arrangements as they assume in the embryo may often be induced under cultural conditions by reactions to solids, there is a presumption in favor of the view that this type of reaction is a potent factor in normal development also.''

IV. ON FORM AND FORMATIVE PROCESSES   397

tensions, exerted on this membrane in post-embryonic life by sound waves, happen to coincide with the embryonic growth tensions, the final structure of the organ looks as if it were functional by origin (Benninghoff, 1931). A more irregular contour causes a more complicated pattern of tension trajectories. For this Benninghoff (1931a) brings a very instructive example. After implanting the dead scapula of a human fetus under the skin of a rabbit, he observed the formation of a fibrous capsule around the graft. The arrangement of the connective tissue of this capsule was very similar to the fiber pattern found in normal scapular periosteum. During the normal development, the tensions determining the orientation of fibers arise from the growth of the scapula; in the experiment, they arise from the contraction of the connective tissue which coats the graft. Since, however, the distribution of tensions is determined by the form of the contour, and the contour being in both cases the same, the pattern of tension trajectories and, hence, the pattern of fibers is similar in both cases, too. We are confronted here with one of the simplest manifestations of a "Gestalt" principle, since we are dealing with a system of typical configuration in which the arrangement of every part is strictly dependent upon the whole. The external shape of the organ is not brought about by mere apposition of elementary parts in a definite arrangement, but, on the contrary, the arrangement of the parts, *i.e.*, the internal structure, is determined by the shape of the entire system. This is further evidenced in cases where an organ, after having undergone some pathological alteration during morphogenesis, shows internal structures which are entirely different from the normal, but which fully correspond to the new shape the organ as a whole has assumed. Cases of this kind have been described in chick embryos with malformed bones by Landauer (1929).

Like growth, the mere change in the shape of an organ will, of course, alter the existing tensions and occasionally establish new ones. Furthermore, it is obvious that the growth, or change in form, of an organ not only affects the ultrastructure inside but affects the surrounding ground-substances as well. Imagine, for instance, the growth of a vesicular or tubular organ. It is easy to realize that by increasing in diameter it must create a steady tangential tension in the surrounding medium which, eventually, leads to the formation of concentric fibrous capsules and sheaths. A tubular organ growing in length and width is bound to exhibit both longitudinal and circular tensions on the surroundings which may result in a corresponding orientation of the surrounding tissues. Examples are found everywhere in the body.[5] Carey (1922) has called attention to the fact that the arrangement of muscle tracts in the embryonic limb can be explained on the basis of tensional stresses exerted on the surrounding mesenchyme by the differential growth of the skeleton. A similar explanation was claimed for the arrangement of the muscular layers in the intestine (Carey, 1921). Considering all existing evidence, it seems that this idea is perfectly substantiated, as far as merely the spatial arrangement of the tissue is concerned. On the other hand, there has not yet been offered any convincing proof to show that the differentiation into muscle tissue is likewise determined by tensional strain. On the contrary, whereas there is some indication that stretch of some kind is essential for the maintenance, and maybe also for the differentiation of muscle fibers, the facts revealed by experimental embryology carry sufficient evidence to emphasize that the factors determining a cell to transform into a myoblast are of a more specific nature than mechanical stress can ever be.

From what we have stated above, it becomes clear that even the epithelial forerunners of organ formation, by

[5] It is obvious that the tangential stretch which leads to the formation of a fibrous capsule around an organ is not always produced by increase in size of the organ, but may as well be due to a gradual contraction of the surrounding substance, as is the case in the formation of capsules around foreign bodies.

IV. On Form and Formative Processes

their moving, folding, stretching, swelling, and so on, must have a marked effect on the ultrastructure of the underlying ground substance, in later stages on the mesenchyme. It is however, hard to say whether this phenomenon is of significance in the process of organ formation or is just a transitory incident. Observations on the development of glands by Flint (1903) suggest that the epithelial formations may play some rôle in the organization of their supporting or surrounding stroma.

The fact that I have been dealing at some length with the effects of mechanical tension may give a wrong impression about the real share which those tensions take in organization. As a matter of fact we frequently find their importance overrated by many authors. Of course, we do not know very much about the other orienting forces besides tension which intervene in morphogenesis. But we know, at least, a little, and this may be outlined as follows: One non-tensional factor which can be assumed as certain to exhibit organizing activities by stamping ultrastructures in the ground-substance is the displacement of fluids, both by slow diffusion and by faster circulation. You all remember having seen brooks streaming over grass or weeds. You remember how the blades were oriented by the current, as if combed. This gives an illustration of how the circulation of liquids in the ground-substance causes the formation of an ultrastructure oriented along the stream lines. Steady diffusion potentials work, of course, in the same way. Undoubtedly, many of the whorl-like formations in organisms can be explained on this basis. I have frequently observed typical stream-lined structures in the semi-fluid regeneration blastema. The morphogenetic action of the displacement of fluids is, of course, most pronounced in those cases where there is excessive resorption of water during growth. Triepel (1911) was already aware of the fact that the fiber tracts of the connective tissue in a tadpole's tail mark those directions in which water from outside penetrated into the tail, and those along which it shifted therein. Once aware of these factors one will probably detect them in many places. It is, for instance, very likely that a difference in the water content of different organs has effects of the kind described. To a certain extent, I was able to reproduce and analyze even in tissue culture the directive effect of fluid displacement. A growing tissue culture, by sucking fluid from the surrounding medium, establishes a steady diffusion gradient. If, now, two cultures are put in the same medium, their reciprocal suction will result in a movement of liquids along their line of connection. This movement results in the formation of a corresponding ultrastructure bridging the two growth centers, and the bridge becomes visible, as soon as the cells grow over it (Weiss, 1929). In the embryo, pictures of this kind, showing strands of cells forming straight connections between two centers of high formative activity are a familiar occurrence. If there is a mechanical barrier somewhere in the ground substance, currents from the surroundings will be deviated in a direction parallel to the wall. This causes the formation of stream-lined ultrastructures which sometimes can be easily mistaken for tensional structures. The transverse growth of a tendon regenerated along a silk thread inserted at right angles to the direction of the tendon stump, as found by Levy (1904), is apparently due rather to the flow of liquids along the thread than to tension, as the author was inclined to believe.

Another factor which during morphogenesis undoubtedly is involved in the formation of ultrastructures is the electric field. Electric fields may act either directly by orienting the polarized micellae or indirectly by causing cataphoretic effects. These latter consist in the establishment, in the capillary spaces, of an oriented migration of electrically charged particles along the lines of electric force, and this polar migration has, of course, the same orienting effect on the ground substance as a streaming fluid. Increased chemical activity in a circumscribed area leads to structural effects both by causing diffusion streams and electric potential differences.   The chemical activity may furthermore alter the qualities of the ground substance itself, may cause dehydration and fibrillation or, on the other hand, by means of proteolytic enzymes, cause some liquefaction and destruction of existing structures.   And if we try to survey all possibilities, the situation becomes so intricate that, at least, one obvious objection against our conception, namely, that it might be too simple, certainly can not be upheld.

As an example of how the principle of structural organization can be applied to special problems in embryology, I may briefly outline its explanatory value for the problem of nerve patterns, both peripheral and intracentral.   As Harrison (1910) has concluded from his tissue culture experiments, outgrowing nerve fibers show a similar affinity to solid structures, as do mesenchyme cells.   So, there was good reason to believe that ultrastructures in the ground substance might be responsible for the orientation of the growth of the nerve fiber.   This assumption is substantiated by the observation of Held (1909) who saw the first processes of the embryonic neuroblast extend along the extremely fine filaments of the mesostroma connecting medullary tube, myotomes, notochord and ectoderm.   Thus, the facts support the idea that the immediate factors in the orientation of the nerve fiber are correspondingly oriented guiding structures.   On the other hand, however, various other kinds of factors have been proved to influence the orientation of nerve fiber growth.   The theory of neurotropism as a form of chemotaxis has been advanced by Cajal,[6] electric fields have been claimed by Kappers (1927), Child (1921) and Ingvar (1920) and Detwiler (1926), finally, have clearly shown the attractive influence on the outgrowing nerves which is exerted by a developing organ rudiment. The principle of ultrastructural organization offers a common and uniform explanation for all these cases, by

[6] *Cf.* Tello, F. Vortr. u. Aufs. u. Entwicklungsmech. H. 33, 1923.

assuming that neither the chemical nor the electrical agents act directly on the nerve fiber, but that both, primarily, produce ultrastructures in the ground substance which, secondarily, serve as guiding paths for the nerves; chemical centers by diffusion currents, electric centers by cataphoresis, both resulting in structures converging toward the respective centers. The occurrence of a peripheral nerve plexus in the limb region could be explained by assuming that there is some metastructural barrier arising from the meeting of two fields of activity, one being the developing limb bud, the other being the axial organs of the embryo. An interlacing of nerve fibers, corresponding to what in the organism occurs as plexus formation, could also be observed in the tissue culture in places where there was some sudden change in the mechanical condition of the medium.[7] The formation of intracentral fiber tracts may obey similar rules. According to Kappers, Coghill (1929) and others, central fiber tracts always connect such parts of the brain as differentiate simultaneously. This reminds us exactly of the conditions of the experiment described above where a cell bridge connected two growing cultures. Furthermore, Bok's (1929) demonstration of the dependence of the internal architecture of the brain upon its curvature almost forces us to consider this architecture as trajectorial. These few remarks may suffice to show the applicability of the principle.

Having established and demonstrated the principle of ultrastructural organization its general bearings may now be pointed out. It is clear, at first glance, that the principle tends to replace a number of different agents acting at a distance which have been hitherto considered as orienting and organizing factors. If cells arrange themselves in definite directions this should not be explained any longer by specific attractions nor by trophic stimulation, but by the establishment of oriented pathways in the ground substances, these latter being the

[7] P. Weiss—Unpublished results.

IV. On Form and Formative Processes   403

common playground of all kinds of forces released during development.[8] Although the principle apparently applies only to moving cells, leaving the cause of their displacement unexplained, a consideration of the changes in surface tension relations, in distribution of liquids, etc., may possibly lead to an understanding of this latter point, too. A certain difficulty seems to arise, however, from cases where in one and the same organ or embryonic district one part shows definite orientation, while another part does not, or does, but differently; for instance, nerves running across·muscles. It is, however, possible to account also for these cases if one keeps in mind two things: First, a possible specificity of the pathways; and second, what we may call the time factor. What specificity of the pathways means was expressed above when we said that differences in the surface tension relations at the solid-liquid interfaces of the medium may well explain why one type of protoplasm extends along the interfaces, whereas another does not. As to the time factor, the rôle of which in development has been especially emphasized by Goldschmidt (1927) and by Brandt (1928), we must be aware of the fact that the directive agents during embryogenesis are very unstable, most of them acting only temporarily, during a definite period. A center of higher developmental activity will, of course, act as an organizing agent on the ground substance only as long as its activity lasts. As soon as its activity decreases and other areas rise to increased activity, the previous ultrastructural pattern will gradually be replaced by a new pattern. Structures developed on the basis of the new pattern will, then, no longer show relations to the structures originated at an earlier time. In this respect the principle agrees with the general conception of tem-

---

[8] There is evidence to show that not only the orientation and arrangement but also the shape of the cells is materially influenced by the organization of the ground substance. It might, therefore, well be that in the loose connective tissue the lamellar structure is not, as Laguesse believes (Arch. de Biol. 31, 173, 1921), the result of, but is the cause for the flat shape of the cells contained therein.

porary gradients of activity as conceived by Child (1929). Taking into account the typical sequence of developmental processes, the time pattern, so to speak, we recognize the developed organism as being, in some way, comparable to a photographic plate on which different pictures have been developed successively, and the difficulty mentioned above ceases to exist. One more thing, however, has to be assumed. That is, that the ultrastructures are plastic and labile enough to give way to new ones, if this is required by the situation. This point is the main reason why so much emphasis has to be laid on proclaiming ultrastructures and not manifest fibrillar structures as the general ·guiding principle. Ultrastructures, indeed, are plastic and reversible and thus conform to the requirements of embryogenesis; fibrillar structures, generally, are not. Fibrillar structures, too, are not altogether unmodifiable, but their breaking down seems to be too slow to keep pace with the rush of developmental changes. Since, however, ultrastructures tend to transform steadily into coarse fibrillar structures, provided they are not stirred up by continual interferences, a gradual loss of plasticity during development may be expected. The degree of plasticity found in any part of the embryo will, then, essentially depend upon how long an organizing factor has had time to act on the surroundings. After a certain duration of continuous uniform action, the ultrastructure will be found to be almost irreversibly transformed into a manifest structure no longer susceptible to new influences; such solidified structures, however, will, on the other hand, be able to serve, henceforth, according to their previously attained organization, as a typical guiding system, even if the original organizing factors do not persist or have changed in character. We will say of such a system that it has undergone "irreversible determination" and has changed from dependent differentiation to self-differentiation. This remark should, however, not be mistaken. I do not think we can expect that the general problem of deter-

IV. On Form and Formative Processes   405

mination could at some time be reduced to so simple a basis. Past experience has warned us of being too prompt in generalizing. So, instead of proclaiming the principle of ultrastructural organization as a universal clue for the problems of embryonic organization, we may content ourselves, for the time being, with the recognition that it is *one* of the principles or mechanisms of morphogenesis. Its best feature is that it opens innumerable ways for further experimental attack.

LITERATURE CITED

Ambronn and Frey
    1926. "Das Polarisationsmikroskop und seine Anwendungen." Leipzig.
G. A. Baitsell
    1915. *Jour. Exp. Med.*, 21: 425.
    1917. *Am. Jour. Physiol.*, 44: 109.
A. Benninghoff.
    1925. *Ztschr. f. Zellforschg.*, 2: 783.
    1931. *Anat. Anz.*, 72 (Erg.-H.): 95.
    1931a. *Anat. Anz.*, 71 (Erg.-H.): 62.
S. T. Bok
    1929. *Ztschr. f. Neurol.*, 121: 682.
W. Brandt
    1928. *Roux' Arch. f. Entwicklungsmech.*, 114: 54.
E. B. Carey
    1921. *Anat. Rec.*, 21: 189.
    1922. *Jour. Morphol.*, 37: 1.
C. M. Child
    1921. "The Origin and Development of the Nervous System." Chicago.
    1929. *Roux' Arch. f. Entwicklungsmech.*, 117: 21.
G. E. Coghill
    1929. "Anatomy and the Problem of Behavior." Cambridge.
S. R. Detwiler
    1926. *Quarterly Rev. Biol.*, 1: 61.
V. v. Ebner
    1906. *Sitzgber. Akad. Wiss. Wien. Math.-Naturwiss.*, Kl. III, 115: 1.
J. M. Flint
    1903. *Arch. f. Anat.*, jg. 1900: 61.
R. Goldschmidt
    1927. "Physiologische Theorie der Vererbung." Berlin.
R. G. Harrison
    1910. *Jour. Exp. Zool.*, 9: 787.
    1914. *Jour. Exp. Zool.*, 17: 521.

H. Held
   1909. ''Die Entwicklung des Nervengewebes bei den Wirbeltieren.''
      Leipzig.
Th. Huzella
   1929. *Anat. Anz.*, 67: 36.
S. Ingvar
   1920. *Proc. Am. Soc. Exper. Biol. and Med.*, 17: 198.
C. U. Äriens Kappers
   1927. *Jour. Comp. Neur.*, 27: 261.
W. Landauer
   1929. *Roux' Arch. f. Entwicklungsmech.*, 115: 911.
O. Levy
   1904. *Arch. f. Entwicklungsmech.*, 18: 184.
F. R. Lillie
   1909. *Biol. Bull.*, 16: 54.
A. Maximow
   1929. *Ztsch. f. mikr.-anat. Forschg.*, 17: 625.
J. Nageotte
   1922. ''L'organisation de la Matière dans Ses Rapports avec la Vie.''
      Paris.
J. Nageotte and L. Guyon
   1931. *Arch. de Biol.*, 41: 1.
L. Rhumbler
   1914. *Vhdlg. dtsch. Zool. Ges.*
Wilh. Roux
   1895. *Gesammelte Abhandlungen*, Vol. 1. Leipzig.
W. J. Schmidt
   1924. ''Die Bausteine des Tierkörpers im Polarisierten Licht.'' Bonn.
P. Snessarew
   1932. *Ergebn. d. Anat. u. Entwicklggesch.* (*Ztschr. ges Anat.*, III)
      29: 618.
H. Triepel
   1911. *Arch. f. Entwicklungsmech.*, 32: 477.
   1922. *Ztschr. f. Konstitutional.*, 8: 269.
P. Weiss
   1929. *Biol. Ztrbl.*, 48: 551.
   1929. *Roux' Arch. f. Entwicklungsmech.*, 116: 438.
J. Wolff
   1892. ''Das Gesetz der Transformation der Knochen.'' Berlin.

IV. On Form and Formative Processes   407

Reprinted from *Experimental Cell Research, Suppl.* 8, 260–281 (1961)

CHAPTER 19

GUIDING PRINCIPLES IN CELL LOCOMOTION AND CELL AGGREGATION

PAUL WEISS

The Rockefeller Institute, New York, U.S.A.[1]

INTRODUCTION

THE invitation of the Editor to write a summary comment on the symposium of cell interactions and cell locomotion held at Noordwijk, Holland, made me ponder whether or not the subject matter, as presented, would lend itself as yet to summing up. The conference has been outstanding as an occasion for reviewing on a common platform the fragmentary information bearing on a subject of crucial significance to development, pathology and cellular biology in general. Beyond documenting the fact that here is one of the large blank areas on the map of biological knowledge, it has pointed to a number of constructive analytical approaches which at least promise to fill the vacuum with factual content. It is remarkable that in an age which strives successfully to pinpoint vital activity on tangible mechanisms of molecular interactions (for instance, gene action, contractility, immune reactions, nerve excitation, etc.), one still acquiesces in "explaining" the directiveness of cell locomotion by such purely symbolic and unanalytical labels as chemotaxis, thigmotaxis, and the like, without much evidence of the intellectual discomfort which one would expect to arise from the lack of technological understanding of just how locomotion is produced and guided. After all, a moving cell is a physical body undergoing displacement in space, a process for which there must be tangible mechanisms, that is, a properly timed and spaced sequence of physical and chemical events. The labels we have invented for this mechanism can be no substitute for its description in objective terms, and such an objective description is, in the main, still wanting. This is no minor matter if one contemplates that it implies deep ignorance as to how a cancer cell metastasizes, how a parasite reaches a predestined location in its host after invasion, or how, even in normal development, the primordial sex cells, which originate outside of the gonad, manage to arrive at their eventual functional positions in the erstwhile sterile beds.

[1] The original work referred to in this paper has been supported in part by grants from the American Cancer Society and the National Cancer Institute (National Institutes of Health of the Public Health Service).

DYNAMICS OF DEVELOPMENT

Evidently, there has been a growing awareness of both the dearth of knowledge in this area, as well as of the opportunity to do something about it. The Noordwijk Symposium has given proof of this growing realization of the problem and of the effective, if sporadic, current efforts to bring modern experimental techniques to bear on its solution. In collating the scanty and scattered contributions, it has created a nucleus for greater cohesion and concert of future efforts. In documenting the disproportion of speculation over tested facts, it has presented a new challenge to imaginative investigation, and in publishing the proceedings, it will, so we may hope, serve as a sounding board to amplify the call for much more vigorous and disciplined experimentation.

In this sense, the conference may some day be looked back to as an inaugural event. It certainly has been in essence forward-looking, projecting and open-ended, rather than integrating or conclusive, either by tendency or in effect. It was essentially a stock-taking operation, and perhaps the main conclusion I, for one, would draw is that, for some time to come, we shall have to concentrate more on propagating, hybridizing and improving the stock of ideas than on carrying them to the market as if they were ripe for general consumption.

Recognizing this state of flux, I concluded that it might be of greater service if instead of trying to summarize what, for the lack of crucial data, does not as yet add up to a critical sum, I were to limit myself to a brief account of my own and my collaborators' modest efforts over the last nearly forty years to contribute, first, to the clarification of the issues and, second, to their elucidation in terms of tangible processes. Far from any pretense at comprehensiveness or finality, I shall simply state the gist of the facts and ideas that have emerged in decades of preoccupation with phenomena in which cellular locomotion plays a part. I shall do this in reportorial form without attempt at systematic coverage, for the latter task would far exceed the scope of a brief essay.

There is another point that makes me choose the form of reportage, and it is this. No one who has lived through this last half-century of scientific development can have missed the change in the manner of scientific publication that has occurred in this period. It is marked by the progressive bankruptcy of faith and the introduction of legalistic principles of accountancy and documentation, however spurious, for the accreditation of scientific data and conclusions. The printed record has become a measure of validity. Unprinted observations do not count. This tends to bar from recognition that vast store of accumulated experience which an investigator develops in years of concentrated occupation with a given problem and much of which is not precipitated in itemized reports on specific projects. Only a fraction of the work I have ever done in the laboratory has been published in printed detail, while the rest has merely served as a private store of information and judgement. But does the fact that uncommunicated experience has no legalistic standing make it illegitimate? I certainly cannot disown myself of its possession and therefore shall draw on it in the following account whenever pertinent; leaving it to the reader whether or not to give credence to statements not specifically certified by reference to chapter and verse of prior publications.

The problems of locomotion has, as the term indicates, two components—motion and dislocation. Motility refers to the power to move, whereas dislocation means a shift from one place to another. There can be motility and motion with the body remaining in dead center, that is, in stationary position, and conversely, there can be displacement of the body without motility, as in the case of passive convection by external force. Active locomotion, in turn, may be random, that is the number of excursions sampled over a sufficiently long period of time may yield no resultant direction, or there may be directiveness to the total displacement of varying degree and strictness of orientation.

I shall exclude from consideration those cell types which have preformed instruments of propulsion, such as cilia or flagella, as well as the various kinds of blood cells capable of floating and swimming inside a liquid. This leaves those cell types which cannot actively advance in the interior of a liquid medium or, to put it positively, which require a substratum—more correctly, an interface—for their locomotion. This condition is all there is implied in the principle of "contact guidance" to which the first part of this essay will be devoted.

CONTACT GUIDANCE

A naked cell, suspended in a fluid, is essentially spherical. In locomotion, this spherical shape becomes distorted. Our first consideration, therefore, must be given to the deforming forces.

A drop of water suspended in oil, or a drop of oil in water, if wholly undisturbed, tends to assume a spherical shape as a result of surface tension. In groping for models of cell locomotion, Rhumbler pointed out that any local reduction of surface tension on a drop will cause an "amoeboid" deformation comparable to the effect of a local weakening in an elastic membrane around a liquid mass: the intact portion of the contractile surface forcing content to herniate through the aperture of lowered resistance, until the system becomes temporarily equilibrated in some sort of dumbbell shape (for moderate inequalities). If a drop of oil (*a*) is deposited in the interface between two different media (*b*, *c*) with which it is immiscible, the sphere will be flattened to a lentoid of such a contact angle along the circumference that the sum of surface tensions *ab* + *bc* + *ac* becomes a minimum. In other words, the work required to enlarge the surface of the original sphere to that of a lentoid is produced by the potential energy inherent in the combined triple system. Thus, in both of these examples, a sphere becomes deformed as a result of inequalities in its environment.

Since cells, other than those enclosed in rigid envelopes or cell walls, likewise tend to assume nearly spherical shape, when they are unrestrained, they have often been treated as drops subject to the laws of surface tension (e.g. by Rhumbler, d'Arcy Thompson [15], and J. Z. Young [33]), and their flattening out along the interface, for instance, between a slide and a liquid medium has been compared to the spreading of a film of oil along the interface of air and water. In formal regards, the proposition is correct, but closer familiarity with the properties of cell surfaces rules out as oversimplified the notion that the expansion of cell surface in contact with an interface is basically a passive process.

The cell, in fact, is involved quite actively, furnishing some of the energy required to enlarge its surface from its metabolic resources. This is evident from some observations on the "bubbling" of cells freely suspended in liquid media. Not only have rounded-up postmitotic cells long been known to "boil" over their whole surface with little blisters, but the same phenomenon can be seen in many types of ordinary tissue cells enzymatically liberated from their confinement in tissues or organs and suspended in liquid media. The surfaces of such cells are under considerable contractile tension, which manifests itself in the protrusions of numerous little blebs at what must be judged to be temporarily weaker spots along the surface. Taken by itself, this phenomenon of a boiling surface would, of course, only express the statistical inequalities of the surface, making the latter a composite mosaic of patches of different contractile and elastic strengths, for which, in line with our previous comments, surface tension alone could readily account. But there are additional facts that force us to credit metabolic energy for the work involved. One is the fact that this surface activity can be modified by adenosine triphosphate (Lettré [7]), and another is that the contractility which provides the protrusive force appears to move about the cell in a coordinated wave. These and other considerations suggest that the cell expends energy to put its surface layer into a contractile state, yet one which does not result in uniform static compression, but rather in a travelling dynamic oscillation between higher and lower contractile states synchronized over a large fraction of the cell surface and sweeping over it in the manner of a pulse.

We have made numerous observations on this rhythmicity of surface activity in cells isolated from tissues *in vitro*, but have gotten no further than to establishing the reality of the phenomenon. It evidently belongs in the same class as the equally obscure processes of cyclosis in plant cells, radiolarians, and slime molds; of the pulsation of the yolk mass in fish eggs; and perhaps many other propagated slow rhythms in organisms not yet considered under a common viewpoint. Since the substance of the cell is essentially incompressible, the surface necessarily is the region in which regional variations of internal pressure can express themselves; for instance by less intensive "bubbling" in sectors where the crest of the wave passes. This must not be interpreted, however, to mean that the contractile force is confined to the cell cortex, although the extent to which subcortical cytoplasm is actively engaged in the coordinated contractile wave remains to be determined. The whole phenomenon has the aspect of a circling excitatory process which causes those portions of the contractile continuum to contract which have recovered from their refractory condition in the wake of the preceding sweep.

The feature that permits this expenditure of energy to engender protoplasmic motion is the fact that activity is co-ordinated over a large area of the protoplasmic continuum. If the force were produced in random distribution, the result would be an unorganized convulsion, heating rather than moving the cell; but by synchronizing the tension in one large cohesive portion of the system, while at the same time correspondingly relaxing another contiguous portion, a systematic gradient of pumping force results. Various mechanisms invoked to explain amoeboid motion conform in general terms to this principle. Their shortcoming seems to be that they localize the source of the motive force in a circumscribed area of the protoplasmic continuum, albeit in different regions (see Goldacre *versus* Allen in this symposium), whereas a general theory of protoplasmic motion that would embrace under a common principle amoebic motion, plant cyclosis, cell gliding, and perhaps even movements of nuclei and mitochondria within the cytoplasm might have to concede motive power to the whole mass in motion. It should be borne in mind that the effect of such motility is the displacement of cell content relative to a stationary substratum, hence the advance of an amoeba or other naked cell along a solid surface and the streaming of the protoplasm of a plant cell along the inside of the cell wall are essentially the same phenomenon, turned inside out.

When an originally spherical cell flattens out to disc shape along an interface, the contractile wave likewise courses mainly in the plane of flattening and registers as rhythmic fluctuation of activity circling in the protoplasmic fringe of the disc-shaped cell. Depending on the kind of cell, this fringe consists either of a continuous ruffled "undulating" membrane or of separate sectors of such a "membrane"; the term "membrane" in this connection being an historic relic, for one is dealing simply with the thinned-out margin of the cytoplasm itself. If the sectors are relatively narrow, we speak of "pseudopodia", which may end in fan-shaped, knobby or pointed tips, representing increasing degrees of geometric restriction from planar to predominantly linear configuration. Concomitantly, the contractile waves become restricted from their circumferential route into linear courses along the radial pseudopodial extensions, which gives them the configuration of peristaltic waves. We have actually observed such peristalsis in tubelike projections of the character of myelin figures, which are thrust forth from the surface of postmitotic liver cells. In fact, our prior demonstration of a continuous translatory movement of axoplasm in nerve fibers (Weiss and Hiscoe [31]) makes it necessary to postulate such a mechanism.

In conclusion, there is rather striking evidence that metabolic energy is channelled through the cell in coordinated fashion so as to produce rhythmic contractions and expansions of the surface, but the mechanism of this action and the manner of its coordination are still undetermined. Whatever contributions surface tensions and interfacial tensions may make toward the passive deformation of a spherical cell, the cell itself has the power to expand its surface actively beyond any passive deformation by environmental forces.

The described contractile waves deform the spherical shape of a freely suspended cell but little. They lead to some heaving and bubbling of the surface with the protrusion of relatively short projections, but the cohesive and viscous-elastic forces of the system maintain its compact rounded shape. For a cell to expand more extensively, it must be lodged at an interface. Innumerable observations on single cells and nerve fibers in tissue culture have shown unequivocally that active extension requires some adhesion between the peripheral fringe of the cell and an appropriate substratum. This makes the area of contact appear as the effective motor of the cell (note the convincing demonstration of this fact by Ambrose at this symposium). Whereas the unattached protrusions of the freely suspended cell are retracted again by contractile and cohesive forces, these same forces will pull the cell content centrifugally in protrusions whose tips have taken a firm hold on the substratum. The configuration of the substratum thus limits decisively the directions in which a cell may expand.

The expansion of a cell can therefore be understood only in terms of the competition between two opposing sets of forces. Cohesion, viscous-elastic forces and surface tension, including the contractility fibrous networks pervading the cortex of the cell, make for a massive shape of minimal surface, ideally a sphere. Adhesive forces at the margin, contraction in the radial direction and inertia of flow, to be described presently, make for centrifugal expansion, which may be aided in special cases by the formation of crystalline reinforcements in the axial direction—a sort of transient microscopic or sub-microscopic endoskeleton. Whether any particular area of the cell surface will become enlarged or reduced, therefore, depends on the net balance of these forces at that point. This balance naturally varies along the surface and from moment to moment, depending on the local variations in the constitution of the surface itself, on the inhomogeneities of its immediate environment, particularly the substratum, and on the fluctuations of the propulsive force, the latter force becoming greatest for any given point when the crest of the contractile wave is in the antipodal position.

Since, as we said above, expansion or retraction in a given radius depends on whether or not the tip of the projection adheres to the substratum, it is clear that the configuration of the contour of an expanding cell depends on the number and relative strengths of environmental contacts. The eventual outcome of this deformation is what we register as cell shape. Considering the variability of conditions along the border of a cell, both within and without, and the resulting inequalities in the balance of forces from point to point along the border, one realizes that problems of shape cannot be dealt with in terms of *average* surface parameters, but on the contrary, that the very *inequalities* are the critical determinants.

With these facts as background, we can now interpret a wide spectrum of cell shapes in terms of graded deformations of spheres containing motive energy and interacting with interfaces. Let us first consider the simplest case, namely, that of a spherical cell attaching itself to a very smooth surface, such as glass. Adhering firmly to the substratum, the circular line of attachment expands in all radial directions, thus flattening the cell to the shape of a pancake (see the presentation by Taylor at this symposium). Many cell types in tissue culture show this pattern, although the degree of flattening varies with cell type, substratum and medium.

Let us now assume a change, either in the cell surface or in the substratum, such that their mutual adhesiveness would be lowered, which increases the incidence of detachments along the cell margin. The detached stretches become immediately subject to centripetal retractive pull, which forces them to sag in catenary form between the nearest two points at which the margin has remained stuck to the substratum. In this manner, the cell acquires a multipolar shape. Just as the disc-shaped cell can be derived from the spherical cell by the restriction of freedom of radial expansion from three to two dimensions, with all radii in the plane still essentially equivalent, so the multipolar cell can now be derived from the disc-shaped cell by a further restriction of the degrees of freedom of expansion from an infinite to a definite number of radii. The weaker the adhesive force between margin and substratum, the fewer adhesive points for radial elongation will persist. The extreme of this condition, short of total detachment, is reached when there are only two anchor points left between which the cell becomes axially elongated. This yields the bipolar "spindle" cell, which thus marks the ultimate restriction of freedom of expansion to a single dimension.

In this perspective, the problems of cell shape and cell orientation merge into one. Axial orientation is the resultant direction in which the spherical cell has undergone elongation. Reorientation of an established spindle cell, however, requires an additional qualification due to some degree of structural inertia in an already deformed cell. The surface configuration of a cell necessarily affects the movement and arrangement of the macromolecular and particulate populations in the interior, hence in due course produce a dynamic framework which tends to perpetuate with some residual inertia the once initiated cell form. Evidently, in order to be able to change that form, countervailing forces must be the stronger, the longer the prior consolidating phase has lasted.

A related qualification applies to the consideration of a freely suspended cell as essentially isotropic. Any cell has come from a mother cell, and in metazoans, has been a member of a cellular community, hence has had a history of asymmetric exposure to its total environment. How much of the polarized structural residue of this anisotropic exposure endures in a subsequent state of free suspension, is a question which must be settled for each type of cell and for each length of period of isolation empirically.

For most metazoan cells thus far studied, however, polarizing effects of the immediate environment, if of sufficient strength, determine the shape and orientation of the cell. Perhaps the most crucial demonstration of this fact has come from the following experiments (see Weiss [25]). Rounded single cells of various tissues, which on a smooth glass surface would flatten out to disc shape, can be forced to assume spindle shape in a direction imposed by the experimenter by being placed on a glass surface engraved with parallel microgrooves. Such cells at first attach and extend processes in several directions, but extensions along the groove attain a firmer hold than do those on smoother portions of the glass and gradually the former win out in the competitive pull for cell content; eventually, after some tug-of-war, the initially multipolar cell becomes relatively stabilized in the bipolar spindle shape, its two ends pulling in opposite directions along the groove. The concept presented above, reducing cell shape to a problem of differential adhesivity around the free margin of the cell, has thus been substantiated. We do not know as yet whether this differential adhesion is due entirely to the difference in roughness of substratum between the grooved and ungrooved portions of the glass or whether a more subtle oriented feature of the substratum, acquired in the course of the scoring procedure, might be involved; for whereas a number of our scored slides proved consistently successful in orienting all cells of a suspension, others were wholly ineffective.

On the other hand, absolutely reliable results are obtained if one uses a substratum whose oriented inhomogeneities consist of what might be considered the negative of grooves, namely, fibers, The interior of a fish scale, for instance, furnishes a preformed fabric of this kind, as each layer contains a grid of collagen fibers in strictly parallel alignment. Single round cells deposited on such a sheet promptly assume spindle shape in the common direction of the collagen fibers, the whole suspension shortly resembling a school of fish.

Although this may seem to be merely a repetition *en masse* of the original classical experiment of Harrison, in which connective tissue cells in tissue culture were shown to cling to and follow threads of spider web placed in the medium, there is this major difference. Whereas the spider threads traversed a liquid medium and furnished the only possible tracks for cells bound to use a solid substratum, both the grooved glass slide and the fish scale constitute continuous solid systems along which cells could crawl in any direction, as they actually do in some of our glass slides. The explanation of cellular orientation, therefore, can no longer be sought in simple "thigmotaxis" to solids, but evidently involves a directional response to directional properties within that solid.

Moreover these directional cues do not stop on the microscopic scale commensurate to the dimensions of the cell body as a whole, but extend downward into the submicroscopic realm. Historically, this has been the source of my concept of "contact guidance", which implies a more subtle mechanism than the crudely mechanical one for which it has sometimes been misconstrued. Back in 1927, I found that spindle cells in a blood plasma clot would orient themselves in the direction of tension lines prevailing in the coagulum. It was easy to show that tension acted primarily not upon the cells, but rather on the medium, since in a two-step operation, in which a cell-free clot was first subjected to tension and then secondarily, in an unstretched condition, was charged with cells, the latter would still take courses corresponding to the former tension pattern. The explanation was given in terms of guidance of cells by micellar pathways: tension aligns and orients filamentous fibrin molecules in a common direction, thereby facilitating their aggregation into bundles, and the latter then act as tracks for the moving cells. As has been confirmed later by electron microscopic studies of plasma clots, the network of micellar aggregates is mostly of sub-microscopic dimensions, and yet it has the faculty of drawing cells into its course. The cue being submicroscopic, the initiation of the response of the cell must likewise be an event of submicroscopic scale, which only subsequently is amplified to an effect on the whole larger cell body.

We may visualize the situation as follows. The mobile cellular fringe can be assumed to be covered, like the rest of the cell surface, with a network of long filamentous molecules in the plane of the cell membrane. On contact with an extraneous submicroscopic fiber, this molecular layer would be drawn out along the fiber surface, creeping in the direction of the fiber axis like a wetting substance on the wall of a capillary. A fiber thus creates a momentary outlet for cell content, and if this channel persists long enough, it may be widened gradually to microscopic dimensions by the pumping forces described above. Whether or not such a projection becomes established, will depend not only on the force of thrust but also on the number and distribution of competing surface projections initiated in other directions. A simple hydrodynamic consideration will show that this competition between simultaneous protrusions disappears when they lie so close together in a common direction as to become confluent. Since this is precisely the condition along that part of the cell margin which crosses a set of parallel microfibers or microgrooves, one can understand, at least in principle, how a submicroscopic orientation is translated into the orientation of a microscopic cell.

But even this may be too simple. Electron microscopic observations of single cells of various types in culture have revealed that most of them, when spread out on a plane substratum, are bristling on their surfaces with cylindrical projections of a rather uniform diameter of about 1000 Å and a length of up to several microns, composed of an electron dense core in a

18 – 60173256

thin sheath of ectoplasm, which between neighboring processes forms a web. It is conceivable that an extraneous fibril on contact with the cell surface might align the nearest group of these "tentacles" and thereby create an internal outflow channel for cytoplasm in that direction. In view of its obvious bearing on the problem of selective contact guidance, to which we shall refer below, it would seem promising to pursue this possibility further.

If the fact that a microscopic cell can be oriented by submicroscopic cues reveals the inadequacy of mere "thigmotaxis," that is, of plain surface application, as the principle of cell guidance, the case is further strengthened by observations at the opposite end of the scale, namely, the orientation of microscopic cells by macroscopic bodies. For instance, spindle cells in culture, attached to a macroscopic fiber of glass or plastic, orient themselves in the direction of the fiber axis, although, if nothing more than simple surface application were involved, they could have assumed any other positions and configurations in the cylindrical surface. The actual explanation was found to be as follows (Weiss, [20]). Cultured cells produce macromolecular exudates ("ground mats"), which spread along the interface at which the cells reside. On a cylindrical fiber, the margin of this coat, advancing distally, stretches the meshes of the micellar network behind it lengthwise, and the leading cell processes then trace these longitudinal micellar tracks.

While these studies have stressed the importance of macromolecular exudates as integrating mechanisms in the morphogenesis of cell populations —a feature recently reemphasized by Moscona—little is known as yet beyond the sheer existence of the principle. And yet, it seems plausible that if such exudates are cell type-specific, they would offer a basic clue to the understanding of selectivity in contact guidance and cell association (see below). Recent studies in our laboratory (Rosenberg [11]) have confirmed quantitatively that cells of different types in culture release macromolecular units which coat the surrounding interfaces with monomolecular films, the properties and kinetics of which vary with cell type, substratum and medium. While the relation of these microexudates to the micellar and fibrous ground mats is still undefined, they do caution us not to consider the mechanism of contact relations and contact guidance in over simple terms.

Inasmuch as we are focussing here mainly on the orientation of cells by existing geometric singularities in their environment, the origin of the latter is not really up for discussion. Yet, a few cursory comments may be added to round out the picture. As we have seen, linear fiber tracks for cells can be generated by tension, which aligns chains of filamentous molecules in a common direction. The generating tensions may be imposed upon the medium from the outside, but they may also be self-generated by cellular activity, e.g., in the case of spreading exudates. A particularly potent effect of this latter kind results from strong syneretic activity of proliferating cells, which causes the surrounding colloidal network to shrink considerably. Such local shrinkage gathers the meshes of the fibrous continuum towards the shrinking center, thus setting up a pattern of radial tracks, which orient immigrating or emigrating cells correspondingly. When two such centers arise in a colloidal continuum, their combined contractions lay down a preferential pathway of fibers along the connecting line and thereby generate automatically a straight traffic channel from one center to the other ("Two-center effect"; Weiss [23, 24]). This is a striking example of scalar chemical activity resulting in vectorial, movement- and form-directing effects.

Model experiments *in vitro*, as well as observations on developing tissues in the body, have furnished numerous concrete examples of how chemical inhomogeneities can translate themselves into the physical effects that underlie cell orientation in general. Yet the detailed shape, orientation and fate of any given cell can only be statistically predicted, as the details will vary with the unpredictable fluctuations within the cell itself and with the contingencies of its immediate micro-environment, both substratum and medium. The variability of cell shape and cell orientation in any single tissue culture illustrates this point. Individual cell shape can be treated quantitatively only in terms of probabilities, but not of microprecise determinacy. This proposition has been tested as follows (Weiss and Garber [30]).

In line with our earlier comments, we may assume that contacts of extraneous fibrils with the cell surface create outlets for local protrusions, some of which will be abortive, while others will succeed. Now, since concurrent projections must compete for cell content, the more of them will be able to coexist, the more equal they all are in draining power. On general grounds, larger fibers can be assumed to cause a stronger outflow of cell content, hence to suppress other weaker protrusions in their vicinity. Therefore, the greater the inequality of fiber size and the higher the incidence of larger fibers in the medium, the lower should be the average number of successful processes which cells in that medium could protrude. The fibrin network of the blood plasma clot consists of fibrous aggregates of various sizes. Since the degree of aggregation is greater at lower than at higher pH values, one can produce a graded series of clots of increasing diversity of fiber sizes. Cells reared in them should then show a corresponding gradation in the average number of processes, with a dominance of multipolar forms at the alkaline end of the scale, and of bipolar forms at the acid extreme. A similar series could be expected from cells reared in clots with different fibrin concentrations, the higher concentrations containing larger fibers. Our observations, carried out on thousands of cells, have fully confirmed these expectations. The average number of cell processes decreased linearly with decreasing pH values (during clot formation), as well as with increasing fibrin concentration.

A summary view of our discourse thus far reveals the futility of any attempt to deal with contact guidance by reference to some general and overall "adhesivity". Rather we are faced with a composite mechanism in which three major sets of separate and partly separable variables are involved. These are (1) motive power from metabolic energy of the cell; (2) inhomogeneities in the constitution of the motile fringe of the cell, due (*a*) to random or systematic inequalities in the structure of the immediate environment, which in turn are partly attributable to prior modifications of the environment by cellular activities and exudates, and (*b*) to residual inequalities in the cell stemming from previous environmental influences; and (3) dependence of general adhesivity of the cell surface upon the composition of the medium, the substratum and the outermost cell covering. Given a complete knowledge of these factors, it ought to be possible to predict statistically the shape any given type of isolated cell should assume under given conditions, as long as it has a freely reactive surface. In many instances, shapes freely assumed in a primary reaction to the environment will gradually become consolidated in various degrees by the formation of rigid internal or surface structures. It seems that both fibrous protein systems and lipid lamellae can subserve this harnessing and stabilizing function, of which the mature nerve fiber or the spermatozoa are notable examples. However, in the following we shall continue to concern ourselves only with the free and plastic cell and its behavior. This behavior is subject to major modifications and restraints if a cell does not remain in isolation but encounters other cells with which it can interact. In one regard, it is just such restraints that lead to locomotion, paradoxical though this sounds, it will be documented in the following section.

LOCOMOTION

In the preceding account we have dwelled on the development of cell shape as a result of cellular motility, in the sense that some parts of the cellular content undergo internal shifts relative to other parts, ending in deformation. However, the system as a whole remained stationary. The center of its mass, even though not necessarily static, oscillates about a fixed position without net translocation. For a flattened disc-shaped cell, this appears as an erratic straying about within a relatively narrow range, while for a bipolar spindle cell, it means a shuttling back and forth on a linear track, due to the random inequalities of adhesive and protrusive force between the two mobile ends straining in opposite directions. In other words, motility does not of itself lead to locomotion. Locomotion results only if the center of the mass is in progressive translocation; and if the translocation follows a steady course in a given direction, we designate the locomotion as oriented.

Theories of cell locomotion have in general been classed under four headings; namely, chemotaxis, galvanotaxis, phototaxis, and mechanotaxis (thigmotaxis). Except for mechanotaxis, all of these terms refer to directional signs provided by asymmetric distributions ("gradients") of chemicals, electric charges, or radiations in the environment, while remaining non-committal as to the mechanism by which the external asymmetry is translated into an oriented locomotor response of the cell. Of course, no external asymmetry other than sheer passive convection can serve as cue to orientation unless it is "perceived" by the cell; that is, establishes a corresponding state of asymmetry within the cell itself, beginning with the surface and involving the interior in various degrees. Omitting the trivial and essentially unbiological instances of passive convection by currents, mechanical pull or electrophoresis as well as the reflex orientation of cells having a specialized locomotor apparatus, such as ciliates or flagellates, the proposition is as follows.

Random fluctuations in the local concentration of chemicals and distribution of charges within the immediate micro-environment of a given cell lead continually to corresponding inequalities, hence asymmetries, along the cell surface of the same order as those discussed in the previous section. A temporary gradient of concentration or potential therefore entails a momentary axiation of the cell in the direction of the gradient. However, such an effect cannot become manifest unless the external gradient is steep enough to supersede the internal differentials within the cell and retains its position long enough for the cell to give an effective reaction. For a chemical gradient, this would require that the difference of concentration between the side of the cell facing the source and the opposite side be of such a magnitude that it constantly exceeds any of the internal gradients that arise within the cell from its autonomous activities; and for an electric gradient, it would presuppose a sufficiently steep and steady gradient of electric potential between the two sides.

If one keeps in mind these prerequisites for effective orientation, one realizes immediately that in many situations in which sources of chemical emanation or electric charges have been considered as either "attractive" or "repulsive" loci for moving cells, those premises are not fulfilled even in a medium fully protected against mechanical agitation and perturbation. The fraction of an external concentration gradient along the short stretch of the diameter of a given cell, which latter is in itself not homogeneous, would not seem to be sufficient to serve as a guiding cue, except perhaps statistically over long periods of time, in which case it would register as a very gradual drift, rather than an oriented movement. Considering further that inside the organism fluids are constantly stirred up by muscular contractions, pulse waves, currents, and so forth, and that the motion of the cell itself stirs its surrounding medium, one realizes that the old primitive concepts of chemotaxis are far from realistic. And yet, aside from many cases of misinterpretation, we do know of such phenomena as the migration of white blood cells toward a focus of inflammation, of slime molds toward a source of acrasin, and of similar well-attested effects in which a dispersed cell population was found to be at least eventually aggregated at, if not necessarily directly guided towards, a chemically distinguished locus in the environment. In the slime mold, a relay action cell-to-cell, initiated from a given locus, but then propagated chain-wise by each responding cell serving as a new amplifying source, and further accentuated by rapid decay of the product (Bonner [3], Schaffer [12]), offers one type of explanation; but even here the intervention of some microstructural devices oriented by the more patent chemical activities (similar to the "two-center effect" mentioned above) has not been definitely ruled out.

Even so, the possibility that cells can polarize each other directly by chemical means without any structural intermediation, becomes a very real one if the distances between the interacting cells are short; for in this case, each cell forms a physical barrier to the rapid dissipation of the discharged products of the other, hence these products will accumulate between them and thus establish the head of a rather steep gradient between the "inner" and the "outer" sides of the confronted cells. Now, if the products in question are of the sort that would alter the surface constitution of a cell exposed to them, whether by causing molecular reorientation (Weiss [26]) or ionic changes or otherwise, their concentration gradient across the cell would create a corresponding difference in surface strength on opposite sides, as a result of which cell content will be protruded on the side that is relatively weaker; whether this is in the uphill or downhill direction of the gradient depends on whether the agent is one that loosens or one that tightens surface structures. In the former case, the new protrusions will point toward the other cell, giving the illusion of "attraction," while in the latter case, they will be directed outward. Striking effects of this kind have actually been observed in the protrusion of rootlets from seaweed eggs grouped close together (Whitaker [32]) and in the outgrowth of processes from nerve cell groups in tissue culture (Stefanelli [13]).

Next, if the protrusions have the properties of pseudopods, that is, are adhesive instead of free at their tips, their one-sided extension will, of course, drag the rest of the cell body after them in the same direction. As a result, the two or more cells in interaction will in one case move closer together and in the other case move farther apart. As I have pointed out on earlier occasions, the latter mechanism accounts readily for the observation by Twitty and Niu [16] that pigment cells in a capillary tube tend to recede from each other, without the need of invoking any mutual "repulsion" or "negative chemotaxis".

Electric gradients can similarly give rise to surface asymmetries, which limit the directions in which cell content can be propelled. As colloids on the anodal side become stiffened (Anderson [2]), locomotion in any but the kathodal direction is ruled out; the kathodal drift of nerve fibers in tissue culture (Marsh and Beams [8]) is an obvious illustration of this fact.

In conclusion, no asymmetric condition in the environment can exert any guiding influence on a cell unless it can evoke a corresponding axial differential in the cell, which confines the cell's intrinsic motility to an outlet in a single direction only instead of the all-round expansion open to it in an isotropic space. An isolated cell on an interface can be regarded as having innumerable degrees of freedom of expansion and therefore, being pulled out equally in all directions, remains stationary in the center of the stalemated radial strains. However, as soon as its motility is reduced or totally suppressed in one sector, the remaining portion, now given competitive advantage or true monopoly, has full sway in pulling the cell in its direction. In this sense, the above statement that locomotion results from the restriction, rather than the addition, of degrees of freedom of a motile cell, is no longer paradoxical. And a brief reflection will show that in this light, the problems of cell shape, cell orientation, and now cell locomotion, appear simply as various expressions of a single common principle: the channeling of cellular motility from random into definite courses singled out by physical and chemical differentials in the environment, of which other cells and their products form integral parts.

IV. ON FORM AND FORMATIVE PROCESSES

Contact guidance, as a critical limiting condition for all cells bound to interfacial support, delineates the limited tracks still open to a given cell at a given point in a given environment; in the extreme, on a strictly oriented medium, this leaves the individual cell, now a bipolar one, with only a single track. Yet, even on a single track, there are still two alternative directions left in which to advance, and evidently contact guidance remains ambiguous as to which one will be taken. This to decide, is a matter left to those differentials and gradients just outlined before.

A pertinent analogy is that of a railroad. The rails and switches limit the possible routes that can be taken, while the engine provides the motive power; but even then it still depends on which way the engine is headed whether the train will move forward or backward on the tracks. And if two engines are hitched to the train pulling with equal force in opposite directions, the train will remain standing as did our single bipolar spindle cells on grooves or fiber tracks.

CONTACT INTERACTIONS

Now, as I stated in the preceding section, a cell does not become a significant polarizing factor in the environment of another cell until they both come within close enough range for their discharges (chemical or electric) to sustain telling inequalities in each others' surfaces. The critical distance will vary according to the type and state of activity of the cell and the condition of the environment (composition of medium, configuration of space, degree of perturbation, etc.) but can scarcely exceed the order of magnitude of the diameter of the cell. The short-range extreme of this type of cell-to-cell interaction is reached in cases in which the critical distance is so short as to appear as "contact". We shall take the word here in its primitive meaning on the microscopic scale, unmindful of a certain fuzziness it assumes in terms of molecular dimensions (Weiss [25]). If cells were stated above to be able to paralyze each other's confronted sides as they come within close range, so that thereafter they will move apart again (for example, in the cited experiment of Twitty and Niu), a far more general expression of the partial immobilization which cells can impose on each other's surfaces is observed when cells actually make physical contact. The sequelae of such contact are quite diverse for different cell types, but they all seem to have one feature in common, which I have stressed and illustrated ever since 1941 (see review in Weiss [21]) and which is as follows.

In a freely mobile cell, that part of the mobile fringe which comes in contact with another cell ceases to extend actively. In an epithelial cell, this causes cell content along that border segment to become diverted from its formerly radial trend into a tangential course, remaining available for extension along the still unrestrained part of the edge. That free portion diminishes progressively, as contacts are established with ever more cells, ending with the full loss of mobility, though not of motility, in the completely surrounded cells— with specific qualifications to be outlined presently. This immobilization is by no means due to crude mechanical confinement, but is a biological response to contact. It seems to be of a similar nature to the stabilization of the bubbling surfaces of floating (or post-mitotic) cells when they become attached (or reattached) to a substratum. It must be emphasized, however, that in cell-to-cell contacts, only the fraction of the border touching the other cell is immobilized, but not the whole border, let alone the entire cell. Epithelial cells thus acquire common borders, and if their association endures (see below), they constitute cohesive sheets. Any later disruption of such a sheet creates again a free edge, liberates the cell borders along that edge, hence restores to them one-sided mobility, which automatically leads the free cell front into the lesion; this is how wounds heal (Weiss [27]).

Whereas epithelial cells tend to stay combined, the reaction to mutual contact in bipolar or multipolar cells, especially of mesenchyme, takes a different course. Again, the primary response is one of mutual immobilization at the contact points. But lacking any tendency to stick together, the cells are moved apart by the continuing advance of their free sides. This process has been observed in time lapse motion pictures by both Abercrombie *et al.* [1] and ourselves (Weiss [25]; it has been termed "contact inhibition" by the former—a term which is illustrative provided one keeps in mind that the "inhibition" pertains to a localized fraction of the cell only. In many of our cultures, the immobilized end does not just come to rest, but looses its hold on the substratum and is vigorously retracted towards the main cell body. But whatever the details, the outcome is a temporary asymmetry in the motile apparatus of the cell, giving the lead to the residual uninhibited pseudopodia. In our simile, a railroad train, stalled by being hitched to two engines pulling in opposite directions, can run again if one of the engines is disengaged.

While clarifying the nature of directive interactions between proximate or contacting cells in the manner here outlined, all observations so far have failed to show any trace of direct "attractions" or "repulsions" between cells, which would seem to make the continued loose usage of these terms not only meaningless, but truly misleading and reprehensible. The outward

IV. On Form and Formative Processes

migration of cells from a tissue fragment in culture is not due to any repulsive action of the center, but is again simply the outcome of an environmental asymmetry, namely, the gradient of cell population density; the mere fact that statistically, the freely roaming cells meet more contacts, hence are subject to more contact immobilizations, in the central than in the peripheral directions, leads to a general centrifugal drift.

SELECTIVITY

Without going further into these interesting problems of cellular population ecology, there is one major facet which I have tried to evade so far in this essay, but which is too fundamental to be wholly bypassed, namely, the feature of specificity and selectivity, both in contact guidance and cell interaction. It signifies that a given cell can distinguish between different kinds of contact substrata, as well as between different kinds of cells, and can respond discriminately.

My preoccupation with this topic dates back to my earliest experimental work (Weiss [17]), in which I noted that a larval amphibian limb transplanted in place of an amputated limb would suppress the regeneration of the latter only if it was inserted in the correct orientation, in which case it was accepted by the host body as a true substitute. Discovering subsequently the specificity of neuro-muscular relations, the "case of the specific match" as a basic biological principle thoroughly captivated my interest and has held it quite firmly ever since. In 1924 and 1930 [18], I reviewed some of the scanty hints at tissue selectivity under the purely descriptive heading of "affinity," a principle which was soon to receive much more solid factual substantiation by the classical studies of Holtfreter [6] in the 1930's. In 1941 [19], the earlier concept of contact guidance was expanded to encompass "selective guidance," based on "selective adhesiveness," largely on the strength of the demonstration of the selective tracking by different classes of nerve fibers of different preneural pathways, and later of fibers of the corresponding type (HAMBURGER [5], Taylor [14]). At the same time, I also introduced a hypothetical molecular model of contact selectivity in general, which, though it may not hold the answer to the problem, has led at least to a more rigorous articulation of the questions the facts present to us. In 1946, I summarized these considerations of specificity of cell interactions in growth and development (Weiss [21]), and in 1949 [22] proposed their applicability to the problem of the alienation and metastasis of the cancer cell. This was followed by the demonstration of selective "homing" at their proper destinations of random disseminated embryonic cells (Weiss and Andres [28, 29]) and by the evidence that epithelia will merge only with epithelia of the same or related kinds, but not with unrelated ones (Chiakulas [4]). Meanwhile, the Mosconas [9, 10] perfected the technique of dissociation and reaggregation of cells from embryonic tissues *in vitro*, which made possible the elegant demonstration that scrambled mixtures of cell types sort themselves out rather cleanly according to type (Moscona [9]).

It was a logical step to study the manner of this selective segregation directly by time-lapse cinemicrography, and during the last five years, we have (with A. C. Taylor and A. Bock) compiled a large stock of film records of cell encounters *in vitro* in many different combinations of tissue origin, species, ages, densities, treatments, media, substrata, etc. A sample film was shown at the Noordwijk symposium. However, no more than the most general conclusions can be reported here. They are as follows (see also Weiss [25]).

A free epithelial cell on a flat surface, when making chance contact with another cell, may form temporarily a local attachment to the latter. If both are of the same type, they mostly remain associated and gradually enlarge their mutual border line. Even so, their surfaces are by no means firmly bonded to each other, but get detached in spots and reattached in others, so that their conjugation must be explained by a statistical-dynamic, rather than static, principle. As larger numbers of cells of the same type join up, their collective outline tends to assume minimum possible length, i.e. become near-circular. But whether this is the result of a condition of maximum possible internal boundary lengths among the constituent cells or of enveloping action of an exuded ground mat, cannot be decided on existing evidence. On the other hand, if the two meeting cell margins belong to cells of unrelated types, they separate again, much as if the other cell were just part of the ordinary substratum. All this applies to cells freshly liberated from tissues, for after prolonged cultivation some of the discriminative reactions are lost. In this connection it is worth noting that Abercrombie (see this symposium) has found the normal and malignant descendants of mesenchyme to have lost their mutual recognition of kinship, as manifested by the loss of "contact inhibition".

The nature of the mechanisms by which cells recognize each other's kinds as well as the kinds of substrata they want to follow, still escapes us. My suggestion [19] of antigen-antibody-like bonding, even though favored by some, does not seem to be adequate if interpreted statically, although a more statistical and dynamic version of it cannot yet be ruled out definitively.

IV. On Form and Formative Processes 427

But one fact stands out clearly: whatever the mechanism of selective "recognition", the response to it by the cell is definitely more comprehensive than a plain surface reaction. The problem is further complicated by the fact that the selectivities in question show gradations, rather than absolutely sharp definition; for instance, in contact guidance, a preference was noted for cells to stay on glass rather than pass on to a fibrin thread [20]. So, taking all in all, it would seem best to leave the matter of specificity open till further analytic work provides some more compelling clues to explanations than we can muster at the present. One cautionary note: some familiarity with the real phenomena that are to be explained is indispensable, for our supply of explanations for imaginary concepts of cell movement and cell interactions is already superabundant.

CONCLUSION

As indicated in the introduction, it would have been presumptious and futile in this essay to attempt a rounded and systematic portrayal of our knowledge, such as it be, of the principles of cell locomotion. The best I could try to do was to record the gist of years of experience with the relevant phenomena, conveying perhaps some semblance of an emerging unifying concept. The field being in flux, any such concept can be considered as no more than *in statu nascendi,* and even then as but a partial contribution to a common goal. Other participants have offered many illuminating contributions of great value to the same cause. The views are starting to converge. No ultimate answers are as yet in sight. But the problems, which pose the questions, have at least gained in sharpness and consistency; and if the conference were to have achieved nothing more, it has scored a major gain as guide post for further research efforts.

REFERENCES

1. ABERCROMBIE, M., JOHNSON, M. L. and THOMAS, F. A., *Proc. Roy. Soc. London B* **136**, 448 (1949).
2. ANDERSON, J. D., *J. Gen. Physiol.* **35**, 1 (1951).
3. BONNER, J. T., The Cellular Slime Molds. Princeton University Press, 1959.
4. CHIAKULAS, J. J., *J. Exptl. Zool.* **121**, 383 (1952).
5. HAMBURGER, V., *Wilhelm Roux' Arch. Entwicklungsmech. Organ.* **119**, 44 (1929).
6. HOLTFRETER, J., *Arch. exptl. Zellforsch. Gewebezücht.* **23**, 169 (1939).
7. LETTRÉ, H., *Cancer Research* **12**, No. 12, 847 (1952).
8. MARSH, G. and BEAMS, H. W., *J. Cellular Comp. Physiol.* **27**, 139 (1946).
9. MOSCONA, A., *Proc. Soc. Exptl. Biol. Med.* **92**, 410 (1956).
10. MOSCONA, A. and MOSCONA, H., *J. Anat.* **86**, 287 (1952).
11. ROSENBERG, M. D., *Biophys. J.* **1**, 2, 137 (1960).

12. SCHAFFER, B. M., *Am. Naturalist* **91**, 19 (1957).
13. STEFANELLI, A., *Acta Embr. Morph. Exp.* **1**, 56 (1957).
14. TAYLOR, A. C., *J. Exptl. Zool.* **96**, 159 (1944).
15. THOMPSON, D'ARCY W., On Growth and Form. University Press, Cambridge, 1942.
16. TWITTY, V. C. and NIU, M. C., *J. Exptl. Zool.* **125**, 541 (1954).
17. WEISS, P., *Wilhelm Roux' Arch. Entwicklungsmech. Organ.* **99**, 168 (1923).
18. —— Entwicklungsphysiologie der Tiere. Steinkopff, Dresden and Leipzig, 1930.
19. —— *Growth* **5**, 163 (1941).
20. —— *J. Exptl. Zool.* **100**, 353 (1945).
21. —— *Yale J. Biol. and Med.* **19**, 235 (1947).
22. —— Differential Growth. *In* Chemistry and Physiology of Growth, p. 135. Ed. A. K. PARPART, Princeton University Press, 1949.
23. —— *Proc. First Natl. Cancer Conf.* **50** (1949).
24. —— *Science* **115**, 293 (1952).
25. —— *Inter. Rev. Cytol.* **7**, 39 (1958).
26. —— *Proc. Natl. Acad. Sci. U.S.* **46**, 993 (1960).
27. —— *Harvey Lectures 1959–1960*, p. 13, Academic Press, New York, 1961.
28. WEISS, P. and ANDRES, G., *Science* **111**, 456 (1950).
29. —— *J. Exptl. Zool.* **121**, 449 (1952).
30. WEISS, P. and GARBER, B., *Proc. Natl. Acad. Sci. U.S.* **28**, 264 (1952).
31. WEISS, P. and HISCOE, H. B., *J. Exptl. Zool.* **107**, 315 (1948).
32. WHITAKER, D. M., *J. Gen. Physiol.* **20**, 491 (1937).
33. YOUNG, J. Z., Growth and Form, p. 41, Clarendon Press, Oxford, 1954.

IV. ON FORM AND FORMATIVE PROCESSES

Reprinted from THE SYMPOSIUM ON THE BIOCHEMICAL AND STRUCTURAL BASIS
OF MORPHOGENESIS. Arch. Neerland. Zool., 10, 165-176, 1953 (Suppl.).
Swets and Zeitlinger, Publishers, Amsterdam.

CHAPTER 20

SUMMARY COMMENTS AT THE CONCLUSION OF THE SYMPOSIUM

by

PAUL WEISS

(Professor of Zoology, University of Chicago, Chicago, Ill.)

In summarizing the proceedings, it would be hopeless to attempt to abstract and review the abundance of valuable data and ideas presented in the various papers. It would be equally hopeless to try to bring the conclusions to common denominators or to single out any one presentation that might, more than others, be presumed to become in the future the master-key to the understanding of developmental phenomena. If anything, the conference has been a powerful reminder that no single approach, but only the concerted combination of the greatest possible variety of approaches, will bring the desired understanding. At the same time, the more the individual approaches call for concentration and specialization, the more imperative it becomes to bring the various groups together in meetings such as this for an occasional exchange of views and clarification of facts and ideas. The value of the gathering has been amply documented by the lively and pertinent discussion that followed the individual papers.

Rather than giving a factual review of the meeting, I shall, therefore, attempt merely to bring together certain general observations that can be distilled from the proceedings, as they reflect the trend of our current thinking on problems of morphogenesis. In surveying this trend, one cannot help but be impressed by the great advances that have been made in recent years both in the discovery of pertinent facts and in the clarification of issues and interpretations.

In general, the meeting gave proof that we have essentially outgrown the era in which general platitudes about the fact that chemistry has something to do with morphogenesis were the order of the day. We have become sufficiently conditioned to taking this fact for granted without reasserting it on every possible occasion. There was a time when it was quite necessary to stress that morphogenesis is not a phenomenon sui generis with which chemical activities are at best only in some vaguely defined interplay. But the advances of the last half century have pretty well disposed of the notion that development is anything but physics and chemistry provided we mean by the latter designations all properties and all possible interactions of the 90-odd elements in all possible combinations in space and time; not just the ones we already happen to know.

All the more surprising is the fact that certain residues of the old opposition between morphology and chemistry still creep into modern discussions. Evidently we still have not quite outlived our past. We still hear occasional remarks, and have heard them at this symposium, to the effect that there is a chemical ontogeny in addition to a morphological ontogeny, as it were, as if the morphological structures which we study as the products of ontogeny were anything but the mere visible expressions of anteceding chains of physico-chemical events. Such remarks obviously are the last traces of an outdated habit of thought, and it seems pretty plain that at least those who attended this meeting will no longer feel the need for asserting that chemistry is related to morphology, but will pass on to the next and more arduous task, namely, to specify just what precisely this relation is.

One of the main difficulties in this task, it seems, lies in the fact that we have not yet been very successful in showing how chemical action can be translated into morphological effects. The appearance of diffuse characters such as colours or pigments, as well as the suppression of structures because of metabolic defects, have been satisfactorily analyzed, but on the constructive side, just how to get from chemical activity to highly specific cytological and histological structure and architecture, is not only unknown but rarely even stated explicitly as a pressing problem. Here part of the blame may be laid right at the doorstep of the morphologist who often has failed to provide a sufficiently detailed clear and precise description of the actual events which occur in histogenesis and morphogenesis. Lacking these data, it is understandable that the biochemist then must operate with rather abstract and generalized notions of "growth and development" instead of with the concrete problems presented by the real objects. The dearth at this conference of explicit references to the "biochemistry → structure" problem (except for TWITTY's pigmentation patterns and examples of egg surface structures; see below) is a fair indicator of the neglected state of this field.

Electronmicroscopy, properly evaluated, has made a promising start in the further analysis of structure formation. However, one must remember that just as in light-microscopy, the pictures reveal only the eventual structural order attained by virtue of antecedent ordering events, but still leave the latter in obscurity. While the orderly distribution and segregation of different chemical systems within the cell space can perhaps be partly referred to the ordering influence of particulate cell inclusions (see below), the elaboration of such characteristic and delicately structured products as the scale of a butterfly, the feather of a bird, the organ of Corti or the oral implements of a ciliate protozoan are still a far cry from physicochemical interpretation, and for such structural over-all regularities as polarity, axiation and symmetry, we have as yet obtained no more than formal descriptions in terms of fields or crystal analogies or the like.

Nor is it likely that we shall make much conceptual progress toward an understanding of these phenomena if we continue to give most of our attention to biochemical reactions in solutions, particularly in homogenized and randomized systems, in which many of the very prerequisites for the development of formed structures have been destroyed. If our success in isolating, identifying and even synthesizing biologically active substances were nearly as well matched by our understanding of how they operate, the situation might change for the better in short order. Even at this meeting there has been evidence that we are more concerned, in a somewhat legalistic frame of mind, with apprehending the criminal than with reconstructing the crime. Undoubtedly, a biochemistry of the living system, particularly in morphogenesis, must not aim merely at a listing of all chemical compounds present and their possible interactions but must deal with the selective combinations and very specific constellations which these chemical systems assume in their integrated and interdependent coexistence in the living system. Only by raising our eyes from the elements to their interrelations can we hope to make the conceptual transition from the biochemistry of homogenates to that of the living system.

It seems that as we do this, stereochemical interlocking and other specific group conformances among molecular complexes, perhaps equilibrium conditions denoting a greater stability of such dynamic complexes than characterize their separate constituents, will increasingly come to the fore. For the organized living system, biochemistry will have to become what I have called "molecular ecology," that is, a science of the behaviour of molecular species and groupings in response to conditions which they set up for one another. Evidently each biochemical reaction presupposes a rather narrowly circumscribed set of conditions in order to be able to proceed. Hence, in an orderly aggregate of molecular realms, such as a cell, only a limited number of interactions would be permitted in any one region, depending wholly on the constellation in the rest of the system. This fact of itself, therefore, sets the various partial processes into an organized interdependency relation which marks the collective activity of the molecular species as something quite different from the random behaviour of the individual components operating in isolation.

For the time being, there seems to be little we can do but keep this general concept in mind and for the rest proceed on a strictly empirical basis and gather more crucial information not just about the correlation between chemical processes and morphological structures but about the actual processes through which the former are converted into the latter.

Another sign of maturity at this conference was the absence of arguments about thermodynamics in development. Evidently, the experience of the last decades has sufficiently proven the point that developmental phenomena do not violate the laws of thermodynamics. If differentiation at any level – from chromosome duplication to the production of the specialized structures of the adult – is tantamount to the elaboration of increasingly more improbable systems, i.e., a process of unrandomization, we are convinced that extra energy is needed for this process even though the amount may be small in comparison to the amounts needed for the sheer maintenance of the cells of the developing system. The old problem of "energy of shape" is still with us, presumably because of the fact that the energy requirements for the slow transformations in growth and differentiation may be greatly overshadowed by the energy requirements for the continuous anabolic renewal of the protoplasmic systems, which proceeds at a much faster rate. For the sake of clear research orientation, it must be pointed out that we do not even know whether or not the metabolic fuel sources for the endergonic processes of growth and differentiation are the same as those serving cellular maintenance. At any rate, even after we have learned all about the fact that metabolic energy is expended in a particular morphogenetic process, and where it comes from, whether we measure it by gas exchange or cytochemically by the demonstration of phosphorylated energy carriers, this still will tell us nothing about the type of specific synthesis or transformation for which this energy is being utilized; and no doubt, the type of processes engendered or supported is of more interest to the student of development than is the mere fact that they require energy.

It was encouraging, therefore, to see evidence at this conference of the growing awareness of the need for more studies on specific synthesis during development. This has come out not only in the pioneering work of the Swedish school, trying to identify actual metabolic patterns involved in different morphogenetic areas of the sea-urchin germ, but from quite a different approach, in the studies on the appearance of specific protein systems during embryogenesis, as tested by the immunological methods referred to by WOERDEMAN and Mrs. CLAYTON, as well as HERRMANN's program on the biochemistry of muscle differentiation. To be sure, much of this work is no more than sequential

IV. ON FORM AND FORMATIVE PROCESSES

descriptive chemistry, just as old-time embryology was sequential descriptive morphology; but we may expect that once a complete record of the biochemical history of a particular ontogenetic series of stages has been obtained, the succession of events in this chain of transformations may furnish a clue to the underlying dynamics of differentiation. Already these studies have proved to be much more than a mere record of the biochemical ontogeny of given organs, in the following sense. For one thing, they have made it possible to trace back specific organ properties, as in the case of lens, into an appreciably earlier stage than that of morphological distinctiveness, thus giving the concept of "determination" a tangible content, and then they have also brought a chemical confirmation of the principle of epigenesis by indicating the emergence of new macromolecular compounds, such as specific proteins, during the development of the germ. As for the latter proposition, it must be admitted, of course, that present techniques are not refined enough to exclude the possibility that compounds which appear to arise de novo may have pre-existed in the system either in such minute amounts as to be undetectable or else in a conjugated or otherwise masked form.

At any rate, it is instructive to note that when it comes to the identification and discrimination of more subtle characters of biochemical units, we still have to resort to the indicator mechanisms of the living organism. The superiority of the immunological tests of protein specificity over the comparatively crude methods of chemical diagnosis makes this fact abundantly clear. Some of the still subtler differences of biochemical constitution, which are basic to the understanding of the processes of differentiation, still escape all but the strictly biological tests of specificity. For example, we have been told that there is really no appreciable difference between the muscles of some invertebrates and the muscles of man when they are subjected to routine biochemical examinations. On the other hand, we know that even the different muscles of the same animal possess specific biochemical distinctions of a qualitative kind which reveal themselves with the greatest clarity in biological tests but defy any straight biochemical identification. The same is true of other tissues. For instance, while the collagen and the mucopolysaccharides of the connective tissue may be chemically the same throughout the body, we know from their responses to hormones, drugs and epithelial influences that the cells in different parts of the mesenchyme have specifically different properties which it would be difficult to visualize otherwise than as distinctions of biochemical constitution. Different parts of the vascular tree, different areas of the epidermis, different groups of neurons in the central nervous system, and even different types of cartilage taken from different sites, prove to be biochemically specialized beyond the power of direct detection by present strictly chemical techniques. If anything, the proceedings of this meeting have only confirmed me in the conviction that, at least conceptually, biochemistry of morphogenesis will progress faster if it frees itself from the compulsive preoccupation with those aspects which are already within the grasp of our exceedingly limited chemical techniques.

The tendency to such confinement has been particularly obvious in most of the discussions of enzyme activities during development. Since the best studied enzyme systems, besides digestive ones, are those engaged in the metabolic cycles of cellular maintenance, they have gained a certain favourite position in studies on development likewise. However, since they are also ubiquitous and essential for the normal life activities of any cell, they lose much of their significance in regard to the problems of differentiation proper, that is, the appearance of properties distinguishing one cell type from another, except, of course, in their subsidiary role in the energy supply for differentiation. The fact that some of these systems have been studied with particular success in the sea-urchin egg does not necessarily make them relevant to the problems of differentiation as such. There have been many encouraging signs at this meeting of a growing realization of this situation; the various attempts to test new substrates and trace changes in enzymatic patterns during development constitute such signs. However, just as in the case of the supposed epigenesis of proteins in general, one could make the valid objection that what is described as the new appearance of an enzyme may merely signify our inability to detect it earlier because it may have been masked, inoperative or perhaps just present in inadequate concentrations to be demonstrable. While it is important to bear these theoretical considerations in mind, they should not be allowed to detract from the appreciation of the impressive array of facts presented to which they refer.

One of the greatest strides recorded at this meeting was the progress made towards localizing biochemical events within the cell space by assigning certain reactions either to the surface or to intracellular particulates. As for the cell surface, the very fact that it is that part of the cell which is directly exposed to the environment gives it not only regulatory power over what goes in and out of the cell, but also a decisive role in establishing contact relations with the neighbouring cells. RUNNSTRÖM's emphasis that it behaves as an active part of the living protoplasm, rather than as a mere passive barrier, is particularly relevant to developmental processes because, as I have tried to emphasizy in earlier discussions, it is through surface contact that such processes as induction, cell association and even polarization are presumably

IV. ON FORM AND FORMATIVE PROCESSES   435

made comprehensible. Another important fact which I have stressed but which is not always duly taken into account is that the surface of the egg, like any interface surrounding a liquid system, has a much greater mechanical stability than the enclosed liquid space, permitting surface particles to retain their relative positons even as the liquid interior is completely reshuffled by the convection currents of mitosis or by artificial stirring. The surface can, therefore, function as the seat of that specific topographic mosaic of physically and chemically distinct areas which furnish the ground plan for the orientation of the subsequent developmental events in interplay with the appropriate reactions of the genic complement.

The same principle of organizational stability of interfaces is found repeated in innumerable examples in the organized particulate cell inclusions. Many of these which had escaped microscopic detection because of their small size are now easily demonstrated under the electronmicroscope. It would be unfair to consider the recent spurt of interest in these various bodies as merely a revival of the old particle theories of life which charged all properties of the living system to hypothetical "biogens," "plasmones," etc., shirking accountability, but solving nothing. The great difference is that we are now able to connect some of these particles with specific functions, e.g, the mitochondria with high enzyme activity, the microsomes with stores of ribonucleotides, etc. Again one could ask, however, how relevant these functions are to the specific phenomena of differentiation, as against their mere participation in ordinary cell life. The observations of GUSTAFSON and those of HÖRSTADIUS seem to have demonstrated that there is a definite relation at least between the mitochondrial population and the morphogenetic fate of a cell group, and the known participation of the mitochondria in actual structure formation, as in the formation of spermatozoan tails, is likewise an established fact.

The electronmicroscopic evidence of a core-crust differential in the structure of such particles, however, cautions us against considering them as "elementary" in the sense of biochemical homogeneity. Indications are that they have a fair degree of organization in themselves so that they cannot help us to resolve the problem of organization as such. Presumably they constitute large-scale coacervates of diverse molecular populations in which each of the constituent molecular species occupies a characteristic place or layer. As the surface of each such particle can again serve as locus for the concentration by selective adsorption or stereochemical bonding of other molecular species from the cell interior, they may become instruments of the progressive unrandomization which we call differentiation.

The phenomena of selective association between molecular groups by coacervation and polymerization represent one of the few tangible cases in which we understand the manner in which chemical action is translated into physical structure. RANZI's interpretation of the effects of lithium on protein systems is probably a case in point. It is irrelevant in this connection that many of the phenomena in this category which were formerly described in terms of colloid physics are now being viewed rather as instances of chemical bonding, nor does it matter that when the term "viscosity" is used as a measure of the degree of association in such a heterogeneous system as protoplasm, it obviously is only a shorthand expression for *average* resistance to internal displacement. The fact remains that such processes of molecular aggregation are involved in the consolidation and structure formation of cells during development, hence deserve our interest. If induced precociously, as perhaps by lithium action, they may preclude the mobilization and transport of intracellular compounds to such an extent that any subsequent developmental process that depended on undiminished mobility would be correspondingly delayed or fully inhibited. Certainly in the painstaking studies of the Swedish school on echinoderms, those of RAVEN on snails, of LEHMANN in amphibians, etc., this course of studies inaugurated by HERBST is still yielding profitable returns. As on other occasions, experience has again shown that things are not as simple as they were originally conceived of, as lithium evidently produces a multiplicity of effects depending on the stage and state of the affected systems.

Of course, the lithium experiments have in common with other methods of chemical interference with development the underlying hope that by finding out more about what blocks a given process, we might eventually be able to reconstruct the process itself. I suspect that the increasing popularity of metabolic inhibitors as a tool in the analysis of developmental processes is motivated by this hope. As a technique for the detection of local or chronological differences in biochemical make-up, they are certainly far more subtle than general inhibition by anoxia, radiation or cold. In helping to detect metabolic distinctions between different tissues arising during development, they might furnish further valuable criteria of the process of differentiation. But whether they will also contribute to our insight into the nature of the process of differentiation itself is still a matter of speculation.

My conclusions up to this point add up to the net impression of remarkable progress made during the last several years; if not yet in the deeper biochemical understanding of the processes of morphogenesis, so at least in the widening of certain avenues along which such understanding may be sought. As in every young scientific branch, we recognize the pardonable traits of oversimplification, premature genera-

IV. ON FORM AND FORMATIVE PROCESSES   437

lization and over-magnification of the little that is known as compared to the vastness that is still unknown. These, in due time, will find their own correctives. What of the future? Undoubtedly the tempo of work along the lines laid out and so competently sampled at this meeting will increase steadily. As the techniques will become more facile, workers less gifted with critical sense and imagination than the pioneers will be attracted to the field. Much of their work will be sheer routine. We shall presumably witness a repetition of the era of purely morphological description in embryology, only this time in chemical terms. The disquieting thought is that this process is really interminable unless somewhere along the line certain general concepts emerge which will lead to a real understanding of the chemistry of morphogenesis and will render further itemized recording of specific ontogenies unnecessary.

However, if we expect a sheer compilation of data gradually to lead to insight, we must not delay this development by tolerating a certain conceptual laxity which often blurs the issues and which has not been wholly absent from our discussions. I find it necessary to call attention briefly to a few signs of this weakness for even though it may not matter too much to the experts, it might befuddle less experienced investigators.

In the first place, we give in to what STUART CHASE has called "the tyranny of words." With all our precision in measurement, all our concern over purity of chemicals and statistical tests of validity, we often couch our conclusions in a phraseology that not only leads to no clarification or further experimental tests but actually bars any profitable approach to such clarification. We are using terms that are meaningless, ambiguous or plainly misleading.

Let me cite a few examples. We have heard statements to the effect that chemical substances are "doing" this or that, are "acting" in one manner or another, are "controlling" a process or "causing" an effect. Such phrases are customary and widely in use without much opposition. On occasions like this, however, their shortcomings should be pointed out. Clearly, all that a compound can "do" is either to combine with another compound or to dissociate, either with the consumption or liberation of energy, or to change the conditions under which other compounds will interact. Thus, all supposed "action" is in reality "interaction". While it might be convenient at times to focus on one partner in the process only and speak of a chemical "agent", it is unthinkable to deal with it except in reference to the "reagents" with which it interacts. The unscientific rating often given in this connection to one member of the pair as more "important" than the other, can easily be countered by asking in return which one is more important in the formation of water, the hydrogen or the oxygen? It is almost universally true in contemporary biology that we know a great deal more about the "agents" than we do about the reacting systems. We know the constitution of hormones, drugs, nutrients, and so forth, but have little understanding as yet of the mode of their interaction with the living system by whose changes we assay their "actions". We know all kinds of chemical "inductors" in amphibian development, but what do we know about the crucial reaction they unleash?

Thus to say that a given chemical "controls" gastrulation or differentiation or growth is really saying no more than that its presence or absence in some way alters the processes referred to. The word "control", therefore, is pretentious by implying more information than it actually conveys, and it would be scientifically sounder to state simply the facts: describe just what changes in the living system are correlated with the administration of what tested chemical, and then go on from there to investigate just what the interactions consist of. I do not doubt that all of those present at this conference have been fully cognizant of this situation, but I do want to submit that we ought to be a little more careful lest we cover up for the uninitiated workers in other areas the gaps in our knowledge and prevent them from formulating practical ways to the solution of the problems which we conceal behind verbal smoke screens.

In the same class, we note the habit of careless mislabelling of phenomena, which misleads others. I refer to labels which a closer look at the real situation or even some reflection would immediately reveal as untenable. The liberal use of the term "diffusion" is a case in point. In dealing with developmental systems, it is quite rash to suppose that diffusion could occur as freely as it does in ordinary solutions; more careful attention given to the structural properties of the system, its colloidal state, adsorptive surfaces, convection currents, local electric potentials, etc., would often lead to greater restraint in ascribing an interaction off-hand to plain diffusion. It is instructive, for instance, to consider how extensively the familiar "induction" phenomena in amphibians have been dealt with in terms of diffusible agents although it has been suspected for forty years and verified by recent experimentation that these effects are transmitted only by cellular contact.

Similarly, it is becoming almost a fashion to refer to any unknown biochemical reaction as "presumably enzymatic in character" even in cases in which the conditions are so little known that even a guess would be gratuitous. This substitution of suggestive terms for unknowns, pardonable in a young science as a sign of immaturity, had yet perhaps better be discouraged. Not only will the progress in the field be faster, but it will be easier to follow, if the facts are stated clearly and plainly

IV. On Form and Formative Processes   439

in sober, unambiguous, objective wording stripped of anthropomorphisms and unwarranted allusions. That this sobering purge of our vocabulary is well under way, is perhaps best documented by the fact that the "organizer" has made no personal appearance in our discussions, nor has there been any recurrence of earlier attempts to personify unknown chemicals hypothetically related to "organ formation".

Another conceptual confusion arises from the fact that we often apply collective statements as if they were valid for each individual component of the collective sytsem. Living systems being complex and heterogeneous and not wholly dissociable into their component parts without destruction of the very processes of life, we always make our measurements and determinations on complex objects, but after having made them, we then tend to "homogenize" them in our minds; that is, we often act as if those measurements and properties could be apportioned among all the members of the measured group in equal measure. When we read about the growth rate of a cell, we have a mental picture of every single constituent of the cell growing at a commensurate rate. When we hear figures of enzymatic activity, we tend to think of every single enzyme molecule of the particular kind as being active, and so forth. Of course, this attitude is gradually corrected, as we acquire more detailed knowledge of the facts; e.g., learn to assign "contractility" to contractile fibrils; respiratory enzymes to mitochondria; etc. But where this has not yet been achieved, or when the knowledge has not been properly communicated and absorbed, for instance, when we speak of the "growth" or the "differentiation" of a particular cell or a particular tissue, we still equate our statements over the whole measured mass as if each part participated equally in the result. For certain purposes, this is quite immaterial. But it certainly is apt to lead to false conclusions when used as the basis for comparisons among different systems.

Finally, due caution is indicated in drawing general conclusions from biochemical relationships discovered in one group of living forms. We have heard, for instance, some mention of the suspected role of insulin in the synthesis of specific proteins in mammals. It does not detract from the value of such observations to point out that specific protein synthesis, after all, goes on in all living systems, most of which possess no pancreas or insulin.

In conclusion, these last few critical observations were neither intended, nor must they be allowed, to overshadow the positive achievements of this conference, which have been quite splendid. Among all similar conferences which I have had the privilege af attending in recent years, this has been one of the very best. It was well focussed, productive and instructive. The objectives were clearly in view all the time, the presentations had logical continuity, and the desired integration of views was admirably accomplished either in the presentations themselves or in the subsequent discussions. All participants deserve our sincerest thanks for giving us such a highly nutritious meal of facts and ideas. We owe a special debt of gratitude, moreover, to the organizing committee for having done such a superb job in planning this meal so that it would give us a balanced diet; for arranging the courses in such rigorous order; and for providing us with the best possible catering service down to the last detail, including table decorations.

More even than by the tangible results, which have been sampled in the preceding printed pages, all members came away enriched by the intangible lessons of the meeting – a new critical outlook, a broadened perspective, a sense of mutual interdependence never before realized so vividly, pride of achievement, but what is more, humility in the face of the enormous job that still remains to be done, and an enthusiastic determination to go ahead with it.

BIBLIOGRAPHY

WEISS, P., 1949: Differential growth. In: Chemistry and Physiology of Growth pp. 135–186. Princeton University Press.

WEISS, P., 1950: Some perspectives in the field of morphogenesis. Quart. Rev. Biol., **25**, 177–198.

V. ON THE DYNAMICS OF THE NERVOUS SYSTEM

Reprinted from *Growth, Third Growth Symposium*, 1941, 5, 163–203.

CHAPTER 21

NERVE PATTERNS: THE MECHANICS OF NERVE GROWTH[1]

PAUL WEISS

Department of Zoölogy, University of Chicago, Chicago, Illinois

Compared with the steady flow of comparative-neurological research, analytical studies into the "how" and "why" of nerve pattern formation have remained a mere trickle. Whenever pressure for dynamic understanding of the development of the nervous system became irresistible, the comparative neurologist turned to speculation rather than to direct experimental attack. Kappers' "neurobiotaxis," Bok's stimulogenous fibrillation, even Child's gradient theory of nerve development, are noteworthy examples. There were sporadic attempts at analytical experimentation, but a deliberate drive did not really get under way until Harrison stepped in and gave the field an aim and a method; an aim, in that he broke down untractable generalities into tangible problems, and a method, in that he developed new techniques of transplantation and tissue culture for the solution of these problems.[2] His work, aptly extended by his students, notably Detwiler ('36), has been the greatest single step towards an analytical understanding of nerve development. Much of what I am to present today has had its roots in his work.

Harrison's experiments have settled a number of controversial issues concerning the elements of the nervous system, which we may list here as points of departure: (*a*) The neurone doctrine of nerve development was confirmed (Harrison '10): the nervous system arises from the differentiation of discrete cells. (*b*) The filamentous nerve processes, axons and dendrites, arise as extensions of the protoplasm of the nerve cell. (*c*) The cell strands enveloping the peripheral nerve fibers, or sheath cells, are later additions of central origin (Harrison '24). They play an accessory but not a formative rôle.

Nerve patterns, accordingly, do not emerge from a protoplasmic continuum, as was at times suggested (Hensen, Held), but are gradually built up in a true synthesis by the activities of individual neurones in constant interplay among each other and with their non-

---

[1]Experimental investigations reported in this paper have been aided by the Dr. Wallace C. and Clara A. Abbott Memorial Fund of the University of Chicago.
[2]For the most recent review of Harrison's work, consult his *Croonian Lecture* (1935).

nervous environment. More specifically, the neurones multiply, differentiate, and aggregate in definite distributions, grow in definite directions, branch in a definite order, assemble into definite bundles, connect with definite organs, assume definite size relationships, and are enveloped by definite types of other, non-nervous, cells. Occurring according to a definite order in space and time, these processes produce the final nerve pattern of the adult, which is highly intricate. But the task of resolving it into simple developmental components is not as arduous as it may seem, for the following reasons: Firstly, we must keep in mind that the final pattern is of composite origin. That is to say, the pattern-determining factors, while varying from stage to stage, may at any one stage be relatively simple in themselves, the final complexity resulting from the superposition of the successive imprints left behind by each pattern in the form of definite cell groupings and fiber connections. A developed nervous system compares, in a sense, with a photographic plate upon which innumerable pictures have been exposed in succession. The second fact to be kept in mind is that the manifest features of nerve patterns are only in part active achievements of the nervous elements; partly they are the result of purely passive dislocations. Neuroblasts may actively aggregate by migration or be passively crowded together by the expansion of neighboring areas. The contorted course of a nerve fiber may indicate that the fiber took a crooked path in growing, but it may also be due to later distortions of an originally straight fiber.

Thus, resolved into its component steps, nerve development is really not so complicated that it would discourage analytical approach. With this in mind, let us now examine how far this approach has taken us.

The differentiation of a neurone proceeds roughly as follows: The neuroblast, after ceasing to divide, grows considerably in size. Next, a localized protrusion of protoplasm appears, destined to become the nerve sprout. Later similar sprouts appear on the opposite part of the surface. In the organism the first sprout usually develops into the axon, the other sprouts under considerable arborization into the dendrites. In nerve cells cultivated in vitro the distinction between axons and dendrites is frequently quite arbitrary. A nerve process elongates by amoeboid activity of its tip (Figure 3). This free end is in a constant state of unrest, sending out pseudopodia in various directions, of which one usually establishes itself, while the others are hauled in.[3] Then from the end of this pseudopodium new competing

---

[3]Instructive illustrations of the amoeboid growth of nerve fibers can be found in Speidel's work (1933).

feelers are stretched out into the surroundings, one of which again takes hold, and so the fiber extends farther and farther, until it is eventually arrested by connecting with other cells, by unsurmountable mechanical obstructions, or by nutritive exhaustion. In advancing, the amoeboid tip spins out the nerve fiber from the body of the nerve cell, which stays behind, anchored in the ganglionic tissue.

All nerve fibers arise in this way.[4] The first fibers to sprout out are fittingly called "pioneering" fibers. When they develop, the body is still of very small dimensions, and all tissues lie fairly close together. The greatest distances to be spanned by the pioneering fibers until they reach their destinations are at best of the order of the millimeter. Once connected with a peripheral tissue, epidermis cells or muscle buds, the roaming life of the pioneering fiber is over. Attached now at both ends, its further course is no longer one of its own choosing: the terminal tissues take it in tow and drag it along during their own extensive shifts.

The number of nerve fibers maturing early enough to act as pioneers is relatively small. Nerves at that stage consist only of a few fibers. New fibers are added as additional neuroblasts differentiate. Thus the fiber complement of a nerve is gradually built up. The road problem facing these later fiber generations is, however, quite different from that of the pioneering phase. Pioneering fibers must find their way in strange country and get orienting cues from a non-nervous environment. Subsequent fibers, on the other hand, need simply cling to the older ones to reach the same destinations. *Free* outgrowth has given way to growth by *application*.

Thus the establishment of a peripheral nerve involves three overlapping phases (Figure 1): First, the *free* outgrowth of a group of pioneering fibers through non-nervous surroundings. Second, the *bound* outgrowth of subsequent fiber generations along the line laid down by the pioneering fibers. And third, the *towing* process in which the nerve is drawn out by the growth and dislocations of its terminal tissues.

This developmental history of the neurone presents a number of clear-cut problems. Do the pioneering fibers move at random or do

---

[4]The idea of a peripheral rather than a central origin of nerve fibers has longest survived in the field of nerve regeneration. The autogenous regeneration of peripheral fiber fragments, advocated mainly by Apáthy and Bethe, has found adherents as recently as 1920 (Spielmeyer). However, all existing evidence is definitetly against any such view (Boeke, '35). It is true that abortive attempts at regeneration have been observed in distal nerve fragments in vitro (Levi, '34), but they are short-lived and of no practical significance.

V. On the Dynamics of the Nervous System

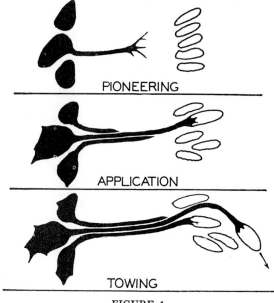

PIONEERING

APPLICATION

TOWING

FIGURE 1

GROUP OF NEUROBLASTS IN THREE SUCCESSIVE STAGES

they follow definite courses? If directed in their outgrowth, what makes them find their way? Are they aimlessly following casual routes for better or worse, or are they aiming towards definite destinations? Do they connect with any terminal tissue, right or wrong, or does selectivity prevail in the process? Why do post-pioneering fibers follow the older fibers? And which ones do they follow?—And many more such questions.

Although the answers to some of them are beginning to emerge in their outlines, our analytical information is on the whole still very inadequate. Therefore, if I am to present to you a coherent story rather than loose scraps, I shall have to call rather liberally on hypotheses to do the cementing. Some gaps can be filled by borrowing analogies from related fields, others by thoughtful evaluation of purely descriptive data. But even so the picture will remain sketchy.

We exclude from consideration the factors that create the initial pattern of neuroblasts, as we find it in the early nervous system. The question of differentiation, i.e., of what makes embryonic cells different among one another, is one that cannot be solved for the nervous system separately. It is one of the most fundamental and most obscure problems of all biology. We find neuroblasts of different potencies in different typical locations. These then transform into the different species of nerve cells. We do not even know whether they are already ear-marked for certain specializations when they are proliferated from the germinative (ependymal) layers, and then migrate each according to its type into a specified location, or whether presumptive nerve cells are still equivalent as they migrate into the various central districts, and then differentiate under the influence of local factors to which they become subjected in their final positions. We do not know, furthermore, what causes the proliferation of nerve cells to be confined to definite loci which shift from stage to stage and have been said to move in waves along the axis of the spinal cord (Coghill, '26, '36).

We, therefore, exclude from our discussions the problems presented by the differentiation of the early nervous system and begin at that later phase in which neuroblasts are already segregated, but nerve fibers have not yet appeared.

The first question of interest is, how does the neuroblast get polarized and oriented? It had been suggested by Child ('21) and Kappers ('17) that neuroblasts are polarized by the potential gradient of a surrounding electric field. Kappers viewed the case as one of direct galvanotropism, while Child suggested differential physiological response of the neuroblast at the anodal and cathodal poles. According to Kappers, the anodal outgrowth would become the axon, and the cathodal outgrowth the dendrite. All this is pure speculation. There simply are not sufficient data on hand to reach any decision. Péterfi and Williams ('33), who attempted to test the theory directly by placing neuroblasts into an electric field in vitro, failed to produce convincing evidence either way. One obvious difficulty of the whole theory lies in the frequent occurrence of tripolar and multipolar neuroblasts; it is hard to visualize how electric polarization could ever produce better than bipolar forms. The points at which the nerve processes emerge from the neuroblast are probably determined by some local changes in the cell surface. However, these changes need not be of electric origin. In fact, they need not even be of a single kind. In this connection Whitaker's ('40) splendid analysis of the polarization of the Fucus egg, reported at last year's symposium, deserves full attention as a possibly pertinent analogy. Similar experiments on neuroblasts are needed. Pending the outcome, we must suspend judgment.

V. On the Dynamics of the Nervous System

This is not serious for our present purpose, because the further course of the nerve fiber does not directly depend upon the polarity and orientation of the nerve cell. Once the amoeboid tip of the fiber has left the cell body, it follows a course of its own. Our interest, thus, turns to the factors determining that course.

Several theories have been developed to account for the oriented outgrowth of nerve fibers. Strasser (1892) suggested the electric field as the orienting factor, and Child ('21) and Kappers ('17) considered the same potential gradients which had supposedly polarized the neuroblast as responsible for the further direction of the axonic outgrowth. However, aside from one positive claim by Ingvar ('20), which can be accounted for by deceptive experimental technique (P. Weiss, '34, p. 426), repeated attempts to demonstrate any orienting effect of the electric field on outgrowing nerve fibers, either in the embryo or in tissue culture, have consistently failed (P. Weiss '34; Karssen and Sager '34; Levi '34, p. 611; Williams '36; Gray '39). This negative evidence is strengthened by the fact that one frequently sees two nerve fibers grow out along a common path, but in opposite directions (Speidel '33). Such reciprocal fiber growth has recently been reproduced on a mass scale in experiments which will be reported below. Other experiments, also to be detailed later, have demonstrated that nerve fibers from the same source grow with equal ease in headward and tailward direction. Thus, whatever axial electrical polarity there may be in the body, has obviously no decisive effect on the direction of fiber growth. Consequently, the theory of a direct orientation of nerve fiber growth by potential gradients of an electric field finds no support in the facts.

Turning to the second group of theories, the chemical theories, or, more specifically, the concept of Cajal ('08), Forssman (1898, 1900), Tello ('23), that nerve orientation is produced by some sort of chemotropism, the existing evidence is overwhelmingly against them, too. Chemotropism implies a directive movement of an organism or part of an organism toward a distant source of chemical emanation. In order to operate, such a mechanism requires: (a) a constant source of chemical diffusion with a steady concentration gradient; (b) selective sensitivity of the growing part for that particular chemical; (c) an ability of the growing part to orient itself along the lines of the concentration gradient; and (d) a stagnant and homogeneous medium in which the concentration gradient can remain stationary during the whole process of oriented growth. These conditions are not realized in the growing organism. There are, of course, numerous centers of chemical diffusion in the growing organism. However, nerve fibers do not follow the direction of concentration gradients, and even if they did, the body would not provide them with gradients of sufficient steadiness to serve as guides, as the agitation of the interstitial fluid between the tissues would blur and distort the original concentration gradients beyond recognition. Furthermore, the fact of reciprocal fiber growth between two points is as fatal to the theory of chemotropism as it is to that of electrotropism, because it would mean that while one fiber runs up the concentration slope, the other one runs down. Also, in spite of several direct experimental attempts it has never been possible to demonstrate a direct attraction of nerve fibers towards a source of chemical emanation in vitro (P. Weiss, '34), not even towards degenerated nerve tissue, according to Cajal the most potent source of such action.[5]

Both the galvanotropical and chemotropical theories proceed from the assumption that the growing tips are oriented by factors acting from a distance. Nerve fibers are seen to *travel towards* certain localized regions and, hence, are assumed to have been *attracted by* those regions. For brevity, let us refer to this kind of orientation as a "distance effect." Now, before looking further into the problem, let us first check the solidity of its premises. In other words, are nerve fibers really moving towards definite destinations, or are they merely running along some preformed traffic routes?

The growth of the segmental nerves along the boundaries of the myotomes, the growth of the lateral line nerve down a longitudinal groove along the trunk musculature, the coursing of the limb nerves along the main blood vessels, and similar facts, would seem to make nerve orientation a matter of the road rather than of the destination. His (1887) has stressed the predilection of nerves for mechanical roadbeds, and Vanlair (1885) has gone so far as to claim that nerves simply take the way of least resistance. Exceptions, in which nerves seem to have gone out of their way, so to speak, in order to reach distant organs, such as the long backward loop of the vagus from the

---

[5]Centanni ('14) once reported "nerve growth" in vitro as oriented towards explanted tissue fragments of supposed chemotactic activity. His statement, however, contains a gross misrepresentation, inasmuch as his "nerve growth" came from isolated fragments of peripheral nerves, in which, as we know, all nerve fibers degenerate rather than regenerate. The so-called "nerve growth," therefore, can have reference only to emigration of spindle cells from the fragments, which has no bearing on the problem of nerve growth at all.

V. On the Dynamics of the Nervous System   451

head to heart and intestine, or the long forward swing of the lingual nerves into the tongue are deceptive, in that these courses reflect extensive shifts of the terminal musculature of these nerves in the towing phase, rather than directive free outgrowth of the nerves. So, one cannot dismiss the possibility that nerves connect with their proper peripheral areas simply because these areas happen to lie at the end of those lines which the nerves find open when they start to grow out. And the problem of whether nerves are destination-bound at all, is a real one.

Experimental evidence on the point is meager but suggestive. Cajal ('08) and Forssman (1898, 1900) have tried to prove under a wide variety of conditions that nerve fibers in the process of regeneration grow actually toward a goal rather than merely along a path. Let me cite one of their classical experiments, exemplary for many others. When a nerve is cut, the distal fragment transforms into a non-nervous protoplasmic cord which persists as such, while the proximal fragment gives rise to abundant outgrowth of new fibers (cf. Boeke, '35). Whether the two stumps are directly apposed or whether they are dislocated, in either case a large number of the new sprouts is finally found trapped inside the old degenerated tubes (Figure 2). The orientation of the regenerating fibers is by no means strict. Numerous

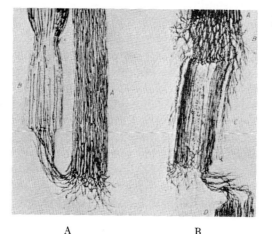

A   B

FIGURE 2

NERVE REGENERATION INTO DISLOCATED PERIPHERAL STUMPS (AFTER CAJAL)

A, Nerve fibers from a severed nerve (*A*) have partly found their way into a degenerated nerve fragment lying some distance behind the cut surface.
B, Nerve fibers from the proximal nerve stump, *A,* after pervading the scar, *B,* have partly entered, partly by-passed, the degenerated nerve segment, *C*; part of the fibers emerging from *C* have traversed the gap to the dislocated fragment *D* and grown into it.

fibers can be seen to have strayed about at random. Nevertheless, a large proportion has reached the peripheral nerve stump. In order to do this, they had to take a somewhat contorted course. Doubtless, the degenerated peripheral stump has exerted some potent influence on the pattern of regeneration.

However, in specifying this influence, we must be aware of two possibilities: One is, that the sprouting fibers were attracted by some factor emanating directly from the open end of the distal fragment. This is the interpretation given to the results by both Cajal and Forssman. In view of experiments which we shall take up in a moment, however, a second possibility seems more likely.

Suppose, the fibers stray out at random. Some strike the peripheral stump by accident. This puts an end to their period of vagrancy. Then something happens to them: As a result of their having become settled, their condition changes; changes in a fashion that will render them sticky or otherwise attractive for other nerve fibers growing out later. In this way, more and more fibers will get trapped along this preferential course. Thus, what originally was accidental encounter, becomes gradually a systematic pursuit. The successful bridging of the gap would be a matter not so much of directive fiber outgrowth, as of the fact that an accidentally successful nerve sprout becomes the center for the building up of a nerve cable. We will call this process *"selective fasciculation"* and return to it in greater detail presently. For the moment, we merely mention it to show that Cajal's experiments do not necessarily prove the case of destination-bound nerve orientation.

Experiments on the embryo, reported by Detwiler (reviewed '36), seem to be more conclusive. The nerves of a vertebrate limb are of plurisegmental origin and converge upon the limb base, where they intertwine in a plexus. When a limb bud is transplanted to a more anterior or posterior position, the innervating centers shift likewise, but less, and there is a certain tendency of nerves to slant more anteriorly in the former, and more posteriorly in the latter case. In spite of great individual variations, limb nerves, as a rule, show a more headward slant if the limb bud was transplanted more anteriorly, and a more tailward slant in the case of a posterior shift. Since the limb plexus converges likewise upon a transplanted nasal organ (Detwiler '28) a tail, or an eye (Detwiler and Van Dyke '34), the attraction, if such it is, is certainly not very specific. Nevertheless, a general diversion of nerves toward growing organs seems to be indicated.[6]

Unfortunately, only the end stages of this process are known, and since secondary events, such as extensive shifts of musculature, plexus formation and, possibly, selective fasciculation, have distorted the original situation as it existed at the moment of the outgrowth of the nerves, it remains unknown just how potent the supposed attraction of embryonic nerves by growing organs is. However, since Burr ('32) has shown that nerves growing out from transplanted nasal organs are likewise showing definite preferences as to the point of ingrowth into the embryonic brain, and since, furthermore, Coghill ('29) has come to the conclusion that in the development of the brain nerve fiber tracts tend to grow towards centers of simultaneous differentiation, we may take it for granted, pending proof to the contrary, that growing regions of the body exert a general influence, which in purely descriptive language may be termed "directional attraction."

The acceptance of this fact puts us into a real dilemma. We admit that nerves can be attracted towards distant organs, and at the same time deny the validity of the best known mechanisms of distance action—galvanotropism and chemotropism. The alternative to distance action is what, with a general term, we may call *"contact action."* All present evidence points to the fact that nerve fibers are conducted on their way solely by "contact action," and in order to extricate ourselves from our dilemma, we would have to prove that distance effects can actually be resolved into contact actions.

This we can prove. But before doing so, we must give some closer attention to the mode of progression of the growing nerve fiber. This takes us into a much wider biological field—the problem of amoeboid movement. Amoeboid movement is produced by the protrusion of lobular or filamentous processes from the protoplasmic surface, so-called pseudopodia, which take hold on the substratum and cause the cell content to stream after. Some cells can project pseudopodia into a homogeneous liquid. Nerve fibers cannot. The nerve protoplasm needs a surface or interface along which to extend. Perhaps this need is imposed by the great length of the fibers. Since the tremen-

---

[6]Cases of extensive nerve detours, as if aiming to reach nerve-free limbs, described by Hamburger ('29), owe much of their impressiveness to the secondary distortion which the nerve courses have suffered during the towing phase. In view of the close apposition of the young limb buds and the closeness of their respective nerve fiber pools, contralateral innervation occurs under infinitely simpler conditions than the final developed product would make one suspect.

dous expansion of relative surface in elongation requires a large output of energy, it is possible that the task would go beyond the capacity of metabolic energy production by the cell, so that external physical structures would have to be called on for support.

However, whatever the reasons, nerve fibers positively do not grow when suspended in a homogeneous liquid medium. When cultured in liquid media in vitro, they cling to the surface of the coverslip or to the surface of the drop (Lewis '12, Levi '34). The occasional observation of nerve fibers inside the spinal canal, which is filled with cerebro-spinal fluid, is no exception, as such nerve fibers have most likely grown along the inner wall of the tube and merely become swept into the liquid in the act of fixation. Let us repeat, therefore, the pseudopodia of the growing tip of a nerve fiber extend along phase boundaries, and only along such.

These phase boundaries need not be of the gross macroscopic kind, such as coverslip–liquid, or liquid–air. In fact, they rarely are. Much more commonly they are of microscopic or submicroscopic order. In the embryo the growth of a nerve fiber along minute microscopic threads has been repeatedly described (cf. Held '09). But since, what in a microscopic preparation appears as a fibril, is often an artifact, resulting from the clotting together of identically oriented ultramicroscopic structures, it is most likely that even in these cases nerve growth in the living occurred along some ultramicroscopic rather than microscopic phase boundaries. The same is true of tissue culture in a blood plasma clot. The plasma coagulum is essentially a spongy gel consisting of a more solid skeleton of fibrin micellae in various degrees of aggregation, imbibed with a continuous phase of blood serum. Interfaces between fibrin aggregates and serum thus are present everywhere in the clot, and they furnish the necessary substratum on which the pseudopodia of growing nerves may take hold.

The question has been raised of how a microscopic body, such as a nerve fiber, can follow an ultramicroscopic interface, which is of lower order of magnitude. The answer is simply that it is not the microscopically visible nerve fiber, which plays the active part in the creeping process, but its delicate terminal pseudopodia, which may very well taper to submicroscopic dimensions.[7] We know that what

---

[7]Grossfeld ('34) has overlooked this fact when taking issue with the explanation of oriented fibrocyte growth in terms of the ultrastructure of the medium. Obviously, what we have stated here about nerve fibers, applies equally well to fibrocytes, which likewise advance by means of fine filamentous pseudopodia.

V. ON THE DYNAMICS OF THE NERVOUS SYSTEM 455

under the microscope appears as the tip of the pseudopodium of a nerve fiber, is not its real end; Giuseppe Levi was able to trace the filaments far beyond that point in dark-field illumination, and we have, of course, no assurance that they may not even extend beyond the limit of dark-field visibility. This makes the pseudopodial tips and the interfaces to which they apply themselves appear as of commensurable dimensions. Once established, the pseudopodium is soon enlarged to microscopic dimensions by the inflow of protoplasm, while new pseudopodia issue from its extreme tip and penetrate into the medium (Figure 3).

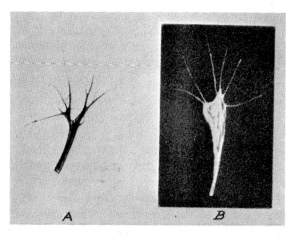

FIGURE 3

PSEUDOPODIAL RAMIFICATION OF THE GROWING TIP OF A NERVE FIBER CULTIVATED IN VITRO (AFTER LEVI)

A, Fixed and stained preparation.—X 950. *B,* Dark-field view of living specimen.

The tendency of organisms to cling, as they grow, to solid structures, is commonly referred to as *stereotropism*. This is a naturalistic term. It classifies the phenomenon, but reveals nothing about its nature. The movement of a nerve fiber falls into this class. Our analytical insight into the phenomenon does not seem to have advanced far beyond where it stood when Quincke, in 1888, explained protoplasmic movement as the work of surface forces, comparing it with the spreading of oil on a water-air interface. It is to be hoped that such studies of the mechanism of amoeboid motion as have been carried out by Mast ('31) on amoebae, by Lewis ('39) on macrophages, and by Fauré-Frémiet on amoebocytes, may lead to information which can be applied to the nerve fiber. For the time being, however, our ideas about the process are still highly speculative.

We are probably safe in saying that interfacial tensions between the nerve fiber protoplasm and the micellar aggregates along which it extends are at work, but just how they operate is still obscure. Of the whole protoplasmic system only that fraction counts which takes part in the surface and is in actual contact with the medium. Reversible gelation of the cortical protoplasm endowing the surface film with adhesiveness, as suggested by Lewis, may be an essential element of the mechanism. But this "adhesiveness" itself can bear further analysis. Are we simply dealing with a purely colloid-mechanical phenomenon, or might we not have to postulate molecular affinities of greater specificity? I have observed some nerve fibers in vitro following a pathway which was wholly ignored by others. This and many similar experiences suggest selective "adhesivity" rather than a common kind of stickiness.

In an unoriented gel of isotropic constitution phase boundaries extend from any one point in several directions. This favors the simultaneous protrusion of several pseudopodia along different lines. (Figure 3). Which one of them is to endure, depends probably largely on local accidents. But it is reasonable to assume that the inertia of the centrifugal protoplasmic flow in the growing sprout will generally enhance the repletion of those pseudopodia that lie more or less in direct line of the axis of the sprout. We have little accurate data about protoplasmic movement inside the fiber, but Levi ('34) has observed some agitation, and Speidel's films show some kind of peristaltic motion. Once a pseudopodium has established itself and caused the mass of protoplasm to flow into it, the drain thereby exerted on the other pseudopodia produces their automatic withdrawal. Only if two pseudopodia happen to be of approximately equal strength, they may divide the protoplasmic influx between each other and thus initiate a terminal branching of the fiber (Figure 11). We shall return to this point. In a medium lacking definite orientation, the nerve processes will thus follow random courses. This is actually realized in ordinary tissue cultures as well as in many parts of the central neuropil.

However, from what we have said, it can be anticipated that if nerve fibers are confronted with a medium in which the interfaces follow a common instead of random direction, their growth, bound to retrace those interfaces, would become likewise oriented, just like vines, growing along parallel stakes. This expectation has been borne out by tissue culture experiments. Since these have been reported at length (P. Weiss '34), we need not go into details. When nerve fibers are made to grow out in a blood plasma clot which had been stretched or otherwise put under tension, the course of the fibers follows the pattern of tension. The tension has served to turn the polarized fibrin micellae into a nearly parallel orientation, and consequently all pseudopodia were likewise drawn into a nearly parallel course.[8] To obtain the effect, the tension need not be exerted from without. Living

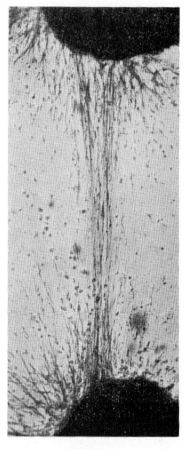

FIGURE 4

ORIENTED NERVE GROWTH CONNECTING TWO SPINAL GANGLIA CULTIVATED IN A THIN MEMBRANE OF BLOOD PLASMA IN VITRO (x 48, P. WEISS, '34).

[8]For illustrations of the effect, see Figures 1-4 in P. Weiss, '34.

proliferating tissue by itself has a contracting effect on the surrounding medium and thereby produces tension. Two such centers of contraction in a common medium automatically produce a pathway of tensionally oriented micellae, which, if nerve cells are present in the centers, becomes soon populated by parallel nerve fibers growing out from both ends in opposite directions (Figure 4). Incidentally, this is another example of reciprocal fiber growth, defying galvanotropic and chemotropic interpretations.

The fact that in these experiments the agent to produce micellar orientation happened to be tension, is irrelevant. For any ultrastructural orientation of the medium, no matter how attained, may have the same effect on the nerve fibers. Convection of liquids through the medium can produce such orientation, and vectorial aggregation of micellae may even occur spontaneously. On the whole, it is possible to explain the oriented outgrowth of individual nerve fibers by taking into account the colloid-physical structure of the medium, and the tendency of the nerve protoplasm to comply with it. The stereotropism of nerve fibers with regard to solid microscopic structures, as assumed by His, and emphasized by Harrison ('14), seems to be only a special case of the more general principle of contact guidance of the nerve tip by the ultra-structure of the medium. To call this contact principle "mechanical," misses the point, the more so as we do not yet know of how much selectivity the adhesiveness to the contact structure admits. It is quite conceivable that differential chemical impregnation endows certain ultra-structures with greater affinity to nerve pseudopodia than others (see below).

Nerve fiber orientation was thus explained, at least in principle. The attraction of nerves towards growing organs, nerve branching, the formation of nerve plexus, and the reciprocal outgrowth of nerves between simultaneously differentiating centers, could be reproduced in vitro. More than that, the experiments have definitely proven that nerve fibers are actually guided by contact and not attracted by agents operating from a distance. So, when nerve fibers are seen to move towards a remote destination, one may take it for granted that that goal was instrumental in laying down an oriented pathway in the medium, rather than acting on the fibers directly. Established in tissue culture, all this certainly holds true for nerves in vitro. But does it likewise apply to the living organism? And if so, does it help to explain the nerve patterns as they actually arise in the course of development?

V. On the Dynamics of the Nervous System   459

As I have pointed out on earlier occasions ('33, '34), the conditions between tissue culture and organism are fundamentally comparable in many respects. However, the actual testing ground is the organism, and this test has thus far remained wanting. Moreover, there are certain features of nerve growth in the organism which cannot be repeated in tissue culture. There was urgent need to clarify the situation in the organism directly. During the last few years I have tried to approach this task by devising methods specially adapted to the problem. The problem was to let nerve growth occur under conditions infinitely simpler and more transparent than those under which it normally occurs and whose complexity defies direct analysis, but still in the organism.

Two different techniques were found to be successful. In both amphibian larvae are used. The first consists of producing an extensive bed of simple granulation tissue and allowing nerves to regenerate into it under various controllable conditions. Locations chosen for the purpose were the *skull cavity*, from which the brain down to the anterior end of the medulla oblongata had been removed, and the *orbit of the eye* after enucleation of the eye ball (see Figure 9, *A*, *C*).

Even large tadpoles survive the removal of brain of the indicated extent for six weeks or more. The skull cavity fills with a soft, loose connective tissue, which provides our testing ground for nerve growth. This nerve growth comes from four sources, the two stumps of the olfactory nerves and the two optic nerve stumps, all four belonging to nerves whose cells lie in the periphery and grow centripetally (Figure 9, *C*). Similarly, after enucleating the eye under certain precautions, the orbital space becomes filled with a simple vascularized connective tissue, and a source for nerve regeneration is made available by cutting the trigeminal nerve, which runs along the inner wall of the orbit (Figure 9, *A*). These experiments have furnished instructive information concerning the relation between nerve growth and connective tissue orientation, relations between nerves and capillaries, the non-existence of any orientation of nerve fiber growth with regard to the general body axis, as well as the non-selectivity of nerve associations (see below), but lack of space prevents us from discussing these points in greater detail at this time.

The second method may be called *deplantation*. It takes advantage of the fact that the amphibian larva contains one extensive and readily accessible tissue which in itself is so elementary that it compares in simplicity almost with a tissue culture, namely, the gelatinous connective tissue of the fin. It consists (Studnicka, '38) of a gelatinous keel-shaped mass extending over most of the dorsal midline, sparsely settled with cells, and pervaded by a moderate amount of capillaries and sensory nerve fibers. Into this tissue we transplant large fragments of developed spinal cord or brain from another larva (P. Weiss, '40, '41, '41a). These fragments are to furnish outgrowing nerves. The transplanted mass undergoes partial resorption, but what is left is sufficient to serve as a potent nerve source. This method has the advantage that other organs can be transplanted into the vicinity of the deplanted center so that the effects of different organs on nerve growth can be analyzed, and, in fact, directly observed, because the fin tissue is sufficiently transparent to permit direct observation of the gross events.

Let us see what happens when a brain fragment and a limb are incorporated in the fin at some distance from each other. (Figure 5).

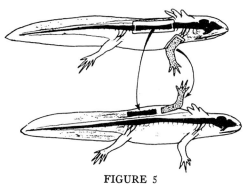

FIGURE 5
DIAGRAM OF DEPLANTATION EXPERIMENT
A portion of the spinal cord and a fore limb are excised from the donor animal (top) and implanted at some distance from each other in the dorsal fin of the host (bottom).

The result is quite impressive: Within two to four weeks after the operation a strong nerve cable forms between brain graft and limb (Figure 6). Inside the limbs these nerves follow the regular nerve channels and make functional connections with the musculature. Innervation completed, the whole grafted complex begins to exhibit spontaneous functional activity which is highly interesting from a physiological standpoint, but which to discuss is beyond the scope of this paper.[9]

Now, in these experiments we have witnessed nerve formation in

---

[9]The functional results have been summarized in P. Weiss, '41.

an almost diagrammatically simple form. Given: an isolated nerve center and an isolated limb in a common gelatinous matrix. Result: a direct nerve connection. That the limb is the actual objective of the outgrowth, is proved by various facts. If no limb is transplanted, only stray fibers leave the brain fragment, but no nerve cables form. Depending on where the limb was inserted, whether behind or in front or on top of the grafted nerve center, the nerve cable grows posteriorly or anteriorly or dorsad, which again proves that the polarity of the body is immaterial in nerve orientation. If two limbs are transplanted, one anteriorly and the other posteriorly, nerve connections form in both directions. Histologically, these nerve cables

FIGURE 6

LATERAL VIEW OF THE TRANSPARENT FIN OF AN EXPERIMENTAL ANIMAL CONTAINING DEPLANT OF SPINAL CORD (*G*) AND GRAFTED LIMB (*L*): A STRONG NERVE CAN BE SEEN TO HAVE GROWN FROM *G* TO *L*
(13 weeks p. op.—cca 6 x.)

consist of non-anastomosing fibers, which, interestingly enough, are devoid of sheaths and sheath cells, just like a central fiber tract (Figure 7). It is noteworthy that they are fully functional, proving the physiological adequacy of naked peripheral nerves as conductors.

As for the analysis of the phenomenon, one technical detail must be taken into consideration. In order to introduce the deplants, a longitudinal channel was made in the fin gel; the brain fragment was deposited in the distal part of this pocket, and the limb was usually inserted into its orifice. Residues of this channel would present a road of least resistance between the deplant and the limb graft and favor nerve growth between the two structures. As you will see, this explains part but not all of the phenomenon. For, even if the limb and the brain grafts are introduced into separate pockets, without open communication, nerve connections between the two are still established. These differ, however, from the channel connections in one notable respect. Instead of a single straight nerve cable, there are usually several nerves, and instead of being straight, their courses are somewhat crooked.

Nerve cables, though of smaller size, may form also between the deplant and underlying trunk musculature. Here, too, the presence of a connecting channel facilitates the connection, but is by no means essential.

If we focus, for the moment, on the cases without common channel, we realize from the crooked course of the nerve connections that they have not come about by directive outgrowth, but rather by differential fasciculation. More specifically, this is the sequence of events. Nerve fibers, singly or in small groups, radiate from the deplant in all directions and stray through the surrounding fin tissue. Some

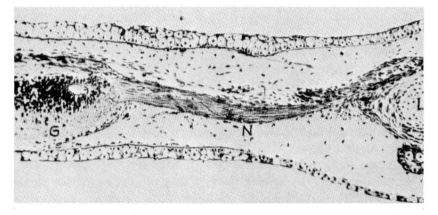

FIGURE 7

FRONTAL SECTION THROUGH FIN OF EXPERIMENTAL ANIMAL CONTAINING DEPLANT OF SPINAL CORD (G), TRANSPLANTED LIMB (L), AND CONNECTING NERVE CABLE (N)
(18 days p. op.—x 72.) Impregnated according to *Bodian*.

have reached the grafted limb, others have not. Only those nerves, however, which have entered the limb, have subsequently become filled up into a sizable bundle. This leaves only one interpretation. Obviously, those pioneering fibers which had accidently struck the limb, had thereby acquired some contact property which made their surface sticky, or otherwise a pathway of preferential application, for other fibers growing out subsequently. This is precisely what I had suggested before as an alternative to Cajal's chemotropism in the regeneration of nerves; only here we have more direct evidence. Just how the fasciculation of "successful" fiber courses comes about, has not yet been decided. Nor do we know whether the successful fiber

owes its rallying power merely to the fact that it has come to a halt
as a result of terminal connection, or rather to some specific chemical
impregnation which it may have received from its new end organ.

The straightness and singleness of the nerve cable in the channel
experiments is easily accounted for as a special case of selective
fasciculation. For the channel scar establishes a straight connective
tissue bridge between the limb and the deplant, along which pioneering fibers are actually guided toward the limb, as in our tissue culture
experiments. For the rest, the gradual building up of the cable occurs
according to the same rule as in the other series, in which the encounter with the limb is purely a matter of chance.

A deeper analysis of the change, presumably in the surface, which
distinguishes successful fibers from unsuccessful ones, is definitely
within the limits of our methods. But nothing has as yet been done
in that direction. One point, however, seems to emerge. That is, that
in order to obtain fasciculation, the immediate vicinity of the nerve
fiber must be in a fairly liquid state. This can be observed directly
in tissue culture. While nerve fibers inside the solid plasma clot grow
out individually, they anastomose and associate easily in liquefied
parts of the medium. Nerve fibers in a liquid medium stick together
because their surfaces have greater adhesivity to each other than to
the surrounding serum. This rôle of liquidity for fasciculation in tissue culture was noted by myself ('34, p. 441) and Levi ('41, p. 177).

In the organism the strong fasciculation of nerves in the liquid
spaces along the large blood vessels furnishes a suggestive example of
the same situation. We can almost generalize and say: Inside the
solid tissues—epithelium, muscles, central gray, and so forth—nerve
processes tend to take individual courses. In between these tissues,
however, where the fibers pass through liquid-filled spaces, they are
grouped into nerves. Now, if liquidity of the medium is a prerequisite
for fasciculation, part of the task of a successful pioneering fiber may
consist of liquefying its immediate surroundings by proteolysis.

Indications of facultative proteolysis by nerve fibers have been observed in tissue culture.[10] Nerve fibers in old cultures, that is those

---

[10]This remark does not refer to the proteolysis of the plasma clot commonly observed
in brain cultures (cf. Olivo '28, Levi '34, P. Weiss '34), which has been shown
(P. Weiss, '34a) to be due to the secretory action of ependymal *cells* (and which, incidentally, has been wrongly interpreted by Levi, '41, as plasma syneresis). We are referring here to proteolytic effects of the *nerve processes*. A few examples were observed
during earlier work (P. Weiss, '34, p. 434). More extensive evidence has been obtained
since, but has not yet been published.

in which growth has essentially come to a standstill, frequently begin to dilute the surrounding plasma medium and, at the same time, combine into bundles (Figure 8). In favorable cases, the formation

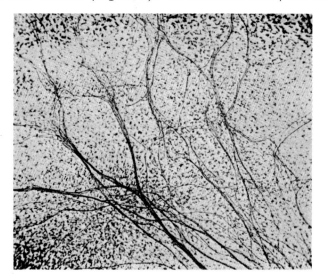

FIGURE 8

PERIPHERAL GROWTH ZONE OF TISSUE CULTURE OF SPINAL GANGLION (CHICK EMBRYO) AFTER SEVEN DAYS' CULTIVATION IN PLASMA CLOT

Extensive bundling of nerve fibers. Impregnated according to *Bodian's* method. x 56.

of blisters in the vicinity of a nerve can be immediately observed. However, whether these cases have a real bearing on the mechanism of fasciculation in the body, remains to be seen.

But whatever the details may be, nerve fibers which have "arrived" somehow bear the stamp of success on their exterior, which induces other nerve fibers to follow them, while unsuccessful ones, lacking similar appeal, usually remain unattended. In the light of these results, the significance of primarily oriented fiber outgrowth for the development of nerve patterns seems to be greatly reduced. The problem obviously exists only during the pioneering phase of early fiber growth and regeneration; during that phase it is apparently well covered by the principle of *ultrastructural orientation*. But from there on all further development of peripheral nerves is a matter of *selective fasciculation*.

This immediately raises two questions: Firstly, the question of what degree of selectivity there is in the application of one nerve fiber to

V. ON THE DYNAMICS OF THE NERVOUS SYSTEM

another. And second, the question of what kind of numerical control limits the size of a nerve cable to its normal proportions.

Let us consider the problem of selectivity first. Selectivity may assume two forms: selectivity of pathway and selectivity of termination. Therefore: can nerve fibers select their courses, and can they choose the tissues with which to connect? To take up the latter point first, the notion that nerve fibers from a given source would possess a prerogative on a corresponding pre-assigned peripheral organ, is wholly untenable. Any number of cross connections between nerves and terminal areas strange to them have been successfully effected.

A brief list may indicate the range of non-selectivity. First, with regard to regeneration: Motor fibers can connect with sensory organs, and sensory fibers can form terminations in muscles (Boeke, '17). It has been shown that such connections may even be physiologically adequate: a dorsal spinal root (P. Weiss, '35) or a sensory nerve (P. Weiss '34b) forced upon a muscle can mediate its contraction. Any motor nerve will connect with any muscle. It even connects with spinal cord when inserted into it (P. Weiss '32). In the deplantation experiments, just discussed, innervation of limb and trunk muscles was obtained from all spinal cord, medulla oblongata, midbrain, and forebrain, all of which had reached functional differentiation at the time of deplantation (P. Weiss, '41a). Turning to the embryonic phase, cranial nerves were shown to penetrate limbs (Harrison, '07; Braus, '05, Detwiler, '30; Nicholas, '33) or trunk muscles (Hoadley, '25), grafted within their domain; and in fact, central fiber tracts can do the same (Nicholas '29, '30).

I can supplement this array further with an example from our recent experiments on nerve growth in the cleared orbital or cranial cavities, which I mentioned before. In the empty skull cavity the regenerating fibers of the optic and olfactory nerves pervade the granulation tissue in all directions, forming an unorganized neuroma. But in most cases there were also strong fascicular connections between at least some of the nerve stumps. Quite aside from their bearing on the problem of fasciculation, they interest us here for the weird combinations they have produced. For instance, the optic and the olfactory nerves had met head-on, merged into a common cord, whereupon the optic fibers have travelled on towards the nose, and the olfactory fibers reciprocally towards the eye (Figure 9, D). Their ultimate fate has not yet been studied, but the gross fact is clear. Similarly, in the enucleated orbit regenerating fibers from the severed

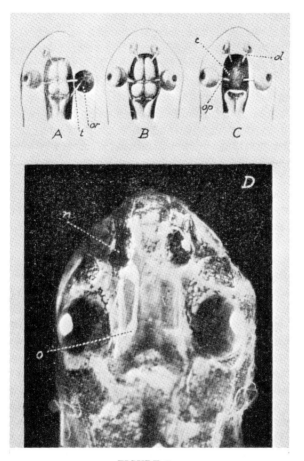

FIGURE 9

METHOD OF STUDYING NERVE GROWTH IN EXPERIMENTALLY PREPARED CAVITIES AS BEDS

A, Orbital bed (*or*), cleared by enucleation of bulb, with trigeminal nerve (*t*) as fiber source.

B, View of normal brain of frog tadpole after dorsal opening of skull.

C, Cranial bed (*c*), cleared by removal of all brain parts anterior to cerebellar lamina, with olfactory (*ol*) and optic (*op*) nerves as fiber sources.

D, Experimental animal, operated according to C, showing fusion of left optic (*o*) and olfactory nerve (coming from nose, *n*) into common cable.

trigeminal nerve have penetrated into the central stump of the deserted optic nerve and grown back into the brain.

In spite of this demonstrated non-selectivity of terminal connections, the fact remains that in normal development highly specific relations are somehow achieved. The simplest expression of these is

V. ON THE DYNAMICS OF THE NERVOUS SYSTEM 467

the fact that ventral root fibers connect with muscles, while fibers growing out from the spinal ganglia connect with sensory end organs, mostly in the skin. Since both travel for some distance along a common road before segregating into muscular and cutaneous branches, they present us with a really puzzling problem. This problem has not yet been brought anywhere near its solution. One might submit that there might first be indiscriminate outgrowth, followed later by the reduction of functionally inadequate connections. This contention, however, does not find support in the facts, as the relations are established very early, long before there can be any question of functional effects.

Mr. Taylor, of our laboratory, has made a thorough investigation of the innervation of hind limbs of frog larvae, which had developed in the absence of either their sensory or their motor nerve sources. Purely motor innervation was obtained by the removal of the sensory ganglia prior to the outgrowth of the limbs, and purely sensory innervation by the early extirpation of the whole spinal cord of the limb area, leaving the ganglia undisturbed. Even the very early limb bud, long before differentiating muscle and skeleton, is already pervaded by sprouts from the nerves which had been waiting at the limb base in large numbers. Later, as the limb elongates, the various nerves can be identified. In animals with purely motor centers most fibers are then found in muscular nerves, while purely sensory sources have sent most of their fibers along cutaneous pathways. The separation is not strict. Motor fibers get into the skin and cutaneous fibers into muscles, but a statistical predilection of each kind is nevertheless indicated. This bears out an earlier observation of Hamburger ('29). The fact seems clear, its explanation less so. We are still collecting evidence, and I do not want to be too positive. But one thing seems to emerge: If there is selectivity in these early stages, it seems to refer to the pathway rather than to the terminal areas. Sensory pioneering fibers would show affinity to a cutaneous nerve path rather than to skin; motor fibers to muscular nerve paths rather than to muscle. These paths might be viewed, in line with our earlier discussion as ultrastructures with differential chemical impregnation. Sensory and motor nerve fibers would have to react differentially to the two types of structure. In assuming this, we are on firmer ground. We know from anatomical and physiological experiences that motor and sensory fibers are constitutionally different. This differential might very well predispose them for different contact affinities. Of course, once the pioneering fibers have been sorted into their proper channels, the rest may be merely a matter of selective fasciculation: Guided by surface affinities, sensory, and motor fibers would apply themselves each to the corresponding type.

The fact of surface affinities among identical kinds of nerve fibers seems to be well supported. For what else can it mean when the neurologist describes the fibers which mediate special sensations, such as pain, proprioception, cutaneous sensation, and so forth, as running in their intraspinal course as separate fiber bundles? Or the fact, that even in a peripheral nerve the sensory and the motor fibers hold pretty much to themselves, although imbedded in a common nerve tract? Each bundle contains fibers of a wide range of ages, which have developed in succession; if they lie together, this must mean that the later ones have grown out along the path of the earlier ones. Oppenheimer ('41) has recently described interesting observations on the course of supernumerary Mauthner's fibers in the fish brain. The extra fibers, easily recognized by their size, have shown a marked tendency to follow the course of the original Mauthner's fiber. This is strong evidence of selective surface application. Thus, it may become necessary to concede to nerve fibers the attribute not only of being just adhesive, but of *selective adhesiveness*. This affinity would presumably be obliterated by the appearance of a myelin sheath.

The problem of selective fasciculation brings up a question of much wider biological significance, namely, that of tissue affinities in general. The intimate fusion or adhesion between tissues of different origin presupposes a compatibility between the contact surfaces which thus far has not found sufficient elucidation. Whether two protoplasmic systems can adhere to each other, depends primarily on some contact affinity which must hold at least until mechanical ties, such as basement membranes, or other fiber structures, have cemented the surfaces. Holtfreter ('39) has reported some highly instructive observations on how this contact affinity among different tissues changes in the course of ontogeny; how tissues which at one time have adhered, later separate off, as their biochemical divergence becomes greater.

I am inclined to regard this merely as one special case of a very general biophysical problem, which finds its expression in such diverse phenomena as the selective fusion of protozoan pseudopodia, depending on whether they belong to the same animal or to different animals; the fusion of individualized cells with a common syncytium and later

V. ON THE DYNAMICS OF THE NERVOUS SYSTEM   469

re-formation of individual cells within the syncytium, depending probably on variations of biochemical constitution of the respective nuclear
territories (see P. Weiss, '40a); the different rate and ease with which
grafts can be incorporated depending on their character and orientation; the differential agglutination of sperm in the presence of different types of eggs; and many other similar phenomena. We may view
all these phenomena in a common light, as follows (Figure 10).

Tissue affinity may be based on the presence in the contact surface
of identical (B, C) or complementary (A) types of proteins, plus
the fact that these will be oriented by surface forces in such regular
manner that they can interlock. Molecular attraction forces among
the two films would seal the surfaces to each other.[11] Furthermore,
different degrees of adhesion might be explained by differences in the
spacing of the molecules on either side of the phase boundary (D).
Any change resulting in divergence of the biochemical characteristics
or the physico-chemical parameters of the two systems in contact
might upset either the orientation (E) or the compatibility (F) of
these films and produce a release of one system from the other. This
may likewise be attained by direct proteolytic action of enzymes
present at the surface.

The association by surface contact between two nerve fibers, or even
between a nerve fiber and a specifically impregnated non-nervous
pathway, could then be viewed in the same light. The fact that we are
seemingly dealing with a problem of very general biological applicability, may give renewed impetus to its analysis on the molecular
level. So much about orientation and fasciculation of nerves.

The second problem to arise in this connection, as we have said, is
one of numerical control. For the size of the fiber complement of a
given nerve follows a definite norm, which raises the specific question
of what determines the final size of a nerve? Four factors enter into
consideration: Size of the nerve source; peripheral amplification by
fiber branching; peripheral reduction by anastomosing; and the size
of the periphery to be innervated. Each one of these factors contributes to the number of nerve fibers ultimately found in operation.

The size of a nerve source anticipates grossly the size relations of
the future nerves; that is to say, the amount of neuroblasts disposed
to send out peripheral nerve fibers varies in different parts of the
central nervous system, in a certain forward reference to the later

---

[11]Compare the paper on *Protein Patterns in Cells* by F. O. Schmitt presented at this
symposium.

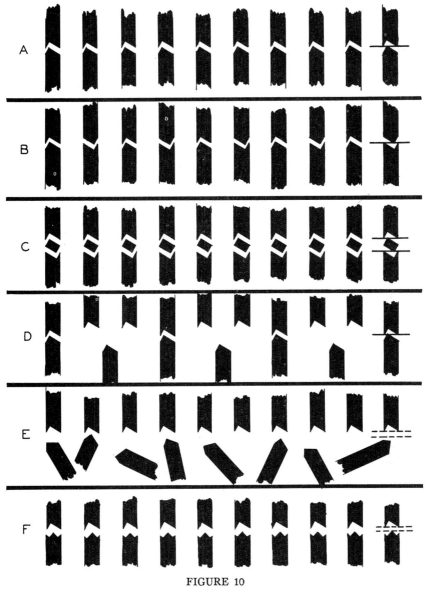

FIGURE 10

SCHEMATIC REPRESENTATION OF SELECTIVE ADHESION BETWEEN TWO ORGANIC SYSTEMS
ON THE ASSUMPTION OF SPECIFIC PROTEIN CONFIGURATION

Each rod represents the polar end of a molecule. The notches and correspondingly
shaped triangular protrusions symbolize complementary steric configurations in the
molecules resulting in selective interlocking, according to the hypothetical analogy of
antigen-antibody union. Vertical position of rods indicates parallel orientation of

molecules in surface films. Full lines on the right indicate surface adhesion, broken lines non-adhesion.

A, Surface application of one system to another containing molecules of complementary configuration (e.g., of nerve fiber to impregnated non-nervous pathway).

B, Surface adhesion between two identical systems, assumed to contain both types of mutually complementary molecules in identical ratios.

C, Surface adhesion between two identical systems, containing only one common type of molecule. Adhesion is mediated through a film consisting of an appropriate cementing substance (cf. Schmitt's presentation at this symposium). (Example: Selective adhesion of nerve fibers to their own kind.)

D, Weakened adhesion owing to different spacing (statistically speaking) of molecules in the apposed surfaces. The surface union in *D* would be only one third as strong as in *A.*

E, Lack of adhesion due to molecular disorientation.

F, Surface detachment, owing to change in the molecular configuration in one of the apposed surfaces.

size of the corresponding periphery. For instance, the limb segments of the cord are intrinsically larger than the trunk segments, even before the limbs have developed (Coghill, '36), and, in fact, even when limb development has been suppressed experimentally (Detwiler, '24). As we have said in the beginning, little is known about the factors which determine this typical spatial pattern of proliferation and differentiation and thereby produce the crude cast of the early central nervous system. Owing to these initial inequalities, the density of nerve outgrowth in different segments varies somewhat from the very start. But the size of the source and the rate of its increase determine only the amount of fibers leaving the centers, while the amount of terminal connections is much larger, owing to the extensive peripheral branching of the fibers.

Branching occurs in two ways: by terminal bifurcation and by collateral sprouting. As we have mentioned before, *terminal* bifurcation presumably results whenever two simultaneous pseudopodia of the growing tip turn out to be of equal strength, so that neither will succumb to the draining effect of the other (Figure 11, *A*). The tendency for profuse peripheral branching can be expected to be the greater the more intersections a fiber finds on its way. Therefore, branching will be much more extensive in a medium with unoriented ultrastructure than in one with definite micellar orientation. This expectation is fully borne out by tissue culture evidence (P. Weiss, '34, p. 440). According to Levi ('41), irradiating a culture with radium may greatly increase the incidence of branching, but the manner in which this occurs has not yet been analyzed. The rôle of the medium is likewise illustrated by the extensive branching which occurs in the unorganized scar around the cut end of a regenerating nerve. Perhaps the different density of the neropil in various parts of the brain (Herrick, '34) will one day be correlated with the different density and micellar organization of the respective colloidal matrices.

Collateral branching, on the other hand, is the result of a new budding process from the consolidated stem of a fiber (Figure 11, *B*). Experiments of Peterfi and Kapel ('28) suggest that local mechanical irritation of the fiber can produce it. Speidel ('33) has shown that the agitation caused by a dividing cell on or near the nerve fiber may act similarly. Obviously, any irritant, which is sufficiently strong to remobilize the local fiber protoplasm may, if given further support

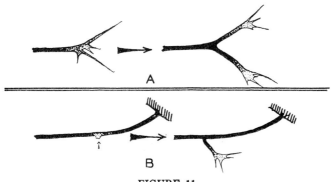

FIGURE 11

MODE OF BRANCHING IN NERVE FIBERS

A, Terminal branching, due to the persistence of two contemporary pseudopodia.
B, Collateral branching, due to lateral activation of stem of fiber.

by external conditions, lead to the establishment of a local outgrowth which then proceeds just like any other new fiber. The conditions under which this occurs remind us strongly of the rules established by Child ('27) for lateral regeneration in Hydroids; that is, in order to be effective, the wound stimulus must exceed a certain strength, and the distance from the existing apical end must be sufficiently great to insure to the new growth center what Child ('41) has called "physiological isolation."

Branching of peripheral nerve fibers occurs chiefly inside the terminal tissues. Each motor axon is eventually connected with a great number of muscle fibers through corresponding peripheral arborizations. There is evidence that muscle fibers are innervated long before they ever become muscle fibers, that is, in the state of myoblasts. This being the case, it is probable that whenever a myoblast or a young muscle fiber divides, this stimulus by itself provokes the formation of

V. ON THE DYNAMICS OF THE NERVOUS SYSTEM

a collateral branch from the nerve fiber, so that each new muscle fiber, as it emerges, takes its share of the neurone along. The fact that the average number of muscle fibers supplied by a single nerve fiber varies systematically for different muscles, would thus be an expression of the different rates of multiplication of the muscle fibers in different muscles.

The question of peripheral *anastomosing* of fibers by protoplasmic fusion is less clear. Boeke ('33) has emphasized the syncytial nature of the peripheral sympathetic plexus, and Stöhr ('35) has taken an even more extreme stand. Boeke ('30, '38) has also stressed the syncytial anastomosing of somatic nerves in the early phases of regeneration, a transitory condition which gradually gives way to fiber individualization from within the common mass. It seems that simple and thin fibers may merge peripherally, while the larger and more differentiated fibers always retain their individuality. This distinction is substantiated by nerve cultures in vitro. Levi ('41, p. 193) points out that the finest fibers frequently anastomose, while larger fibers never do. Might it not be that fusion is again contingent on full biochemical identity, which would exist among the primordial fibers, but disappear with their progressive divergent differentiation as they mature? This would explain the gradual individualization of fibers in regenerated nerves. However, these processes have not yet received enough attention to permit us to evaluate their significance for the final configuration of the peripheral nerve pattern.

Besides affecting nerve orientation and branching, the periphery regulates the volume of its innervation in two other ways. One of these has become familiar from the comprehensive studies of Detwiler and his co-workers (reviewed in Detwiler '36). Their experiments have shown that in urodele amphibians the size of the spinal ganglia is adjustable within limits to the actual extent of the periphery which they supply. An experimental increase or decrease of the peripheral tissue produces an augmentation or reduction in the number of spinal ganglion cells in the corresponding segments. May ('33) in the frog, and Hamburger ('34, '39) in the chick have demonstrated a similar relation between size of periphery and size of the motor nuclei of the cord. While the manner in which the expanse of the periphery reflects upon the volume of the centers is still wholly obscure, it is certain that the periphery exerts some influence on the number of fibers it is to receive right at their source.

Recently, however, we have come to learn about a second regulative influence by the periphery which does not affect the production, but rather the admission of fibers, in that it determines what proportion of the total fiber production is actually admitted into final functional connections. It was noted that in the case of transplanted full-sized limbs innervated from a reduced nerve source (P. Weiss, '37), as well as in the case of undersized regenerating limbs innervated from a superabundant nerve source (Weiss and Walker, '34), the total amount of peripheral nerve branches bore a definite relation to the size of the limbs rather than to the size of the nerve sources. These experiments, amplified later by Litwiller ('38, '38a) have shown beyond doubt that the tissues of the limb exert a controlling influence on the admission of nerve fibers into the peripheral field. The experiments can be best summarized in a diagram (Figure 12). Faced with

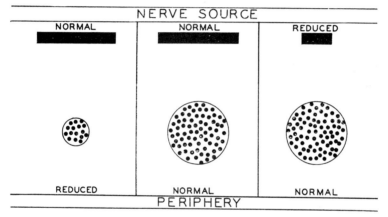

FIGURE 12

DIAGRAM, INDICATING LACK OF DEPENDENCE OF PERIPHERAL NERVE FILLING UPON SIZE OF THE NERVE SOURCE

an oversized nerve supply, the tissue can somehow prevent its becoming oversaturated with nerve fibers. Faced with an undersized nerve source, it can yet draw its full quota of peripheral branches. How it does this, is still unknown. A direct experimental demonstration of the effect has thus far been furnished only for regenerated nerves. However, the remark by Harrison ('35, p. 148), that a giant limb grown on a small body may contain a motor nerve complement proportional to its size without corresponding increase of the innervating centers, indicates that a similar mechanism of control operates in the embryo.

V. ON THE DYNAMICS OF THE NERVOUS SYSTEM 475

From these studies it would seem, that an innervated area spreads some influence which prevents the penetration of further innervation into its domain. It stakes off a territory. This may account for a well known observation. When a skin nerve is cut, the denervated area is slowly invaded by collateral nerve fibers from surrounding intact areas: the denervation has apparently removed some restraining influence previously exerted by the former occupants of the territory.

Thus receptor and effector tissues regulate the density of their innervation. The mechanism need not be the same for all tissues, and each one will have to be accorded a separate investigation. In only one case do we have more concrete information, that of the muscle. It is a well established anatomical fact that the large majority of muscle fibers receive but one single motor nerve branch each. Harrison ('10) has pointed to the analogy between this fact and the monospermy of eggs. Just as an egg, after receiving a single spermatozoan, would produce a surface reaction through which additional spermatozoa would be kept from penetrating, so the muscle fiber, after admitting its first nerve branch, would have become immune to further impregnation. This subject has recently been studied more extensively by Fort ('40). It was confirmed that muscle fibers, even when confronted with a superabundant supply of nerve terminals, as a rule, cannot be forced or conditioned to accept functional connections with more than a single nerve terminal. There are some indications, although not yet very convincing, that the reason why an innervated muscle fiber becomes resistant to further nerve impregnation may lie in a change of its surface constitution. The muscle, however, is the only case in which at least a beginning has been made to analyze the nature of what we may call the "saturation factor" of peripheral innervation.

Lack of space as well as of reliable information prevents us from going into the subject of nerve fiber *size,* the diameter of the individual fiber. One observes fibers in all gradations, from down at the limits of microscopic visibility up to those visible to the naked eye. They can be grouped into different size classes with different physiological properties (Erlanger and Gasser, '37), and different nerves contain fibers of the different classes in different proportions. These proportions seem to be essentially re-established in regeneration (P. Weiss, '37, p. 517), and the question of what determines the diameter of a nerve fiber seems to offer considerable interest. However, I know of no single analytical examination devoted to this problem. Correlations have been established from observational data between the length of fibers and their size, between innervated area and size, and the like, but with the possible exception of observations by Detwiler and Lewis ('25), indicating that motor neurones may show size reduction in the case of an experimental reduction of their periphery, no analytical experiments are available to decide in which way intrinsic and environmental factors contribute to the determination of the final dimensions of a neurone. We may mention that in tissue culture nerve cells and fibers occur in all sizes, but their variation is too unsystematic to furnish any clues.

Let me stop here with the analytical part of the discussion. The picture I have given you has covered only the most prominent features of nerve patterns. If I were to be exhaustive, I would have to go on and discuss the finer differentiation and physiological distinction among fibers, the arborization of the dendrites in the gray, the problem of the stratification of the brain, the problem of the relation between nerve fibers and capillaries, the differential susceptibility to hormonal agents which may be of behavioral significance, the causes of plexus formation, the association with sheath cells, the production of myelin, and many other features. But even the very incomplete picture, which you have received, will have impressed you with the multiplicity and complexity of factors which participate in turning out the final intricate fabric as which the nervous system confronts the anatomist and physiologist. If, in spite of this multiplicity and diversity of agents, the end product turns out to be, on the average, a nervous system fully capable of coördinating, controlling, and integrating the functions of the body, the credit for this achievement must be ascribed to three main factors.

Firstly, to the existence in the embryo of a definite spatial and chronological organization, according to which the individual events are seriated. Secondly, to the fact that the initial organization provides only for the gross outlines of the future development, leaving sufficient latitude for direct adjustive interactions to allow for a certain amount of variation among individuals. And thirdly, to the fact that the operation of the nervous system does not require absolute structural stereotypism, so that only statistically speaking must there be something like constancy and repetition in the final developmental product.

Just a few words in amplification of these three principles. The

V. On the Dynamics of the Nervous System 477

standardized dosing and timing of the morphogenetic processes is something the nervous system has in common with other parts of the body, and need not be discussed here (cf. P. Weiss, '39, pp. 104, 319, 486, 558).

The second point, however, deserves illustration because it gives a tangible conception of what we usually refer to as "regulation." Let us take a specific example. Remember that the formation of a straight nerve connection between a center and a peripheral organ is due to a dual process: Oriented outgrowth of the pathfinders, and the subsequent filling up of the cable by selective fasciculation. Suppose now, that owing to a genetic mutation or to developmental accidents, the pioneering fibers grew out precociously, before their supposed leading structure is ready: they will stray about and take devious routes, but some of them have still a good chance to reach the periphery. After that, selective fasciculation will do its part and produce a fair-sized nerve connection after all. The initial size deficit of the nerve may be partly corrected by the stimulative influence of the periphery which will mobilize additional neurones. But even with a constantly undersized nerve source, the peripheral field can still obtain its full quota of innervation, owing to the control it exerts over the amount of peripheral branching. The fact that each neurone will now have to carry a heavier load, is immaterial, because the size of the innervating centers is no measure of functional perfection, as is illustrated by the perfect functioning of transplanted supernumerary limbs innervated from only a fraction of the normal number of motor neurones (P. Weiss, '36). We realize, thus, that quite a few steps in the complex process of nerve formation may deviate from the norm without endangering the essential adequacy of the final product.

This situation again is characteristic not only of the development of the nervous system, but of development in general. One must remember that no developmental process is a unitary event. Innumerable independent and interdependent partial events share in the formation of an embryonic part. So, if one among the many fails to coöperate properly, the result may be only slightly off normal. This is merely translating the conclusions at which the geneticist has arrived in his multiple factor analysis from the symbolic language of genetics into the concrete terms of embryodynamics (cf. P. Weiss, '39, p. 479-486).

The latitude left to developmental processes leads us to the third point. No two nervous systems are identical or even nearly identical, if we concentrate on details. Yet, many of the most popular schemes to interpret the physiological action of the nervous system in terms of neurone relations, are based on the assumption that every detail of design and measurement in each individual neurone is significant as such. The number of collaterals, the length of collaterals, the points where they branch off, the number of end feet, their density on the cell body, the spacing between individual endings—all such details are treated as if there were agents in the organism which could wisely see to it that each individual neurone would really be as precisely constructed as physiological theory demands.

Now, while we are still far from complete insight into the factors which operate in establishing the finer structural details of the neurone architecture—and there is no saying what surprises the future may hold—we do know certain things; and one of them is that those factors which we have discussed in this report and which certainly do operate, are not of the kind that could produce the envisaged precision machine. The finished machine can be of no greater precision than the agents by which it has been constructed. We do not mean to deny here the possible intimate correlation between structural aspects and physiological performance altogether, but we must point out that, if the neurophysiologist wants to take into account the lessons of embryology, he will have to substitute *statistical* considerations for his overemphasis on *systematic* traits.

For instance, the *average* density of dendrites or collaterals in a certain neurone group may be functionally relevant, even though the number of dendrites or collaterals on each element may fluctuate considerably about that average number. As we have said before, the statistical amount of branching is a function of the colloidal consistency of the medium, and could, therefore, be very well understood on embryological principles. On the other hand, the assumption that each and every nerve cell of the group might possess a fixed number of dendrites or collaterals is positively discouraged by what we know about developmental mechanisms.

Here is another example of how statistical properties can play as good an explanatory rôle as is usually ascribed to structural details. The phenomenon of central irradiation, first described by Sherrington, implies that, as the excitatory influx into a center increases, more and more cells become engaged, leading to a proportional increase of the response. This is usually explained on the assumption that the central cells are connected in definite chain arrangements. In

spreading from one cell to the next, synaptic resistance must be overcome. Thus, the stronger the initial charge of the first excited cell,
the more hurdles it will be able to take, and the more cells of the chain
will be set off. But since each cell is viewed as a link in several
intersecting chains, the structural provisions prerequisite for an orderly gradation of the response would have to be infinitely complicated.
In contrast to this view, I have recently obtained evidence of graded
responses, increasing with the strength of the stimulus, in deplanted
centers (see p. 179), in which every trace of systematic organization
had vanished. In these deranged neurone pools, cells are still intercon-
nected, but at random. If graded responses can still be obtained, there
seems to be only one explanation: That is, that central cells develop
as a population in which the *thresholds* of excitability vary *at random*
(P. Weiss, '41). In such a population the number of elements of a
given threshold class, plotted against the thresholds, would follow a
near-normal distribution curve (Figure 13, *A*). A stimulus of given

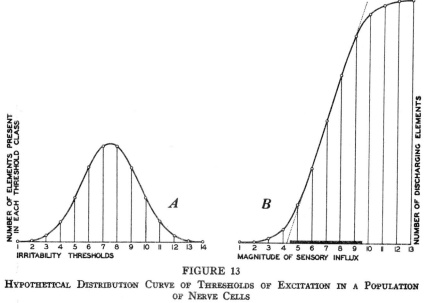

FIGURE 13

HYPOTHETICAL DISTRIBUTION CURVE OF THRESHOLDS OF EXCITATION IN A POPULATION
OF NERVE CELLS
(Explanation in text.)

strength would set into action all cells of the corresponding and lower
thresholds. The number of cells activated by stimuli of varying
strength would be expressed by the integral of the normal distribution

curve, and this, a sigmoid curve, very nearly approaches the course of a straight line for about 80 per cent of its range (Figure 13, *B*). Thus, within that range, any increase of the stimulus would bring in a proportionally larger number of cells, without the need for any special structural provisions. It is interesting to note, that v. Brücke, Early, and Forbes have actually demonstrated that the distribution of thresholds among the α fibers in a peripheral nerve follows a normal distribution curve, in other words, expresses random fluctuation within a population of living elements, rather than any systematic provisions specially devised in development. I think Hecht ('26), in his treatment of the visual apparatus, was the first to suggest such statistical considerations.

It has been my intention in this review not only to point out the multiplicity and complexity of factors that enter into the making of a nervous system—and of any organic system for that matter—but also to indicate the analytical insight that may be gained by resolving the complex situations into their simpler components and treating these one by one. In doing this, we have seen that the resolution can sometimes even be carried to the point where the biophysicist proper may take over. Once a vital process has been reduced to terms of molecular events, nothing remains to be done by the biologist in that particular direction. Thus, after having deposited the problems of adhesivity, selectivity, ultrastructure, etc., at the doorstep of the biophysicist, we may withdraw expectantly to see what further elucidation he may be able to provide from his own province. Or better still, we may combine forces with the biophysicists to create teamwork as proficient as that of that well-known symbiosis of the fable, the blind and the lame.

REFERENCES

1. BOEKE, J. 1917. Studien zur Nervenregeneration. II. Die Regeneration nach Vereinigung ungleichartiger Nervenstücke (heterogene Regeneration), und die Funktion der Augenmuskel- und Zungennerven. Die allgemeinen Gesetze der Nervenregeneration. *Verh. Konin. Akad. v. Wetensch. Amsterdam*, **19**, 1.

2. ———. 1930. De- und Regeneration des peripheren Nervensystems. *Deut. Z. f. Nervenheilk.*, **115**, 160.

3. ———. 1933. Innervationsstudien. V. Der sympathische Grundplexus und seine Beziehungen zu den quergestreiften Muskelfasern und zu den Herzmuskelfasern. *Z. mikr. Anat. Forsch.*, **34**, 330.

4. ———. 1935. Nervenregeneration. *Handb. d. Neurol.*, **1**, 996.

5. ———. 1938. Über die Verbindungen der Nervenzellen untereinander und mit den Erfolgsorganen. *Verh. d. anat. Ges. Anat. Anz.*, **85**, 111.

V. ON THE DYNAMICS OF THE NERVOUS SYSTEM

6. BOK, S. T. 1915. Die Entwicklung der Hirnnerven und ihrer zentralen Bahnen. Die stimulogene Fibrillation. *Fol. Neuro-biol.,* **9**, 475.

7. BRAUS, H. 1905. Experimentelle Beiträge zur Frage nach der Entwicklung peripherer Nerven. *Anat. Anz.,* **26**, 433.

8. VON BRÜCKE, E. TH., MARIE EARLY, & ALEXANDER FORBES. 1941. Fatigue and refractoriness in nerve. *J. Neurophysiol.,* **4**, 456.

9. BURR, H. S. 1932. An electro-dynamic theory of development suggested by studies of proliferation rates in the brain of Amblystoma. *J. Comp. Neurol.,* **56**, 347.

10. CAJAL, S. R. 1908. Studien über Nervenregeneration. Leipzig.

11. CENTANNI, E. 1914. Sulle colture affrontate dei tessuti in vitro nello studio della polarita di accrescimento. *Pathologica,* **6**, 305.

12. CHILD, C. M. 1921. The Origin and Development of the Nervous System. Chicago: Univ. Chicago Press.

13. ———. 1927. Experimental localization of new axes in Corymorpha without obliteration of the original polarity. *Biol. Bull.,* **53**, 469.

14. ———. 1941. Patterns and Problems of Development. Chicago: Univ. Chicago Press.

15. COGHILL, G. E. 1926. Correlated anatomical and physiological studies of the nervous system of Amphibia: V. The growth of the pattern of the motor mechanism of Amblystoma punctatum. *J. Comp. Neurol.,* **40**, 47.

16. ———. 1929. Anatomy and the Problem of Behavior. Cambridge: Cambridge Univ. Press.

17. ———. 1936. Correlated anatomical and physiological studies of the growth of the nervous system of Amphibia: XII. Quantitative relations of the spinal cord and ganglia correlated with the development of reflexes of the leg in Amblystoma punctatum Cope. *J. Comp. Neurol.,* **64**, 135.

18. DETWILER, S. R. 1920. On the hyperplasia of nerve centers resulting from excessive peripheral loading. *Proc. Nat. Acad. Sci.,* **6**, 96.

19. ———. 1924. The effects of bilateral extirpation of the anterior limb rudiments of Amblystoma embryos. *J. Comp. Neurol.,* **37**, 1.

20. ———. 1928. Further experiments upon alteration of the direction of growth in amphibian spinal nerves. *J. Exper. Zoöl.,* **51**, 1.

21. ———. 1930. Observations upon the growth, function, and nerve supply of limbs when grafted to the head of salamander embryos. *J. Exper. Zoöl.,* **55**, 319.

22. ———. 1936. Neuroembryology: An Experimental Study. New York.

23. DETWILER, S. R. & R. W. LEWIS. 1925. Size changes in primary brachial motor neurones following limb excision in Amblystoma embryos. *J. Comp. Neurol.,* **39**, 291.

24. DETWILER, S. R., & R. H. VAN DYKE. 1934. Further observations upon abnormal growth responses of spinal nerves in Amblystoma embryos. *J. Exper. Zoöl.,* **69**, 137.

25. ERLANGER, J., & H. S. GASSER. 1937. Electrical Signs of Nervous Activity. Philadelphia: Johnson Foundation Lectures.

26. FORSSMAN, J. 1898. Über die Ursachen, welche die Wachstumsrichtung der peripheren Nervenfasern bei der Regeneration bestimmen. *Beitr. z. pathol. Anat.,* **24**, 56.

27. ———. 1900. Zur Kenntnis des Neurotropismus. Weitere Beiträge. *Beitr. z. pathol. Anat.,* **27**, 407.

28. FORT, W. B. 1940. An experimental study of the factors involved in the establishment of neuromuscular connections. Dissertation, Univ. Chicago.

29. GRAY, P. 1939. Experiments with direct currents on chick embryos. *Roux' Arch. f. Entwicklungsmech. d. Org.,* **139**, 732.

30. GROSSFELD, H. 1934. Zellstreckung und Kohäsionskräfte im gallertigen Wachstumsmedium. *Roux' Arch. f. Entwicklungsmech. d. Org.*, **131**, 324.

31. HAMBURGER, V. 1929. Experimentelle Beiträge zur Entwicklungsphysiologie der Nervenbahnen in der Froschextremität. *Roux' Arch. f. Entwicklungsmech. d. Org.*, **119**, 47.

32. ————. 1934. The effects of wing bud extirpation on the development of the central nervous system in chick embryos. *J. Exper. Zoöl.*, **68**, 449.

33. ————. 1939. Motor and sensory hyperplasia following limb bud transplantations in chick embryos. *Physiol Zoöl.*, **12**, 268.

34. HARRISON, R. G. 1907. Experiments in transplanting limbs and their bearing upon the problems of the development of nerves. *J. Exper. Zoöl.*, **4**, 239.

35. ————. 1910. The outgrowth of the nerve fiber as a mode of protoplasmic movement. *J. Exper. Zoöl.*, **9**, 787.

36. ————. 1914. The reaction of embryonic cells to solid structures. *J. Exper. Zoöl.*, **17**, 521.

37. ————. 1924. Neuroblast versus sheath cell in the development of peripheral nerves. *J. Comp. Neurol.*, **37**, 123.

38. ————. 1935a. Heteroplastic Grafting in Embryology. The Harvey Lectures, 1933-1934, p. 116.

39. ————. 1935b. The Croonian lecture on the origin and development of the nervous system studied by the methods of experimental embryology. *Proc. Roy. Soc. London, Ser. B*, **118**, 155.

40. HECHT, S. 1926. A quantitative basis for visual acuity and intensity discrimination. *Skand. Arch. Physiol.*, **49**, 146.

41. HELD, H. 1909. Die Entwicklung des Nervengewebes bei den Wirbeltieren. Leipzig.

42. HENSEN, V. 1903. Die Entwicklungsmechanik der Nervenbahnen im Embryo der Säugetiere. Kiel and Leipzig.

43. HERRICK, C. J. 1934. The amphibian forebrain: IX. Neuropil and other interstitial nervous tissue. *J. Comp. Neurol.*, **59**, 93.

44. HIS, W. 1887. Die Entwicklung der ersten Nervenbahnen beim menschlichen Embryo. Übersichtliche Darstellung. *Arch f. Anat. & Physiol.* (Anat. Abt.), *Jahrg.*, 1887, 368.

45. HOADLEY, L. 1925. The differentiation of isolated chick primordia in chorioallantoic grafts: III. On the specificity of nerve processes arising from the mesencephalon in grafts. *J. Exper. Zoöl.*, **42**, 163.

46. HOLTFRETER, J. 1939. Gewebeaffinität, ein Mittel der embryonalen Formbildung. *Arch. f. exper. Zellforsch.*, **23**, 169.

47. INGVAR, S. 1920. Reactions of cells to the galvanic current in tissue cultures. *Proc. Soc. Exper. Biol. & Med.*, **17**, 198.

48. KAPPERS, C. U. A. 1917. Further contributions on neurobiotaxis: IX. An attempt to compare the phenomena of neurobiotaxis with other phenomena of taxis and tropism. The dynamic polarisation of the neurone. *J. Comp. Neurol.*, **27**, 261.

49. KARSSEN, A., & B. SAGER. 1934. Sur l'influence du courant electrique sur la croissance des neuroblastes in vitro. *Arch f. exper. Zellforsch.*, **16**, 255.

50. LEVI, G. 1934. Explantation, besonders die Struktur und die biologischen Eigenschaften der in vitro gezüchteten Zellen und Gewebe. *Ergebn. Anat. & Entwicklung.*, **31**, 125.

51. ————. 1941. Nouvelles recherches sur le tissu nerveux cultivé in vitro. Morphologie, croissance et relations réciproques des neurones. *Arch. de Biol.*, **52**, 133.

52. LEWIS, W. H. 1939. The rôle of a superficial plasma gel layer in changes of form, locomotion and division of cells in tissue cultures. *Arch. f. exper. Zellforsch.*, **23**, 1.

V. ON THE DYNAMICS OF THE NERVOUS SYSTEM 483

53. LEWIS, W. H., & M. REED LEWIS. 1912. The cultivation of sympathetic nerves from the intestine of chick embryos in saline solutions. *Anat. Rec.*, **6**, 7.

54. LITWILLER, R. 1938. Quantitative studies on nerve regeneration in Amphibia: I. Factors controlling nerve regeneration in adult limbs. *J. Comp. Neurol.*, **69**, 427.

55. ————. 1938a. Quantitative studies on nerve regeneration in Amphibia: II. Factors controlling nerve regeneration in regenerating limbs. *J. Exper. Zoöl.*, **79**, 377.

56. MAST, S. O. 1931. Locomotion in Amoeba proteus (Leidy). *Protoplasma*, **14**, 321.

57. MAY, R. M. 1933. Réactions neurogéniques de la moelle à la greffe en surnombre, ou à l'ablation d'une ébauche de patte postérieure chez l'embryon de l'anoure, Discoglossus pictus, Otth. *Bull. Biol.*, **67**, 327.

58. NICHOLAS, J. S. 1929. An analysis of the responses of isolated portions of the Amphibian nervous system. *Roux' Arch f. Entwicklungsmech. d. Org.*, **118**, 78.

59. ————. 1930. The effects of the separation of the medulla and spinal cord from the cerebral mechanism by the extirpation of the embryonic mesencephalon. *J. Exper. Zoöl.*, **55**, 1.

60. ————. 1933. The correlation of movement and nerve supply in transplanted limbs of Amblystoma. *J. Comp. Neurol.*, **57**, 253.

61. OPPENHEIMER, J. M. 1941. The anatomical relationships of abnormally located Mauthner's cells in Fundulus embryos. *J. Comp. Neurol.*, **74**, 131.

62. PÉTERFI, T., & O. KAPEL. 1928. Die Wirkung des Anstechens auf das Protoplasma der in vitro gezüchteten Gewebezellen: III. Anstichversuche an den Nervenzellen. *Arch f. exper. Zellforsch.*, **5**, 341.

63. PÉTERFI, T., & S. C. WILLIAMS. 1933. Elektrische Reizversuche an gezüchteten Gewebezellen: I. Versuche an Nervenzellen. *Arch f. exper. Zellforsch.*, **14**, 210.

64. SPEIDEL, C. C. 1933. Studies of living nerves: II. Activities of amoeboid growth cones, sheath cells, and myelin segments, as revealed by prolonged observation of individual nerve fibers in frog tadpoles. *Amer. J. Anat.*, **52**, 1.

65. STÖHR, P. JR. 1935. Beobachtungen und Bemerkungen über die Endausbreitung des vegetativen Nervensystems. *Z. f. Anat. u. Entwicklungsgesch.*, **104**, 133.

66. STRASSER, H. 1892. Alte unde neue Probleme der entwicklungsgeschichtlichen Forschung auf dem Gebiete des Nervensystems. *Ergebn. d. Anat. u. Entwicklungsgesch.*, **1**, 721.

67. STUDNIČKA, F. K. 1938. Die weichen Gewebe der Mesenchymreihe (Gallertgewebe und Bindgewebe) bei den Larven von Pelobates fuscus Laur. *Z. f. Zellforsch. u. mikr. Anat.*, **28**, 414.

68. TELLO, F. 1923. Gegenwärtige Anschauungen über den Neurotropismus. *Vortr. u. Aufs. über Entwicklungsmech.*, **33**.

69. VANLAIR, C. 1885. Nouvelles recherches expérimentales sur la régénération des nerfs. *Arch. de Biol.*, **6**, 127.

70. WEISS, P. 1932. Versuche über die Wirkung der operativen Einleitung motorischer Nerven in das Rückenmark (Parabioseversuche an Kröten). *Arb. d. ungar. biol. Forschg. Inst.*, **5**, 131.

71. ————. 1933. Functional adaptation and the rôle of ground substances in development. *Amer. Nat.*, **67**, 322.

72. ————. 1934a. Secretory activity of the inner layer of the embryonic midbrain of the chick, as revealed by tissue culture. *Anat. Rec.*, **58**, 299.

73. ————. 1934b. Motor effects of sensory nerves experimentally connected with muscles. *Anat. Rec.*, **60**, 437.

74. ————. 1935. Experimental innervation of muscles by the central ends of afferent nerves (establishment of a one-neurone connection between receptor and effector organ), with functional tests. *J. Comp. Neurol.*, **61**, 135.

75. ————. 1936. Selectivity controlling the central-peripheral relations in the nervous system. *Biol. Rev.,* **11**, 494.

76. ————. 1937. Further experimental investigations on the phenomenon of homologous response in transplanted amphibian limbs: II. Nerve regeneration and the innervation of the transplanted limbs. *J. Comp. Neurol.,* **66**, 481.

77. ————. 1939. Principles of Development. New York.

78. ————. 1940a. The problem of cell individuality in development. *Amer. Nat.,* **74**, 34.

79. ————. 1940b. Functional properties of isolated spinal cord grafts in larval amphibians. *Proc. Soc. Exper. Biol. & Med.,* **44**, 350.

80. ————. 1941a. Further experiments with deplanted and deranged nerve centers in amphibians. *Proc. Soc. Exper. Biol. & Med.,* **46**, 14.

81. ————. 1941b. Autonomous versus reflexogenous activity of the central nervous system. *Proc. Amer. Philos. Soc.,* **84**, 53.

82. WEISS, P., & R. WALKER. 1934. Nerve pattern in regenerated Urodele limbs. *Proc. Soc. Exper. Biol. & Med.,* **31**, 810.

83. WHITAKER, D. M. 1941. Physical factors of growth. *Growth,* **4**, Suppl. 75.

84. WILLIAMS, S. C. 1935. A study of the reactions of growing embryonic nerve fibers to the passage of direct electric current through the surrounding medium. *Anat. Rec.,* **64**, *Suppl.,* 56.

V. ON THE DYNAMICS OF THE NERVOUS SYSTEM 485

Reprinted from COMPARATIVE PSYCHOLOGY MONOGRAPHS.
The Williams and Wilkins Co., Baltimore, Md., 17, 1-96, 1941.

CHAPTER 22

SELF-DIFFERENTIATION OF THE BASIC PATTERNS OF COORDINATION*

PAUL WEISS

Hull Zoological Laboratory, The University of Chicago

CONTENTS

THE PROBLEM OF COORDINATION

Introduction

Motor behavior is effected through the coordinated operation of the musculature. The problem of the ontogenetic origin of behavior, therefore, resolves itself essentially into the problem of the ontogeny of coordination.

Whether "coordination" is a *constitutional*, i.e. pre-functional, faculty of the central nervous system or an *acquired* property, gained by experience, has long been a matter of dispute. The

* The experimental investigations were aided by the Dr. Wallace C. and Clara A. Abbott Memorial Fund of the University of Chicago.

main objective of this paper is to present direct experimental proof that *the basic patterns of coordination arise by self-differentiation within the nerve centers, prior to, and irrespective of actual experience in their use.* In addition, experimental data will be reported indicating what coordination consists of, and how it is laid down in the centers. So specific and articulate has this experimental information been that we would forego the full benefit of its instructiveness if we were to report it in the general inarticulate language in which the problems of coordination are conventionally treated. Since the experiments have produced answers much more differentiated and detailed than any of the questions commonly being asked, we must prepare the ground for their presentation by reformulating the questions with greater precision. One cannot discuss "coordination" profitably so long as the term is kept on the abstract level; ill-defined, and non-committal in regard to concrete implications. We, therefore, shall try to dissect the general concept of "coordination." By breaking it down into more tangible issues, we make it tractable and give experimental analysis a chance to substitute knowledge for conjecture.

Thus, with a view to discontinuing the practice of speculating about "coordination" without a clear mental picture of just what it implies, we shall first review the facts and reformulate the problems, and only then proceed to present the experiments and, in their light, scrutinize the existing theories of coordination. The notable confusion about terms and facts in this field would justify a much more thorough reconsideration of the whole subject than can here be afforded. However, such an ambitious attempt had better be postponed until more concrete building stones for a good and sound theory of coordination have been gathered than are now available. It is only a few of these building stones that the present paper aims to contribute.

There is a striking disproportion between our knowledge of the physiological properties of nervous *elements* and our understanding of the operation of the nervous *system* as the coordinator of those elemental activities in the service of the organism. While the combined efforts of electro-physiological, histological, biochemical and mathematical studies have produced a great wealth of data concerning the *elemental* activity of the neurone, our conception of the *systemic* activity of their organized totality has essentially remained pegged to the level to which it had been raised by *Sherrington's* classical work on the "Integrative Action of the Nervous System." We do not mean to imply that some progress has not been made here and there. Brain physiologists and neurologically-minded psychologists, in their efforts to interpret behavior in terms of the function of the nervous system, naturally had to focus on the system as such. From their studies, they were led to conclusions which partly supplemented, partly discredited, the synthetic conception of the nervous system to which preoccupation with the nervous elements had led. Yet, although they succeeded in pointing the direction toward a more adequate theory of central functions, the actual progress made thus far appears small when contrasted with the spectacular growth of our information about the nervous units during the same period. A mere comparison between the volume of attention currently paid to the issue of synaptic transmission—whether chemically or electrically mediated—on the one hand, and the almost complete neglect of the problem of how central transmission has come to be so discriminatory and selective as to lead to coordinated responses, rather than to unorganized convulsions, on the other, puts the situation into sharp relief. A restoration of sounder proportions should be attempted, of course, not by detracting from the current vigorous trend toward the isolated elements, but by reviving interest and revitalizing research in matters which concern the integrative aspects of the nervous system. If such course is to be followed with profit, it will pay, at the outset, to examine the possible reasons of its lag in the past.

A few explanations suggest themselves quite readily. The most obvious one is the infinitely greater difficulty and complexity of the task facing the student of the nervous system. This may be a challenge to inquisitive minds, but it certainly does not predispose the subject for mass attack by routine methods. So long as one clings to the study of elements, one is dealing with well-circumscribed units, a well-defined subject, presenting clear-cut problems, and one can call on familiar and approved methods of analysis. As soon as one raises the eye from the unit to the whole system, the subject becomes fuzzy, the problems ill-descript, and the prospect of fruitful attack discouraging in its indefiniteness. This may explain why a considerable number of able experimental workers prefer to circle around the focal problems at a respectful distance rather than heading straight at them. It also explains why discussions of central nervous function operate so much more liberally with words than with facts; for it is remarkable how general the tendency is in this field to cover up factual ignorance by verbalisms. The average attitude is somewhat like this: the "whole" gets a large share of one's thought and talk, but the elements get all the benefit of one's actual work; here the problems seem to be so infinitely more tangible. *Adrian* (1932) has expressed this very plainly at the conclusion of his lectures on "the mechanism of nervous action." He says (p. 93): "The nervous system is built up of specialized cells whose reactions do not differ fundamentally from one another or from the reactions of the other kinds of excitable cell. They have a fairly simple mechanism when we treat them as individuals. Their behavior *in the mass*[1] may be quite another story, but this is for future work to decide." The search for that "other story" deserves encouragement.

However, a second point which has detracted from a vigorous pursuit of this search is not to be overlooked. This is the reiterated expression by highly competent students of the nervous system of their conviction that a thorough understanding of what is going on in the isolated peripheral nervous units, will eventually explain the operation of the centers, too. *Gasser* (1937), for instance, in his Harvey lecture on "the control of excitation in the nervous system," states this belief quite explicitly (p. 171): "Admittedly the nervous system can be understood only as it is operating as a whole, but it is equally true that an insight into its working can be gained *only*[1] by a detailed analysis of its parts. If the isolation of a part results in the sacrifice of some of its qualities, the loss is compensated for by the acquisition of a degree of sim-

[1] Not italicized in text.

V. ON THE DYNAMICS OF THE NERVOUS SYSTEM   489

plicity making the part more amenable to investigation. The organization of the nervous system is such that an understanding of the mode of activity of any part of the ganglionic apparatus would mean a long step forward, for all parts of the nervous system are fundamentally alike. We can, therefore, proceed, confident in the belief that when the parts are understood, they *can be added together into larger units*;[2] and that, as the addition takes place, the lost qualities will again emerge and be recognized. Bit by bit it should be possible in the end to build back to the elaborate patterns of activity which are characteristic of the intact organism." This stand seems justified if one accepts the premise proposed by *Keith Lucas* and quoted and amplified by *Gasser*, "that the phenomena taking place in the central nervous system could be explained without the assumption of any properties which could not be experimentally identified in peripheral nerve." "If the reactions in the central nervous system are to be explained on the basis of occurrences in peripheral nerves, . . . it is evident that the starting point of all discussions must be a thoroughgoing understanding of the physiology of nerve fibers" (p. 173).

At the same time, the opinion is growing that this precept of *Lucas* has not worked, and cannot work, in bringing us real understanding of the centers, in spite of numerous ingenious hypotheses designed to make it work. *Forbes*, for one, once an outstanding exponent of *Lucas'* scheme, expresses this change of opinion poignantly, when he declares (1936): "In developing these schemata no attempt was made to offer a final theory of the workings of the central nervous system, but merely to see whether research had yet brought to light any facts which were utterly incapable of explanation in terms of the phenomena of peripheral conduction. As newer information has come to light, the subsidiary hypotheses needed to explain the facts in terms of the nerve impulses alone have made it increasingly improbable that these working hypotheses would be adequate, until now they have become of little more than historical interest" (p. 164). Without arguing the issue any further, we merely present it as a symptom of the discouragement not uncommonly held out to

[2] Not italicized in text.

those who might have wanted to give the central nervous system an independent examination as a system in its own right.

Consequently, a large group of workers who did concern themselves primarily with the nervous system accepted the thesis of the "elementarians," that inasmuch as central activity can be conceived of as merely a proper linking together—"association"—of individual neurone activities, understanding of central activity can be pieced together from bits of knowledge about the elements. Thus the theory that "associations" form the basis of behavior, which is the psychological version of the theory of the "reflex" as the basis of central nervous action, played into the hands of the elementarians and greatly increased their prestige. Now, we still are not saying that they may not in the end turn out to have been right; but we do want to point out that the question as to whether there is more in the centers than there is to be found in the peripheral elements is a purely empirical question which cannot be solved on *a priori* grounds and which should not be prejudicated by recommendations for procedure which by their very nature preclude any but the anticipated answer. Nobody is likely to do prospecting in an area where he is constantly assured by experts that nothing worthwhile can be found; nor is he going to be encouraged by the repeated warning that whatever he is looking for has already been found—or, at least, will most certainly be found—by his neighbor. In this sense, the denial that the central nervous system presents problems *sui generis*, has undoubtedly been a potent deterrent from a vigorous attack on those problems.

It is not to be questioned that the attempt to identify in the central nervous system properties familiar from peripheral nerve elements has met with spectacular success. In fact, faith in the fundamental identity of both has been rewarded by discovery in the peripheral elements of properties which had formerly been known only to occur within the centers (see *Gasser*). However, by confining attention to those phenomena which the peripheral and central systems have in common, we plainly relegate the specifically central phenomena to continued obscurity; by "specifically central" we mean the ones that are not recoverable from peripheral

V. On the Dynamics of the Nervous System   491

investigations.   *Coordination* is a case in point, and this brings us to our immediate subject.

Even though we can interpret other central phenomena, such as reaction time, summation, rhythmicity, inhibition, irradiation, fatigue, etc., in terms of known properties of neurones, the specific *order* in which the units are brought into play so as to produce effects serviceable to the organism, is nowhere accounted for in this scheme.   It is this order that is commonly referred to as "coordination," with implications that are not always clearly realized.   Coordination means the *selective* activation of definite groups of units in such combinations that their united action will result in an organized peripheral effect that makes sense.   But what principle is there in operation in the centers to make the appropriate selection?   And in what terms is the choice being made?   Here is a question aimed at a "specifically central" phenomenon, evidently of fundamental importance, and yet one that is hardly ever asked explicitly, and still more rarely answered in anything but the most general of terms.   Really only a very few have taken the trouble of penetrating beneath the surface of the problem, and although, as we shall show below, none of their efforts have as yet yielded wholly acceptable results, they have at least emerged with some definite suggestions that can be put to test and serve as points of departure for further clarification.   In addition to the few who have given the matter mature thought, practically every biologist and psychologist carries in his mind some sort of notion, specific or hazy, of the mechanism of coordination, which he has usually acquired unconsciously and by accident.

Disregarding their various shades, we can class these notions, both rational and instinctive ones, into three different groups: the *preformistic*, the *heuristic*, and the *systemic* theories.   All three start from the fact of the transmissibility of excitation from one nerve element to another, and go on to explain why transmission in the normal nervous system does not occur indiscriminately, ending up in mass contraction of all muscles, but remains confined within certain channels, yielding an orderly and differentiated response.   For the time being, we may ignore the fact that by stamping the problem as purely one of controlled transmission we subject our search to an unwarranted limitation from the very start. Inasmuch as most speculations on coordination have tacitly accepted this limitation, they do not essentially differ on that point. From here on, however, they diverge.

The theories of coordination

The preformistic-structural concept

The first and most popular interpretation of coordination is the reference to stereotyped inherited anatomical neurone connections in the centers. It is based on the study of reflexes, the observation that in many simple reflexes there is a fairly definite and constant relation between the point of stimulation and the nature of the response ("reflex-arc"), and the assumption that the chain of events leading from stimulus to response is anatomically preformed in a chain of neurones leading from the sense organ through the centers to the effector. The biological adequacy of the response, according to this concept, is a result of the correct construction of the anatomical apparatus, that is, among other things, of the suitable distribution of the peripheral and intracentral neurones, suitable arrangement of the central switches among neurones, suitable arrangement of the muscles on the skeleton and suitable form of the joints; "suitable" in the sense of making the whole response come out as of service to the organism. In other words, the body has its coordination built in.

In designating this concept as "preformistic," we do not use the term "preformation" in the sense in which it is used in embryology. There it refers to the existence of organized patterns in the egg prior to the onset of *development*, while in the present connection it merely implies the presence of definitely organized innervation patterns in the centers prior to the onset of their actual *operation*. Developmentally speaking, these patterns have, of course, been differentiated according to the same principle of progressive (epigenetic) determination which dominates embryonic development in general (*P. Weiss*, 1939). Only in functional regards may we call them "self-differentiated," that is, differentiated in their essential characteristics independently of the actual intervention of function.

Movements with only one degree of freedom give an excellent illustration of this kind of anatomically preformed coordination. All a clam can do with its shell, is to open and close it, the contraction of the adductor muscle effecting the closure, and the elastic ligament over the hinge effecting the opening when the muscle relaxes. The whole performance is rigidly determined by the construction of valves and hinge. The situation would not be fundamentally different, if the opening of the hinge were effected by another muscle, instead of an elastic ligament, as is the case in bivalve brachiopods. In this case, the "suitability" of the situation implies, in addition to the hinging of the joint, the insertion of the two muscles on opposite sides of the joint (mechanically antagonistic action) and the ability of the nervous system to contract one muscle while the other relaxes (reciprocal innervation). If one admits the possibility that, for instance, tactile receptors are connected with the shutting muscle, and chemoreceptors with the opening muscle, the differential response of the animal—closing after mechanical disturbance and opening in the presence of food—would be accounted for on purely anatomical grounds.

Cases have been reported in which the anatomical predisposition to produce a given peripheral effect is even visibly expressed by the nervous elements so predisposed. The giant nerve fibers of the earthworm are the mediators of certain fast responses only, while other fiber systems serve slower reactions of different pattern (see *Prosser*, 1934). Making use of the fact that thicker fibers conduct faster (*Erlanger* and *Gasser*, 1935), the anatomy of these animals has provided for a system of central superhighways for undelayed through traffic. Still more spectacular is the case of the innervation of the mantle of the squid. The giant nerve fibers to the mantle muscles, which radiate from an anterior ganglion, increase both in length and thickness in anteroposterior order (*J. Z. Young*, 1938). In view of the proportionality between thickness and conduction speed, this provision enables a simultaneous discharge from the ganglion to arrive at all muscles at approximately the same time, producing a powerful synchronized over-all contraction in spite of the graded lengths of the supplying nerves (*Pumphrey* and *Young*, 1938). But for this anatomical provision, the result could be achieved only by having the impulses go off staggered in definite regular intervals, beginning with the longest fibers.   Here, too, coordination is efficiently preformed in the anatomical structure of the nervous system.   Similarly, the fast and slow reactions of the claws of crustaceans are mediated by different sets of nerve fibers, either set being predisposed for its function by appropriate constitutional properties (*Wiersma* and *van Harreveld*, 1938).

There can be no doubt, therefore, about the existence of built-in coordination.   The only question is whether these observations can be generalized and made the basis for a theory of *all* coordination.   In point of fact, they have been generalized without much opposition.   For instance, the flexing and bending of a knee joint has been viewed much in the same light as the opening and closing of the valves in a brachiopod.   However, we must reiterate that the brachiopod exemplifies a system with only a single degree of freedom of motion.   The only two movements compatible with the rigid mechanical limitations occur both in the same plane, and the muscles merely determine whether the joint will move one way or the other.   It is after this sort of model that the anatomical theory of reflex action has been fashioned, and one recognizes immediately the close resemblance with the conventional description of the stepping movements of a vertebrate limb.   The main action of the joints is represented as occurring in one plane, with two groups of antagonistic muscles producing excursions of opposite sign, which, depending on whether they reduce or increase the angle of the joint, are designated as "flexion" and "extension." To represent a limb movement in this manner involves a deliberate abstraction, in that we confine our attention to those joints which, in crude approximation at least, can be considered to conform to the model, that is, to true hinge joints (fig. 1, *a*).   Under these conditions, simplified to the extreme, peripheral coordination would appear simply as the alternating contraction of two antagonistic muscle groups, and the central basis of that coordination might pass for merely an oscillation between the excitation of a "flexor" and that of an "extensor" center, each attended by inhibition of the other ("reciprocal" innervation).[3]   An underlying anatomical neurone set-up is conceivable, and several such schemes have been promoted in the past.   They have become so firmly ingrained in our thinking that most textbooks deal with them as realities.

This is no place to evaluate the merits of the various anatomical switchboard concepts which have been advanced in explanation of simple type reflexes, and, in further consequence, of coordinated behavior in general.   But it will be well to keep in mind what

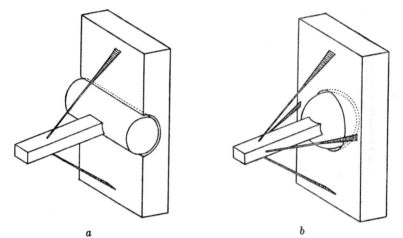

a   *b*

Fig. 1. Models of Joints

a, hinge joint with single degree of freedom; moved by a pair of antagonistic muscles; *b*, ball joint with free rotational motion, operated by four muscles.

Herrick, that most judicious student of the anatomy of the central nervous system, had to say in this connection (1930, p. 645): "No complication of separate and insulated reflex arcs, each of which is conceived as giving a one-to-one relation between stimulus and response, and no interconnection of such arcs by elaborate switchboard devices can conceivably yield the type of behavior which we actually find in higher vertebrates. . . . These facts are

[3] It is irrelevant in the present context whether this oscillation is regarded as due to alternating stretch-reflexes (Sherrington) or to autonomous rhythms of the centers (Graham Brown, 1914).

regarded as incompatible with the traditional dogmas of reflex physiology, with its precisely localized and well-insulated reflex arcs and centers of reflex adjustment. . . . The mechanisms of traditional reflexology seem hopelessly inadequate." However, even if the concept of a rigid anatomical neurone linkage were adequate to explain the single pendulum-like action of a hinge joint, it must utterly fail as a basis of coordination in any more general sense. For most joints are constructed so as to allow of more than one degree of freedom and, therefore, require more than two sets of muscles, inserting and acting in different planes.

A ball joint, such as the shoulder joint, can be moved in any plane laid through its center (fig. 1, b). The direction in which it actually moves at any given moment, is determined by the resultant of muscular tensions acting from all sides. Depending on the combination of active muscles and the relative strength of their contraction, an infinite variety of positions can be assumed. Variety of movement is thus made possible by varying the combination of muscles called into action. While it is easy to separate the muscles of a hinge joint sharply into agonists and antagonists, such classification is no longer applicable to a ball joint. Any two muscles may facultatively operate as agonists or antagonists. The following example, which refers to a specific case dealt with later in this paper, will help to make this clear.

Let us consider the shoulder joint of a tetrapod. Ignoring the finer details of the distribution of its muscles, we recognize four main groups converging upon the humerus from four different directions, schematically along the four edges of an imaginary pyramid (fig. 1b, 2). Acting individually, these muscles—listed in counterclockwise order—would pull the humerus upward, forward, downward, and backward, respectively, and they may be designated accordingly as elevator (El), abductor (Ab), depressor (De), and adductor (Ad). Through their graded contraction in proper combinations the humerus can be made to describe a full circle, eight representative stages of which are reproduced in figure 2. To bring the humerus into any of these eight positions, the following muscle combinations must enter into action (table 1).

According to the table, any one muscle may be engaged either in phase or out of phase with any other muscle. The diagram (fig. 3) expresses for each pair of muscles the phases when they act synergically (convergent arrows) and antagonistically (di-

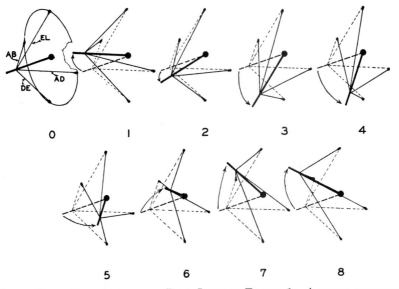

FIG. 2. EIGHT POSITIONS OF THE BALL JOINT OF FIGURE 1B, ASSUMED THROUGH THE CONTRACTION OF ITS MUSCLES IN THE COMBINATIONS LISTED IN TABLE 1

The posture in which all muscles are evenly contracted is pictured in phase O; it is indicated throughout the following phases by dotted lines, the arrows showing the direction of the excursions.

TABLE 1

PHASE	MUSCLES IN ACTION	RESULTING MOVEMENT
1	El, Ab, De	Craniad
2	Ab, De	Cranio-ventrad
3	Ab, De, Ad	Ventrad
4	De, Ad	Ventro-caudad
5	El, De, Ad	Caudad
6	El, Ad	Caudo-dorsad
7	El, Ab, Ad	Dorsad
8	El, Ab	Dorso-craniad

vergent arrows). Thus, in contrast to the hinge joint with its single direction of motion, the ball joint presents the centers with a problem of multiple choice: the grouping of the mucles changes with the movement to be effected. The problem of coordination, therefore, is no longer simply one of alternative innervation of one out of two muscle groups ("flexors" and "extensors"), but it involves the selection of a definite combination of muscles out of a large number of possible combinations. If we add that a change in the intensity of the contraction of any one muscle necessarily changes the direction of the resulting movement, we realize that

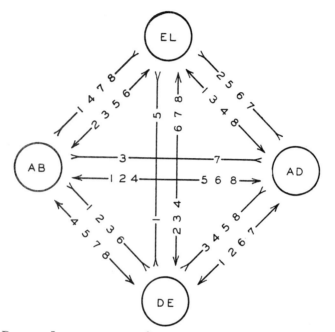

Fig. 3. Diagram Illustrating the Changing Agonistic (›—‹) or Antagonistic (↔) Relations between any Two Muscles in Executing the Eight Movements of Figure 2

none of the schemata developed for hinge joints are applicable to this more general case, and, particularly, that this type of coordination must defeat any interpretation in terms of monotonous central connections. Here the problem of coordination presents itself in its full meaning: *What determines the choice of muscles to be engaged in a given movement, and how is this selective activation being put into effect?* Viewing the problem in the light of the ball joint, sets it into the right perspective, while the overemphasis of the flexor-extensor pendulum essentially misses the problem and thus helps to side-track rather than to solve it. We shall return to this subject later in the paper.

From a correct appreciation of the implications of the problem, several theories have arisen, striving to replace, or, at least, supplement, the thesis of pure structural preformation of coordination. We may call them the *heuristic* and the *systemic* theories.

The heuristic concept

Any theory which submits that the common appropriateness for the body of motor effects has been developed ontogenetically under the molding action of practice and experience, may be called *heuristic*. The animal is thought to be capable of activating each one of its numerous motor effectors independently. Through repeated trials with constantly changing combinations it is supposed to produce a kaleidoscopic variety of peripheral effects, of which some are successful from the standpoint of the body and its needs, some are failures, and some are indifferent. Combinations leading to useful effects are somehow preserved in the central organization and fixed and improved by repetition, while those leading to useless or adverse effects are eliminated. According to this view, patterns of coordination arise through the accumulation of nervous "associations" which have proved their usefulness for the body. The essential point is that the *effectiveness* of a response is thought to confer selective value, and hence stability, upon the originally wholly tentative grouping of the muscles through which it is brought about. Whether the trials in this "trial-and-error" procedure are entirely random, or whether they show some method and direction; whether the drive to move is produced within the organism or furnished by external stimulation; whether the "associations" are to be viewed as nerve fiber connections, or whether—following behavioristic maxim—one had better refrain from such attempts at visualization; all these are relatively minor matters compared with the basic tenet in which all heuristic theories agree: that the central nervous system is a plastic mold upon which experience gained in actual performance gradually inscribes the patterns of coordinated behavior, with the adequacy of the effect for the organism as a whole serving as the standard of rating. This concept has been advanced for the lowest (*Jennings*, 1931) as well as for the highest forms of animals (*Pavlov*, 1927), and it has been variously applied to the development of nervous coordination from the highest cortical acts down to the most elementary motor functions.

The systemic concept

The systemic theories of coordination have in common with the heuristic theories the assumption of practically unlimited *plasticity* of the nervous system. However, instead of letting coordination become built up bit by bit through trial and error methods, they concede to the nervous system a *primordial dynamic ability* to respond to any change in the external stimulus situation by a total response of maximum adequacy for the organism as a whole. According to this view, entirely novel stimulus situations, neither provided for in the organization of the animal nor previously experienced, can be met by a primary response of great suitability. This view, shared by many Gestalt psychologists, has been particularly elaborated by *Bethe* (1931) and *Goldstein* (1939). The contrast between the heuristic and systemic concept can perhaps be expressed as follows: according to the former, partial reactions (elementary senso-motor responses) of no definite directiveness are variously recombined until they finally compose a chain the resultant direction of which has affirmed its value for the organism; whereas, according to the latter, resourceful dynamics of the central nervous system lay down the general direction of the total response as a sort of frame through which the partial reactions necessary for its execution are forced in channels consonant with the general "intention." In this view, the drive toward a certain real or visualized goal would directly produce the proper muscular innervation necessary to attain that goal.

Summary

Reduced to the terms of the preformistic-anatomical (1), heuristic (2), and systemic (3) theory, respectively, the coordinated advance of an organism toward a desirable goal, and its coordinated retreat from a harmful situation, could be expressed as follows:

(1) Beneficial and nocuous stimuli enter through different adequate receptors, activate each a system of separate pre-arranged lines which, in turn, engage a pre-arranged selection of muscles in a pre-arranged time order, the combined action of which then becomes manifest as a motion of advance or withdrawal. The appropriateness of the response is based on the appropriateness of the inherited pre-arrangements; the individual itself deserves no credit for it.

(2) Either kind of stimulus evokes ubiquitous random reactions, including excursions of limbs and trunk of continuously varying patterns, which are tried and discarded and repeated and altered, until eventually the correct composition and sequence is discovered; the animal is to be given credit for its resourcefulness in producing ever changing assortments of undirected responses, as well as for its faculty to choose and retain those that prove to lie in the right direction.

(3) Any stimulus produces a general response the character of which is directly determined by the constellation of the external field of stimuli and the internal state of the centers, resulting in a primarily directed movement; the centers get credit for their ingenuity.

To reduce the various theories of coordination to such simple formulae, admittedly involves a great deal of abstraction and oversimplification. Moreover, many authors, in discussing these matters, have failed to take an explicit stand, which puts it up to the interpreter to extrapolate their basic beliefs from casual remarks; which is a doubtful task. Again, some have taken compromise attitudes, contaminating one theory with admixtures from another, which makes a strict classification of their views impossible. Hence, no more than practical significance should be attached to the attempt of the preceding pages to group all existing theories of coordination into the three outlined categories. The justification for that attempt lies merely in that it serves to crystallize the issues. It puts into specific form the premises and implications of the various current theories and thus prepares them for the experimental tests to which they must be subjected for verification.

The central hierarchy

To decide between the performistic and the heuristic concept of coordination, is an empirical problem. Here is the alternative which the experimental investigator faces: If coordination is preformed in self-differentiated central impulse patterns, which yield adequate peripheral effects only by virtue of what may be called evolutionary precedent and in the individual case amounts to predesign, they should prove stable and conservative even if experimentally prevented from producing appropriate functional effects. If, on the other hand, functional effectiveness is all that counts in shaping the patterns of coordination, one should expect any experimental reduction of that effectiveness to be followed by corrective modifications of the impulse patterns—evidence of plasticity and of lack of intrinsic organization.

It is evident that the decision cannot be reached by even the most intimate study of the *normal* organism, with its inherited stereotypism of central structures, peripheral structures and nerve connections. For this stereotypism means that the same standard central patterns and the same standard effects appear always in conjunction. A crucial experiment, therefore, must aim at disrupting the monotony of central-peripheral correspondence. It must upset either the discharge pattern of the centers or the play of muscles or the distribution of nerve connections in such a manner as to make the established central impulse patterns yield incongruous effects for the body. If, thereafter, the body recovers more efficient use of the affected part—either instantaneously by systemic reaction, or gradually by heuristic procedure—, the systemic or heuristic theories would score. If, on the other hand, corrective changes fail to occur and the nervous system continues to operate the part according to the old standard scheme of innervation now rendered inadequate, this would be incontestable proof of the *preformation* of coordination in form of definite central impulse patterns which do or do not produce appropriate effects, depending on whether the effector system for whose operation they are predesigned is intact or disarranged.

All this seems so plain that one might expect the issue to have long been settled one way or the other. In fact, the indicated experimental course has been followed by some authors in the past. If, nevertheless, there is, even at this date, a basic lack of agreement, this suggests that either the experimental results or their interpretations have remained inconclusive. The reasons for this failure will become increasingly clear in the course of this paper. Two of them may be specifically mentioned here: preoccupation with higher mammals, particularly man; and injudicious generalization of concepts of "learning" or "conditioning."

The substitution of a healthy muscle for a paralyzed one is standard practice among orthopedic surgeons. The replacing muscle is sutured to the tendon stump of the incapacitated muscle and thus the mechanical part played by the lost muscle is taken over by the substitute. To be physiologically effective, the transposed muscle must, of course, be operated according to a new time schedule. Its former functional associations must be dissolved and replaced by new ones in accordance with its new function. To be sure, tendon transposition as such, even without retiming of muscle actions, produces some degree of improvement, simply owing to the restoration of a more normal balance of tensions around the joint. For the shift of a muscle from the vigorous to the frail side not only reduces the bulk of muscle left on the intact side, but also cancels part of the remaining muscle power of that side by the opposing action of the shifted portion contracting simultaneously. Further postoperative adjustments are brought about by compensatory changes in the strength with which other, unimpaired, muscle groups are being engaged, still according to the original time pattern.

However, there is incontrovertible evidence to show that the gradual restoration of relatively efficient limb coordination in patients with translocated tendon insertions is not merely due to changes of the kind just mentioned, but involves an actual modification of the original time pattern of innervation, the transposed muscle assuming functionally—i.e., with regard to its phase of innervation relative to other muscles—the place formerly held by the muscle which it has replaced mechanically. While there seem to be limitations to this adaptive change of coordination (*Scherb*, 1938), there is consensus of opinion that some "re-education" of the play of muscles can be attained by proper training in all human beings.

By sheer extrapolation it was then conjectured that what is true of man, would likewise hold for other animals. A broad experimental foundation of this assumption seems to have been neither sought nor offered. Some pertinent experiments were made on the coordination of eye movements after muscle translocation in mammals (*Marina*, 1912), but the results have remained controversial (*Dusser de Barenne and de Kleyn*, 1929; *Olmsted, Margutti and Yanagisana*, 1936). An isolated report of reorganization of limb coordination in the frog (*Manigk*, 1934) was shown to have arisen from a faulty interpretation of the underlying experiments (*Taylor*, 1936). But in spite of this lack of convincing experimental proof, the view seems to prevail that the locomotor apparatus of an animal can undergo essentially the same kind of adjustive "re-education" which has been demonstrated in man.

Critical experiments on amphibians, however, have contradicted this view decisively. As will be reported below in greater detail, these animals show no trace of re-adjustment of muscle coordination under comparable circumstances. To avoid misunderstandings, it may be added that while the basic coordination mechanisms through which all locomotor acts must be executed are in themselves quite unmodifiable in amphibians, the total behavior of these animals can be somewhat modified by training. They can learn to advance or retreat on different occasions, but they cannot learn to change their manner of walking or retreating. Similarly, recent experiments on the rat have definitely shown that the time pattern of coordination of the hind limb muscles of these animals, too, is rigidly fixed and remains incorrigible even after the crossing of antagonistic muscles resulting in permanent reversal of movements (*Sperry*, 1940). There may be a trace of re-adjustment after transposition of muscles in the *fore* limb of the rat (*Sperry*, see below), and as we extend the examination to higher and higher mammals, we may expect to find a growing faculty for such corrective measures of the nervous system. The essential point, however, remains that this faculty is a very late evolutionary acquisition of the central nervous system, practically still absent in as high an animal as the rat, and consequently entirely unfit as a model of the principle of coordination in general. (See p. 80.)

We must once and for all renounce the idea that the type of muscular control with which we are most familiar, namely, our own, or at least that part of it of which we are consciously aware, represents the fundamental type of vertebrate coordination. Man can learn to engage individual muscles independently, but most animals cannot. This is why the anthropocentric approach to the general problem of coordination is misleading and has failed to produce results of general applicability whenever attempted.

The issue has often been further obscured by ill-defined and unverified generalizations of the concept of "learning." "Learning," that is, an adaptive modification of behavior in response to recurrent stimulus situations, has been demonstrated to occur, at least in traces, in most branches of the animal kingdom, from the lowest forms up. However, the strict constitutional limitations of this learning ability do not seem to be generally realized, or if so, have certainly not received due emphasis. There is agreement that the total motor performance of an animal can be modified by experience, but since the total performance is an integrated act, involving shifting combinations of partial performances of more elementary character, it remains to be demonstrated whether the modification concerns those elementary acts—the building blocks of behavior, as it were—as such, or merely their combination into more complex actions on a higher level. The mere assertion that the response mechanisms of the animal as a whole are not absolutely rigid but provide for some degree of adaptation, does not reveal whether this plasticity extends to all parts of the behavioral mechanism alike or is a privilege of certain components only, and if so, of which. Adaptive behavior presupposes functional reorganization somewhere; but where? Is the whole nervous system one vast pool of equivalent elements whose functional relations can be infinitely varied by experience, or is adaptability confined to some of its divisions or some of its functions only, while the rest are immutable?

The question is no longer whether learning is a common faculty or not. The answer to this has become a matter of course. What we need, is to know precisely what functions are amenable to change and what others are not; further, what functional elements, or groupings of elements, remain constant and unmodifiable even as the behavior pattern of which they form integral parts changes. Behavior results from the activities of a hierarchy of functional levels, each of which may or may not be adaptable. Plasticity on any one level neither implies nor precludes plasticity on any other level; the only means to test their capacities is by way of experiments.

Let us briefly illustrate the various levels.[4] Confining ourselves to metazoans possessing a differentiated nervous system, we may dismiss subcellular entities and start right at the cellular level. There we find as the lowest recognizable elements to which a measure of functional stability may be conceded, the individual *motor units*, the term signifying, according to *Sherrington*, a motor neurone with its attached muscle fibers (level 1).

The orderly contraction of a whole *muscle* is the result of collective action of its constituent motor units, and the characteristics as well as the grading of the resultant contractile effect depend largely on the proportion of active to inactive units, on their rate of alternation, on the time relations of their activation (synchronization or temporal dispersion), on the frequencies of their discharges, and several other factors. In other words, in order that a muscle may function properly, its motor units must be definitely under common control. What happens when motor units act without control and at random, is impressively demonstrated by the functional inefficiency of a muscle in the state of

---

[4] This list is more pertinent than one published on an earlier occasion (*P. Weiss*, 1925).

pathological fibrillation.   The level of integration of the motor units of a given muscle may be designated as level 2.

Next, we must remember that every *joint* or other movable part (e.g., eyeball) is operated not by one, but by several muscles. The relative strength and timing of their contractions determine the direction and speed of the movement and the duration and stability of the resulting position.   Thus the simple muscular actions of level 2 are integrated into orderly functions of muscular complexes relating to a single joint (level 3).

Again, in order to obtain an efficient movement or to maintain a definite posture of a segmented structure, such as a limb or the spine, containing *several joints*, the activities of the various muscle groups of level 3 must be finely correlated among one another, a task which gains in complexity in those cases in which a single muscle spans two joints, moving either one or the other or both, depending on the degree to which it is opposed at the time by other muscles.   Thus, limb movement requires a higher level of integration (level 4) than does single joint action.

On the next higher level (level 5), we find movements of the various locomotor organs (limbs, trunk segments, tail) combined in orderly fashion so as to yield a definite act of the entire *locomotor system*, such as ambulatory progression and regression, jumping, swimming, or the like.   The integration of widely separated muscle groups in the act of breathing (laryngeal, intercostal, abdominal, diaphragmal muscles) is obviously of the same order.

Finally, on the highest level common to all animals (level 6), the various motor acts are put into the service of the *animal as a whole* under the control of the sensory apparatus, whose reports are evaluated by the centers in accordance with the external stimulus situation, the internal state of the animal, the inherited response mechanisms of the centers, and such modifications of the latter as past experience may have brought about.   It is in their bearing on this level that all motor acts of the lower levels gain *biological significance*.   Viewed from this level, "progression" becomes an instrument of "preying," "regression" of "escaping," "eye movements" of "orientation in space," and so forth.

For abbreviated reference to the various levels, we shall use the following symbols:

Level 1, i.e., level of the Neurone............................... N
Level 2, i.e., level of the Muscle................................ M
Level 3, i.e., level of the muscle Group......................... G
Level 4, i.e., level of the Organ................................ O
Level 5, i.e., level of the organ System......................... S
Level 6, i.e., level of the organism as a Whole.................. W

Now, after this survey, let us repeat the question: On which one of these levels does "learning" occur? Possibly on all of them? Or on the highest (W) only? Or the lowest (N) only?

To those who are either unaware of the hierarchical principle of nervous function or may think to have grounds for denying it, these questions must seem utterly senseless. For them there exists no central organization on a level higher than that of the neurone, and as they would describe all behavior merely in terms of connections among individual neurones, so they would naturally be disinclined to conceive of "learning" otherwise than as of a free rearrangement of individual neurone connections. In the terms of our question this amounts to asserting that "learning" occurs on the level N, and exclusively there. There is no room in this concept for stability and unmodifiability of functions of lower order in an organism whose behavior has been demonstrated to be adaptable. It is either plasticity all the way, or rigidity all the way. "Learning," according to this view, proceeds by tentatively engaging, disengaging, and re-engaging independent efferent neurones in varying constellations, letting the biological value of the results for the organism decide which ones of the tried combinations are to be preserved, and which to be discarded. If this were true, the problem for an animal would be the same, whether it faces the necessity to substitute one established pattern of locomotion for another established pattern (e.g., hopping for running; see *Bethe and Fischer*, 1931), or to make modifications *within* a standard pattern (e.g., longer strides in trotting; swimming in circles instead of straight), or to use its healthy legs for unusual tricks, or finally to return limbs with arbitrarily disarranged muscles to their usefulness as instruments of locomotion.   There would seem to be no reason why an animal which can change the rhythm of its several limbs in the acts of locomotion should not be equally adept at changing the rhythm of the several muscles within the limb, if in last analysis it all comes down simply to rearranging neurone linkages.

However, this view is strictly contradicted by the facts.   We have already quoted evidence to show that in amphibians, and even in the rat, the time order according to which muscles execute a limb movement is unalterably fixed, while at the same time the total behavior of these animals is amenable to reconditioning by training and other regulatory adjustments.   In the light of these facts, the distinction between rigid and plastic functional levels assumes great significance, as indeed the neglect of the hierarchical principle would lead, and has led, to serious confusion.   In brief, the fact that an animal can learn (on levels S and W) to use its limbs differently in moving the body, does not necessarily imply that it can likewise learn (on levels O and G) to use its muscles differently in moving a limb.   Adaptive functional reorganization is a prerogative of certain functional levels only.   Therefore, in raising the question of learning separately for each level, we merely give expression to realities.

The preceding pages may suffice to bear out our contention, that progress in the study of coordination has been held up both by lack of restraint in extrapolating from higher mammals to animals in general, and by lack of precision in the application of the principle of learning.   In this state of affairs it is not surprising to find the changes observed after tendon transposition, muscle transplantation, nerve crossing, sectioning of central tracts, destruction of brain portions, and similar interferences, lumped together indiscriminately under the common headings of re-education, regulation, functional restoration, reconstitution, reparation, re-organization, re-adjustment, and the like.   If in the future more discretion will be exercised in the use of these terms and if the mere statement, that a behavioral change has occurred, will be amplified by precise information as to what this change has consisted of and where it has taken place, the gain for our understanding of nervous function will be enormous.   Then, all the mentioned interferences, instead of merely serving to tell whether or not "functional recovery" can occur, become discriminative assay methods, revealing the degree in which the various functional levels participate in the noted "adjustment." It is in this assaying capacity that the transplantation experiments to be discussed below have been used, and since this method invites much wider application, a few comments on general methodology seem appropriate.

Experimental methodology

As all biological experiments, those dealing with the nervous system fall essentially into three classes: *defect* experiments, *isolation* experiments, and *recombination* experiments (compare *P. Weiss*, 1939, p. 147 f.). Given a system Y, consisting of parts A, B, C, D, and so forth, the experiment aims at establishing the relations between the system and its parts, as well as among the constituent parts, by severing the existing relations. Singling out, for instance, part A, the *isolation* experiment determines the properties and capacities of A, when completely released from the rest of the system, while the *defect* experiment, complementary to the former, ascertains the properties and capacities of the remainder of the system (Y minus A). In both cases the relations between the system and the part are permanently interrupted. It is left to the *recombination* experiment to supply the positive part of the story by restoring connection and relations between the severed components, however, with such added variations from the original condition that it will be possible to discern whether and in what respect the relation between the system and part A differs from its relation to parts B, C, etc. Part B is supplanted for part A, and vice versa, and the subsequent conduct of the altered system is studied. If the system behaves as before, we conclude that A and B are equivalent; if it behaves differently, the change is ascribed to the differential between A and B.

In the past the defect experiment has been by far the predominant method in the study of nervous function. The value of the isolation experiment is increasingly appreciated; witness the work on isolated nerve fibers and isolated brain parts. The recombination experiment, however, has been largely neglected. Yet, its instructiveness greatly exceeds that of the defect experiment. In what respect, can be easily shown.

Let us quote an example. We cut a tendon and note subsequently that limb movements are changed. Now, the operation has altered a number of conditions in one stroke: It has caused trauma, produced a gap in the elastic continuity of the tissues, interrupted the transmission of pull from muscle to skeleton, abolished stretch reflexes from the affected muscle, and, as a result, changed the mechanical and innervatory balance of other muscles. What each one of these factors contributes to the common defect, cannot be immediately discerned. Their effects can be separated, however, by resuturing the tendon stumps in various modifications—under slack or shortening; to the old muscle or to an antagonist; with or without concomitant denervation of the muscle;—in other words, by restoring certain, but not all, of the severed relations.

Similarly, peripheral nerve section leads to a complex functional disturbance the net effect of which—impairment of motility—does not reveal its composite nature. The locomotor apparatus suffers changes which are partly due to the trophic and mechanical by-products of muscle denervation in general, partly to the fact that a particular nerve (not just innervation in general) has been lost, partly to compensatory reactions of other muscle centers not directly affected by the operation, with every one of these effects telling on all levels, from the simple movement of a joint up to the aimed behavior of the animal as a whole. We know that all these factors enter into the result, but only systematic recombination experiments, consisting in this case of the replacement of old peripheral nerve connections by new ones, with or without rearrangement of tendons, can help us to disentangle them. Only then can we learn what difference it makes whether a muscle is merely supplied with nerves or actually receives impulses; whether it is just innervated or innervated by one particular nerve rather than another; whether it merely contracts or actually moves the skeleton; whether it operates the skeleton to good use or in a mechanically inefficient manner; whether it contributes to an act of biological significance or one that runs counter to the interests of the organism as a whole.[5]

The rest of this article is essentially a detailed account of how problems of this kind, refractory to other methods, could be solved by recombination experiments. It is hoped that the advantages of the method will become sufficiently evident to encourage its extension to problems not yet hitherto tackled, but entirely within its reach. In view of the fact that endocrinology owes much of its rapid progress to the introduction of routine transplantation methods, it is surprising that neurology has not yet adopted analogous methods to any appreciable extent.

AN ANALYSIS OF COORDINATION IN AMPHIBIA

The principle of myotypic response ("resonance")

Transplantation experiments carried out after the scheme just outlined have led to an astonishing discovery concerning the manner in which the central nervous system controls the musculature. They have revealed a series of phenomena, commonly referred to, somewhat vaguely, as *resonance* phenomena, which have provided us with an assay method of nervous function of much greater discriminatory power than any other method available. Since the results obtained with this method have been amply reported and reviewed on previous occasions, only those points will be recapitulated here which have an immediate bearing on the problem of this paper. For the latest comprehensive review, see *P. Weiss*, 1936b.

Reduced to the simplest formula, the results have shown that muscular control is based on a principle of selective correspond-

---

[5] The lack of distinction in these matters is clearly reflected in the anatomical nomenclature of muscles. Some muscles received their names from some constitutional characteristics, such as shape, (e.g., m. trapezius; m. piriformis), some from their topographic relations (e.g., m. intercostalis; m. subscapularis), some from their skeletal connections (e.g., m. ileo-fibularis; m. coraco-brachialis), some from their kinetic effects (e.g., m. levator scapulae; m. extensor carpi; m. corrugator supercilii), some from their biological function (e.g., m. masseter [from μασσα εσθαι, to chew]; m. risorius, the "laughing" muscle; m. vocalis, the "vocalizing" muscle).

ence between nerve centers and individual muscles, which enables the centers to identify and engage any given muscle by its *name*, irrespective of its mechanical effects, or the biological effects which the latter have for the body as a whole. Each individual muscle owns some distinctive constitutional characteristic through which it is differentially distinguished from all other normal muscles. By virtue of this distinctiveness, in a manner not directly discernible from the experiment, the central nervous system can discriminate muscle from muscle, regardless of where they are attached and how they act. Each muscle is centrally represented by units of corresponding specificity, and these, in turn, are activated by the centers selectively. The basic elements of motor function are specific central "calls," one for each muscle, and each so organized that it will affect just those motor units belonging to the appropriate muscle, and no others. The totality of "calls" at the disposal of the centers represent the code or vocabulary, as it were, of which all central messages are necessarily composed.

All of these statements contain no hypothetical implications, but represent merely the net result of a large series of experimentally established facts. Presented in a nut shell, the underlying facts are the following.

In amphibians[6] it is possible to graft an extra muscle, or group of muscles, or even a whole limb, to the body wall and provide motor innervation for these transplants by diverting to them some motor nerve branch from one of the normal host limbs. The amount of deviated nerve fibers can be held to such insignificant proportions that no perceptible change in the function of the host limb results. This small nerve source is fully adequate to assure complete re-innervation of the transplant, inasmuch as nerve fibers in the course of regeneration undergo profuse branching. The experiments were so devised as to insure that the muscles of the transplants would be re-innervated for the most part or wholly by nerves with which they formerly had no relations.

---

[6] All statements in this chapter relate only to young amphibians, for which direct experimental proof is at hand, while the possibility of extending the results to other groups will be discussed later.

After transmissive connections between the regenerated nerve fibers and the grafted muscles had been restored, the supernumerary muscles began to exhibit regular and strong contractions whenever the host limb, from which their nerve supply was derived, moved.

The stress, however, lies not so much on the fact that the transplanted muscles had become re-engaged in functional activity, but on the peculiar time order in which they were found to operate. Extensive studies, under a great variety of conditions, of the precise times when a transplanted muscle starts to contract and ceases to contract, as well as of the degree of its contractions during that active period, has revealed a principle of such definiteness and constancy that it amounts to a law. *The phases of activity of an extra muscle correspond precisely to the phases in which the muscle of the same name, or synonymous muscle, is found to be active in the host limb innervated from the same plexus of the spinal cord.* Whether the transplant consists of a single muscle, or a group of muscles, or a whole limb, each individual muscle as such duplicates the action of the synonymous muscle in the normal limb nearby. This phenomenon has been described, and ever since been referred to, as *"homologous response"* of *synonymous muscles.* The term signifies that if a body district is provided, instead of with a single muscle of a given kind, with two, three, or even four homologous muscles of the same kind and name, all of them will act in unison, the contractions beginning at the same moment, developing the same proportional tension and subsiding at the same time; the only prerequisites being, that all of them receive their nerve supply from the same side and the same general level of the spinal cord (e.g., limb level in the case of limb muscles), and that the transplantation be done in young, preferably premetamorphic, animals.

Now, what does this phenomenon of "homologous response" actually mean? In spite of the ample attention given to it in the past, it does not seem that the majority of authors have succeeded in seeing it in the correct light. Many authors, while reporting the phenomenon correctly, yet have missed the point which it so clearly proves, namely, the existence of correspondences of a specific kind between nerve centers and individual muscles. Thus, the phenomenon was variously described as demonstrating almost unlimited "plasticity of coordination," "learning capacity" even at the lowest level, "adjustments" of the nervous system to the introduction of a new organ, integrative action of the spinal cord, and so forth. Some of these interpretations are strictly incorrect, others merely besides the point. It would be idle to try to fix the blame for these misinterpretations. Part of it can probably be ascribed to lack of clarity in the earlier descriptions of the phenomenon, as well as to misleading terminology, part to the fact that the problem of coordination, to which the phenomenon offered a clue, was not usually presented in the correct light. It thus becomes necessary once more to explain the intrinsic meaning of this phenomenon of "homologous response" of synonymous muscles.

To begin with, to emphasize the fact that the transplanted muscle and the synonymous muscle act in unison, is already putting the wrong slant on the phenomenon, because in stressing the association of the two peripheral parts, we give prominence to a rather irrelevant aspect. We make it appear as if the transplanted muscles, or rather their centers, had in some way learned to imitate the synonymous normal muscles; even though it would be difficult to find any plausible reasons why they should have done so. To avoid this misconception, it must be stressed that in all these experiments the normal muscles simply serve as *indicators* of the hidden activities of the central nervous system, and that their actual presence is in no way required for the appearance of the phenomenon. Even if all muscles of the normal limb are removed, the transplanted muscle keeps on functioning at precisely the phases when the removed synonymous muscle would have functioned if it were still present. But so long as a normal limb is available, we use it as a detector to tell us which combination of muscles the central nervous system tends to activate in any given phase of a locomotor act.

Let us now forget for a while that effective limb movements are of service to the animal, and let us consider them simply in their service to the observer as convenient instruments for the visible registration of the content of the varying central commands. To be used for this purpose, movements must be resolved into muscle actions. This can be done directly or indirectly—directly, by connecting the muscles with mechanical or electrical recording devices, as is commonly done in the study of simple reflexes with fairly constant stimulus-response relations; indirectly, by taking cinematographic records of the movements and reconstructing the muscular activity from the measurable changes in the angles of the various joints. The former method, besides being limited to animals above a certain size, has the disadvantage of interfering with the free execution of the movements, while the latter falls short in two respects: firstly, it fails to register isometric contractions, which produce tensions to overcome resistance without producing excursion of the joints, and secondly, the individual muscular contributions to the movement, instead of self-registering, are only indirectly recoverable from the record. In the work on homologous response both kymographic registration and slow-motion picture analysis have been employed. However, experience has shown that the former method is by far less suitable in the study of coordination, because the technical measures necessary for direct muscle registration (anaesthesia, strapping, fixing of joints, tendon dissection, etc.) interfere with the execution of most of the regular locomotor repertoire of the animal to such a degree that little insight into the normal, unrestrained, performance can be gained.

By slow-motion picture analysis, a complex movement can be resolved into the constituent muscular actions: the sequence in which different muscles become dominant and the duration of their phases of activity can be determined. We thereby obtain a time record—a chronological "*score*," as it were,—of the central activities through which the various muscles are engaged. Thus, when the forearm is bent against the upper arm, we take this to signify that the central nervous system has set into action chiefly that group of neurones which innervates the biceps muscle. As the flexion slows down and finally reverts into extension, we interpret this as central activation of the triceps neurones. Doing this for all muscles involved in a given movement, we compose a master chart of the central timing mechanism effecting the movement. Such a master record of coordination may be compared to the score of a piece of orchestral music in which onset, intensity, and duration of each instrumental part are recorded separately, except that in the case of our muscular orchestra no instrument can produce more than a single tone: contraction of the particular muscle. It is herewith proposed to call this master time record of all muscles participating in a given movement, the *myochronogram*.[7]

As a concrete example on which to carry on our further discussion, we reproduce here the myochronograms of the fore limbs of a salamander in the act of walking on solid ground (fig. 4). For the sake of simplicity, only the shoulder and elbow joints are included, while the wrist has been ignored. Furthermore, only six major muscles have been selected as representatives. These muscles are shown in a dorsal view of the left fore limb in the inset (upper left of the figure). The shoulder muscles chosen are the same as those of figure 2, namely, an elevator ($\triangle$), an abductor ($\bigcirc$), a depressor ($\times$), and an adductor ($\bullet$). The upper arm muscles are represented by a flexor ($\square$) and an extensor ($\blacksquare$) of the elbow. The two central strips of the picture reproduce in diagrammatic outline six phases of one complete walking cycle, both in dorsal and rear views. Double rings indicate the fixed points on the ground in which the animal sets down its wrist and around which as pivotal points it swings the body forward. The muscles in action are marked by their respective symbols.

In phase 1, the right hand takes hold on the ground ($\circledcirc$), and presently (phases 2 and 3) depression ($\times$), adduction ($\bullet$), and extension ($\blacksquare$) of the right arm swing the animal forward, while

---

[7] Although we have used myochronograms as the simplest method to represent movements for years, none were published until 1937 (*P. Weiss*, 1937d). Coincidentally, it was then found that a Swiss surgeon, *Scherb* (1938), had been using the same method to symbolize movements under the name of "myokinesiogram." If we give preference to the term "myochronogram," it is only beause "*kinesiogram*" means record of motion, whereas "*chronogram*" simply means "time record," thus providing for the inclusion of non-motile, isometric, muscle contractions.

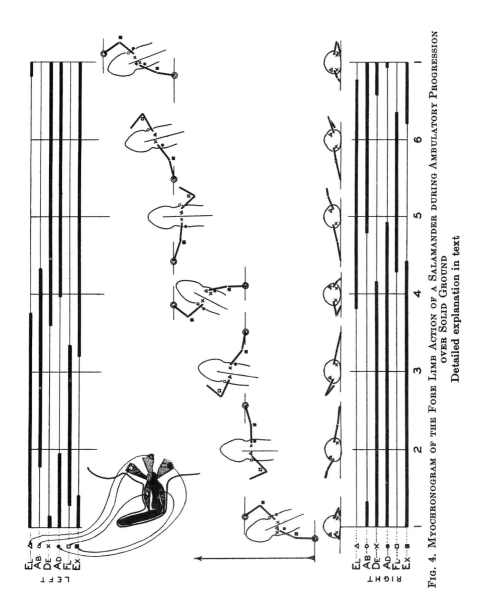

FIG. 4. MYOCHRONOGRAM OF THE FORE LIMB ACTION OF A SALAMANDER DURING AMBULATORY PROGRESSION OVER SOLID GROUND

Detailed explanation in text

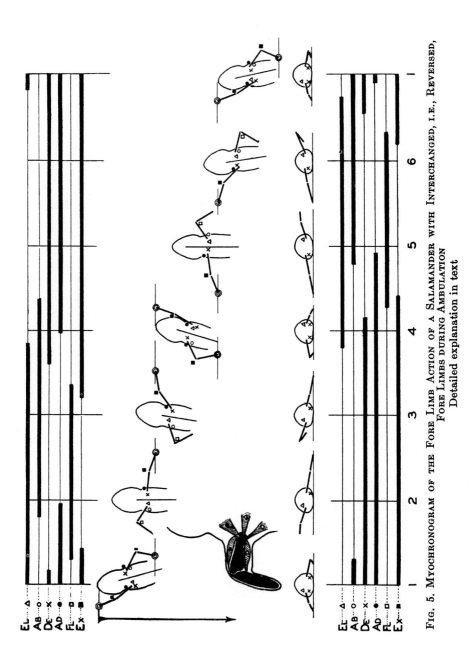

FIG. 5. MYOCHRONOGRAM OF THE FORE LIMB ACTION OF A SALAMANDER WITH INTERCHANGED, I.E., REVERSED, FORE LIMBS DURING AMBULATION
Detailed explanation in text

at the same time the left arm is lifted ($\triangle$) and brought forward ($\bigcirc$, $\square$). At point 4, the left hand is then set down and serves as pivot around which the body is swung forward through phases 5 and 6, while the right arm in turn is lifted from the ground and brought forward. The top and bottom records represent the myochronograms of the left and right fore limbs, respectively, as each goes through the illustrated phases. In order to be truly representative of the resulting movements, these records should include an account of the varying intensities of the muscular contractions. Since these cannot be directly observed, we confine ourselves to the score of time relations.[8]

With sufficient practice it might be possible in reading a myochronogram to visualize the resulting movement just as an orchestra conductor reads a musical score. To determine how much variation in detail there is in the play of muscles in moving a limb, would require much more intimate studies. However, all observations thus far concur in demonstrating that the essential chronology expressed in the above myochronograms is typical. Walking is always effected by the same general sequence of muscular activities in all animals of the same species. In other words, the muscular integration on the level O of our hierarchical scale (p. 24) follows a stable chronological pattern.

Compared with this firmly set pattern for a given limb, the association between left and right fore limb, or that between fore limbs and hind limbs is much looser. That is to say, while usually in undisturbed and vigorous walking the alternation between left and right limb is as strict as is indicated in the diagram, amounting to a shift of the lower myochronogram against the upper one of just one half phase of the whole cycle, there is also frequently independent action of the two limbs, either one moving without the other, or both moving in phase, or even both entirely out of turn. Similarly, in vigorous walking there is usually an intimate correlation between the movements of hind limbs and

[8] The overlap between antagonistic muscles in these diagrams has not been directly observed, but put into the records in accordance with the work of *Wachholder* (1923), showing that the contraction of antagonistic muscles actually sets in some time before the reversal of the movement becomes visible.

V. On the Dynamics of the Nervous System

fore limbs in that the adduction phases, and likewise the abduction phases, of diagonal limbs coincide.   However, at other times, the hind limbs operate independently of the fore limbs, and dissociation between the two pairs of limbs occurs even more commonly than between the two partners of the same pair.   Simple observation thus demonstrates that integration on the level S of the hierarchical scale is subject to much greater variation than that found on the lower level O.

Now, let us return to the animals with supernumerary limbs. The movements of their normal limbs serve us to construct the central myochronograms, that is, to reveal what muscles are centrally being called up for action at any given moment.   Suppose we examine an animal with an extra limb attached to the left fore limb plexus.   The myochronogram informs us that in phase 3 the left fore limb centers discharge impulses destined to engage the elevator, the abductor, and the flexor of the elbow. Watching the transplanted limb, we note that in it, too, out of the whole extra set of muscles, just the elevator, the abductor, and the flexor respond, i.e., precisely those muscles provided for in the central score.   Since this holds for all muscles and at all times, we must conclude that the centers, in a sense, "call up" the individual muscles by their names; further, that when the name of one muscle is called, all muscles of that name react, which implies that each muscle is endowed with some peculiar property enabling it to respond to the calls of its own kind selectively.   Thus, the "homologous" response of supernumerary muscles signifies a selective correspondence between central impulses and peripheral effectors rather than a tie-up between synonymous muscles as such.

Obviously, the designation of the phenomenon as "homologous response" has been misleading in that it places the emphasis on a technical rather than on an essential feature.   For this reason, it would seem more to the point to speak of a principle of "*myotypic response*," which means "muscle-specific response."   This change of terminology is herewith proposed.

The factual content of the phenomenon of myotypic response can be reduced in essence to two points: (1) The protoplasm of each individual muscle has a specific and distinctive constitution, distinguishing it from all other muscles. (2) This constitutional specificity is instrumental in establishing a selective relation between the centers and any muscle of that particular kind of specificity.

This formulation merely expresses logical conclusions to be drawn from the observed facts. It contains no reference whatever as to just how those specific relations between muscles and centers are effected. If we are to consider this latter question, we find ourselves no longer on the same solid ground as before. Here the experiments fail us. While they have set up a definite frame within which any explanation of central-peripheral correspondence must hold itself, they carry no further positive suggestions. The tentative explanation currently favored and presented in the following pages should, therefore, be considered as entirely hypothetical. Whether or not it will ultimately prove to be correct, does not affect the validity of the principle itself. The reality of the myotypic principle remains a fact, its mechanism a matter of further research.

Three possible mechanisms suggest themselves.

(1) One might assume that there are as many specifically different types of motor cells in the spinal cord as the corresponding peripheral district contains specific individualized muscles, and that in development, as well as in regeneration, each muscle receives exclusive and selective innervation by fibers of the corresponding type. This would presuppose either selective attraction by the muscle of its appropriate kind of nerve fibers from a distance, or some scheme by which nerve fibers, though growing out indiscriminately, would be admitted for functional connection only into those muscles which are of the corresponding type. As one can readily see, this assumption would match the qualitative differential among muscles by a corresponding inherited qualitative differential among the central cells. It would further postulate an unfailing capacity of the two predestined partners of each nerve-muscle pair to find each other. The possibility that such strict selectivity might prevail in the innervation and re-innervation of muscles has been definitely ruled out by ex-

tensive studies on nerve regeneration in general, and specifically in reference to the experiments on supernumerary muscles and limbs (*P. Weiss*, 1937b). We know for sure that regenerating nerve fibers coming from whatever source can form functional connections with whatever kind of muscle. Thus, any explanation of myotypic response on the basis of selective fiber regeneration can be definitely excluded. We, therefore, pass on to a second possibility.

(2) The excitation produced in the centers might be of composite character, that is, have a different pattern depending on which muscles it is to set into action. The resultant excitation would be dispatched over all motor fibers of the district, delivered to all the muscles and there be analyzed by the end organs, each one picking out its proper component. This interpretation of myotypic response, tentatively introduced in the early reports on the phenomenon (*P. Weiss*, 1924) had to be later abandoned as not consonant with the subsequent development of nerve physiology, in that it has since been fairly securely established that no excitations are electrically demonstrable in motoneurons unless the muscle, too, is active (see *Wiersma*, 1931). In other words, the motor impulses are already assorted before they leave the centers, and the idea of an analysis of the impulse pattern by the peripheral end-organs can no longer be entertained. It is somewhat disconcerting to see that some authors, ignoring the explicit renunciation of this idea (*P. Weiss*, 1934, 1935a, 1936b), still insist on presenting it as the official version, as it were, of the resonance principle. At present, the view best in agreement with the facts, although not entirely free from difficulties, is the following.

(3) Each muscle exerts, by virtue of its individual protoplasmic specificity, a correspondingly specific influence upon its motor nerve fibers. As a result, the nerves acquire specific differentials which match and centrally represent the variety of muscles. The biceps muscle, for instance, would gradually transform any nerve fibers connecting with it into strictly biceps-specific fibers; that is, it would impress upon them some biochemical tag through which they become centrally recognizable as belonging to the

biceps muscle. This hypothetical process of appropriation of the motoneurons by their muscles has been called *"modulation"* (*P. Weiss*, 1936b).

According to this view, a nerve fiber which has been severed from its erstwhile connections and switched to another muscle, would lose its former specificity and acquire the new one. However, it seems that nerve fibers, as they grow older, lose their plasticity and become irreversibly ingrained with the specificity of the muscle with which they had been connected up to then. This is the reason why myotypic response of transplanted muscles is not observed unless the transplantation is performed below a certain critical age of the animal. If a nerve is switched to a new muscle after that critical period has passed, the muscle will always contract in the phases in which the old muscle, which used to be at the end of this nerve, would have contracted. In other words, the response has become *neurotypic* (nerve-specific), the nerve specificity being the residue of muscle-specific effects received during an earlier phase of life. A *young* nerve, which after connecting with a biceps muscle had become biceps-specific, can still be transposed to a triceps muscle and assume triceps-specificity, thus changing its central tune. However, after having been under prolonged influence of its original biceps contact, transfer to the triceps has no longer any retuning effect, and the nerve responds centrally as if the muscle at the end were still a biceps.

That the muscular specificity is conveyed to the spinal centers through "modulation," i.e., an ascending process extending over the motor neurones directly, rather than through the mediation of sensory fibers, is proved, apart from other evidence, by the fact that it occurs just the same after the sensory innervation has been eradicated (*P. Weiss*, 1937c). It seems reasonable to assume that modulation embraces the whole motor neurone, including the ganglion cell. But whether it extends farther centrally, and how far, remains to be determined.

Modulation itself being largely hypothetical, it would be idle to indulge in speculations about its nature. Merely as a matter of personal preference, I like to view it as belonging to the class of specific molecular adaptations of which the immunological reactions are the most familiar example.   Just as an antigen calls forth in the cell the production of a correspondingly shaped antibody, so the specific biochemical factor of each muscle may provoke in the nerve protoplasm complementary molecular configurations, beginning at the myo-neural junction and from there spreading through the neurone much in the fashion of virus reproduction.   The molecules thus molded may be essential links in the chemism of the "propagated disturbance," or they may be merely indirectly related to it.   In either case, their biophysical and biochemical parameters would determine the selectivity of the response of the whole unit to the biophysical and biochemical activities of its central environment.   Evidently, this is one way to explain the fact that the central response of each motoneurone is selective in accordance with the identification mark of its muscle.   However, there is no evidence that this hypothesis comes anywhere near the truth, and we mention it only to prove that a rational explanation of modulation is not beyond reach.

The process of modulation sheds no immediate light on the central mechanism of coordination.   Modulation merely furnishes the centers with necessary clues without which central coordination could not become peripherally effective.   For if central coordination, as we now realize, operates in terms of individual muscular specificities, the centers need means to identify the various muscles.   Modulation provides them with such means of recognition.   However, the fact that the motor ganglion cells, after undergoing modulation, represent an assortment of elements with distinctive characteristics, in no way explains the central mechanism through which these elements are being activated in the varying combinations provided for in the central "scores" of coordination.   Is it that dendritic links develop among ganglion cells of the same "tune" through which all cells belonging to the same muscle would become connected into a unitary system responding to impulse patterns from higher stations as an entity; or is the association based on purely physiological affinities of a sort that would force all those elements which have the same selective specificity to respond in unison to a common activating agent of corresponding specificity? This is a problem which furnishes much food for speculation but little hope of solution in our present state of knowledge.

It seems preferable, therefore, to refrain in this place from a further consideration of these hypothetical matters and to concentrate on the consolidated factual content of the principle of myotypic response, which is that *the central nervous system, in dealing with the musculature, utilizes specific means through which each muscle can be called into action independently, in accordance with its individual constitution.*

Myotypic response as assay method

Having established the general validity within known limits of the principle of myotypic response, we may pass on and use it to assay central coordination. How, will become clear from the following example.

Let us consider what is conventionally described as a *flexion reflex*. A stimulus applied to the toes results in the withdrawal of the foot and leg. This reaction occurs essentially on the level O of the hierarchical scale, implying operations on the subordinate levels G, M, and N. In terms of the top level (W) it can be rated as part of an escape reaction from a harmful stimulus. In terms of the organ level (O) it means approximation between base and tip of the limb. In terms of the muscle group level (G) it amounts to reducing the angles of the ankle and knee joints, commonly designated as "flexion"; and on the level of the individual muscle (M), it simply means contraction of those muscles which happen to be inserted on the flexor sides of the joints (e.g., hamstring at the knee; tibialis at the ankle). Our problem is to decide in terms of which of these levels the coordination of the withdrawal reflex is laid down.

It is here that the assay function of a transplanted muscle can prove its value. For we can transplant a flexor muscle in such a fashion that it will have the mechanical effect of extending the joint instead of flexing it, as it did before. Thereby we alter the relation between the M level and all higher levels. When a "flexion" reflex is now elicited, will the response still be "flexion," or will it be a contraction of what used to be the "flexor" muscle now producing extension? Or we can change the insertion and orientation of a whole limb with regard to the body in such a manner that, while the muscles will continue to produce the normal kinetic effects within the limb, the net result of the limb action for the body as a whole will become quite different from what it was before. Thereby we upset the relation between level O and the higher levels S and W. Will coordination patterns within the limb thereafter remain as they were before, ·perpetually in discord with the needs of the body, or will they be remodeled and re-integrated with the levels S and W so as to restore harmonious operation of the whole?

Applying these experimental tests, it was found that a stimulus which normally yields a "flexion" reflex will invariably lead to a contraction of the hamstring muscles and the tibialis group, no matter whether the resulting movement comes actually out as flexion, or, owing to transposition of the muscles, as extension, rotation, or any other joint excursion, irrespective also of whether or not the resulting flexion or extension, as the case may be, leads to an effect which can be considered adequate from the standpoint of the body.

Now, let us go one step farther. Let us change the nerve supply of a "flexor" muscle, either one that still flexes, or one that has been transposed to the extensor side, by substituting an "extensor" nerve for the original "flexor" nerve. As we have outlined before, the result will vary with the age of the animal at the time of operation. Late operations will lead to neurotypic response; that is, the muscle will contract during "extensor" phases only and, hence, not take part in the "flexion" reflex (cf. *Sperry*, 1941). Early operations, however, will result in continued myotypic response, that is, a "flexion" reflex will bring in the "flexor" muscle even though it is now innervated from an "extensor" nerve and may have been switched over to the extensor side so as actually to produce extension. Whatever we do to it, the muscle with the "flexor" constitution will be the one to respond in the so-called "flexion" reflex (see table 2).

Such being the situation, it would seem much more to the point to speak of a "tibialis-semitendinosus-semimembranosus" reflex, rather than of a "flexion" reflex, and to describe the "flexion" reflex about as follows: A stimulation of sensory fibers from the skin of a toe sets off a central discharge pattern, which selectively engages all motor neurones, however much scattered over the central district, which bear the specific "tibialis," "semitendinosus," and "semimembranosus" tags previously acquired from

TABLE 2

Effect of a "flexion" reflex on a "flexor" and "extensor" muscle before and after tendon crossing or nerve crossing or both

OPERATION	OPERATED MUSCLE	INSERTION		INNERVATION		REFLEX EFFECT	
		Left on	Transferred to	Original	Transposed	Contracting muscle	Kinetic effect
Control	Flexor	Flexor side		Flexor nerve		*Flexor*	Flexion
	Extensor	Extensor side		Extensor nerve			
Tendon crossing	Flexor		Extensor side	Flexor nerve		*Flexor*	Extension
	Extensor		Flexor side	Extensor nerve			
Nerve crossing*	Flexor	Flexor side			Extensor nerve	*Flexor*	Flexion
	Extensor	Extensor side			Flexor nerve		
Tendon and nerve crossing*	Flexor		Extensor side		Extensor nerve	*Flexor*	Extension
	Extensor		Flexor side		Flexor nerve		

* Nerve crossing prior to loss of ability of re-modulation.

their respective muscles, with no regard to the actual kinetic and biological effects of the resulting contractions. The fact that the contraction of the hamstrings produces flexion, which has given the reflex its name, is, physiologically speaking, pure coincidence; fortunate from the standpoint of the animal and, of course, fixed by virtue of that very fact during the phylogenetic evolution of the species, but entirely dependent on the skeletal attachments of the muscles being and remaining what they are. If we disrupt this anatomical wisdom, we note no tendency of the centers to maintain the integrity of the response in terms of its effect ("flexion," "withdrawal"), but a blind continuance of the inherited central impulse scheme, delivered in terms of muscle-specific calls, in spite of the adversity or, at best, indifference to the individual of the resulting effects.

We have chosen a reflex as our first example because reflexes are usually conceded to be sufficiently rigid to fit into this picture. Therefore, the statements of these last pages do not exact much revision of current thinking, except in so far as they show that the response called for in a given reflex is not due to firmly set central connections, but that the nerves are conditioned for their response by their muscles. All the other conclusions could have been reached without knowing about the myotypic principle. It is only on the level of the more complex motor activities that uncertainties arise which it might not have been possible to clear up without the aid of the myotypic test.

A transplanted supernumerary limb can be of no use to the body unless possibly in the very special case where it has been inserted exactly in the same orientation as the near-by normal limb so that the pair can execute parallel action. In all other cases the actions of the transplant are sheer waste from the standpoint of the body. Conditions can even be created in which the action of the transplant is distinctly harmful in that it counteracts the normal limb (*P. Weiss*, 1937a). No adjustment or elimination of the wasteful action has ever been observed. It was suggested by *Bethe and Fischer* (1931, p. 1119) that the disturbance caused to the animal by the extra limb might not have been sufficiently vital for the centers to do something about it. It was argued that so long as the host limbs could continue in their normal function, the incentive to change the functional pattern might not have been strong enough. This criticism, however, has been invalidated by later experiments in which the nuisance value of the transplant was so aggravated that it created a serious predicament for the animal. Since these experiments illuminate the problem of coordination most clearly, we shall recount them here briefly, adding a number of comments that were not contained in the original publication (*P. Weiss*, 1937d).

Unmodifiability of locomotor scores

In larval salamanders possessing developed and functional limbs, the two fore limbs were mutually exchanged under preservation of their original dorsoventral orientation. Since the two limbs are mirror images of each other, this operation amounts to replacing one limb by another limb which has the same assortment of muscles but in exactly the reverse arrangement. A comparison between the insets of figures 4 and 5 explains the situation. Of the six muscles which represent the limb in our myochronograms only the elevator ($\triangle$) and the depressor ($\times$) have retained their normal positions relative to the body, while the adductor and abductor of the shoulder, and likewise the extensor and flexor of the elbow, have traded places. Adductor ($\bullet$) and extensor ($\blacksquare$) now lie at the anterior instead of the posterior border of the limb, and abductor ($\bigcirc$) and flexor ($\square$) lie on the posterior instead of the anterior side.

After being re-innervated by regenerating nerves, these limbs resume function. The characteristics of this function are outlined in figure 5, in which six phases of a full walking cycle have been reproduced diagrammatically. Strips of the moving pictures from which these diagrams were reconstructed have been reproduced previously, and the reader may be referred to the earlier publication (*P. Weiss*, 1937d) for further details. The functional effect of the anteroposterior reversal of the whole muscle apparatus was so obvious that it seems hardly necessary to add much to the story, as it unfolds itself in a comparison between figures 4 and 5 (pp. 34, 35).

All movements of the trunk, hind legs, and other parts which have been left untouched by the operation, are identical with those of a normal animal in the act of progression. Hence, we can use these normal parts to identify the successive phases of locomotion and to line them up with the corresponding phases of the normal animal in the diagram. This being done, we realize immediately that the positions of the transposed fore limb (fig. 5) and those of the normal fore limb (fig. 4) are precise mirror images for each corresponding phase of the body movement. If we resolve the movement again into its component muscular contractions, we note that at any one moment the combination of muscles active in the reversed limb is identically the same as the one that would be active at that particular moment if the limb were a normal unreversed one. Using the myochronogram as index of the central impulse pattern, we thus learn that the centers have continued to call up the individual muscles in the same rhythm, sequence and intensity as they had done when they were still operating normal legs with unreversed musculature. In doing this, however, they lead to peripheral effects which are exactly the opposite from what would serve the organism: instead of progression, they produce regression. This is explained in the diagram (fig. 5).

In phase 1, the reversed limb on the right side has taken hold on the ground. During phases 2 and 3 the extensor (■) and adductor (●) muscles contract— the same muscles which are active during phases 2 and 3 in the normal animal (fig. 4). This swings the body backwards (see arrow), while the arm on the left side reaches backward owing to the contraction of its elevator (△), flexor (□) and abductor (○). In phase 4, this free arm, in turn, takes hold on the ground, and the following contractions of its adductor (●) and extensor (■) bring the body still further backward through phases 4, 5, and 6. Thus, the muscles of the reversed limbs, while going through precisely the same cycle of innervation which their synonymous muscles would go through in the normal limbs, move the body backward instead of forward.

Actual regression occurs only if other means of progression, such as the tail and hind limbs, have been removed or paralyzed. If the hind limbs are present, however, the resultant effect is a constant struggle between the hind limbs and the fore limbs, the former striving to advance the body and the latter cancelling the effect by moving the body backwards by the same amount. The net result is that the animal swings back and forth without ever moving from the spot. It is almost pathetic to see how helpless the animals are about their predicament, and although some of them have been kept for more than a year, long beyond metamorphosis, their behavior has never changed.

From these results it must be concluded that the centers operate in terms of individual muscle calls which are combined into definite groupings so pre-arranged as to yield suitable effects in an animal with normal distribution and normal attachments of its muscles, and that the centers continue to operate according to the old scheme even when the peripheral anatomy is no longer normal and the central design no longer yields the desired peripheral effect. Paraphrasing the situation, one might say that the centers continue to act under the illusion that they are still operating a normal limb with consequential results.

In these experiments, the objection certainly no longer holds that the disturbance of behavior was not sufficiently crucial·for the animal to attempt an adaptive change. If they could not improve under these conditions, then they surely cannot under any circumstances. There is one possible objection, however, that deserves some consideration. It might be submitted that the anatomical conditions in the reversed legs might have been such as to preclude, for purely mechanical reasons, an effective participation of these limbs in forward locomotion, in which case even the highest power of central re-education would have been able to do no good. To this one could simply reply that in that case the animal might, at least, have learned to suppress the activities of the useless fore limbs altogether, which undoubtedly would have been of some help. However, there is more pertinent evidence on hand to invalidate the mechanical argument. It lies in the observation that animals with reversed fore legs can actually exert forward traction through these limbs. This happens whenever the animal as a whole tends to retreat.

Figure 6 explains the case. The upper half of the picture shows the fore limb coordination of a normal animal which tries to recede from a repulsive stimulus; for instance, strong ammonia vapor, or a moving object of threateningly large dimensions. The essential mechanism consists of adducting ($\bullet$) and flexing ($\square$) the elevated ($\triangle$) arm, then setting it down on the ground ($\times$), and finally extending ($\blacksquare$) and adducting ($\bigcirc$) it, with the result that the body is thrust backward. Usually the right and the left arm alternate, as is shown in the myochronograms at the top and bottom of the figure. A comparison of these myochronograms with the records of forward locomotion in figure 4 reveals that the essential difference between the two types of movement is the change in the phases in which the abductor and adductor muscles

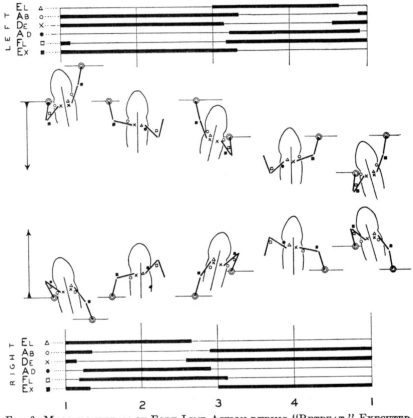

FIG. 6. MYOCHRONOGRAM OF FORE LIMB ACTION DURING "RETREAT," EXECUTED BY NORMAL SALAMANDER (UPPER CENTER) AND SALAMANDER WITH REVERSED FORE LIMBS (LOWER CENTER)
For details, see text

come in. In both movements the elevator and flexor muscle, and likewise the depressor and extensor, operate approximately in phase, while the abductor works with the former group in the case of progression and with the latter group in the case of retreat, the reciprocal holding for the adductor. There are other minor differences, but this is the most conspicuous one. Incidentally, this alternative association of the shoulder muscles with either one or the other elbow muscle group is a good illustration of the diversity of muscle combinations possible on level O.

Now, if the myochronogram of retreat is projected into the musculature of an animal with reversed forelegs, a movement results such as the one illustrated in the lower half of the center strip of figure 6. As one can see, the effect is that the body is being pushed forward. This the animals have actually been seen to do in the face of a repelling stimulus, and, biologically speaking, the result is as absurd, if not even more so, as in the case of forward locomotion: In their attempts to recede, they bring themselves closer and closer to the stimulus which they tend to avoid. For us the observation proves that even reversed fore limbs can efficiently contribute to forward locomotion of the body if only their muscles are activated in a time pattern appropriate for the purpose. In purely anatomical regards, the reversed fore legs are, therefore, as adequate for forward as for backward motion; hence their persistent failure to cooperate in the total locomotion of the body cannot be ascribed to mechanical incompetence.

A comparison of the myochronograms of ambulation and retreat at the same time permits us to define precisely what changes in the pattern of locomotion would have been necessary in order that the animals with reversed fore limbs might have learned to employ their limbs more judiciously. There are two ways in which the functional incongruity between the normal hind limbs and the reversed fore limbs could have been removed. One would have been to make the adductor and abductor phases trade places in the central time score of progression at the fore limb level, and the other would have been to combine the hind limb fraction of the time score of progression with the fore limb fraction of the time score of retreat. Both changes would have led to essentially the same net results, namely, a transformation at the fore limb level exclusively of the myochronogram of figure 5 into the myochronogram of figure 6. This would have restored harmony between the fore and hind limbs in that it would have enabled the fore limbs, too, to take part in body propulsion (lower row of fig. 6).   In other words, it would not even have been necessary to rebuild the whole locomotor pattern de novo.   Most of it could have been left unaltered, with a simple shift of the abductor innervation from the elevator-flexor phase to the depressor-extensor phase, and conversely, of the adductor from the depressor-extensor phase into the elevator-flexor phase.   This would have involved the time schedule of one muscle pair only.   Or the retreat pattern might have been divided into its fore limb and hind limb parts, and the fore limb part alone substituted for the part normally assigned to the fore limbs in progression.

These would seem to be relatively minor changes, and if the amphibian central nervous system had any tendency and power to take into account and to repair inadequate peripheral results, the emergency of the reversed fore limbs should have proved to be a minor problem.   As it is, however, it proved to be insurmountable, and neither were corrections effected nor any tendencies at correction, however abortive, ever observed.

The conclusions to be drawn from the reported results are the following: *The chronological scores*, according to which muscles are called into action when a limb is supposed to move *are rigidly fixed*.   The centers contain a definite repertoire of such fixed and discrete scores; for instance, one for ambulation, one for retreat, one for swimming, one for righting, one for turning, and so forth, each of which can be displayed only as a whole or not at all. The nervous system cannot recombine for simultaneous execution parts of one score with parts of another score, nor can it alter the sequence and associations among the individual muscles within a given score.   In other words, coordination patterns from level S down are ingrained in the centers and are not "effect-determined."

Basic coordination is thus revealed to deal exclusively with the central representatives of muscles, regardless of what effects these dealings will entail.   So far as the basic scores are concerned, the muscles might be non-existent.   Amphibian coordination operates "blindly," reeling off available central scores evoked by the stimulus situation.   In fact, it can be predicted that they would continue to do so even after the interruption of all motor nerves, or the amputation of all limbs. If it were technically feasible to dissect each muscle free, fully protecting its nerve supply, and then to attach them individually to writing levers,[9] we should expect to obtain a myochronogram which would in its major lines coincide with the myochronogram reconstructed from the muscle play in a smoothly moving normal limb with all muscles in place. In fact, the result should not be essentially different if we cut all nerves and registered oscillographically the activity of all central stumps. Pieced together, the records should again present the myochronogram of one definite movement or another. This would be true not only of type reflexes, for which our statement is not likely to be questioned because it refers to a standard practice in reflex registration, but also for the much more complex and highly coordinated movements which form the locomotor repertoire of the species.

It will be noted that in this description a possible determining influence of sensory innervation has been completely left out of consideration. This is fully justified by the facts. While a more detailed discussion of this problem will be presented below, we may already in this place point to the fact that in amphibians the basic patterns of coordination are not disturbed by the radical removal of sensory innervation from the muscles executing those patterns.

THE CENTRAL SELF-DIFFERENTIATION OF COORDINATION PATTERNS

Reversed locomotion after pre-functional limb reversal

All experiments thus far reported were done in animals which had already exercised locomotor function for some time prior to the operation, i.e., during a period in which they had been anatomically normal. Thus, while the results have definitely shown that coordination patterns, once developed, remain firmly ingrained in the centers, they have not eliminated the possibility of a constructive influence of experience on these patterns during the embryonic and early larval phase.

[9] This feat was once actually attempted by *Warren Lombard*, but with little practical success.

V. On the Dynamics of the Nervous System   537

Coghill (1929) has emphasized the autonomy of the early motile patterns of the amphibian embryo.   His view is corroborated by the fact that embryos reared under narcosis develop the ability for coordinated swimming (*Harrison*, 1904; *Carmichael*, 1926; *Matthews and Detwiler*, 1926).   However, it cannot be overlooked that none of these observations have been extended beyond the early phases, in which motility is confined to orderly mass movements of the trunk, much simpler and less differentiated than the limb movements with which we are concerned.   One might, therefore, concede autonomy of origin to the former and still deny it for the latter.   This has, in fact, been done.

Some experiments were reported which seemed to indicate that the development of coordinated limb movements depends upon the presence of sensory innervation in the limbs (*Nicholas and Barron*, 1935; *Detwiler and Vandyke*, 1934; *Chase*, 1940).   This could be interpreted to mean that the central patterns of motor coordination are gradually built up in the larva under the guidance of sensory control.   As we shall explain below, this interpretation cannot be accepted.   Nevertheless, one must give credit to these experiments for having called attention to the problem of the origin of coordination in the embryo and for having raised it above the purely verbalistic treatment.   No further pertinent experiments on this vital problem were at hand.

Under the circumstances, it seemed promising to adapt the assay method of the transplanted muscle to the study of the developmental phase in the same sense in which it had been applied previously to the post-developmental phase.

Detwiler (1925) had already shown that supernumerary limbs, developed from transplanted embryonic limb buds, exhibit homologous response, just as do limbs transplanted in the functional stage.   This demonstrates that muscles which have developed from the start in abnormal positions make just as reliable indicators of myotypic response as do muscles which have fully differentiated in their normal location.   As for the origin of central coordination, these experiments were, however, inconclusive since the animals had retained their full complement of four normal legs, which one could assume to have furnished to the central nervous system the clues which it needs in organizing its motor patterns.

A more crucial situation could be created by exchanging the right and left fore limbs in the stage of buds prior to the onset of function. This operation would produce animals which could never have had experience with normal limbs, but which, according to what was said above, might very well have learned to use the reversed limbs to good advantage, provided the young centers were plastic enough and not occupied by inherited self-differentiated patterns.

The method of transplanting limb buds in amphibian embryos has been carried to such perfection by *Harrison* and his school that the operation meets with no technical difficulty. If the transplantation is carried out after a certain critical stage, the limbs differentiate true to their origin; that is, a left limb bud on the right side will form a left limb. There is one after-effect of the operation, however, which tends to defeat our purpose. As *Harrison* (1918, 1921), and later particularly *Swett* (1937), have shown, the transplantation of a limb bud which is in disharmony with its surroundings on the body is usually followed not only by the differentiation of the transplant, but also by the production on the part of the body of an accessory limb, which then usually is a mirror image of the transplant. Therefore, upon transplanting a limb bud from one side of the body to the other, we would obtain a reversed limb, as intended; but, in addition a secondary limb would sprout out which, being a mirror image of the (reversed) graft, would be a duplicate of the removed original limb. Hence, again an undesirable normal limb would force itself into the picture. In order to evade this difficulty, it seemed indicated to transplant the limbs at a fairly late stage of differentiation, because the ability to suppress local regeneration of extra appendages increases with the age and size of the limb graft. Even so only a small number of cases were obtained in which the operation was wholly successful.

The species used for the experiment was *Amblystoma punctatum*. The eggs were reared in the laboratory. All operations were done in chloretone anaesthesia, the animals being immobilized in a strong solution (saturated solution diluted in a ratio of 1:8) and then kept for several hours in a solution diluted 1:24. All host animals were chosen at stages in which the fore limbs had begun to show signs of differentiation, but had not yet attained motility. The larvae ranged in age between *Harrison*'s stages 39 and 43. In these stages, the fore limbs project from the body wall as rigid, cylindrical, and immobile rods with the contour of the hand plate and the elbow just beginning to appear. The host limbs were extirpated, including as much of the shoulder girdle and shoulder muscles as possible. They were then replaced by limbs of the opposite side taken from donors of slightly older stages. In some cases the donor limbs had already been functional when they were transferred to the younger non-functional host. In removing a future transplant from its donor, a good-sized disk of shoulder girdle and shoulder musculature was taken along with the free extremity.

The transplants were adjusted at the vacated sites of the extirpated host limbs, care being taken to maintain their original dorso-ventral orientation in order to enable them after differentiation to apply their plantar surfaces to the ground. All transplants were at the time of operation far beyond the stage of definitive determination, so that one could be sure that they would continue after transplantation to self-differentiate in all major respects as if they had been left in their original places, that is, with an anatomy exactly the reverse of that fitting the location.

The further differentiation of the transplants occurred as expected. A right fore limb developed at the left flank, and a left limb at the right flank, flexor sides facing posteriorly instead of anteriorly. However, in many cases the transplant had not been successful in suppressing the regeneration of a secondary limb from the host body, and consequently one, and sometimes even two, additional limbs arose near the base of the graft. Fortunately, these accessory regenerates did not affect the conclusiveness of these cases for our purpose, for the following reason. Since the transplants had a head start over the regenerates of several weeks, they became functional and exhibited well coordinated movements long before the regenerates attained motility. Thus they can be

counted along with those other cases in which regeneration had
failed to occur.

The operation was successful in twenty-five cases. Thirteen
of these died prematurely or had to be discarded for other reasons.
The remaining twelve animals with bilateral operations were
carried through and form the basis of our further discussion.

a   *b*

Fig. 7. *a*, Salamander Larva With Reversed Fore Limbs (right limb on left side
and left limb on right side) developed after the pre-functional limb rudiments of
the host (stage 43) had been replaced by limbs of the opposite sides from an older
donor (stage 45). Five months after operation. (Gills have been removed for a
better view of the limbs.) Note that extensor side of the fore limbs faces headward (compare inset of figure 5). *b*, normal control animal (gills removed).

Only in one of these cases, illustrated in figure 7, had regeneration
failed to occur on both sides, so that this animal had no other
fore limbs except the two reversed transplants. In 4 cases the
transplant of one side remained single, whereas the transplant
of the opposite side was later joined by either one (fig. 8) or

two regenerates.   In the 7 remaining cases regenerates appeared on both sides, leaving both transplants single only during the earlier stages.

The transplants received innervation from the severed brachial plexus of the host, but since the conditions under which such re-innervation occurs have been amply studied (*P. Weiss*, 1937 b),

FIG. 8. SALAMANDER LARVA DEVELOPED AFTER OPERATION AS IN FIGURE 7A
Behind the reversed transplant on the right flank, the host has regenerated a right limb.   (Donor: stage 46; host: stage 43; fixed 5 months p. op.)

a detailed anatomical study of the present cases was omitted. The earliest functional contractions were seen in the transplanted limbs about two weeks after the transplantation, which interval may then be taken to be sufficient for successful regeneration of the nerves.   Even in older larvae two weeks are sometimes adequate for the re-innervation of transplanted limbs.

The observations on the function of the reversed limbs were so uniform that they can be summarized in a single sentence:

From the very first stages of motility, the limbs moved in reverse.
This is as true of the stages of incipient function, in which the
limb moves in association with the trunk (*Coghill*, 1933), as it is of
the later stages, during which the limb exhibits independent and
more differentiated action.   In some cases the excursions of the
shoulder joint were of reduced extent.   The movements of the
elbow, the wrist, and the fingers, however, were always smoothly
executed and of normal amplitude.   The restriction of the shoulder
movement is to be ascribed to slight anatomical deformities of
the transplanted shoulder girdles, and in a few cases may also be
due to some remnants of the host shoulder muscles counteracting
those of the transplants.   The shoulder movements were more
seriously affected in those cases in which a regenerate had developed and become fused to the base of the transplant.

For the rest, however, the transplants exhibited precisely the
same form of coordination which we described above for fore
limbs reversed in the functional stage.   As in those earlier cases,
the same combination of muscles was active at any given moment
in the reversed limb which would have been active in the same
phase of movement in a normal unreversed limb innervated from
the same central district.   The effect was again a persistent effort
of the fore limbs to move the body backwards whenever the rest
of the animal attempted to progress.   Some of the animals were
kept up to six months after the operation, but they never showed
any sign of modification of this peculiar locomotor coordination.
Of course, no such adaptation was to be expected from what the
earlier experiments on older larvae had taught us.   Besides, the
point to be stressed here is not that the animals have retained their
reversed locomotion, but that they had developed it in the first
place.

In those specimens in which one or two regenerates arose next
to the transplant, the resulting limb duplet or triplet exhibited
again "homologous response."   Since cases of this kind have been
amply described in previous publications, we merely mention
them without further comment.

V. On the Dynamics of the Nervous System   543

The inherited motor repertoire

Let us now briefly consider the implications of these experiments. The reversed forelimbs have from the beginning functioned in a manner which precluded their useful participation in the locomotion of the body. All the points discussed in connection with limb reversal in older larvae hold for these younger stages as well. Just as in the older cases, the objection that the reversed movements might have been enforced by mechanical limitations is invalidated by the fact that when the animals retreat the reversed limbs do exert satisfactory forward traction; but then, of course, the hind limbs march backwards. Combination of the retreat pattern for the fore limbs with the advance pattern for the hind limbs would have straightened out matters, as we explained above. But the early developing nervous system was no better at solving the problem than were the more mature centers of the older larvae.

These animals have never in their lives experienced normal locomotion. Nevertheless, they have developed the same typical central patterns of locomotor coordination which make an anatomically normal individual walk. In other words, the centers have come to "walk" correctly, even if the limbs have not. Surely, therefore, the experience of actual walking can play no part in molding in the centers those patterns of coordination through which walking is effected, as the heuristic theories would have it. Nor does it seem to make any difference for these patterns whether they produce effective walking or—as in the case of our experiments—some wholly incongruous performance, which is quite contrary to the systemic theory, according to which the total effect counts. Consequently, *neither a heuristic nor a systemic explanation of the origin of coordination can be harmonized with the facts.* Neither experience gained through trial and error in performance, nor autonomous regulative powers of the nervous system can subordinate the timing of the muscles to the needs of the body as a whole.

Of far greater significance, however, than this negative statement is the positive information contained in the experiments, namely, the proof, firstly, that the centers develop and elaborate the patterns of coordination by pure *self-differentiation* and, secondly, that this differentiation occurs in the *myotypic code*.

It must be kept in mind that our experiments have altered the normal anatomical relations not only between muscles and body, but also between spinal cord and muscles: In removing the host limb to make room for the transplant, the young limb nerves were interrupted, which must have abolished any original point-to-point correlation between spinal cord and musculature, if ever such a point-to-point correlation existed. Then, the transplant was inserted, and the severed nerves grew into it. The facts of nerve regeneration are such that, statistically speaking, these nerve fibers growing out for a second time must have come to land on entirely different muscles from where they had ended before. Thus, each muscle of the transplant is finally connected with an entirely different constellation of ganglion cells than is the corresponding muscle of an unoperated limb. That this does not confuse the muscular control, is to be credited to the principle of myotypic response.

Any more concrete idea of how this may occur, depends on what particular explanation of the mechanism of myotypic response one favors. As we have said before, this is still largely a matter of conjecture. If we accept, for instance, the suggestion that each muscle after its own differentiation "modulates" its motor nerve fibers, this retrograde action would project the various muscular specificities into the motor nuclei and there set up a true representation of the effector system, independent of the peripheral nerve topography. Thus, even if the nerves are distributed over the limb at random, the ganglion cells of the motor columns of the spinal limb segments will at the end of modulation consist of approximately two dozen different varieties, each type corresponding to a particular muscle. How closely the ganglion cells belonging to an individual muscle are assembled, is still a controversial question. But it is obvious that even if a strict topographical projection of the musculature in the cord existed in a normal animal, it would be abolished in the course of peripheral nerve regeneration. Consequently, in our experimental animals the various motor cells of the two dozen varieties must have been profusely intermingled.

These cells now are the ultimate executives of the centers. In dealing with them, the centers seem to have no additional means of discerning just how they are connected with their muscles, how the muscles are connected with the skeleton, what kind of movements the contractions of these muscles will produce, and finally, what the outcome of these movements will be for the body as a whole. Thus, when the central score requires the activation of, for example, the pectoral muscle, which in a normal animal adducts the arm, the pectoral-specific "call" is emitted within the limb cord and is responded to by the pectoral-modulated ganglion cells, even though the pectoral muscle, as is the case in the reversed legs, may lie at the anterior instead of posterior border of the limb, abducting instead of adducting the upper arm, and thereby producing a backward instead of a forward swing of the body. Moreover, the pectoral-modulated cells are likely to be quite a different group from what would have been the pectoral center in the normal animal.

While these facts are merely a confirmation of the general validity of the principle of myotypic response, the new point brought out by the experiments concerns not so much the *response* of the motor units as their *selective activation* according to firmly set time patterns, self-constructed by the centers in forward reference to, but without the constructive aid of, actual function.

How are we to visualize these patterns? It must be admitted that the prospect of identifying their material basis and dynamic properties is not yet very bright. It seems that all we can do at present, is to reconstruct them from their manifestations. We know that when we provide them with a system of detector muscles, they will produce a definite myochronogram for each particular act, such as exemplified in figures 4, 5, and 6. The myochronogram, therefore, may be regarded as a peripheral projection of the time schedule according to which the various modulated ganglion cell groups are being activated. If we could let all motor ganglion cells register their phases of activity in separate tracings, this would give us a direct record of their time schedule—a *"neurochronogram."* The neurochronogram, in turn, is only an expression of the time order in which the principle which activates the motor cells becomes effective, a record of what we have called the "central score." This is as far as our factual knowledge will take us, and unless we want to enter the field of hypothesis, the nature and localization of the agents in back of it all must at present remain unaccounted for. Even so, however, we can make certain definite statements about their reality, characteristics, development, behavior, and effects.

(1) The central score remains constant amidst the reported experimental changes forced upon the periphery. This stability of organization, therefore, justifies the assumption of an underlying central state or process of correspondingly definite organization.

(2) The organization of the scores on level O (coordination within a limb; "intra-appendicular" coordination) is stable only for a given motor act. The fore limb score for ambulation, for instance, provides for the combination of the adductor, depressor and extensor in one team, and of the abductor, elevator and flexor in the other. The score of the same center, however, for the act of retreating combines the abductor, depressor and extensor in one phase, and the adductor, elevator and flexor in the other.

(3) The intra-appendicular scores on level O are indissoluble in their composition within any given act. Even urgent biological necessity cannot modify the chronological patterns exemplified under (2).

(4) The scores on level S (coordination among limbs; "inter-appendicular" coordination) for a given act (ambulation; retreat; etc.) operate through the subordinated intra-appendicular scores provided for those acts according to (2). But the time relations among the individual limb actions are more variable than the time relations of muscle actions within the limb (see p. 36).

(5) In spite of greater temporal elasticity, the combination of partial scores of level O executing interappendicular coordination of level S is indissoluble within a given act. Fore limb scores of retreat cannot be combined with hind limb scores of progression even under the pressure of biological necessity.

(6) The scores provided for the different basic biological acts appear to be discrete entities with no intergradations. Each one represents a ready mechanism to execute one vital function with great adequacy, provided the animal is equipped with just the kind of anatomical apparatus for which the score is designed, including proper nerve modulation.

(7) The basic scores develop by self-differentiation within the centers, molded by the pre-functional agents of embryonic differentiation. Not all of them develop at the same time, and many are not completed until late in post-embryonic life. Some are permanently in operation (e. g., respiratory movements), others may remain latent under ordinary circumstances. The totality of preformed scores owned by a species may be called its *"motor repertoire."*

(8) Since different scores are executed by the same muscles and ganglion cells, only in different assortments and sequences, it becomes impossible to view them in terms of rigidly established neurone connections after the fashion discussed on page 8. Consequently, the definiteness and constancy of their structure must be based upon definite dynamic properties (time parameters; chemical affinities; or the like) of the central agents, rather than on unique morphological connections.

According to these points, the amphibian central nervous system self-differentiates a definite and limited repertoire of discrete and strictly circumscribed scores upon which the animal must draw for all its performances. Animals of the type here discussed have never shown any ability of adding to this inherited repertoire in their later life by experience. *The elementary mechanisms of coordination are inherent in the centers,* and *the units of coordinated behavior are integrated complexes of the kind reflected in a myochronogram,* rather than individual neurone activities. This is the basic fact which all theories of coordination will have to keep in view, and compatibility with which will be their test.

Muscle center vs. impulse pattern

In order not to vitiate the factual content of the experiments by speculative admixtures, we have refrained from trying to translate into more concrete terms the essentially formal conclusions to which our experiments have led. We have derived from our observations as much insight into the origin of coordination as the method employed has been able to yield. But our formulations have not gone beyond evaluating the phenomena and furnishing a definite conceptual frame which any theory of coordination will have to fit. On the negative side this has led to the elimination of certain theories of coordination favored in the past, for the reason that they are in conflict with the facts. On the positive side, however, we have not been able to establish anything more concrete than the fact that the central agents to which the chronological scores refer, have real existence, strong individuality, and develop their properties, as well as maintain their identity, in the face of all kinds of peripheral changes. To study their nature, remains an object for future research. Our experiments have merely put this objective into clear view. As a first move toward this goal, a few speculative comments will here be added. But in offering them, it should be understood that, whatever criticism they may arouse, would not shake the factual content of the experiments thus far discussed.

The problem is plain: The motor columns consist of a fixed number of nerve cells, modulated into a few dozen varieties, and susceptible to activation by agents of corresponding specificity. What is it, then, that makes those agents appear in such standardized chronological order as to yield the stereotyped pattern of activation represented by the "score?" One is tempted to think of a piano in which keys are operated in definite combinations and sequences.[10] However, in a piano, any combination of keys can be played. Possibly in the amphibian nervous system, too, any combination of motor cells *can* be activated; but the fact is that only certain definite ones actually *are* activated. Thus, if the piano were to serve at all as an analogy of the central apparatus, it would have to be the automatic kind of piano with its limited repertoire. An electric piano is made to give off a definite tune

[10] Except that in the piano the position of the key in the scale is a clue to the effect (tone) it will produce, whereas in the case of the motor centers the clues consist of the specific moduli acquired by the cells in the course of modulation.

by running a properly cut out stencil over contact points which release a corresponding set of keys. The harmony of the tune is preformed in the pattern of the stencil. Similarly, we understand the operation of moving electric light signs as the result of a properly shaped contact brush sweeping over a field of contacts, each of which has a one-to-one connection with a single light bulb.

But this analogy does not stand up under closer inspection. For one thing, there is no constant topographical correspondence between the muscles and their central representatives, as there is between keys and contacts. Even if individual muscles should be proven to possess each a localized nuclear representation in the motor part of the spinal cord under normal conditions—a fact which, as we mentioned before, is still disputed (see *van Rynberk*, 1908)—random nerve regeneration would upset the scheme without consequence for the peripheral coordination of the muscles. For we have seen that experiments which enforce random connections do not upset the strict functional correspondence between center and muscle. In contrast, an automatic piano with the wires between contacts and keys crossed gives off a tuneless dissonance. Thus, we had better not follow the analogy any further.

In conversations bearing on the homologous response of supernumerary limbs, it has often been suggested that the motor cells newly connected with a certain kind of muscle might through a selective union of dendritic processes have managed to join with the motor cells connected with the synonymous normal muscle, thus tuning themselves in on all calls going to the normal muscle, much in the way of a two-party telephone line. Nothing was said about how the coordination of the normal muscle centers was achieved, but this, of course, was tacitly assumed to be simply a matter of one or the other of those theories (switch-board; heuristic; systemic) currently in favor as explanations of coordination. However objectionable for other reasons, this view becomes untenable in view of the animals with reversed fore limbs. For in these cases, whatever may have normally passed for a "normal muscle center" with topographical identity, is now dissipated at random over the whole extent of the limb segment. Surely, in a physiological sense, one could still speak of "muscle centers" including under this term the totality of ganglion cells modulated to the tune of a given muscle. Dispersed as they are, however, no topographical sense could be connected with this designation.

It is fully conceivable that motoneurones of the same tune might become knit together into functional unions through their widely ramified dendritic processes. In other words, it is quite possible, as I have suggested previously, that the process of modulation extends farther centrally and thus permits the innumerable cell processes of the neuropil to identify each other. However, this does not really bear on the essential problem of coordination, because the question still remains: why do these muscle centers always fall into certain definite patterns and chronological sequences of action, and only into these patterns, instead of being plastically associable, as has been the view of most theories?

One might try to solve the difficulty, or at least remove it by one step, by assuming that the real muscle centers are not composed of the motor root cells, i.e., the "final common paths" of *Sherrington*, but lie farther centrally in the internuntial gray from which the motor neurones derive their excitation. This assumption would save the concept of a separate localized center for each muscle; one would have to conceive of them as biochemically differentiated neurone pools within the central gray with which the efferent neurones would effect dendritic contact selectively, each one with the appropriate focus according to its modulation. These central stations, one neurone removed from the motor effector units, might then be treated according to the well known schemata of reflexology.

This view would deserve more serious consideration if the seat of basic muscular integration were to be found in the brain at supraspinal levels, where a considerable amount of gray matter is still without physiological assignment. But since we know that a spinal amphibian exhibits well coordinated movements of its extremities (last reported by *Gray* and *Lissmann* 1940b; "homologous response" in spinal animal: *P. Weiss*, 1937a), it is the spinal cord alone to which we have to look for the mechanism of basic coordination. The histology of the amphibian spinal cord, however, discourages emphatically the attempt of identifying anything like discrete muscle centers in its internuntial gray.

Furthermore, the question as to why definite combinations of muscles, and only a very limited choice of such combinations, comes into real operation, would still remain as obscure as before. Recourse to the assumption of stereotyped anatomical connections between definite muscle representatives on the one hand and higher centers or sensory nuclei on the other is of no avail. As we have pointed out earlier, coordination on the basis of built-in anatomical connections has been demonstrated in certain cases and might possibly be passable as an explanation of the coordination of hinge joints, the paradigm of conventional reflexological schemata. It is irreconcilable, however, with the existence of multiple-choice reactions, as exemplified by forward and backward locomotion, in both of which depressor and extensor act essentially in phase, while the adductor may be engaged either in their phase or in the antagonistic phase, and so may the abductor. Obviously, the morphological connections are equally suited for either form of association, and, hence, the choice must depend on physiological factors. Other patterns may show the elevator and flexor out of phase, and the depressor dissociated from the extensor; hence, their association in progression and retreat cannot be anatomically fixed. In general, different patterns show the muscles in different associations. Yet, how these functional associations are brought about, and why they are as strict as they are, is beyond our present knowledge. The heuristic theories had an answer for it, the systemic theories at least a claim to an answer. With both these now eliminated, so far as *basic* coordination is concerned, we must start our search anew. It will probably be helpful in this search not to follow the ruts of the now discredited scheme of rigid fiber connections as the basis of coordination, and turn from a topographic-geometric concept to a more dynamic concept in which a "muscle center" would be a *specific central state of activity* characteristic of a given muscle.

Of course, if we concede nothing more to the centers than what is found in the periphery, the task is hopeless, just as all the phenomena we have described would become incomprehensible.

However, if we accord to the centers an unbiased investigation, we might some day discover that the "excitatory state" of *Sherrington,* for instance, may not be after all such a trivial and colorless condition as electrotonus, or the concentration of a chemical substance, but that it may assume a variety of differentiated states in which *specific biochemical configurations* or *specific electric parameters* would appear as keys selectively unlocking for the discharge of excitation peripheral channels of correspondingly specific configuration. Yet, it seems quite futile at the present state of our knowledge to indulge in these vague prophecies. And we wish to repeat that their rejection in toto would not in any way reflect on the experimental facts and conclusions presented earlier in the paper.

THE ROLE OF SENSORY CONTROL

Motility after de-afferentation

The inadequacy of both heuristic and reflexological interpretations of coordination is further underscored by the fact that sensory excitations play a constructive part neither in the origin nor in the maintenance of basic coordination patterns in amphibians.

Ever since *Exner* (1894) advanced his principle of "senso-motility," the importance of sensory control in the execution of movements has been staunchly defended. Observations on motor defects following sensory disturbances in mammals, clinical observations of similar nature, the importance of sensation in the re-education of the muscular coordination in patients with transposed muscles, and finally the emphasis placed on the afferent input as the key to the motor response in all reflexological interpretations of coordination, have contributed to the popular notion that motor coordination would break down in the absence of sensory control. Extrapolated into the embryonic phase, this notion would suggest that without sensory control motor coordination could not be established, either.

While this is not the place to give the whole problem as thorough a consideration as it would deserve, its relation to our subject matter is close enough to require some comment. Heuristic, systemic and reflexological theories of coordination have considered it vital, each for a different reason, to ascribe to sensory innervation the role of a determining agent in the development and maintenance of coordination. According to the heuristic view, the sensory messages provide the centers with the necessary reports of the success or failure of a tried muscular effort, so that the effort may be either repeated or abandoned. According to the systemic view, the total sensory input creates a definitely shaped central state which, in turn, determines the character of the total response. According to the reflexological view, each sensory channel opens into its own peculiar system of central communications, the efferent ramifications of which determine the pattern of the response. But all three are agreed that the sensory influx is indispensable for coordinated motility. So far as the *basic* patterns of coordination are concerned, this is a wholly untenable thesis.

The whole matter is one about which there seems to be still as much divergence of opinion as there ever has been. The controversy rests not so much on discrepancies of experimental results as on divergent interpretations, loose or mistaken formulations, and a heavy dose of anthropocentric thinking of the kind referred to on page 20. In order to avoid further confusion, let us be explicit about the issue: It is not whether sensory functions are "important" or not, and whether or not the elimination of sensory function "affects" the motor behavior of an animal, because nobody would hesitate to answer both these questions in the affirmative. The point with which we are solely concerned in the present connection is whether the *patterns* of coordination, i.e., the organized combinations of muscles into definite chronological scores, owe their *organization* to any effects exerted through afferent nerves. We further confine the issue to the *basic* patterns of locomotion, excluding the finer adjustments, and discuss the problem, for the time being, only as it applies to the *amphibians*.

Within these limits it can be safely stated that sensory control plays no *constructive* part in the development and maintenance of the patterns of motor coordination. To qualify this statement, we may first point to evidence coming from experimental de-afferentation of limbs.   The experimental results are rather convincing because most authors, even though favoring diametrically opposed interpretations, are in essential agreement about the underlying facts.   It has been observed by *Brown-Séquard* (1850), *H. E. Hering* (1896), and *Bickel* (1897), that frogs continue to use their legs in essentially unimpaired coordination after the dorsal roots of the limb segments have been transected. These observations were confirmed and amplified by more recent work on the toad and on salamanders (*P. Weiss*, 1936 a; 1937 c). Still more recent repetitions of the experiments by *Chase* (1940) and by *Gray* (1939) have added no data that would conflict with the earlier findings.   According to all these findings, a limb completely deprived of sensory innervation continues to take part in locomotion, such as jumping, walking, and swimming, as well as in postural and other reactions, such as righting, turning, wiping.

These facts prove that the basic patterns of coordination, at least up to level O, are not acutely controlled by sensory impulses from the peripheral organs through which they are effected.   Just as we have seen the inherited central motor patterns to survive in the face of *altered* peripheral conditions, so we find them to continue under a complete *lack* of clues concerning their peripheral effects.   These same experiments disprove the theory of the rhythmic gait as a series of alternating stretch reflexes.   While stretch reflexes may, and probably do, re-inforce and smooth out various phases of the locomotor acts, they are not instrumental, and not even indispensable, in producing coordinated locomotion.

The undulating swimming movements of the trunk of an amphibian are as little the product of a self-perpetuating chain reflex as is the alternating gait of an individual limb.   Experiments by *von Holst* (1935) have already demonstrated that the swimming mechanism in fishes does not consist of alternating stretch reflexes of symmetrical segments of the trunk musculature, as had formerly been suggested.   After cutting most of the sensory roots, he observed no major impairment of the swimming rhythm, which led him to the conclusion that swimming is effected by autonomous intracentral waves of excitation passing down the length of the spinal cord with a definite phase shift between the left and the right sides. Since *Gray* (1939, p. 42) has questioned the conclusiveness of *von Holst*'s results, because a few dorsal roots had been left intact in the operated fishes, I recently repeated the experiments with amphibians, following a more radical procedure (*P. Weiss*, 1941). In frog tadpoles, *all* spinal dorsal roots were cut on both sides way down into the tail base, and the remaining part of the tail, where de-afferentation is less well controllable, was amputated. Swimming coordination of these animals was as good as that of normal animals.

In view of the existing evidence, which, as one can plainly see, rests on a solid and consistent experimental foundation, it would be hardly necessary to dwell on these points any longer, were it not for the fact that *Gray* (1939) and *Gray and Lissman* (1940a) have recently come out strongly in the defense of the reflex chain theory of ambulation. In point of fact, they confirm that amphibians with one, two, three, or even all four limbs de-afferented are still capable of performing all the typical acts of locomotion. Their description coincides in all major points with the earlier report on the subject by *P. Weiss* (1936a): There is agreement that with increasing extent of de-afferentation, (a) the *spontaneity* of the animals, i.e., reactivity on the level W, decreases; (b) the effects of loss of postural control and possibly tone become increasingly conspicuous in the form of *cruder* and less well balanced movements; neither of which defects, however, can obscure the fact that the *typical coordination patterns* of levels O and S, the former integrating muscles into limb movements and the latter integrating different limbs among one another, have not been abolished and cannot, therefore, be under sensory guidance.

We prefer to place the emphasis on the persistence of the patterns as such, regardless of whether or not their execution is equally easy and equally smooth. That the sensory channel may be necessary in order to start the centers off, is an entirely different problem. It requires a switch to start an electric plant going, and fuel must be fed into an engine to make it operate. In a normal animal, aside from respiratory and highest brain functions, some stimulus is obviously necessary to set off a reaction, but that does not mean that the pattern of the reaction is likewise determined by the input, as little as the distribution of the electric power or the mechanism of the machine are determined by the switch or the fuel. Thus, if *Gray and Lissman* state that ambulation[11] is impossible unless *some* sensory fibers, even though few in number, are left intact *somewhere* in the trunk region, one will agree that even if this statement were to be taken at full face value,[12] it would only confirm the *autonomy* of the central patterns of coordination. For if the stimulation of *any* kind of nerve entails such a highly organized performance as ambulation, and invariably the same standard pattern, no matter which nerve was stimulated, this should satisfy the authors that the *character* of the response cannot have been determined by the stimulated nerve. This is merely a matter of logic.

In a subsequent paper, *Gray and Lissman* (1940b) describe interesting observations on locomotor patterns obtainable in spinal amphibians. In line with what is well known to everybody who has occupied himself with this field, they describe how a stretch reflex may set off a partial pattern of locomotion, that is, a coordinated limb reaction on the level O, frequently entailing a spinal reaction on the level S, engaging all four limbs. How the authors can be so inconsistent as to suggest from this observation that the act of ambulation actually amounts to nothing more than a series of stretch reflexes elicited from the legs, after they have themselves in the immediately preceding paper demonstrated the persistence of the ambulating pattern, even after all four limbs had been de-afferented, is difficult to understand.

Moreover, if the authors had paid attention to the locomotion of our animals with reversed fore limbs, as described and depicted in the 1936 paper, they would have realized that these specimens carry the whole stretch-reflex-chain idea ad absurdum. For in these animals diagonal limbs (e.g., left fore limb and right hind limb) move in opposite directions, rather than in parallel as do

---

[11] The authors are no longer equally positive about other types of locomotion (1940a, page 232–233).

[12] Coordinated locomotor movements have actually been observed in toads in which the whole length of the spinal cord was de-afferented (*P. Weiss*, 1936a).

diagonal limbs in a normal walking animal (see p. 46). If muscle stretch due to the drag of the advancing body were responsible for the functional linkage between the four limbs, it should have operated in the reversed fore limbs in the same sense as in normal limbs, resulting in effective progressive ambulation, which did not happen. Or if the reversed fore limb movements had set the pace, the backward drag imposed upon the hind limbs should then have made the latter cooperate in a coordinated backward movement; which did not happen, either. What did happen, was that animals possessing reversed fore limbs and normal hind limbs remained stationary, with each one of the four legs going through the motions of what would be its part in typical ambulation, but without the body as a whole moving at all. This should have cautioned the authors against vesting the simple stretch reflex with imaginary pattern-determining powers. In fact, a mere consideration of the normal retreat pattern, to which the authors give no thought, but in which, as we have seen, the abductor-adductor associations with the flexor-extensor phases are exactly the reverse from what they are in forward ambulation, would have upset their scheme, the more so as they themselves deny the existence of stretch reflexes in the adductor groups, which are the ones that would be subject to passive stretch in backward locomotion. On the whole, if the losing cause of the reflex chain concept of locomotion needed any further weakening, the inconclusive and contradictory re-interpretation of otherwise sound experiments by *Gray and Lissman* has certainly produced it.

While the suggestion of a *constructive*, that is, pattern-forming, role of the sensory influx must be emphatically rejected, the *regulative* influence of sensory innervation is duly appreciated. While the major characteristics of the locomotor pattern are laid down as autonomous functions of the centers, the minor details of their execution are undeniably under peripheral control. Failure to keep the two issues separated, confuses the discussion, and many misunderstandings of the past can be referred to this failure. To be sure, the distinction between "major" and "minor" is somewhat arbitrary, but no more so than the decision in physical and mathematical matters as to where quantities become negligible.

Plainly there are differences between the movements of a fully sensitive and a de-afferented limb, as well as differences in the behavior of the whole animal, which become increasingly greater as the extent of de-afferentation increases. The finer polish through which the movements become smooth and are kept in harmony with the changing topography of the environment disappears, and only the crude basic structure of the main patterns is left. But the difference between the polished behavior of the normal animal and the cruder performance of the de-afferented animal is so much smaller than the difference between the still highly coordinated function of the latter and a disorganized state of random contractions, which would mark the break-down of coordination, that it becomes practically negligible so far as the problem of basic locomotor coordination is concerned. The step from the irregular twitching of an uncontrolled muscle machine to the coordinated activities observed even in de-afferented animals is so immense, when compared with what sensory control has to add in the way of further accomplishment, that our sense of proportions should revolt against the recurrent attempts to give sensory control full credit for the whole achievement.

However, clarification of the whole issue will be greatly aided, if we abandon such inarticulate utterances about sensory control as that it is "of paramount importance," "dominant," "essential," "vital," or, on the other side of the picture, "irrelevant," "practically insignificant," etc., and replace them by precise statements as to what phases of motor activity depend upon the integrity of sensory innervation, in what respect, and to what degree. Such a program would make no sense unless the hierarchical constitution of nervous functions is recognized. But if we admit that sensory influx may have different effects with regard to some levels of nervous activity than with regard to others, we realize the necessity of a more differentiated rating of those effects, than merely as a point on a scale from "unimportant" to "highly important." Without trying to be exhaustive, here is a brief list of known sensory functions as they affect motor behavior.

(1) The afferent influx initiates responses by releasing central discharges of definite pattern.

(2) It conditions the centers for subsequent excitations by residual effects on central excitability and excitatory state. The total afferent influx thus produces a continuously shifting background of central excitability, which explains much of the latitude of the stimulus-response relation.

(3) It decides which response from among the plurality of latent discharge patterns composing the central "repertoire" is actually to go into effect. It also influences direction, intensity, speed and duration of the elicited response.

(4) Afferent proprioceptive impulses control the degree of muscular tone, and hence, maintenance of posture against gravity.

(5) They also contribute to the precision and smoothness of a movement through local stretch reflexes (myotatic reflexes) acting as "governors."

Since each major item of this list can be still further subdivided, it will be realized how complex the effects of sensory influx are, hence, how futile it is simply to assert their bearing on motor functions without further qualification. A toad with de-afferented hind limbs moves perfectly well over rough ground; but when it happens to land from a jump with its limbs contorted, no postural correction will ensue until the next locomotor impulse automatically returns the limbs to their normal position. This exemplifies the kind of disturbances to be ascribed to lack of sensation. Other shortcomings are the cruder dosing of the muscular contractions, exaggeration of movements, abnormalities of the tonic background, etc., none of them serious enough to mask the essential fact that the basic central scores through which locomotor coordination is effected survive the elimination of the sensory influx.

Many statements to the contrary can be traced to obvious sources of error. The most trivial is primary damage done to the motor centers in the course of de-afferentation. When *Moldaver* (1936, p. 457) states that "the transection of the dorsal roots in the toad results in an immediate deep depression of spontaneous motor activity of the de-afferented leg, which presents the aspects of paresis, or even veritable flaccid paralysis," his result cannot be reconciled with the ample evidence to the contrary, except by assuming traumatic motor damage; for if the de-afferentation is performed with the necessary care, leaving the spinal cord unexposed (*P. Weiss*, 1936a), no such effect is noted, not even immediately after the operation.[13]

Less easy to explain, however, are certain secondary late effects of de-afferentation which seem to hamper insensitive limbs in the execution of movements. In amphibians these effects become distinct several weeks after de-afferentation. While varying in extent among different individuals, their common denominator, as judged from observations of *Brown-Séquard* (1850), *Weiss* (1936a), and *Chase* (1940), seems to be a marked *hypertonicity* of the limb muscles, tending to throw the legs into rigid extension and adduction. Whether this exaggerated tone affects the extensor and adductor muscles selectively, or applies to all muscles, with the extensors and adductors simply dominating in appearance owing to their greater weight, has not been determined. At any rate, while it lasts, this hypertonicity prevents the limb from taking part in locomotion. Yet, one commonly observes that, when the muscles limber up temporarily, normal coordinated movements are resumed (*P. Weiss*, 1936a; *Chase*, 1940), and, reciprocally, that with the onset of a locomotor sequence the hypertonicity may abruptly give way to unhampered movements, only to reappear again during the following resting period.

These observations in themselves demonstrate clearly that we are not dealing with a real deterioration of the patterns of coordination—their integrity has not suffered appreciably—but rather with the development of some new pathological condition which blocks the execution of movements. There is evidence that de-afferentation of an amphibian limb automatically raises the excitability of the corresponding motor segment (*Moldaver*, 1936), thus submitting the motor centers to bombardment by

[13] It is gratifying to the author that a satisfactory clarification of this point could be reached during a personal visit in 1937 to the laboratory of Dr. *Bremer* in Brussels where Dr. *Moldaver* was working.

impulses that would otherwise have remained below the threshold of effectiveness. The central tone, which normally merely furnishes the background upon which the locomotor innervation is superimposed, may thus produce in the hyperexcitable de-afferented region a state of chronic excitation of such extent that no motor units would be left for the execution of coordinated acts. By engaging all motor pathways, the hypertonic state would deprive the mechanisms of coordinated locomotion of their effectors. This would explain the immediate resumption of coordinated movements upon any temporary subsidence of the tonic phenomenon. One could also imagine that the relentless bombardment of the de-afferented motor segments might in due course of time produce some lasting damage, although there is, at present, no real foundation for such an assumption. But whatever the explanation of this interesting phenomenon of hypertonicity after de-afferentation may turn out to be, the outstanding fact remains that the phenomenon merely supersedes locomotor activity and by no means signifies disintegration of coordination. Consequently, the attempt to advance this phenomenon as indirect evidence of a constructive role of the sensory influx in the formation of patterns of coordination cannot be regarded as warranted.

Development of motility without sensory control

The point is of particular significance in view of the fact that similar observations were made in animals whose sensory system had been damaged in the *embryonic* phase, giving rise to the claim that the integrity of the sensory system is a prerequisite for the differentiation of the motor patterns (*Detwiler and Vandyke*, 1934; *Chase*, 1940). The experiments in question were done in larval urodeles and consisted of the removal of the dorsal portion of the neural tube, which includes the neural crest rudiments of the spinal ganglia. Such embryos develop into larvae with defects of varying degrees in both spinal ganglia and spinal cord. Functionally, the larva exhibited the full range from undisturbed to completely paralyzed limb motility. Since the functional disturbances could have been due to either sensory or motor

defects, the authors made careful estimates of the relative degrees
in which either central component was affected in various cases.
As a result of these studies, they have come to the conclusion that
the severity of the peripheral functional trouble is correlated with
the degree of damage done to the sensory rather than to the
motor system.

Whether or not their conclusion is accepted, it is reasonable to
assume that at least part of the motor disturbance was due to
the underdevelopment of the afferent system.   It is noteworthy,
however, that this disturbance was more in the nature of an excess
than of a deficiency, in that the limbs were afflicted by the same
kind of excessive extensor and adductor tone which had been
observed in de-afferented adult limbs.   Thus the question arises:
Did the central motor patterns in these cases remain undeveloped,
or did they develop but become chronically overridden by an exaggerated tone occupying the whole available motor pool, in the
manner discussed above?   The description of some of the
operated cases seems to give a clear answer.   We read, for instance: "Although both legs tended to hyperextend at rest, all
reflex and spontaneous movements were considered to be normal"
(*Chase*, 1940, p. 74).   Obviously, the mechanism for coordinated
reflex and spontaneous movements had not failed to undergo
differentiation.   It was only covered up by the hypertonicity,
which can perhaps be explained in the same way as in the analogous adult cases: hyperexcitability of the motor centers as a result
of the chronic weakness of the sensory influx.   Moreover, the
possibility that the sensory system may exert some trophic influence upon the motor parts of the cord cannot be ignored.   One
could think of the hypertonicity of the de-afferented limbs as a
result of trophic disturbance, but the data are insufficient for any
decision.   At any rate, so long as we do not acknowledge the
hypertonicity as a sign of disintegration of the coordinative
mechanism, its further explanation becomes a side issue.

The embryonic experiments just reported have not brought the
desired clarification of the issue.   If anything, they speak against
a constructive part of the sensory control in the development of
coordinated locomotor patterns.   The main reason why they

were bound to remain inconclusive is that damage was mostly done to both the sensory and the motor systems.   In the urodeles, which were used for the experiments, the limbs develop in the embryonic and early larval period so that prefunctional de-afferentation is feasible only by removing the whole rudiment of the ganglionic column (neural crest) in bulk.   Although this has been done with great skill (*DuShane*, 1938), one has no·as-surance that the removal has not deprived the differentiating neural tube, if not of some cellular constituents, so at least of some embryogenetic and trophic influences possibly required for its elaboration into a normal spinal cord.   This difficulty can be avoided if *anurans* are used instead of urodeles.

The tadpole of the frog develops hind legs not until compara-tively late in larval life, at a time when the central nervous system has reached a very advanced stage of differentiation and the animal has been functional for several months.   The spinal ganglia and dorsal roots are in their definitive positions and can be easily and neatly extirpated without causing damage to the spinal cord.   This was recently done in a number of tadpoles whose limbs had not yet reached the functional stage.   Since the loss of the ganglia precluded regeneration of a sensory supply, these limbs grew up with purely motor innervation, which was confirmed by their lack of sensitivity to stimulation.   While in certain cases secondary complications, not of concern to us in the present connection, tended to obscure the picture, the general result was that a number of these animals developed coordinated locomotor function in hind limbs from which no sensory message had ever gone to the centers.   A full report of these experiments will be presented on a later occasion, but it seemed desirable to refer to them in this place in order to stress that *locomotor co-ordination can develop by central self-differentiation of the motor centers in the complete absence of sensory control.*

On the whole, we see, therefore, that a close scrutiny of sen-somotility has failed to reveal facts that would militate against the theory of the autonomous development of motor patterns by *central self-differentiation*, the primary evidence for which we had adduced from recombination experiments.   In this sense, the results of the recombination and the defect experiment mutually support each other.

COORDINATION IN MAMMALS

If central self-differentiation unguided by peripheral experience is so dominant in the establishment of motor coordination, how is it then possible that the recognition of this fact could so long have remained in doubt? Two reasons may be advanced: First, the lack of adequate experimentation; second, the anthropocentric factor, that is, the tendency, mentioned in the introductory chapter, to interpret animal behavior on the basis of human experience. The prominence of the volume of acquired habits in human coordination intimated that all basic coordination had originated in a similar way. By extrapolation from his own faculties, *man* conceded *experience* a role in the primary modelling of coordination, as well as a capacity to remodel such coordination once it had ceased to serve the needs of the body. In contrast to this view, we now learn of the great rigidity and unmodifiability of coordination patterns in the *lower vertebrates*.

Since it would be entirely unsatisfactory to let the matter rest with this obvious schism between lower and higher vertebrates, it seems desirable to indicate briefly how the gap can be bridged. In *amphibians* learning ability seems to be definitely confined to levels higher than S, with all partial acts below level S being firmly and irrevocably set. Thus an amphibian can be conditioned to exhibit a certain motor reaction, e.g., alarm, preying, retreat, etc., in response to a certain set of sensory stimuli. Yet, in producing these responses, it is bound to use the existing repertoire of preformed motor mechanisms, such as they are.

Proceeding to a higher animal, and choosing the *rat* as representative of lower mammals, the experiments of *Sperry* (1940, 1941) have demonstrated that the basic central scores are strikingly rigid and unmodifiable even on this scale of organization. The tendons of the major dorsi-flexor and plantar-flexor of the foot were crossed so as to cause either muscle to produce an excursion of the ankle joint in the opposite direction from normal. In order to avoid possible compensatory adjustments through normal muscles, all muscles but the crossed ones were extirpated.   After the operation, the rats first showed clear-cut reversal of foot movements, which was evident both in type reflexes and in complex postural and locomotor actions.   If the central nervous system were to have re-integrated the ankle movements with those of the other joints, it would have had to retime the innervation of the crossed muscles so as to excite either of them at such moments when in normal coordination the other would have been activated. That such retiming would have led to mechanically and biologically satisfactory results, was shown by crossing the nerves to the crossed muscles.   As we mentioned earlier, nerves transposed after their modulation has become immutable continue to operate according to the time schedule of their original muscles.   Thus, a *plantar-flexor* muscle, provided with a firmly modulated *dorsi-flexor* nerve, acts in the *dorsi-flexor* phase of each movement (*Sperry*, 1941).   Since in the animals with crossed tendons the effect of a former *plantar-flexor* is mechanically converted into *dorsi-flexion*, the nerve crossing rectifies the reversal produced by the tendon crossing, and movements are again correct.   However, in no case did rats after simple tendon crossing (*Sperry*, 1940) or simple nerve crossing (*Sperry*, 1941) learn to adjust their foot movements.   They persevered in operating the hind limb muscles according to the inherited coordination scores and failed to re-arrange the timing even under crucial training conditions.   In conclusion, so far as plasticity of basic coordination patterns is concerned, the rat possesses none in its *hind* limbs, and, therefore, lines up in this respect with the amphibians.

A repetition of the experiments on the *fore* limb, however, has suggested that there may be a significant difference between fore limb and hind limb coordination (*Sperry*, 1942).   Tendon crossing in the fore limb again led to reversal of all standard movements and postures, in the manner described for the hind limbs. However, some rats seem to have discovered an emergency solution, consisting of locking the extended elbow joint mechanically so that the stiff fore limb can be used as a brace for the support of the fore body.   In this manner the animals can avert the caving in of the elbow joint which would otherwise accompany the supporting phase of all movements and postures because of the translocation of the extensor tendons to the flexor side of the joint. Even while this adjustment is in effect, the transposed muscles continue to contract in their original phases. The adjustment consists of an appropriate twisting of the whole arm by the shoulder muscles rather than of a corrective retiming of the arm muscles. A point to be stressed in this connection is that the locked position is assumed and maintained only for the one specific act for which it has been acquired, namely, the support of the body, while in all other performances the elbow is still moved in reverse. Moreover, frequent relapses occur even during the supporting phase.

In other words, the basic patterns of coordination have not been remodeled, extensor and flexor muscles have not traded their phases of innervation as they would have had to do if the mechanical reversal were to have been compensated for, and in this respect the experiments on the fore limb merely duplicate those of the hind limb. However, in addition and on top of the immutable inherited pattern, a trick performance has been established, the locking reaction, through which the old automatic and inadequate response can be temporarily superseded in a manner profitable to the body as a whole. This new performance is neither a permanent substitute for, nor is it in itself a revised edition of, the old pattern. The old stereotyped automatism continues in existence, only intermittently covered up by the action of another nervous apparatus more responsive to the needs of the body.

The adjustment of the fore limb behavior is of a very crude and primitive nature. However, an adjustment it is, nevertheless, and possibly the first faint trace of that capacity for learned coordination, which has reached such high degree in man. Pending proof to the contrary, one would feel inclined to ascribe this incipient adjustive capacity to the beginning evolutionary efflorescence of the motor cortex. Accordingly, the cortico-spinal system would have to be considered as the mediator in these adjustments (see *Tower*, 1936), and the lack of secondary adjustments of the *hind* limb movements of the rat could be correlated with the fact that in this animal only a small fraction of the pyramidal system reaches the hind limb centers (*Ranson*, 1913). Through its short-cut from the cortex to the spinal efferent neurones, this system is obviously enabled to deal with the muscles directly under circumvention of the whole hierarchy of lower centers. Whereas such motor acts as are produced through the mediation of lower centers will continue to exhibit the stereotyped inherited patterns, responses effected over the cortico-spinal system may engage the muscles in new temporal groupings of varying combinations, to be deleted or retained depending on their ultimate success for the body.

Whether these new patterns are established by a trial-and-error procedure, or by virtue of some intrinsic self-regulatory capacity of the cortical system, is impossible to say and also wholly irrelevant from our present point of interest. The main thing is that this cortical activity, or to put it more cautiously, adjustive capacity of higher centers, is limited to setting up new secondary patterns without power to remodel or abolish the primary patterns. This would seem to imply that *the primary and secondary patterns are operated by different central mechanisms.*

As we go up in the scale of mammals, the wealth of secondary patterns—that is, of acquired performances, learned under the guidance of cortical activity—becomes so enormous that their preponderance tends to obscure the existence of the old primary patterns which dominated the amphibian picture. The presence of basic patterns of the primary unlearned type even in man, has of course been widely recognized. Studies on fetal behavior (*Hooker*, 1939) and child development (*Gesell*, 1929, *Shirley*, 1931) have been particularly suggestive. However, the distinction between primary and secondary patterns was usually based merely on differences of origin: autonomous central maturation of the former, as against peripheral acquisition of the latter by experience. There has been no intimation that the difference may also be one of plasticity. Hence, if we want to homologize the primary innate patterns of man with the basic coordination patterns of the lower vertebrates, we must first prove that they are equally unmodifiable. This is an empirical task which has not yet been accomplished thus far.

The problem is to separate those motor performances in which the chronological scores of muscular contractions are absolutely fixed, and remain so even when they lead to unsatisfactory results for the body, from those in which the muscles can be operated in freely variable combinations so as to yield aimful responses. The most valuable experimental material bearing on the problem is in the hands of orthopedic surgeons, who are studying the recuperation of useful coordination after muscle transplantation in partially paralyzed limbs. Some well analyzed cases have brought to light a real conflict between inherited and unmodifiable patterns on the one hand, and novel patterns learned by experience with the aid of physical therapy, on the other (*Scherb*, 1938). However, no more than the first step towards a really clearcut classification and distinction in these matters has been undertaken.[14]

Another source of valuable evidence lies in the study of the comparatively rare cases of functional supernumerary appendages in man. One such case has been examined and has yielded some instructive data. A girl with three supernumerary fingers, which could be identified as a third, fourth, and fifth finger, when first tested, showed distinct "homologous response" between each extra finger and the corresponding normal finger of the same hand (*P. Weiss*, 1935b). Thus, obviously, the principle of myotypic response is as valid in man as it is in lower vertebrates. Whenever homologous muscles operate in association, this may be taken to indicate that they were activated through the mediation of the spinal mechanism to which the neurones of synonymous muscles respond in unison. However, after continued training with conscious effort, the girl finally managed to produce a very clumsy but, nevertheless, real dissociation between homologous muscles (*P. Weiss and Ruch*, 1936). She had apparently learned to innervate the extra fingers and the homologous normal fingers independently. This lasted, however, only so long as her attention was concentrated on the job. As soon as mental or physical

---

[14] The author is at present conducting a research project jointly with the Orthopedics Department of the University of Chicago, in which oscillographic records of muscle potentials are used in the analysis of "functional restoration" after tendon transposition.

V. On the Dynamics of the Nervous System   569

fatigue or distraction of attention weakened the effort, the old associated movements of homologous fingers re-appeared immediately. Here, too, cortical efforts have been successful in temporarily superseding and circumventing lower mechanisms with the result of greater refinement of movement, but this has not entailed a permanent reorganization of the lower effector patterns in the direction of better adjustment. These results, moreover, suggest that the adjustive higher mechanism (hypothetically identified here with the cortex) does not necessarily operate according to the myotypic principle, as the lower mechanisms of coordination do.

Of course, this is only a single case, and the examination leaves much to be desired. Much valuable evidence is constantly being wasted by not giving natural occurrences, such as the one here described, the critical experimental study which their fundamental significance would warrant. It is hoped that the concrete and differentiated questions raised in this paper may lead to increased interest in phenomena of this kind and to a more articulate evaluation of the information which they present.

CONCLUSIONS

It would be needless repetition to review here the specific conclusions reached from our experiments and deliberations, which have already been summarized in their context throughout this paper.

However, we may briefly examine how the answers to some of the standard questions about coordination will look in the light of our results. The term "coordination," one will remember, is used here strictly in reference to the fact that the central nervous system engages the muscles in such a definite order that, in a normal animal, their combined activities result in orderly movements, which, in turn, yield acts of biological adequacy for the animal as a whole.

It will have become obvious that questions such as: "*Is coordination inherited or acquired?*", "*Is coordination rigid or plastic?*", "*Is coordination under sensory control?*" just cannot be answered in that generality. There are patterns of coordination, as we have seen, that are definitely inherited, of prefunctional and pre-experiential origin; there are others that are definitely "learned." The former are rigid in some regards, but show a certain latitude in other regards. The latter show greater plasticity, but even so within definite bounds. Sensory "control" is not vital for coordination; however, coordination may suffer from its absence, the degree varying from one class of animals to another. We have found it necessary to speak of different degrees of coordination; to distinguish levels of coordination; to separate coordination of muscles in moving a limb from coordination of limbs in moving the body, and the latter from coordination of body movements for the satisfaction of biological needs. These are not all the same thing. They cannot be treated all alike and squeezed into a single formula. What holds for one level or one animal, cannot be applied as a matter of course to all levels and all animals.

Failure to recognize this truth cannot but breed sterile controversies. In fact, if it were not too far a digression from our main subject, it would be an easy matter to trace many a heated dispute of the past back to the fact that two schools, starting from two sets of different but equally valid data, generalized far beyond the legitimate scope of those data, ending up with irreconcilable theories, all in the name of simplicity. We prefer to think of natural principles as of great uniformity and universality, and we are partial to doctrines which present them as such. Accordingly, we would have expected coordination to be either all plastic or all rigid; all preformed or all individually acquired. To learn that it is partly one way and partly the other, is disappointing. Nevertheless, this is precisely what the facts have revealed, and we must acknowledge their testimony. Let us briefly review the evidence.

Is coordination inherited or acquired?

Undeniably, the *basic* patterns of coordination are inherited. That is, the nervous system of every vertebrate about which we have sufficient information, amphibian as well as mammal, develops a certain repertoire of patterns of coordination pre-functionally. These patterns differentiate by virtue of the developmental dynamics of the growing organism in forward reference to their future function, but without the benefit of exercising that function during their formative period. They are laid down in a hierarchy of functional levels, of which the lowest, i.e., the one dealing directly with the muscles, operates in terms of specific signals, one signal for each individual muscle ("myotypic" principle). That the centers should be able to differentiate such a variety of specific signals (specific biochemical processes or specific electric states) is no more surprising than that different glands should be able to produce different secretions.

Once in operation, these basic patterns of coordination act "blindly," unconcerned of whether or not their peripheral effects are of service to the animal. Achievement counts in neither their making nor maintenance. In normal animals they are serviceable by predesign—evolution has taken care of that. When disarrangement of the bodily machine for which they are pre-adapted abolishes their serviceability, they continue unaltered. This, better than any indirect evidence, proves their preformed stereotypism. To this extent the data confirm the *preformistic* concept.

However, there is a second side to the story of coordination that is distinctly *non-preformistic*. The preformed patterns are relatively crude, and only grossly speaking are they stereotyped. The inherited repertoire provides an animal only with what we may call an existential minimum of vital performances. Improvements are called for and occur in varying degrees.

In this connection it should be pointed out that the inherited patterns, of course, do not arise all at the same time, nor all in the embryonic phase. Not only does the metamorphosis of amphibians furnish many dramatic examples of comprehensive behavioral changes during the functional life span, but a progressive expansion of the motor repertoire is plainly observable even in non-metamorphotic animals. There is not the least doubt that this gradual enrichment has the same non-experiential origin as the earlier functional endowment, and is nothing but an external manifestation of the continuous progress in the elaboration of the central coordination systems by self-differentiation. The inability of amphibians to readjust primary coordination at any phase in life seems to dispose of the possibility that coordination patterns first exhibited in later life may, in contrast to earlier ones, have been molded by experience.

The inherent repertoire of an amphibian is fully adequate to carry the individual through life without major changes, and, qualitatively, the animal must get along with its limited repertoire of "scores." However, there is room left for improvement on the quantitative side: in the readiness with which a certain score is activated, the smoothness and speed with which it is executed, and in its competitive rating relative to other scores. That is to say, the behavior of an amphibian can become "conditioned" to the exigencies of its environment by selective facilitation or inhibition of existing motor patterns. Within these narrow limits actual experience with the environment then modifies the structure of behavior—although not the structure of its component scores—, and within these limits the *heuristic* concept finds support. So much for the amphibians.

Vertebrates higher on the evolutionary scale, of more complex organization and more specific in their requirements, face more complex tasks. The elasticity of the inherited scores is becoming increasingly insufficient to meet the accidents of the environment. It is on this level of the scale that a new method of coordination came into being: coordination by individual design and discovery, rather than by predesign and evolutionary tradition. This new development culminates in man. Man can operate his muscles in ever varying combinations, can discover and retain successful effects, eliminate wasteful ones, and thus force his motor apparatus into increasingly better adapted service. Here non-preformed, "invented," coordination patterns become so prominent that they obscure the more ancient stereotyped patterns with which they coexist and overlap. We have tentatively identified this plastic coordination with the activity of the cortex, but there is no definite proof that subcortical functions may not take part in it in higher mammals.

In the lower mammals, this type of coordination is, if present at all, still in a very rudimentary condition. While both the stock repertoire of the species and the ability of the individual to adapt the elastic stock performances to its needs seem to be considerably increased in the rat over what there is to be found in an amphibian, fundamentally the difference seems negligible as compared with the tremendous efflorescence of the ability to "invent" coordination patterns on the way from rat to man.   On the other hand, the primitive trick adjustments of which the fore limbs of the rat are capable (p. 81) may be a true trace of emergent "inventive" coordination.

Whether this type of coordination attains its effects by trial-and-error procedure or by more direct means, as suggested by adherents of a systemic concept (e.g., *Goldstein*), is entirely beyond the competence of the present article to decide.   The only positive statement we can make is that, in contrast to preformed coordination, the adequacy of the effects is the guiding principle in it, and that it employs different mechanisms from the ones through which the preformed coordination patterns are put into effect. As we have indicated above (p. 85), it does not, for instance, operate through the myotypic principle, and sensory control, Exner's "sensomotility," seems to play a much more constructive role than it does in preformed coordination.   For the rest, adequate elucidation of these problems will come from the admirable progress of primate neuro-psychology.

The strictly preformistic concept must be qualified in still another respect.   As it is usually presented, it is meant to imply preformation of function in form of a set neurone architecture: "anatomical" preformation.   Because of this common connotation we have treated it above (p. 8) under the double heading of "preformistic-structural."   But it will have become obvious from our investigations that such an implication does not have to be accepted.   An account of some reasons why minute and systematic anatomical preformation of central functions must be questioned, was presented above in the chapters dealing with the myotypic principle.   While it is needless to repeat the whole argument, we may restate the case.

The function of each motor cell is variable, depending on its modulation.   The central coordination scores, on the other hand, are invariable. Hence, their effect on the motor cells cannot be transmitted through fixed neurone connections. The existence of central states or processes of a high degree of specificity, a corollary of the myotypic principle, would explain the situation on a non-structural basis. In this view, "preformation" of a coordination pattern would mean that those hypothetic specific states or processes are activated in definite chronological sequences; much as in endocrine cycles, in which one gland activates another gland, which in turn, activates a third organ, and so forth, likewise without special anatomical channels to pipe the effect along. To fill the picture with concrete sense, is a task of the future. But we had to stress the point that to acknowledge the preformation of coordination does not mean to acquiesce in the strictly structural interpretation of coordination. Preformation of dynamics is the alternative.

Is coordination plastic or rigid?

After the foregoing remarks this question can be easily rectified. We must add: "Which type of coordination?" And, "Plasticity in what respect?"

If "plasticity" is understood in the sense of elasticity within a given qualitative performance, admitting of quantitative adaptation to what for the given species is a normal range of variability of the environment, both preformed and acquired coordination are "plastic." If, however, one defines plasticity as the ability of an organism to cope with emergency situations, lying beyond the normal range of elasticity, by creating new performances, previously not even latently in existence, the preformed type of coordination is definitely devoid of it. Since most of our experiments bear on this point, we need not labor it further.

Elasticity of performance is a different matter. As we mentioned before, there is latitude in the extent to which an animal draws on its motor repertoire, and in the rate at which it displays it. It may walk *or* swim, and do it either faster *or* more slowly. We also pointed out that the timing of the individual limbs in walking or swimming is not nearly as constant as the timing of the muscles in moving the limb (p. 36). But even so, the pattern according to which the limbs are moved will always be the same for all four—following either the "progression" score or the "retreat" score—without a trace of dissociability and "plastic" re-coordination.

That nervous integration bears all signs of being a *self-regulatory systemic* action, such as we know them from other organic systems, is not to be doubted. As a matter of fact, one of the earlier "organismic" attempts to interpret "animal behavior as the reaction of a system" came from the present author (*P. Weiss*, 1925). To discuss the merits of this view, is beyond the scope of this paper. It is strictly within our province, however, to stress that if the systemic concept is to be reconciled with the facts as they now have come to light, the nervous system will have to be viewed not as if it were *one monotonous whole*, in which what happens to any one part has a direct impact on all other parts, but rather as *a system of systems, each of which consists of subsystems*, and so on; in other words, as a *functional hierarchy* in which the competence, freedom of action, range of variability, etc., of each constituent member is strictly delimited by constitutional—and, on the higher levels, acquired—properties.

Thus, one might conceive, for instance, of the integrative complexes involved in progression or retreat as acting each as a single system, within which the actions of all component parts would be independent. However, the working units of these systems are in themselves systems of a lower order, in the present example elementary limb scores, and the integrative system is bound to operate through these subordinate systems as wholes. In other words, the systemic action on level S of the hierarchic scale (p. 24) might regulate the timing according to which the individual limbs are set in motion, but would be powerless to alter the limb scores themselves which operate on the lower level O. On this basis it would be understandable that after the amputation of a limb the timing of the remaining limbs can be aimfully revised for better service to the body (*Bethe*, 1931), while retiming of the muscles within the limbs is yet impossible in all lower vertebrates thus far studied, up to and including the rat. The assumption that the progression system and the retreat system are each an entity, likewise explains the inability of our animals to piece together half of one with half of the other (p. 50). In a more general sense this view can be well reconciled, it seems, with the ideas of *Lashley* (1937, 1939) on systemic action of the brain, and with the known facts concerning vicarious action and compensatory regulation in integrative function.

However, we do not propose to go into the subject here any further. We merely wanted to indicate that the negation for lower vertebrates of "plasticity" in the sense of an omnipotent faculty to invent novel coordination patterns to meet emergencies outside of the elasticity limits of the inherited repertoire, and the affirmation of systemic properties of the centers, implying free interplay of forces within the limits of the constitution of the central system, are not at variance. There will be no misunderstanding on this point so long as one keeps in mind that the higher integrative functions are bound to operate through the existing lower functional mechanisms and that the latter, qualitatively determined during the developmental, i.e., pre-operational, phase, are inaccessible to reorganization. Only the higher mammals seem to have developed a new superstructure capable of setting up "plastic" coordination patterns by means which are not yet available on the lower levels of the animal scale.

SUMMARY

Experiments are described in which the method of transplantation of muscles and nerves was used to analyze the origin and, in certain regards, the operation of motor coordination in amphibians. The results support a preformistic concept of coordination in these forms. A basic repertoire of primary motor patterns develops during the developmental phase. These arise essentially by self-differentiation within the central nervous system, independent of the benefits of sensory control and guidance by experience. They are so predesigned that, when later projected into an anatomically normal peripheral effector system, they produce biologically adequate effects. If confronted, however, with an anatomically disarranged periphery, they produce correspondingly distorted effects without signs of corrective adjustment. The relation of these facts to the phenomenon of "plastic" coordination observed in man and higher mammals, and their bearing on the theories of coordination is discussed.

BIBLIOGRAPHY

ADRIAN, E. D. 1932. The Mechanism of Nervous Action; Electrical Studies of the Neurone. Philadelphia. University of Pennsylvania Press. Pp. 103.

BETHE, ALBRECHT. 1931. Plastizität und Zentrenlehre. Handbuch der norm. u. pathol. Physiol., Berlin, J. Springer, **15**, 1175–1220.

BETHE, ALBRECHT, AND FISCHER, ERNST. 1931. Die Anpassungsfähigkeit (Plastizität) des Nervensystems. Handbuch der norm. u. pathol. Physiol., Berlin, J. Springer, **15**, 1045–1130.

BICKEL, A. 1897. Ueber den Einfluss der sensibelen Nerven und der Labyrinthe auf die Bewegungen der Tiere. Pfluegers Arch., **67**, 299–344.

BROWN, T. G. 1914. Fundamental nervous activity. J. Physiol., **48**, 18–46.

BROWN-SÉQUARD, C. E. 1850. Des rapports qui existent entre la lésion des racines motrices et des racines sensitives. Compt. rend. Soc. Biol., **1**, 15–17.

CARMICHAEL, LEONARD. 1926. The development of behavior in vertebrates experimentally removed from the influence of external stimulation. Psychol. Review, **33**, 51–58.

CHASE, P. E. 1940. An experimental study of the relations of sensory control to motor function in amphibian limbs. J. Exper. Zool., **83**, 61–87.

COGHILL, G. E. 1929. Anatomy and the Problem of Behavior. Cambridge. University Press. Pp. 113.

——— 1933. Correlated anatomical and physiological studies of growth of the nervous system of Amphibia. XI. The proliferation of cells in the spinal cord as a factor in the individuation of reflexes of the hind leg of Amblystoma punctatum, Cope. J. Comp. Neurol., **57**, 327–345.

DETWILER, S. R. 1925. Coordinated movements in supernumerary transplanted limbs. J. Comp. Neurol., **38**, 461–490.

DETWILER, S. R., AND VANDYKE, R. H. 1934. The development and function of deafferented forelimbs in Amblystoma. J. Exper. Zool., **68**, 321–246.

DUSHANE, G. P. 1938. Neural fold derivatives in the Amphibia. Pigment cells, spinal ganglia and Rohon-Beard cells. J. Exper. Zool., **78**, 485–501.

DUSSER DE BARENNE, J. G., AND DE KLEYN, A. 1929. Ueber vestibulaeren Nystagmus nach Exstirpation von allen sechs Augenmuskeln beim Kaninchen. Beitrag zur Wirkung und Innervation des Musculus retractor bulbi. Pflueger's Arch., **221**, 1–14.

ERLANGER, J., AND GASSER, H. S. 1937. Electrical Signs of Nervous Activity. Johnson Foundation Lectures. Philadelphia. University of Pennsylvania Press. Pp. 221.

EXNER, SIEGMUND. 1894. Entwurf zu einer physiologischen Erklärung der psychischen Erscheinungen. Leipzig und Wien. F. Deuticke. Pp. 380.

FORBES, ALEXANDER. 1936. Conduction in axon and synapse. Cold Spring Harbor Symposia on Quant. Biol., **4**, 163–169.

GASSER, H. S. 1937. The control of excitation in the nervous system. Harvey Lectures, **1936–1937**, 169–193.

GESELL, ARNOLD. 1929. Maturation and infant behavior patterns. Psychol. Rev., **36**, 307–319.

GOLDSTEIN, KURT. 1939. The Organism. A Holistic Approach to Biology Derived from Pathological Data in Man. New York and Cincinnati. American Book Company. Pp. 533.

GRAY, JAMES. 1939. Aspects of animal locomotion. Proc. Roy. Soc. B, **128**, 28–62.

GRAY, JAMES, AND LISSMANN, H. W. 1940a. The effect of deafferentation upon locomotory activity of amphibian limbs. J. Exper. Biol., **17**, 227–235.

———— 1940b. Ambulatory reflexes in spinal amphibians. J. Exper. Biol., **17**, 237–251.

HARRISON, R. G. 1904. An experimental study of the relation of the nervous system to the developing musculature in the embryo of the frog. Am. J. Anat., **3**, 197–220.

———— 1918. Experiments on the development of the forelimb in Amblystoma. a self-differentiating equipotential system. J. Exper. Zool., **25**, 413–461.

———— 1921. On relations of symmetry in transplanted limbs. J. Exper. Zool., **32**, 1–136.

HERING, H. E. 1896. Über Bewegungsstudien nach zentripetaler Lähmung. Arch. Exper. Pathol. und Pharmak., **38**, 266–283.

HERRICK, C. J. 1930. Localization of function in the nervous system. Proc. Nat. Acad. Sci., **16**, 643–650.

VON HOLST, E. 1935. Erregungsbildung und Erregungsleitung im Fischrückenmark. Pflügers Arch., **235**, 345–359.

HOOKER, DAVENPORT. 1939. Fetal behavior. Research Publ. Assoc. Research Nerv. and Ment. Disease, **19**, 237–243.

JENNINGS, H. S. 1931. Behavior of the Lower Organisms. New York. Columbia University Press. Pp. 366.

LASHLEY, K. S. 1937. Functional determinants of cerebral localization. Arch. Neurol. and Psychiat., **38**, 371–387.

———— 1938. Factors limiting recovery after central nervous lesions. J. Nerv. and Ment. Dis., **88**, 733–755.

MANIGK, W. 1934. Umstellung der Koordination nach Kreuzung der Achillessehnen des Frosches. Pflügers Arch., **234**, 176–181.

MARINA, A. 1912. Die Theorien über den Mechanismus der assoziierten Konvergenz- und Seitwärtsbewegungen, studiert auf Grundlage experimenteller Forschungsergebnisse mittels Augenmuskeltransplantationen an Affen. Dtsch. Z. Nervenheilk., **44**, 138–162.

MATTHEWS, S. A., AND DETWILER, S. R. 1926. The reactions of Amblystoma embryos following prolonged treatment with Chloretone. J. Exper. Zool., **45**, 279–292.

MOLDAVER, J. 1936. Contribution a l'étude de la régulation réflexe des mouvements. Arch. Internat. Med. Expér., **11**, 405–476.

NICHOLAS, J. S., AND BARRON, D. H. 1935. Limb movements studied by electrical stimulation of nerve roots and trunks in Amblystoma. J. Comp. Neurol., **61**, 413–431.

V. ON THE DYNAMICS OF THE NERVOUS SYSTEM

Olmsted, J., Margutti, M., and Yanagisana, K. 1936. Adaptation to transposition of eye muscles. Am. J. Physiol., **116**, 245–251.

Pavlov, I. P. 1927. Conditioned Reflexes. An Investigation of the Physiological Activity of the Cerebral Cortex. London. Oxford University Press. Pp. 430.

Prosser, C. L. 1934. The nervous system of the earthworm. Quart. Review Biol., **9**, 181–200.

Pumphrey, R. J., and Young, J. Z. 1938. The rates of conduction of nerve fibers of various diamters in cephalopods. J. Exper. Biol., **15**, 453–466.

Ranson, S. W. 1913. The fasciculus cerebro-spinalis in the albino rat. Am. J. Anat., **14**, 411–424.

van Rynberk, G. 1908. Versuch einer Segmentalanatomie. Ergebn. Anat. und Entwicklungsgesch., **18**, 353–800.

Scherb, Richard. 1938. Zur Sehnentransplantation bei poliomyelitischen Lähmungen. Schweiz. Med. Wochenschr., **68**, 354–360.

Sherrington, C. S. 1906. The Integrative Action of the Nervous System. New York. C. Scribners Sons, Pp. 427.

Shirley, Mary M. 1931. The First Two Years. A Study of Twenty-five Babies. Vol. I. Univ. of Minnesota Press. Pp. 227.

Sperry, R. W. 1940. The functional results of muscle transposition in the hind limb of the rat. J. Comp. Neurol., **73**, 379–404.

———— 1941. The effect of crossing nerves to antagonistic muscles in the hind limb of the rat. J. Comp. Neurol., **75**, 1–19.

———— 1942. Transplantation of motor nerves and muscles in the fore limb of the rat. In Press.

Swett, F. H. 1937. Determination of limb axes. Quart. Review Biol., **12**, 322–339.

Taylor, F. W. 1936. The effect of transposition of the Achilles tendon on the walking and righting movements of the frog. J. Comp. Physiol., **21**, 241–273.

Tower, S. S. 1936. Extrapyramidal action from the cat's cerebral cortex: motor and inhibitory. Brain, **59**, 408–444.

Wachholder, Kurt. 1923. Untersuchungen über die Innervation und Koordination der Bewegungen mit Hilfe der Aktionsströme. II. Mitteilung: Die Koordination der Agonisten und Antagonisten bei den menschlichen Bewegungen. Pflügers Arch., **199**, 625–650.

Weiss, Paul. 1924. Die Funktion transplantierter Amphibienextremitäten. Aufstellung einer Resonanz-theorie der motorischen Nerventätigkeit auf Grund abgestimmter Endorgane. Roux' Arch., **102**, 635–672.

———— 1925. Tierisches Verhalten als "Systemreaktion". Die Orientierung der Ruhestellungen von Schmetterlingen (Vanessa) gegen Licht und Schwerkraft. Biologia Generalis, **1**, 167–248.

———— 1934. Comments on the so-called resonance principle of nervous control. Anat. Rec., **60**, Supplement to No. 4, 30–31.

———— 1935a. Experimental innervation of muscles by the central ends of afferent nerves (establishment of a one-neurone connection between receptor and effector organ), with functional tests. J. Comp. Neurol., **61**, 135–174.

Weiss, Paul. 1935b. Homologous (resonance-like) function in supernumerary fingers in a human case. Proc. Soc. Exper. Biol. and Med., **33**, 426–430.

———— 1936a. A study of motor coordination and tonus in deafferented limbs of amphibia. Am. J. Physiol., **115**, 461–475.

———— 1936b. Selectivity controlling the central-peripheral relations in the nervous system. Biol. Reviews, **11**, 494–531.

———— 1937a. Further experimental investigations on the phenomenon of homologous response in transplanted amphibian limbs. I. Functional observations. J. Comp. Neurol., **66**, 181–209.

———— 1937b. Further experimental investigations on the phenomenon of homologous response in transplanted amphibian limbs. II. Nerve regeneration and the innervation of transplanted limbs. J. Comp. Neurol., **66**, 481–535.

———— 1937c. Further experimental investigations on the phenomenon of homologous response in transplanted amphibian limbs. III. Homologous response in the absence of sensory innervation. J. Comp. Neurol., **66**, 537–548.

———— 1937d. Further experimental investigations on the phenomenon of homologous response in transplanted amphibian limbs. IV. Reverse locomotion after the interchange of right and left limbs. J. Comp. Neurol., **67**, 269–315.

———— 1939. Principles of Development. New York. Henry Holt. Pp. 601.

———— 1941. Does sensory control play a constructive role in the development of motor coordination? Schweiz. Med. Wochenschr., **71**, 591–595.

Weiss, Paul, and Ruch, T. C. 1936. Further observations on the function of supernumerary fingers in man. Proc. Soc. Exper. Biol. and Med., **34**, 569–570.

Wiersma, C. A. G. 1931. An experiment on the "resonance theory" of muscular activity. Arch. Néerland. de Physiol., **16**, 337–345.

Wiersma, C. A. G., and van Harreveld, A. 1938. The influence of the frequency of stimulation on the slow and the fast contraction in crustacean muscle. Physiol. Zool., **11**, 75–81.

Young, J. Z. 1938. The functioning of the giant nerve fibres of the squid. J. Exper. Biol., **15**, 170–185.

V. On the Dynamics of the Nervous System

Reprinted from The Report of the Work Session on Axoplasmic Transport, NEUROSCIENCES
RESEARCH PROGRAM BULLETIN, 5, No. 4, 371–400, 1967.

CHAPTER 23

NEURONAL DYNAMICS, AN ESSAY BY P. WEISS

A. Introduction

The first intimations of axonal flow date back roughly a quarter of a century (Weiss, 1943a, 1944a,b,c; Weiss and Davis, 1943; Weiss and Taylor, 1944). From the very first it has been clear that axonal flow is merely an indicator and measure of the dynamics of the neuron and cannot be separated from that integral context. Studies have revealed the neuron to be an infinitely more complex and more variable system than was implied in earlier days when it was treated essentially as a fixed substrate for the electrochemical processes of impulse conduction. The concurrence in the individual neuron of processes of chemical, physicochemical, ergonic, electrical, and mechanical interest naturally leads to isolating each one of these components and analyzing it in its own right, apart from the others. However, there is a danger that the resulting fragmentary approaches may result in a loss of a cohesive picture in neuronal study and some signs of such a trend have already appeared in the literature on axonal flow. The present occasion thus offers a unique opportunity to present the *phenomenon* of axonal flow in an integrated perspective which combines aspects gained by various specialized techniques into a unified picture of a basic neuronal principle. The pertinent techniques, used concertedly in our laboratory, have been those of microsurgery, microscopy, cytochemistry, electrophysiology, cinemicrography, electron microscopy, mechanics, and autoradiography. I am stressing the presentation of a "phenomenon" rather than of a "theory" for, even though a unified picture is beginning to emerge, many features are still either wholly obscure or, at best, conjectural; thus it seems far more important to set in clear focus the facts that have been established conclusively than to indulge in ambitious generalizations. Even the use of the simple term "axonal flow" carries a risk, for the oversimplified concept of the underlying facts it implies is apt to give rise to all sorts of unrealistic notions, which in turn can lead to spurious research problems if not outright confusion. Therefore, I aim to confine myself essentially to the record of facts, to such interpretations as can be cogently derived from them, and to the listing of alternatives when cogent conclusions are not yet possible, without trying to force the still very spotty array of data into a "theory" with the pretentious aspect of finality.

Axonal flow is essentially a cell biological phenomenon character-

DYNAMICS OF DEVELOPMENT

istic of a cell of peculiar constitution, the neuron, its peculiarity being its enormous elongation and the extreme eccentricity of the nucleated part of the cell body — the "perikaryon." In the molecular communities of most other cells, the processes of synthesis, macromolecular compounding, internal transport, catabolic breakdown, and elimination of metabolites and cell products are so grossly intermingled that they are difficult to separate. Likewise, the neuron used to be taken for granted as a cell in which anabolic and catabolic processes go on ubiquitously, although, for instance, the confinement of Nissl substance to the perikaryon clearly expressed some degree of internal segregation. Axonal flow has radically changed that notion by demonstrating that in the neuron the major production site of cellular constituents and the area of their consumption are not intermingled and coextensive but are neatly segregated, which necessarily requires an organized traffic system from source to consumers — and this is exactly what axonal flow provides. In that sense, the neuron constitutes a uniquely favorable object for basic studies on cell biology in general. In turn, what has been learned through these studies about the dynamics of that specialized cell has introduced a new dimension into our thinking about the nervous system; for the shift of our image of the nerve cell, from that of a rather static fixture to that of a rapidly growing body in constant flux, widens considerably the range of possibilities for explaining the great adaptability of nervous functions throughout life.

B. History

The concept of continuous axonal flow emerged from a fortuitous observation, namely, the damming up of surplus axoplasm at the proximal side of a sudden narrowing of the cross section of a nerve fiber by a chronic local constriction, followed by a corresponding reduction of the fiber diameter at the distal side of the "bottleneck" (Figure 1, A-C). Even though the first descriptions were published in 1943 and 1944 (Weiss, 1943a, 1944a,b,c; Weiss and Davis, 1943; Weiss and Taylor, 1944), the comprehensive account of the results, covering 6 years of experimentation and roughly 100,000 nerve fibers, did not appear until 1948 (Weiss and Hiscoe, 1948). During that early phase, the investigations were concerned largely with a systematic follow-up of the constriction effect. It established in detail (1) the acute effects of the constriction, that is, the displacement of neuronal material in the proximal and distal directions; (2) physiological effects (reversible pressure block of conductivity) (Weiss and

Davis, 1943); (3) the development of edema between the nerve fibers at the proximal side; (4) the origin of this edema from dammed-up endoneurial fluid streaming proximo-distally; (5) the appearance of various characteristic deformations of the individual nerve fibers at the proximal side of a constriction; (6) the demyelinization within the constricted zone (Weiss and Taylor, 1944); (7) the reduction of fiber caliber throughout the part distal to the constriction; (8) the permanent character of the observed proximo-distal asymmetry; (9) the general functional adequacy of chronically affected fibers of this kind (save for rates of impulse conduction); and, finally, (10) the downflow of axonal content piled up proximally into the distal portion after the release of a chronic constriction.

The conclusion from those experiments was inescapably that something on the neuron was constantly moving in a proximo-distal direction and that any local throttling of the progress of that "something" resulted in the piling up of neuronal content. The "something" emanating from the perikaryon might have been construed to be some "growth factor" indispensible for local growth and maintenance of the axon all along its length. This supposition would have been in line with hypothetical suggestions put forth in a prophetic vein by G. H. Parker (1932) and by Gerard (1932). In fact, in my first preliminary report on the phenomenon, I myself ascribed the phenomenon to the damming of "some essential factor produced by the nerve cell body and required for continued growth in width of the peripheral fiber." This clearly referred to transport of something *within* the axon. It soon, however, became evident that this interpretation was incorrect and that the damming actually signified that the entire "column of axonal substance is subject to a steady centrifugal pressure or flow" (Weiss and Davis, 1943); in other words, one is dealing with a movement *of* the axon rather than *in* the axon, the axon growing forth continuously from its root in the cell body.

This obviously raised the question of the fate of this incessant increment in a nerve fiber which, after all, retains stationary dimensions. My answer at that time was wholly conjectural. I assumed that the axon has to be furnished all its macromolecular constituents from its central cell body and that the axonal column consumes itself as it moves along, replacing the catabolically degrading elements of its molecular population by its content of fresh ones in a statistical equilibrium between rates of degradation and of replenishment. Some tenuous support for this contention came from data in the literature that I shall mention later. Major corroboration came from the work of Hydén (1943), done about the same time in Sweden, but unavailable to me during the war years. That work presented direct cytological evidence for the very high rate of protein synthesis within the soma of the neuron. As our inferences from the axonal flow had made us postulate a high rate of macromolecular, particularly protein, regeneration in the cell body, so the direct demonstration of that high rate, of necessity, called for an outlet in a transport system such as we had discovered. This mutual reinforcement dictated the further steps in our experimental program.

Feeling reasonably reassured that the axonal transport mechanism was an actual fact, and having established some of its properties, I thus proceeded to explore some of its major aspects in greater detail, essentially in three directions. One was to extend the morphological evidence on which the technological interpretation of the phenomenon rested from the microscopical to the electron microscopical level. The second was a search for direct chemical markers, preferably isotopic labels for the demonstration and finer resolution of the molecular traffic involved. The third objective was a cinemicrographic recording, if possible, of the mode of movement to furnish clues to the underlying physical mechanism. None of this program could have been carried forth without the extraordinarily resourceful collaboration of members of my staff, particularly Dr. A. Cecil Taylor, Dr. Aiyappan Pillai, and Mr. Bock, as well as the invaluable help of friends from other institutions, above all, Dr. Heinrich Waelsch and Dr. Abel Lajtha.

Whereas the electron microscopic and cinemicrographic parts of the program have largely remained confined to our laboratory, the biochemical pursuit of the axonal flow by means of isotopes has become widespread, and, as the program of this conference indicates, has attained a dominant position. The isotope tracer method was started early (in 1949) in my laboratory (Shepherd, 1951) and the neighboring laboratory of Gerard (Samuels et al., 1951) with the use of radioactive phosphorus, carried on later (Waelsch, 1958) with the more highly diagnostic ^{14}-C-amino acids (Weiss, 1961a), and greatly furthered by the introduction of tritiated amino acid (Droz and Leblond, 1963). It was further refined by strict confinement of the application of label to a circumscribed nerve source (see Section F), and having been adopted ever more widely, has now become a major tool in the gauging of the chemical dynamics of the neuron by axonal flow. The results have well sustained the original proposition of protein involvement.

In combination, these fine structural, cinemicrographic and isotope studies have yielded a picture of the neuronal dynamics far more composite and intricate than a simple term such as "axonal flow" can convey. It seems, therefore, that I can serve the purpose of this Work Session best by dissecting the complex phenomenon into its several and diverse components. (For earlier condensed summaries, see Weiss (1960b, 1961a, 1963).) I shall try to extract the gist of the large array of data amassed by my laboratory, only a small fraction of which has been published, as well as regroup the relevant facts according to a methodological or logical consistency, regardless of their chronological order.

C. The Basic Test: Damming

The crucial test of axonal flow is provided by any sudden narrowing of the flow channel within which the axon is constrained in its extracellular course. The manner in which this diminution of the flow channel is produced is of no consequence. Local constriction by an artificial ligature (Figure 1, A-C) has the same effect as the natural entrance of a nerve fiber from a looser packing into a fibrotic zone, such as a scar. Conversely, a very thin regenerating axon, advancing in an old oversized though partly ligated tube (Figure 1, E-G), continues unimpeded without deformation

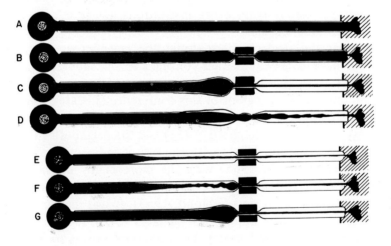

Figure 1. Schematic representation of single nerve fibers subject to chronic constrictions. A, normal mature fiber; B, same fiber as A immediately after application of constricting cuff; C, same fiber as B after assuming new stationary asymmetry: damming of neuroplasm on the proximal side of the "bottleneck" coupled with reduction of caliber distally; D, same fiber as C after removal of chronic constriction: downflow of dammed-up neuroplasm. E, a severed fiber (at taper) that has regenerated a thin axon through a constricted zone; F, same fiber as E after gaining width: beginning proximal damming, distal part remains thin; G, same fiber as F in terminal state: G = C. [A-D, Weiss, 1963; E-G, Weiss and Hiscoe, 1948]

until it has enlarged up to the width of the lumen left open in the pinched portion; only during its further growth in width does excess material begin to dam up at the entrance to the narrow stretch (Figure 1, F). While the distal portion of such *regenerating* fibers simply stops widening (Figure 1, G) the parts distal to the constriction of a *mature* full-sized fiber actually lose size (Figure 1, C). Since in both cases the results are principally the same, I shall deal from here on only with *mature, uninterrupted fibers* in a steady-state condition, so as to dispel any notion that axonal flow and damming might be features peculiar to "growing" embryonic or regenerating fibers with free mobile tips, which obviously must be fed by influx from the cell body.

Since the bulk of the experiments were done by means of active constriction, it is important to learn to distinguish clearly the chronic consequences from the acute short-term effects of the operation. The immediate effect of a constriction is the displacement of substance from within the constricted zone into the free sectors up and down the nerve, both within the axons and between them, the volume of the extrusions varying with the degree and speed of strangulation. (Undistensible ligatures being most harmful, we have used mostly elastic cuffs of artery which contract gradually until the turgor of the nerve and the elasticity of the cuff are in equilibrium.) This acute flanging of the nerve on both sides of a stricture (Weiss and Davis, 1943; Weiss, 1943b) must be taken into account in explaining transitory histochemical modifications near both ends, entrance as well as exit, of a constriction (Lubińska, 1964; Kreutzberg, 1963; Dahlström, 1966). The possibility of electrophoretic effects converging upon the traumatized, supposedly electronegative, region must likewise be taken into consideration. Because of these uncertainties, I shall pass up short-term experiments entirely and confine the following account to nerves under local constriction for at least several months up to more than a year. All chronic constrictions were of a moderate degree, producing a statistical diminution of fiber diameters without complete occlusion, severence, or lasting pressure block. Such specimens displayed no gross functional impairment in impulse transmission or muscle function. Monkeys kept in this condition (histologically verified) for about 10 months, while showing a mild muscular atrophy (15% on the average), showed perfect motor coordination in the use of their legs despite the presumable slowing of impulse conduction in the distal nerve stretches of reduced fiber diameter (Matson et al., 1948; Alexander et al., 1948).

The changes proximal to the bottleneck were of two different kinds. The endoneurial spaces between the nerve fibers were distended by an edematous fluid moving continously in a proximo-distal direction down the nerve (see below in Section F). More significant were the changes of the axons themselves. To characterize them properly, the term "damming," used for brevity, is not sufficiently descriptive, as it suggests a mere quantitative piling up of surplus material. The morphological expression of such a simple congestion would be a massive bulbous swelling tapering off in the proximal direction. The actual changes, however, consist of complicated structural deformations, which denote conclusively that the axonal column is a semi-solid cohesive body and not a liquid-conducting pipe. The major forms of these structural modifications are illustrated in Figure 2. They can be described (from bottom to top) as telescoping, ballooning, corkscrewing, and beading. Partly, the myelin sheath conforms to the contortions and convolutions; partly, the axonal folds become detached from the sheath with liquid filling the interstices.

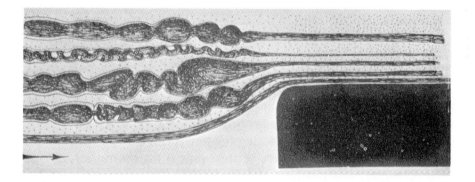

Figure 2.   Tracings of characteristic axonal deformations at the proximal entrance to a constricted zone. [Weiss and Hiscoe, 1948]

In a technological interpretation (see Section E), these deformations denote the resistance met by a viscous cylindrical column propelled in a channel of matching size when passing into a narrower portion of the channel. The sudden throttling of access to the channel by new material jams up the traffic column proximally. Prevented from proceeding at its former rate, the elongation is accommodated partly by folding and spiraling, and partly by local widening, until new stationary equilibrium is attained. Concurrently with these proximal alterations, the parts of the nerve fiber lying distally to the bottleneck become emaciated in proportion to the throttling of inflow by the bottleneck. This attenuated size of the distal length of the fibers persists as long as the constriction is maintained.

After a moderate constriction, the distal fiber portions, though reduced in size, are still wide enough to register the effects of a second constriction applied in tandem further down. In such cases, individual fibers can be shown to have become dammed up at the proximal entrances to both the upper and the lower bottlenecks (Weiss and Hiscoe, 1948). The marked proximo-distal structural asymmetry in the stretch of fiber lying between two bottlenecks, also confirmed by chemical tests (Dahlström, 1967a), proves clearly that the driving mechanism of axonal flow operates unidirectionally and is operative actively at every point along a nerve fiber (see below in Section H).

The chronic proximo-distal asymmetry relative to a constriction, that is, proximal engorgement concomitant with distal emaciation, is not a static, structurally fixed condition, but is the expression of a stationary dynamic configuration of the flow pattern of a highly viscous and relatively form-consistent material, comparable vaguely to the flow of lava with internal reinforcement by semi-rigid fibers. To test the flow properties directly, two series of experiments were carried out. One consisted of removing the constriction after stationary asymmetry had been reached, sometimes as late as after 1 year (Weiss and Cavanaugh, 1959). The dammed up material could then be seen (Figure 1, D) to be gorged down through the formerly constricted region into the distal portion in the form of a tidal wave, eventually restoring the emaciated distal stretch to near-normal dimensions. The advance of this wave front was roughly 1 to several millimeters per day. From this we reached the reasonable, but unsubstantiated, conclusion that the observed rate was a fair measure of the order of magnitude of the normal translatory progress of axonal flow. It was risky to generalize this conclusion in view of the rather limited range of testing, which pertained to limb nerves of rats of a given age measured at a given level, and not even accurately measurable at that. Yet, convection rates in nerve of the order of millimeters per day have in the meantime been rediscovered in such a variety of forms and conditions (see Section F) that the value is beginning to assume an aspect of universality. It refers only to the advance of the axonal column as a whole and not to other traffic that may be going on within the axon and between nerve fibers, as reported below.

Recently, we have resumed the study of axonal consistency and flow properties by a second method, i.e., direct cinemicrography (films were shown at the Work Session). While I reserve the cinemicrographic evi-

V. On the Dynamics of the Nervous System

dence for the intrinsic axonal driving mechanism for Section H, our studies on enforced flow are pertinent in the present connection. Myelinated nerve fibers of mature animals (mice, rats, etc.) were filmed at near-normal speeds under high-power phase-constrast optics while mild compression was applied to the nerve at a point outside the visual field. Our observations confirmed the fact that the axonal column, yielding to the local compression by translatory displacement, moves like a semi-solid body with frictional delay along the wall, yet with considerable resilience, as evidenced by its recoil to the original position after decompression.

In this connection one point must be stressed rather forcefully. Having had extensive first-hand experience with living nerve fibers in both the embryonic and the consolidated mature state, I find the differences in physical consistency and structural properties between the two so profound that off-hand extensions of conclusions from one to the other are gratuitous. It would be like comparing the nucleated and highly mobile mammalian erythroblast with the mature, anucleate, encysted erythrocyte to which it becomes converted.

D. Electron Microscopy

Because the structural deformities in the damming process are decisive for the issue of whether axonal flow represents movement *of* the axon or traffic *in* the axon, we have carried on extensive studies of both constricted and normal nerves under the electron microscope. (Save for a few preliminary notes (Weiss et al., 1962; Pillai, 1964; Weiss and Pillai, 1965), this material, amounting to many thousand electron micrographs, has not yet been published. But, being systematically annotated and ordered, the collection is readily accessible for inspection and study in our laboratory.)

In general, our electron optical data on axonal fine structure are in accord with the observations reported by others in the current literature. The matrix contains the two commonly acknowledged, sharply distinct longitudinal structures—the rectilinear neurotubules, about 200 A wide, and the more wavy neurofilaments, about 70 A thick. (Incidentally, young collateral sprouts of only a fraction of a micron in diameter contain only neurotubules, but no filaments.) Our preparations have shown only sparse stretches of vacuolated strands that could pass for endoplasmic reticulum. One observation rather common in overosmicated preparations, however, deserves special mention here as it has not been previously recorded and

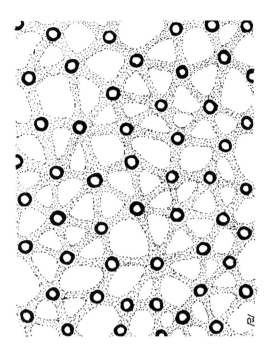

Figure 3. Schematic representation (from an electron microscope view) of part of an axonal cross section. Neurotubules (dark rings) lie in a spongy matrix permeated by seemingly liquid channels. [P. Weiss]

might augment the variety of transport systems with which the interior of the axon is equipped. (It is shown diagrammatically in Figure 3.) In cross sections the rather regularly distributed neurotubules often appear at the intersections of a lattice of rather electron-dense strands, delineating spaces between them, which by their rounded contours intimate liquid turgidity. Those meshes probably represent cross sections of a second can-alicular system in the axonal matrix.

The electron microscopic pictures of the deformities and convolu-tions on the proximal side of the bottleneck were in full accord with our earlier microscopic records. In cross sections of axons at this level, the bundles of neurofilaments and neurotubules showed confused, contorted courses with intersections of whorls and loops. In longitudinal sections, the convoluted course of both neurotubules and neurofilaments con-formed exactly to the contour folds and twists of the whole axon. This seals the proof that the deformities, pictured in Figure 2, reflect the interior structure of the axon faithfully down to the last detail. In the dammed portions, the neurotubules were found more crowded toward the surface, while the filaments occupied mainly the axial core. In many cases, the outline between these two regions was quite sharp, corresponding presumably to the frictional retardation of flow along the axonal wall observed in the motion pictures of enforced flow. The microstructures thus can serve as flow gauges.

The occurrence of strings of vesicles in front of constrictions and at the blind ends of completely severed fibers, first described by van Breemen et al. (1958), has been observed routinely in our preparations of both partially and totally occluded nerves. Their possible bearing on the origin of "synaptic vesicles" calls for more systematic investigations into the pinched-off bulbous swellings at the blind ends of the various canal systems in the axon (see De Robertis, 1967). Significantly, tangential sections through the congested dammed-up portion of an axon have frequently shown a network of true anastomoses among the neurotubules; in those cases, widenings and vesicles were less frequent.

The most conspicuous electron microscopic disclosure was the shift in the mitochondrial distribution (Weiss and Pillai, 1965). A 50-milli-micra-thick cross section through a normal axon contains, on an average, two mitochondrial cross sections—a very sparse population. A similar section through an axon at a chronic bottleneck can show a hundred or more mitochondrial sections. Like the neurotubules, and interspersed with them, they are concentrated in the superficial layers of the axon, in all stages of degeneration from swollen bodies with intact cristae, through lamellated myelin figures and heavily osmiophilic spherules, to vacuoles presumably filled with free phospholipid. At the same time, mitochondria located in the axial region are usually of normal appearance and elongated. From all these observations, a consistent picture has emerged as follows: The damming of flow within each individual axon permits the unimpeded passing of only the axial part of the flow, the normal configuration of the mitochondria there indicating their being carried by the current. The more marginal layers are increasingly retarded. As a result, the mitochondria traveling near the surface are arrested and accumulate like floats washed ashore at the banks of a river. Conceivably, the breakdown products of the stagnated mitochondria lead to the coagulation of an axonal crust, which might explain the sharp interface between a crust and a core portion, harboring neurotubules and neurofilaments, respectively, as described above. At any rate, it is clear that mitochondria are being carried down continuously with the axonal stream. This in no way precludes the possibility of their carrying out additional active local excursions up and down within a narrow range, much as a swimmer can swim upstream or downstream in a current.

Since the density of the sparse mitochondrial population of the axon does not seem to vary along its course, the main and presumably exclusive source of mitochondrial reproduction must lie in the perikaryon. Thus the rate of reproduction can then be estimated. If mitochondria, which have a length of a few microns, are carried down by axonal flow at a rate of a few millimeters per day, the distance of their daily travel is thus about 500 to 1,000 times their own length. Accordingly, in order to maintain this standard rate of export of mitochondria, the cell body would have to reproduce up to 1,000 mitochondria per day. This figure also gives, then, the number of mitochondria going daily past any given level of a nerve fiber. Applied to the level of the entrance to a bottleneck and assuming a case in which only about 80% could move through freely in the axial part of the stream, this would calculate to a local retention and accumulation of the order of 100 mitochondria in front of the entrance, which agrees roughly with our actual observations. This population of arrested mitochondria changes constantly, their number being determined by the ratio of new arrivals over disintegrated earlier arrivals. From these figures, the rate of breakdown of an arrested neural mitochondrion would seem to be very rapid. More systematic calculations are essential for cell biology. As far as the axon is concerned, the mitochondrion serves mainly as a useful marker and index of the axonal stream.

E. Mechanodynamic Analysis

The morphological manifestations of axonal flow described in the preceding sections, in addition to serving as signs of proximo-distal convection of neuroplasm, contain the key to the mechanism of the convective process. As indicated above, they characterize the translatory movement as that of a semi-solid body of structural consistency and not of a liquid system of freely mobile molecules in solution. It is common for purely chemical bulk determinations of proximo-distal shifts of substance in nerve to be referred to as "movement of substance" or "axonal transport," without concern about the physical mechanism of the displacement, much as an economic statistician might list changes in the balance sheet of production over consumption without attention to traffic and the channels of commerce.

This relative disinterest in mechanisms extends to the problems of intracellular traffic in general. As a result, many cell physiological problems are still dealt with as if the dynamics of diffusion in test tubes, driven by the collision frequencies among freely mobile, thermally agitated particles, could be automatically scaled down to the microdimensions of the cell space. In capillary spaces, with lumina or meshes of the order of a few hundred angstroms, convection of macromolecules and particles of sizes no more than one order of magnitude smaller cannot take place by mere molecular buffeting but requires external driving mechanisms of mechanical and electrical nature. Such mechanisms, which channel kinetic energy from random dissipation towards a resultant translatory effect, might be designated as "pumping mechanisms." We still know very little about them in general cell biology; in a few favorable cases, they have been demonstrated to exist (Weiss, 1961b), but for the rest they must simply be postulated. The mechanodynamics of axonal flow are unquestionably a case in point. Lest it become eclipsed, I shall restate in the following the *quantitative* evidence for the fact that neuroplasmic flow involves *translatory movement of the axon as a whole,* in contradistinction to mere fluid transport *within* a static axon. As will be brought out later, this does not rule out separate liquid traffic routes within the axon, for which the more solid axonal column is the basic carrier and reference system.

The mechanical effect of a constrictive force applied locally to a plastic or elastic system grades off from the point of application according to a relatively simple formula. The same is true for the resistance encountered by an elastic system in motion when meeting an obstruction. For instance, a constricting cuff, compressing a bundle of tubes such as a nerve circumferentially, affects the most superficial tubes first by reducing their diameters (note the manometer rise in Figure 4,a). This leaves less force for the subjacent tubes, and so forth. Thus the most central tubes are least affected as each outer layer acts as a "shock absorber" for the next inner one. Figure 4,b illustrates this radial surface-to-interior gradient in the reduction of the cross sections of the fibers in the constricted zone of three individual nerves, as well as the correlated chronic changes in the diameters of the same fibers on the proximal and distal sides of the constriction (Weiss and Hiscoe, 1948). One notes that the proximal size spectrum is exactly the inverse of that of the originally compressed zone, whereas the distal size distribution has become the same as that in the constriction itself. There is thus a direct quantitative relation between the degree of artificial narrowing of the flow channel and the severity of damming on the one hand and peripheral shrinkage on the other. Such results are positive

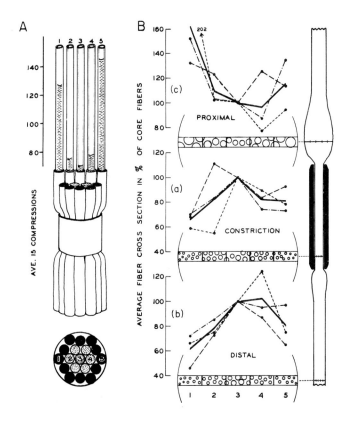

Figure 4. Radial surface-to-center decline of constriction effects in a model or rubber tubes (A) and a cuffed nerve (B). A, manometer tubes connected to rubber tubes registering the actual reduction of diameter of the latter (proportional to the rise of the indicator column) on uniform compression. B, fiber size spectra of axons in three nerves constricted for ± 8 months showing the surface-to-core gradient in the effective reduction of fiber diameters in the constricted zone (a), the corresponding gradient in the free distal zone (b), and the inverse pattern of damming intensity proximally (c). [Weiss and Hiscoe, 1948]

proof of non-Newtonian flow.

Measurements of longitudinal deformation have given even more direct quantitative evidence of the semi-solid nature of the axonal column, as follows. Model experiments with steadily propelled elastic rubber tubes (Figure 5,a) have shown that obstructions (O) in the path of the advance produce deformations (e.g., twists) similar to the ones in the dammed portions of nerve fibers, and that the degree of deformation declines as a linear function of the distance from the obstruction (Figure 5,b). The narrowing of the axonal flow channel at the entrance to a constricted nerve

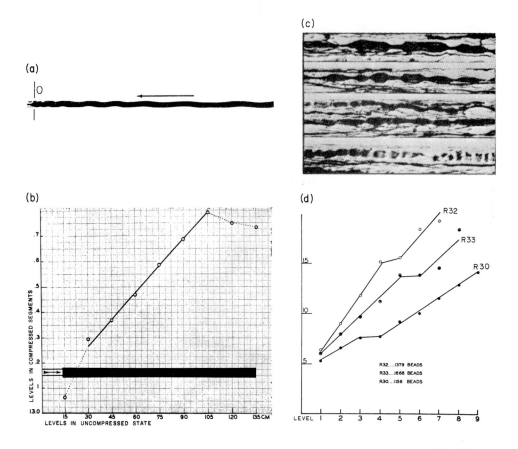

Figure 5. Linear decline from the effect of force (or resistance) applied locally (a and b represent a rubber model; c and d represent a constricted nerve) along the longitudinal axis of an elastic tube. a, twisting of a rubber tube propelled (arrow) inside a glass tube against a block at 0. b, longitudinal shortening (compression) of consecutive segments originally equal in length (15 cm). c, samples from a beaded zone proximal to a chronic constriction (10 weeks), taken from the same nerve at equidistant intervals; proximo-distal direction from top to bottom. d, linear increase of average distance between beads (equals decline of number of beads per unit length) with increasing distance from level of constriction (level 1) in three nerves constricted for 10 weeks.   [Weiss and Hiscoe, 1948]

portion amounts to partial obstruction. Liquid content of Newtonian flow characteristics would remain unaffected by such throttling, except for an increase in the velocity of flow through the narrows. The axonal content, by contrast, is thrown into the configurations exemplified in Figure 2. Now, if one measures the least complicated cases, those of simple beading, one finds that the number of beads per unit length, as a measure of the deforming force, decreases upstream as a linear function of the distance from the bottleneck (Figure 5,d), exactly as in the rubber model. Figure 5,c shows four samples from the same nerve, the bottom one nearest to the bottleneck, the upper ones farther proximally at equidistant intervals.

385

Without elaborating further, this is another proof of the semi-solid, rather than fluid, consistency of the axonal column.

Having dealt with the phenomenon of axonal flow in descriptive terms, we can now turn our attention to the meaning of this chemical transport system.

F. Chemical Interpretation and Isotope Studies

The observations on axonal flow led to the hypothesis that it represented a feeder column carrying materials produced in the cell body for the needs of both the internal household of the nerve fiber and the periphery innervated by it. As I stated in 1944, "mature axons thus seem to grow perpetually from their cells, undergoing commensurate peripheral dissipation." I also indicated at that time that the dissipation might include "the discharge of substances (e.g., acetylcholine and other neuro-humors) from peripheral nerve." I then made the fairly broad jump to postulate that the primacy of the perikaryon for macromolecular synthesis pertained particularly to the *proteins* destined for the axon as replacements for its catabolically degrading enzyme systems. This antedated our knowledge of Hydén's work (1943) which clearly anticipated and corroborated our presumptions by far more direct means. All of this, of course, also preceded by a considerable period the identification of the pathway of synthesis from DNA through transcription to RNA, to eventual translation into primary protein assembly.

My rationale at that time was derived from scattered data in the literature about the high level of ammonia liberation from peripheral nerves. The argument ran about as follows (Weiss and Hiscoe, 1948): If protein is manufactured exclusively in the perikaryon and shipped from there into the peripheral nerve fiber, then its progressive catabolic degradation there, with no opportunity for local re-utilization of the breakdown products, should end in complete deamination, the end product being eliminated as ammonia. By calculating, on that highly tenuous assumption, the average rate of catabolism of proteins in nerve, we arrived at a figure for their half-life time of the order of a month. This rough estimate permitted us to make a tentative calculation of what the rate of replenishment of protein in nerve fibers would have to be if the whole supply were to come from a central source in step with its peripheral degradation. It turned out that this would require a steady supply stream from the cell advancing at a rate of the order of millimeters per day. The correspondence between this figure and the actually observed rate of advance of dammed axoplasm in deconstricted fibers (Section C), suggestive though it was, might also have been sheer coincidence.

It did, however, encourage more direct tests, such as the tracing of the purported transport from its site of origin with radioactive markers. Being then at the University of Chicago, where I had access to some of the high-powered isotopes of the Manhattan project on atomic energy, I took advantage of their availability, first to track down endoneurial flow (Weiss et al., 1945), which the edematous accumulations of fluid in front of constrictions (see above) had intimated. By means of microamounts of isotopes injected into peripheral nerve, the existence of a constant proximo-distal drift of fluid in the endoneurial spaces between the nerve fibers was confirmed and its rate determined as of a few millimeters per hour, that is, about 25 times faster than the axonal flow in the same direction.

Attempts to mark the latter, however, were not at first equally successful. We started in 1949 with radioactive phosphorus, obtaining suggestive but by no means conclusive results (Shepherd, 1951). At the same time, Gerard and his collaborators (Samuels et al., 1951) succeeded in demonstrating a general shift of radioactivity in ^{35}P-labeled nerves, corroborated later by others (Ochs and Burger, 1958). After the more instructive demonstration in 1958 by Waelsch, using ^{14}C-amino acids as labels, of a progressive proximo-distal shift of protein, we embarked, with the collaboration of Waelsch and Lajtha, on a rather diversified series of experiments using ^{14}C-amino acids and ^{14}C-glucose as markers. We made chemical as well as autoradiographic identification of the protein-incorporated label in the nerves of amphibians, mammals, and birds. All these experiments, summarized in an earlier report (Weiss, 1961a), concurred in confirming the cellulifugal traffic of labeled compounds, specifically proteins, but did not go much beyond substantiating the original hypothesis. Introducing the use of tritiated amino acids, Droz and Leblond (1963) then showed unequivocally that ^{3}H-leucine-marked protein in rabbit sciatic nerves shifts in a proximo-distal direction at a daily rate of the order of 1 millimeter. Since these developments are discussed by other participants in this conference, I shall pass over the details.

Prompted by the success of this technique, we took it up and sharpened it by localizing the labeling of the protein source more rigorously to the nerve cells themselves, avoiding the contamination of the rest of the animal which is unavoidable in most injection methods. Moreover, the increased signal-to-noise ratio of "hot" axons over a "cold" background permits the performance of large-scale experiments with scintillation counter recording. Applied to symmetrical nerve sources, it also makes it possible to compare a "hot" nerve with a "cold" control nerve in the same animal, thus circumventing the uncertainties of interindividual variability. The first experiments, the administering of microinjections of tritiated leucine into the vitreous body of the eye in the mouse, fully met the requirement (Taylor and Weiss, 1965). The amino acid was promptly incorporated into the cells of the retina, leaving the rest of the animal, even the coats of the eye, unmarked. The autoradiographic records of optic nerves preserved on successive days revealed two phenomena (Figure 6). The first was that some of the labeling solution seeped into the optic nerve during the first 3 hours after injection and became fixed in the proteins of the glial cells of the nerve near its exit from the orbit (shaded area in Figure 6). In line with

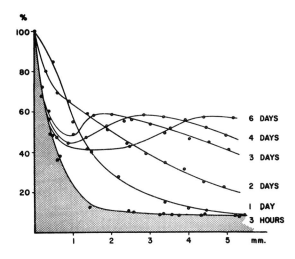

Figure 6. Counts of silver grains over autoradiographs of optic nerves from eyes injected with tritiated leucine, normalized by setting each highest count (at exit from bulb) as 100%. [Taylor and Weiss, 1965]

subsequent observations on the olfactory nerve, the pathway of this seepage must be sought in the extraaxonal spaces. On top of this fleeting primary contamination, the intraaxonal advance of the label from the retina to the brain proceeded in the form of a traveling wave, again at the rate of about 1 millimeter per day. Its crest showed a progressive flattening (see Figure 6), which proved that different points along the wave front advanced at different rates. This signifies that axonal flow proceeds either at different rates in different neurons, or with a core-to-surface velocity gradient within each individual axon,* or, most likely, at a combination of both rates. Symmetrical labeling of both eyes gave sufficiently symmetrical results to accept one eye as a reliable control for the other in future tests of the effects on neuron dynamics and axonal flow rate of metabolic inhibitors, drugs, stimulation (light vs. darkness), and so forth.

The most convenient nerve for the labeling of axonal flow is the olfactory, the only one in higher animals whose cell body lies in the body surface and hence is directly accessible. By applying tritiated-leucine-soaked gelfoam directly to the lining of the nostril, the cells of origin of the olfactory fibers in that surface epithelium could be readily marked and the advance of the labeled proteins toward the brain be followed autoradiographically (Weiss and Holland, 1967). Significantly, no transsynaptic transfer of labeled material beyond the primary neuron has been seen, at least for the first 3 weeks. (The distinction between labeled axons and their completely blank environment being absolutely sharp, this method recommends itself more generally for the accurate tracing of terminal ramifications of neurons; it is comparable to the Golgi method yet is without its capriciousness.)

In order to circumvent the limitations imposed by conditions in the living body and yet attain circumscribed labeling of the neuronal cell bodies, I finally resorted to using excised mouse neuron preparations, that is, spinal nerves removed from the body with their attached spinal ganglia intact (Weiss, 1967). It proved feasible, with certain precautions, to immerse the ganglionic end in the labeling solution without external contamination of the rest of the nerve. The incorporation of ^{3}H-leucine in the ganglionic proteins reached nearly saturation level within 2 hours of submergence. The treated nerve preparations were then either transplanted subcutaneously into host animals or else kept in nutrient solutions *in vitro*. As in the optic nerve experiments, there was an initial endoneurial and vascular infiltration of labeled substance from the treated ganglionic end into the nerve trunk but this again remained confined to stationary non-neural cells within the first few millimeters from the source. All preparations were handled pairwise; after identical treatment through the labeling and washing procedure, the ganglion of one of the members of a pair was cut off, while the other nerve was left with its labeled source. The difference in radioactivity between the ganglionated and deganglionated partners after various periods of transplantation or explantation *in vitro* then

---

* The assumption of a velocity profile with a peak in the center of the axon and the slowest rate at the circumference is almost a postulate if one is to reconcile steady-state axonal dimensions with the longitudinal age gradient of the traveling protein systems. [P.W.]

served as measure of the progressive influx of axonal material into those fibers that had been left connected with their supply source in the ganglion. This translocation of intraneuronal protein from ganglion to nerve amounted to between 6% and 11% per day, depending on the experimental conditions. Assuming a commensurate traffic rate for the non-protein compounds, one can roughly calculate that the whole macromolecular population of the nerve cell is renewed about once every day, which, translated into axonal dimensions, gives an average centrifugal outflow from the cell of the order of 1 millimeter of axonal length per day.

Even though this last series included a total of 255 test nerves, one should still view the quantitative data given here within an uncertainty range of perhaps one order of magnitude. A further complication, not previously taken into account, appeared in the tests of directly immersed ganglia. Autoradiographs of sections of ganglia immersed for 2 hours, when 95% of the maximum attainable uptake of leucine had been reached, showed extreme differences of radioactivity among neighboring ganglion cells. This indicates great disparities among different neurons in either the uptake of amino acid or in their protein synthesis. Whether this fact reflects differences of physiological state or of cell type, it cautions us against treating the ganglion cell population as if it were uniform. Autoradiographs of ganglia, which had been kept as subcutaneous grafts for several days after the labeling and washing, still showed a mottled population; while most of the cell bodies were no longer marked, their labeled content apparently having been drained off into the axons, some were still remarkably radioactive. It is impossible to judge from our sparse data whether these latter are the ones which had initially been the most heavily marked or whether the extrusion from cell to axon perhaps does not go in a continuous sequence, but occurs rather in alternating phases of production, storage, and release.

It will be seen that, on the basis of all the diversified evidence accumulated through many methods and from many different quarters, the general concept of the neuron as one of the most rapidly growing cells of the body, feeding the axonal flow, seems solidly established. The magic number for the daily flow rate of the order of 1 millimeter has likewise been arrived at independently in so many diverse circumstances that one could scarcely doubt its reality. However, to go into much greater detail and try to insist on specific interpretations of the basic phenomenon, save as a private guide to further experimentation, would be rash. The main point is to keep the total complex of observations to be harmonized clearly in view; the rest is up to the future. Solely in the interest of bringing together as many data as might be of potential relevance to the problem, the following comments are added:

G. Supporting Data

The evidence gathered thus far strongly supports the conclusion that the perikaryon is by far the major source of macromolecular synthesis and further assembly of cell products for the whole neuron. On the other hand, whether the cell body is the exclusive source of *all* protein, structural or enzymatic, will be determined by facts and not preconceptions. From the early days of the identification of the Nissl substance in the perikaryon as ribonucleoprotein to the brilliant unraveling in recent years of the role of RNA in the patterning of amino acids in proteins, the primacy of the nucleated cell territory in protein synthesis might have been suspected. However, as this conference will bring out, the data on the distribution and relevance of local RNA in the neuron as a true indicator of protein synthesis have not been unequivocal. Cytochemically, virtually all RNA in the neuron has been found concentrated in the perikaryon. Axons have been described as either devoid of it or containing minimal amounts (with some noteworthy exceptions, such as Mauthner's fiber in lower vertebrates (Jakoubek and Edström, 1965)). *In vitro* tests with tritiated uridine have shown RNA turnover to be sharply confined to perikaryon with no incorporation beyond the axon hillock (Utakoji and Hsu, 1966). Thus far, peripheral protein synthesis has only been claimed for cholinesterase in nerve endings; yet, since the evidence that this is a truly *de novo* production is still rather indirect, it would be unsafe at the present stage to take final monopolistic attitudes.

Far more definite than its role in the internal metabolic household of the neuron is the role of axonal flow as a mechanism for the transport of specialized cell products from the central site of manufacture to peripheral destinations. The whole field of neurosecretion is full of illustrative examples. In fact, as one now recognizes, the cases in which traveling products of the cell body can be tracked readily because of their particulate or vacuolar shape appear now only as special manifestations of the general principle of axonal flow. Originally determined from microscopic observations, the list of pertinent examples has been widely enlarged with the advent of the electron microscope, many of the formed globules being of submicroscopic dimensions. Whenever rates of movement of microscopically visible inclusions have been reported, they seem to fall again in the millimeter class. But this does not seem to hold universally for the sub-microscopic particles.

The transport of substances instrumental in the excitatory and transmitter functions of neurons from cell body to site of action farther peripherally has been postulated or positively demonstrated, particularly for catecholamines (von Euler, 1958). Rate determinations of this transport made by consecutively timed assays after constrictions analogous to those that demonstrated axonal flow, however, indicated speeds significantly faster than those of the 1-millimeter class. Such faster rates were also indicated for the proximo-distal convection of injected phospholipids (Miani, 1964) without, however, establishing the exact route of the convection. Except for acute transitory accumulations at both sides of a constriction, which for the reasons mentioned in Section C must be discounted as unrelated to the continuously operating axonal flow, the direction in all reported observations of sufficient duration has been *cellulifugal*. The fact that the observed slow and fast rates do not form a graded scale, but cluster about distinctly separate modes, strongly suggests a multiplicity of pathways in the system of the nerve, to which we shall return in Section I. Additional experiments on more diffuse labeling of nerve sources in brain and spinal cord have confirmed the proximo-distal convection (Ochs et al., 1962; Rahmann, 1965), but, being of lower resolving power than those reported in the preceding sections, they need not be detailed in this summary account. One further observation presumably related to the internal catabolical degradation of the axonal column on its way is the proximo-distal decline of the axonal content of cholinesterase and cholineacetylase (Lubińska, 1964; Hebb and Silver, 1961).

Lastly, it is of historic interest that Ramón y Cajal (1928) had correctly observed and pictured the "beads" of nerve fibers on the proximal side of a tightly ligated nerve, but misinterpreted their origin. He thought that the beads were residues of the club-shaped ends of regenerating fibers which he assumed had thus been ramming their way through the tissues.

H. Transport Mechanism

The technological analysis of axonal flow given in Section E had led to certain minimum presuppositions about the driving mechanism involved. *Static* pressure from an expanding cell body with increasing turgor (Young, 1945) can definitely be ruled out, for in any tubular system with some degree of elasticity and plasticity, such as a nerve fiber, inflation from the end could only result in a dilation of the base tapering off toward the tip (Figure 7, bottom), which is exactly the reverse of the polarity of the observed configurations of dammed axons. The actual mechanism had to be sought in a *dynamic* drive operating all along the length of the nerve fiber, roughly comparable to the roller belt of an assembly line. The most plausible assumption (Weiss and Hiscoe, 1948) was a microperistaltic wave in the surface of the fiber propelling the enclosed content away from the cell (Figure 7, top). In mechanical devices of this kind, the rate of the traveling wave of the drive need not be directly related to the rate of the advance of the driven core. Observations on rhythmic contraction-relaxation

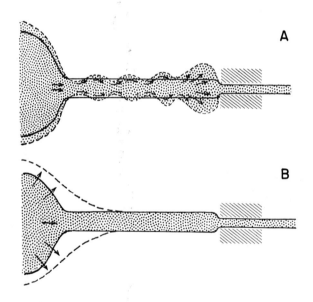

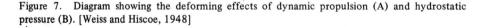

Figure 7. Diagram showing the deforming effects of dynamic propulsion (A) and hydrostatic pressure (B). [Weiss and Hiscoe, 1948]

pulses sweeping over the surfaces of certain eggs and tissue cells (Weiss, 1961b) and the fact that such circling waves, when constrained linearly into a cylindrical path, assume the aspect of a peristaltic wave, encouraged the attempt to obtain direct visual signs of the axonal drive by cinemicrography (Weiss, 1963; Weiss et al., 1962).

Spinal ganglia of the trunk region of young mice were excised with their intercostal nerves attached and placed into specially designed chambers continuously perfused by a carefully balanced nutrient medium, in which they stayed alive for a week or longer. In a part of the nerve, stripped of its perineurium, the nerve fibers were teased apart sufficiently to explore individual fibers under high-power phase-contrast microscopy. Time-lapse cinemicrographs were taken, mostly at intervals of 1 second between frames, which (on projection at 24 frames per second) amounted to an acceleration on the screen of 1440 times, i.e., 1 minute on the screen equalling 1 day in actual life. (Samples of the film were shown at the meeting.) The major features can be summarized as follows.

Both intact fibers and fibers in the early stages of Wallerian degeneration, already segmented into ovoids, show continuous peristaltic motions polarized proximo-distally with regard to the position of the ganglion. Once started, these waves go on without interruption at a relatively constant rhythm of about 16 minutes per single contraction-relaxation cycle. This rhythm was about the same regardless of the fiber diameter and whether the fiber was still continuous or already degeneratively segmented. This evidently confirms the inference drawn from the sum of our experiments that the motive mechanism for the axonal propulsion must be present at every differential of the fiber surface along its length. The wave appears as a conspicuous traveling deformation of both the myelin sheath and the enclosed axon, involving either the whole circumference or rippling only a part of it. The axonal content often moves more slowly than the surface wave; in the presence of obstructions by kinks or convolutions, it may be completely arrested until its continuing longitudinal compression has built up a sufficient pressure head to ram the front across the block. Sheath cells glide up and down along the fiber within the endoneurial tube with great agility, but so erratically that it is doubtful that the axonal beat in its polarization could be correlated with their motility. At the same time, the observation of rhythmic pulsation in explanted glia cells (Pomerat, 1961), extended now to many other cell types (Weiss, 1961b), and the general symbiotic relation between Schwann cell and axon call for a far more systematic investigation of their role. Considering the paucity of mitochondria within the axon, it would seem logical to ascribe the energy supply for the perpetual axonal drive to the mitochondria-laden Schwann cell. The actual motile mechanism, however, must be located somewhere in the myelin sheath, or the axolemma, or perhaps in a cooperative process between the lipid layers and the spiral windings of sheath cell protoplasm sandwiched between them. I have presented a hypothetical model of how such a combined lipid-protein array, surrounding an incompressible content, could act as mechanism for a peristaltic drive in cells in general (Weiss, 1964), but factual data to support the model are still missing. One reason why the Schwann cell, though perhaps energizing the axonal drive, is unlikely to contain also the effectuating machinery is that the frequency ranges observed in the pulsation of glia cells (Pomerat, 1961) and of nerve fibers (Weiss, 1963) are distinctly different.

A lucky incident in one of the films has furnished a significant clue to the understanding of the peristaltic process. A rather quiescent nerve fiber had one local spot on its surface at which the myelin sheath bulged into the axon in the form of a dimple, and this local pit contracted rhythmically and continuously at the usual frequency of about 4 pulses per hour. The rest of the myelin sheath and of the axon remained mostly inactive and smooth. Every once in a while, however, a polarized traveling wave would start from the beating center and run off as a peristaltic movement down the whole stretch of fiber in the visual field. Formally, the picture is similar to that of a local rhythmic excitatory process of subliminal intensity building up to threshold level and actuating then a propagated disturbance. Any more specific contentions about the process would seem unwarranted, particularly as we have no information on the axonal flow mechanism in unmyelinated fibers in which the damming phenomenon in front of constrictions has been fully confirmed (Dahlström and Häggendal, 1966a; Kapeller and Mayor, 1967).

Motion pictures taken at the cut ends of nerve fibers revealed bulbous widenings at the cut upon the arrival of the peristaltic waves, but no actual extrusion of axonal material into the medium. Conceivably, in other media, osmotically or otherwise less balanced, true outflow may occur, as we have also seen it to occur from ruptured fibers.

Despite our extensive studies over many years, more questions remain open than have been answered. For instance, our attempts to demonstrate axonal flow in small subcutaneous nerve bundles in anesthetized living animals have remained inconclusive. Under these conditions we have recorded flow slow enough to be ascribed to axonal flow of the standard type, but we could not rule out the possibility that it may have been actuated by some asymmetry of local pressure across the visual field imposed by the experimental set-up. However, it is possible that the spectacular surface deformations that mark the peristaltic wave in our teased fibers, which are unconstrained, could not occur in intact nerves *in situ* whose fibers are tightly bound by the indistensible perineurial sheath; in the animal, the waves of *isotonic* contractions seen in our liberated fibers *in vitro* would then appear as *isometric* waves of pressure differences. But for this kind of dynamics, adequate engineering models are still lacking.

I. Additional Transport Systems

As indicated earlier, the continuous outgrowth of the axon from its root in the cell body and its cellulifugal propulsion as a semisolid column in no way precludes the presence within that carrier system of separate routes or channels for express traffic. I have already mentioned transport in nerves at rates far in excess of the average rate of axonal flow itself. Some data in the literature referring to such fast transport are ambiguous because of the absence of a clear distinction between intraaxonal and interaxonal (endoneurial) flow. As stated in Section F, we have recorded *endoneurial* flow rates of up to 70 millimeters per day. We therefore shall concentrate here only on instances in which such fast transport was rather definitely identified as intraaxonal. This has been shown most conclusively for the proximo-distal catecholamine shift in the unmyelinated fibers of the autonomic system (von Euler, 1958; Dahlström and Häggendal, 1966a).

Two intraaxonal systems can be considered as candidates for the role of fast traffic channels. They are the neurotubules and the intertubular spaces in the matrix illustrated above in Figure 3. The neurotubules, themselves, are evidently integral constituents of the moving axonal mass, as is convincingly demonstrated by their strict conformation to the convolutions and gyrate deformations of the dammed axonal stretches as described in Section D. It would seem most plausible to assume that they grow continuously from their bases in the cell body by the apposition of new elements to their wall, the older stem being carried along by the axonal stream in which it is imbedded. Either the neurotubules or the intertubular liquid channels in the matrix or both could serve as specialized canals for rapid transit of solutes or small particles suspended in solutes. Evidence suggesting this comes from the fact that the dammed-up portion of axons frequently contains bulbous widenings, sometimes pinched off as vesicles, with a size gradient from surface to center identical with that described above (Figure 4) for the axon population within the nerve as a whole. While our electron micrographs have definitely identified some of these widenings with neurotubules, the larger fluid bulbs are presumably dammed-up intertubular spaces.

In technological interpretation, this radial gradient of damming intensity within the individual axon strongly indicates that the driving force for the proximo-distal propulsion of liquid in intraaxonal channels resides in the surface of the axon rather than in the individual microchannels. Technically, it is plausible for such intratubular and intertubular liquids to assume velocities of flow considerably in excess of those of the semi-solid matrix; whereas the peristaltic contractions can move the cohesive axonal column only slowly, each new contractile wave can impart additional acceleration to freely mobile liquid columns in the interior, somewhat in the manner of a linear accelerator principle. The effective rate at any instant should then be limited only by the degree and rate of the extrusion of the liquid at the distal end. Peripheral depletion would be expected to entail faster transport, and consequently drainage, from the cell body, whereas peripheral oversupply would lead to stagnation in the liquid channels, thus perhaps constituting a negative feedback repression of further central production of the particular product. Whether or not this could become a general principle for transmitter substances and their supply to nerve endings remains to be tested. One realizes how important it is to clarify the relation between the blind ends of neurotubules and the synaptic vesicles. (For further reading, see De Robertis, 1967.)

All traffic in the neural system thus far considered has been in the proximo-distal direction.* Nothing in our observations on axonal flow has given any hint of traffic in the reverse direction. Yet there are facts to indicate the existence of some sort of direct communication between the periphery of a nerve fiber and the central system other than through impulse conduction. For example, neurotropic viruses (lissa, poliomyelitis, etc.) and some toxins (e.g., tetanus), when applied peripherally, are known to penetrate the nerve centers by way of the nerves; but in most cases, the exact pathway, whether intraaxonal or extraaxonal, has not been identified. Even if it were mostly endoneurial, convection by liquid currents is doubtful because of the prevailing centrifugal direction of endoneurial flow (Weiss et al., 1945). Moreover, polio virus has been asserted to move centrad inside the axon (Bodian and Howe, 1941a,b).

In a more subtle way, the problem of ascending communication between periphery and centers arises from the existence of a direct control of the state of the nerve cell body by its terminal connections, as in the chromatolytic reaction of the cell after peripheral severance of its axon, which indicates that the perikaryon has received information about the peripheral disturbance (see Weiss et al., 1945; Cavanaugh, 1951). Even more specific is the "modulating" effect which a peripheral organ, whether muscle or receptor, exerts upon its connecting neuron and which endows the cell body with properties distinctive of the particular terminal connection (Weiss, 1966).

* "Proximo-distal" in axons is always used in reference to the position of the cell body; i.e., synonymous to "cellulifugal." For spinal sensory nerves, it means "toward the receptor" for the peripheral branches, and "toward the spinal cord" for the dorsal roots. There is thus clearly no relation between the direction of axonal flow and that of the physiologic impulse conduction. [P.W.]

It is not even known, of course, whether in these latter cases the critical information is conveyed to the cell body by actual substance transport or by transmission of a wave of molecular changes along the line. If one adds to this the uncertainties of the host of empirical laboratory and pathological experiences about communication in the nervous system other than by impulses, compounded by our relative ignorance about transport and pumping mechanisms in living systems in general, one comes to realize that further speculation on these matters would be gratuitous.

By the same token, however, it is unwarranted to interpret ascending communication in nerve as a sign of a reversal of axonal flow. Our motion picture studies, particularly those on pressure-forced flow, not only have shown no sign of two-way substance transfer in axons, but have positively ruled out the physical possibility for such reciprocal flow in *mature* nerve fibers. It is true, of course, as has been known since the classical observations of embryonic nerve fibers by Harrison and the Lewises, and amplified by several decades of studies, including my own, that the *immature* (developing or regenerating) "outgrowing" axon exhibits in its rather liquid interior bidirectional movements of particles and fountain currents comparable to those of simple protozoan cells; but, as I indicated earlier, these immature or remobilized neurons are so different in state, consistency, and behavior from the mature nerve fiber that unsubstantiated transfer of conclusions from one to the other is quite unrealistic.

Even though pleading ignorance of the nature of ascending traffic in nerve, which in its relation to axonal flow corresponds to upstream navigation in a river, I wish to repeat a suggestion made on an earlier occasion (Weiss, 1963), which might offer a clue to a future solution of the problem. This is *interfacial* transport or "creep" in multiphasic systems. In studying the spread of macromolecules (dyes for microscopy; heavy metal compounds for electron microscopy), for instance, in tendons composed of collagen fibers embedded in a continuous mucopolysaccharide matrix, we found the test substance to concentrate heavily along the fiber surfaces and to spread there in the direction of the fiber axes much faster than in other directions (Weiss, 1961b; Grover, 1966). Since this was observed only for molecules or particles which carried an electric charge, and since clusters formed at the periodically spaced polar bands of the collagen fibers, one is led to conclude that there may be fast saltatory transfers along a structured interface, in the manner of a molecular bucket brigade. It is conceivable to ascribe a similar guiding function to the interfaces between neurofilaments and their surrounding axonal matrix. The participation of neurotubules in the ascending traffic is far less likely, but it cannot be definitively discounted.

J. Conclusion

Having exploited in the preceding sections the factual content of our data as far as their logical interpretation would justify, I might, in closing, indulge in a brief excursion beyond that safe line and list a few of the areas that stand to gain from the new light shed on the growth dynamics of the neuron.

As has been clearly recognized and experimentally verified by Hydén and his group (e.g., Hamberger and Hydén, 1945), the quantitative adjustments of the size of the nerve cell body and its constituents to variations of functional load are both the effect as well as the measure of the dynamics here discussed. Given this fact, far more research is needed with a view to acquiring deliberate control, first experimentally but in the end one hopes therapeutically, over the circumstances that would permit optimization of the trophic state of the nerve cell and its synthetic machinery. Compensatory hypertrophy of residual neurons in a pathologically decimated neuron population due to their taking over the additional innervation of deserted peripheral tissue (Wohlfahrt, 1935, 1941) or to increased functional bombardment (Edds, 1950) is known. Extrapolation from this intrinsic regulation might furnish guide lines for the deliberate enhancement of neural activities. In this connection, it would be important to study now in greater detail whether and to what extent various neurotropic agents and drugs enter, and how they affect, the unsuspectedly high productivity of the mature neuron. This also leads directly into the problem of the so-called "trophic" function of nerves in the maintenance of the integrity and the capacity for repair and regeneration of innervated tissues. It should prove possible to trace "trophic" exudates from nerve in the axonal stream and trap them, much as neurosecretions have been trapped, collected, and identified.

Furthermore, the realization of the intensive and incessant activity of the neuron has an immediate bearing on the problem of specific qualitative adaptability of neurons such as underlies functional plasticity of the central nervous system, as manifested in learning, habituation, acquired hypersensitivities, idiosyncrasies, addictions, and so forth. Changes in the base composition of the ribonucleic acid system of the nerve cell in the course of the acquisition of certain trained performances has recently come into the foreground of interest (see Hydén, 1965). Even if only a few subspecies of the cellular RNAs become altered in the process, the transcription of their altered code to a correspondingly modified protein will henceforth entail perpetually modified protein production, which, at the rate at which the machine works, enormously *amplifies* the initial effect. If such an altered constitution of the macromolecular population of the cell body were to project itself into specific surface properties, and if conformance or nonconformance of such surface properties between cells in mutual contact were to be decisive for the making or breaking of transmissive synaptic junctions, one could readily understand how self-perpetuating changes within a single cell body would affect the functional relation of that cell body to the whole network (Weiss, 1959). In this fashion the molecular and the systemic theories of neural activity would not only be reconciled but would be made to mesh. Yet this is leaving the solid ground of facts, and, since my declared purpose in this presentation has been to emphasize phenomena and not theory, I have reached the point to close this account.

(Research supported by grants from the National Institutes of Health No. CA 10096 (National Cancer Institute), No. NB 07348 (National Institute of Neurological Diseases and Blindness), and the Faith Foundation.)

BIBLIOGRAPHY

The numbers listed in the right-hand margin refer to pages in the essay on which the references are cited.

Page

Alexander, E., Jr., Woods, R.P. and Weiss, P. (1948): Further experiments on bridging of long nerve gaps in monkeys. *Proc. Soc. Exp. Biol. Med.* 68:380-382. 376

Bodian, D. and Howe, H.A. (1941a): Experimental studies of intraneural spread of poliomyelitis virus. *Bull. Johns Hopkins Hosp.* 68:248-267. 397

Bodian, D. and Howe, H.A. (1941b): Rate of progression of poliomyelitis virus in nerves. *Bull. Johns Hopkins Hosp.* 69:79-85. 397

Breemen, V.L. van, Anderson, E. and Reger, J.F. (1958): An attempt to determine the origin of synaptic vesicles. *Exp. Cell Res.* 5(Suppl.):153-167. 381

Cavanaugh, M.W. (1951): Quantitative effects of peripheral innervation area on nerves and spinal ganglion cells. *J. Comp. Neurol.* 94:181-219. 397

Dahlström, A. (1966): *The Intraneuronal Distribution of Noradrenaline and the Transport and Life-Span of Amine Storage Granules in the Sympathetic Adrenergic Neuron.* Stockholm: Karolinska Institutet. M.D. Thesis. 376

Dahlström, A. (1967a): The effect of reserpine and tetrabenazine on the accumulation of noradrenaline in the rat sciatic nerve after ligation. *Acta Physiol. Scand.* 62:167-179. 378

Dahlström, A. and Häggendal, J. (1966a): Some quantitative studies on the noradrenaline content in the cell bodies and terminals of a sympathetic adrenergic neuron system. *Acta Physiol. Scand.* 67:271-277. 395,396

De Robertis, E. (1967): Ultrastructure and cytochemistry of the synaptic region. The macromolecular components involved in nerve transmission are being studied. *Science* 156:907-914. 381,397

Droz, B. and Leblond, C.P. (1963): Axonal migration of proteins in the central nervous system and peripheral nerves as shown by radioautography. *J. Comp. Neurol.* 121:325-345. 374,384, 387

Edds, M.V., Jr. (1950): Hypertrophy of nerve fibers to functionally overloaded muscles. *J. Comp. Neurol.* 93:258-276. 399

Gerard, R.W. (1932): Nerve metabolism. *Physiol. Rev.* 12:469-592. 373

Grover, N. (1966): Anisometric transport of ions and particles in anisotrophic tissue spaces. *Biophys. J.* 6:71-85. 398

Page

Hamberger, C.-A. and Hydén, H. (1945): Cytochemical changes in the cochlear ganglion   399
   caused by acoustic stimulation and trauma. *Acta Oto-Laryngol.* 61(Suppl.):5-89.

Hebb, C.O. and Silver, A. (1961): Gradient of choline acetylase activity. *Nature* 189:123-125.   392

Hydén, H. (1943): Protein metabolism in the nerve cell during growth and function. *Acta*   373,386
   Physiol. Scand. 6(Suppl. 17):5-136.

Hydén, H. (1965): RNA—A functional characteristic of the neuron and its glia. *In: Brain*   400
   Function: RNA and Brain Function, Memory and Learning. Brazier, M.A.B., ed. Berkeley:
   University of California Press, pp. 29-68.

Jakoubek, B. and Edström, J.-E. (1965): RNA changes in the Mauthner axon and myelin   391
   sheath after increased functional activity. *J. Neurochem.* 12:845-849.

Kapeller, K. and Mayor, D. (1967): The accumulation of noradrenaline in constricted sympa-   395
   thetic nerves as studied by fluorescence and electron microscopy. *Proc. Roy. Soc., B* 167:
   282-292.

Kreutzberg, G.W. (1963): Changes of coenzyme (TPN) diaphorase and TPN-linked dehydro-
   genase during axonal reaction of the nerve cell. *Nature* 199:393-394.

Kreutzberg, G.W. (1963): Enzymhistochemische Veranderungen in Axonen des Rückenmarks   376
   nach Durchschneidung der langen Bahnen. *Dtsch. A. Nervenheilk.* 185:308-318.

Lubińska, L. (1964): Axoplasmic streaming in regenerating and in normal nerve fibres. *In:*   376,392
   Mechanisms of Neural Regeneration. (Progress in Brain Research, Vol. 13). Singer, M.
   and Schadé, J.P., eds. Amsterdam: Elsevier, pp. 1-66.

Matson, D.D., Alexander, E. and Weiss, P. (1948): Experiments on the bridging of gaps in   376
   severed peripheral nerve of monkeys. *J. Neurosurg.* 5:230-248.

Miani, N. (1964): Proximo-distal movement of phospholipid in the axoplasm of the intact and   392
   regenerating neurons. *In: Mechanisms of Neural Regeneration.* (Progress in Brain Research,
   Vol. 13) Singer, M. and Schadé, J.P., eds. Amsterdam: Elsevier, pp. 115-126.

Nauta, W.J.H. and Koella, W.P. (1966): Sleep, wakefulness, dreams and memory. *Neuro-*
   sciences Res. Prog. Bull. 4(1):1-103.

Ochs, S. and Burger, E. (1958): Movement of substance proximo-distally in nerve axons as   387
   studied with spinal cord injections of radioactive phosphorus. *Amer. J. Physiol.* 194:499-
   506.

Ochs, S., Dalrymple, D.E. and Richards, G. (1962): Axoplasmic flow in nerve fibers. *Proc. Int.*   392
   Physiol. Sci. 22nd Int. Congr. 2:797.

Parker, G.H. (1932): On the trophic impulse so-called, its rate and nature. *Amer. Natur.*   373
   67:147-158.

Page

Pillai, P.A. (1964): A banded structure in the connective tissue of nerve. *J. Ultrastruct. Res.* 379
11:455-468.

Pomerat, C.M. (1961): Cinematology, indispensable tool for cytology. *Int. Rev. Cytol.* 11:307- 394
339.

Rahmann, H. and Korfsmeier, K.H. (1965): Autoradiographische Untersuchunger über den 392
Eiweisstransport im Zentralnervensystem von Brachydanio nerio Ham. Buch. *Zool. Jahrb.*
(Physiol.) 71:475-488.

Ramón y Cajal, S. (1928): *Degeneration and Regeneration of the Nervous System.* (Trans. and 392
ed. by R.M. May) Cambridge: Oxford University Press.

Samuels, A.J., Boyarsky. L.L., Gerard, R.W., Libet, B. and Brust, M. (1951): Distribution, ex- 374,387
change and migration of phosphate compounds in the nervous system. *Amer. J. Physiol.*
164:1-5.

Shepherd, E.H. (1951): The movement of radioactive phosphorus 32 in peripheral nerve. 374,387
Ph.D. Dissertation, University of Chicago.

Taylor, A.C. and Weiss, P. (1965): Demonstration of axonal flow by the movement of tritium- 388
labeled protein in mature optic nerve fibers. *Proc. Nat. Acad. Sci.* 54:1521-1527.

Utakoji, T. and Hsu, T.C. (1965): Nucleic acids and protein synthesis of isolated cells from 391
chick embryonic spinal ganglia in culture. *J. Exp. Zool.* 158:181-202.

von Euler, U.S. (1958): The presence of the adrenergic neurotransmitter in intraaxonal struc- 392,396
ture. *Acta Physiol. Scand.* 43:155-166.

Waelsch, H. (1958): Some aspects of amino acid and protein metabolism of the nervous sys- 374,387
tem. *J. Nerv. Ment. Dis.* 126:33-39.

Weiss, P. (1943a): Endoneurial edema in constricted nerve. *Anat. Rec.* 86:491-522. 371,372

Weiss, P. (1943b): Nerve regeneration in the rat following tubular splicing of severed nerves. 376
Arch. Surg. 46:525-547.

Weiss, P. (1944a): Damming of axoplasm in constricted nerve: a sign of perpetual growth in 371,372
nerve fibers. *Anat. Rec.* 88(Suppl.):48.

Weiss, P. (1944b): Evidence of perpetual proximo-distal growth of nerve fibers. *Biol. Bull.* 371,372
87:160.

Weiss, P. (1944c): The technology of nerve regeneration: a review. Sutureless tubulation and 371,372
related methods of nerve repair. *J. Neurosurg.* 1:400-450.

Weiss, P. (1947): Protoplasm synthesis and substance transfer in neurons. XVII International
Physiological Congress. Oxford, England.

V. On the Dynamics of the Nervous System

Page

Weiss, P., Edds, M.V., Jr. and Cavanaugh, M. (1945): The effect of terminal connections on   387,397
the caliber of nerve fibers. *Anat. Rec.* 92:215-233.

Weiss, P. and Hiscoe, H.B. (1948): Experiments on the mechanism of nerve growth. *J. Exp.*   372,375,
Zool. 107:315-395.   377,378,
                                                                                  383,384,385,386,393

Weiss, P. and Holland, Y. (1967): Neuronal dynamics and axonal flow. II. The olfactory nerve   389
as model test object. *Proc. Nat. Acad. Sci.* 57:258-264.

Weiss, P. and Pillai, A. (1965): Convection and fate of mitochondria in nerve fibers: axonal   379,381
flow as vehicle. *Proc. Nat. Acad. Sci.* 54:48-56.

Weiss, P. and Rossetti, F. (1951): Growth responses of opposite sign among different neuron
types exposed to thyroid hormone. *Proc. Nat. Acad. Sci.* 37:540-556.

Weiss, P. and Taylor, A.C. (1944): Impairment of growth and myelinization in regenerating   371,372,
nerve fibers subject to constriction. *Proc. Soc. Exp. Biol. Med.* 55:77-80.   373

Weiss, P. and Taylor, A.C. (1964): Synthesis and flow of neuroplasm: a progress report.
Science 148:669-670.

Weiss, P., Taylor, A.C. and Pillai, P.A. (1962): The nerve fiber as a system in continuous flow: 379,393
microcinematographic and electronmicroscopic demonstrations. *Science* 136:330.

Weiss, P. and Wang, H. (1944): Proximo-distal fluid convection in nerves, demonstrated by
color indicators. *Fed. Proc.* 3:51.

Weiss, P., Wang, H., Taylor, A.C. and Edds, M.V.,Jr. (1944): Proximo-distal fluid convection
in nerves, demonstrated by radioactive tracer substances. *Fed. Proc.* 3:51.

Weiss, P., Wang, H., Taylor, A.C. and Edds, M.V., Jr. (1945): Proximo-distal fluid convection
in the endoneurial spaces of peripheral nerves, demonstrated by colored and radioactive
(isotope) tracers. *Amer. J. Physiol.* 143:521-540.

Wohlfahrt, G. and Swank, R.L. (1941): Pathology of amyotrophic lateral sclerosis; fiber analy-   399
sis of ventral roots and pyramidal tracts of spinal cord. *Arch. Neurol. Psychiat.* 46:783-799.

Wohlfahrt, S. and Wohlfahrt, G. (1935): Mikroskopische Untersuchungen an progressiven   399
Muskelatrophien unter besonderer Rücksichtsnahme auf Rückenmarks- und Muskelbe-
funde. *Acta Med. Scand.* 63(Suppl.):1-37.

Young, J.Z. (1945): The history of the shape of a nerve fiber. *In: Essays on Growth and Form.*   392
LeGros Clark, W. and Medawar, P.B., eds. Oxford: Clarendon Press, pp. 41-94.

Reprinted from the A. M. A. Archives of Neurology
June 1960, Vol. 2, pp. 595-599
Copyright 1960, by American Medical Association

CHAPTER 24

Editorial

Modifiability of the Neuron

For nearly a century neuroanatomists and neurophysiologists in concert have sought to understand the nervous *system* in terms of the properties of its elements. Yet, despite spectacular analytical advances, we are still far short of an explanation of its integrative, coordinative, regulatory, and adaptive features that would satisfy those who are mindful of the true problems of behavior and who find no comfort in fragmentary, unrepresentative, or even fictional models or purely verbal surrogates for knowledge. Does this impasse discredit the validity of the attempt to reconstruct the whole from the components? Or does it perhaps just signify that some deficiencies of discernment in dealing with the elements have made us overlook some relevant properties? The decision lies in the future. But resignation from the analytical course would seem premature so long as the elements have not been entered into the account with their full share of capacities, as thus far they have not.

Let us not forget that when we speak of "properties" and "faculties" of any natural object, we actually refer not to its full endowment but only to that empirical fraction which we have come to know from observations and measurements by suitable detector devices, each of which reveals but a limited absorption band from the total spectrum with which the object is endowed. A listening device registers acoustic signals; an electric instrument, electric information; an optical apparatus, pictures. This being so, it would be folly to concede to an object no wider a repertory of properties than that recorded by the limited choice of detection devices which we happen to have favored routinely; the limitations of the student and of his tools must not be projected into the object of his study. Rather, one must constantly be on the lookout for ever-new means of disclosing additional properties formerly undiscerned for lack of proper indicators. The time is ripe—and reasons are compelling—for giving the neuron the benefit of such more generous considerations and treatment.

Our customary treatment has concentrated on structural, electric, and metabolic features. Optical devices, microscopic and submicroscopic, have furnished structural data: numbers, dimensions, orientations, distributions, branchings, groupings, and connections, including their variations. Electric measurements have led to detailed information about potentials, conductances, resistances, capacitances, polarizations, thresholds, and the like, as well as about the time courses, periodic or aperiodic, of their changes during activity. Metabolic studies have given us the balance sheet of energy requirements and exchanges during rest, activity, and

Accepted for publication March 14, 1960.

recovery, as well as insight into the underlying chemical processes. Furthermore, the discovery of chemical transmitters has upgraded neurochemistry to something above a mere supply source of metabolic energy.

Without in any way belittling the achievements of this tripartite methodology, still what it can tell us about the neuron is limited to the restricted vocabulary of the techniques employed. So, as a result of our confinement to these powerful, yet limited, techniques, a correspondingly constrained conception of the nervous system has crystallized in our minds, which unnecessarily and unnaturally denies to neurons a lot of properties that other living cells demonstrably possess. Frankly, in essence we credit neurons with "life" not much more fully than we do "live" wires in an electric circuit. They merely vegetate, subserving the monotonic—and monotonous—function of carrying and transmitting impulses, much as Atlas was thought to be no more than a "live" pillar to support the globe. Except for quantitative gradations of structural, electric, and metabolic parameters, plus a very few gross chemical distinctions, as between "cholinergic" and "adrenergic" units, the neuron population has been considered as all of one kind and character, and, for that matter, static throughout life. At best, the neurons were conceded some minor variations of size and threshold values, but decidedly none of the sharp and discriminatory character distinctions that mark the cells of other tissues; for instance, the cells of different glands which have radically different chemical production plants.

Another methodological self-confinement has helped to consolidate this picture. Neuroanatomy and neurophysiology have, in a sense, accepted delivery of the nervous system into their hands only after it was fully formed, ready-made, in working order. Just how it had gotten to be that way, was a matter for which they did not feel accountable. It was something that one labeled noncommitally as "maturation," plus "conditioning," and then left to other departments to crack. Yet, it was precisely in those other biological departments, where technical conventions were less restrictive, that the neuron was found to have some unsuspected new properties, which, on the one hand, tie it more closely to the rest of the family of truly living cell types, and, on the other, throw new light on its functional performances.

The two major new additions to the list of properties of neurons are the demonstration of (1) a wide range of "specific" type differences within the neuronal population as clue to the establishment and maintenance of supraelemental order; and (2) a state of flux and continuous self-renewal of the body of the neuron. The meaning of these points may here be briefly summarized.

"Specificity," which underlies our highly discriminative sensory perception, used to be relegated to the sense organs, which, as transducers, were thought to reduce diversity to a single code in common to all neurons. Yet the simple experiences of pharmacology, toxicology, and endocrinology, according to which agents of a given molecular structure have a selective action on certain groups of central neurons to the exclusion of others, prove clearly that specificity is not confined to the periphery. "Specificity," in this context, refers to that constitutional property which enables a given unit to respond discriminately to an outer agent, depending on whether or not the latter is of a properly matching configuration—much as a lock yields to a matching key. Such pairwise specific correspondences are among the most general attributes of living systems, as manifested, for instance, in hormone action, fertilization, parasitism, immunology, and enzymology.

DYNAMICS OF DEVELOPMENT

Whatever the underlying mechanism may be—according to some, it lies in the steric conformance of complementary molecules—the only suitable detectors so far at our disposal are the specific biological responses themselves. Specific interactions between two systems, A and B, are of two kinds: Either A imposes its own specificity (or a complement to it) on a more plastic B ("modulation"); or, if both A and B are already fixed in their specificities, the degree of mutual correspondence or resonance ("affinity") will determine whether or not they will become linked, morphologically or functionally.

Ever since 1922, I have presented evidence to show that both individual muscles and individual receptor areas possess distinctive constitutional differentials, presumably biochemical in nature, which they confer upon the neurons to which they are attached ("modulation"), thus creating a corresponding number of distinctive subspecies among the peripheral neurons. The latter, in turn, would impose similar specificities upon penultimate neurons, and in this manner the centers would be briefed on just which central stations are connected in each case with precisely what muscle or receptor.[1] The receptor side of this principle, originally demonstrated for muscular proprioception[2] and corneal exteroception,[3] has since been widely confirmed for vision and other senses.[4] From all this, the conclusion became inescapable that neurons differ much more widely among one another in subtle biochemical characteristics than had previously been recognized, and that these highly specific differentials are instrumental in establishing the selective linkages requisite for the orderly functioning of the nervous system.

The nervous system thus is comparable to an integrated system of closed-circuit broadcasting networks. Whether the individual channels are controlled by a "resonance" system of communication,[5] or are joined by "wire connections,"[4] is debatable, but inconsequential for the general thesis, which is that "specifically matching properties" are a basic principle of neural intercommunication. The crucial points of this thesis are (1) that neuronal specificity can be modified ("modulated") by the terminal effectors and receptors, as well as by fellow neurons; and (2) that by virtue of the acquired "tunes," they can selectively enter into higher-order unions of concordant elements.

It is important to realize that neurons share both these properties with other living cells. Point 1 is illustrated by the various template analogues invoked in cell reproduction, differentiation, growth control, antibody production, induced enzyme formation, etc., in all of which cells undergo some specific adaptive re-shaping of parts of their molecular populations.[6] Point 2 has recently been clinched by the following two discoveries: (*a*) Cells of a given kind, disseminated throughout the body by random routes, have been found "homing" at their specific sites of destination.[7] (*b*) Mixtures of cells of different types, scrambled at random, have been found to sort themselves out actively according to kinds and mutual "affinities"[8]: They recognize each other's kinds on contact and then react discriminately, remaining associated if they are alike, but separating again if they are unrelated.[9] There is thus now available definite proof that the "specific" faculties of neurons outlined above are quite in keeping with demonstrated properties of other living cells. Some more detailed suggestions have been set forth on how the interaction of specific complementary molecular end-groups facing each other across cell membranes at points of cell-to-cell contact (as at synaptic junctions) might force the carrier macromolecules from their "barrier" alignment in the cell surface into radial "open-gate" positions, thereby opening

V. On the Dynamics of the Nervous System   619

wide "pores" and "leaks" between them for the transcellular passage of substances and increased flow of current [9]; but such speculations indicate only the feasibility, rather than the attainment, of compatibility between the principle of specificity in cellular and neuronal interactions and the conventional, "nonspecific" concepts.

Once recognized, the property of specificity presents us with countless new questions: How many types are laid out as such in the mosaic of the embryonic nervous system, and how many additional subspecifications are later imposed upon them by modulating influences of somatic tissues, or even functional adaptation? How far do such secondary modifications become ingrained and indelible, and to what extent can they be superseded by new ones? And so forth. The answers to the What's and When's and How's will vary with the type of neuron. But it seems intriguing to contemplate that some degree of adaptive modifiability may be retained by almost any neuron, so that the plasticity of the nervous *system*, as manifested in behavior, might rest not only on latitude of *inter*neuronal relations, but on *intra*neuronal modifiability as well.

A tangible basis for such a lasting opportunity for specific adaptation is offered by the disclosure that the neuron is in a state of perpetual renewal of its substance, which proceeds from the nuclear territory of the cell. From this supply source, a continuous viscous stream of new cell substance is conveyed down the nerve fiber, where it evidently replenishes the stores of metabolically degraded macromolecules, besides serving as vehicle for neurohumors and other products to be discharged into the periphery.[10,11] The evidence, derived at first from the dynamic deformations of throttled supply streams, has recently been greatly strengthened by electron-microscopic, cytochemical, and isotope-tracer studies. The neuron, in this version, abandons its former aspect of a static fixture and assumes instead that of a stationary process in continuous flux. This obviously creates opportunities for specific adaptive changes in its molecular population, assuming that the reproductive source of cell-specific molecules can be refashioned, template-wise, as in cellular immune reactions, so as to adopt configurations of chemical structures introduced from the neuronal environment. It is not inconceivable that this concept might some day even embrace the explanation of acquired idiosyncrasies and selective sensitization and tolerance to drugs.

To what extent the various instances of "revitalization" of the neuron concept which have been outlined in this essay will open and illuminate new avenues to the obscure problems of coordination, regulation, learning, ageing, and so forth, is unpredictable. But it stands to reason that the described additional degrees of freedom of the neuron will at least have to be taken seriously into account in further efforts to reconstruct a reasonably faithful picture of the nervous *system* from what little is known of its elements—that is, provided we want to let our brain exercise the privilege of adaptive change for which its "living" elements so well predispose it. New methods yield new facts. New facts which cannot be accommodated in existing concepts call for adaptive conceptual revisions. The living nervous system is calling.

PAUL WEISS, Ph.D., M.D. (Hon.), Sc.D.
Member and Professor
The Rockefeller Institute, New York

REFERENCES

1. Weiss, P.: A. Res. Nerv. & Ment. Dis., Proc. (1950) 30:3-23, 1952.
2. Verzár, F., and Weiss, P.: Arch. ges. Physiol. 223:671-684, 1930.
3. Weiss, P.: J. Comp. Neurol. 77:131-169, 1942.
4. Sperry, R. W.: Quart. Rev. Biol. 12:66-73, 1951.
5. Weiss, P.: Biol. Rev. 11:494-531, 1936.
6. Weiss, P.: Yale J. Biol. & Med. 19:235-278, 1947.
7. Weiss, P., and Andres, G.: J. Exper. Zool. 121:449-487, 1952.
8. Moscona, A. A.: Proc. Soc. Exper. Biol. & Med 92:410-416, 1956.
9. Weiss, P.: Internat. Rev. Cytol. 7:319-423, 1958.
10. Weiss, P., and Hiscoe, H. B.: J. Exper. Zool. 107:315-395, 1948.
11. Weiss, P.: A. Res. Nerv. & Ment. Dis., Proc. (1954) 35:8-18, 1956.

TOPICAL INDEX

Numbers following topic entry refer to chapter numbers.

623